Applying Geomorphology to Environmental Management

Applying Geomorphology to Environmental Management

Edited by

Deborah J. Anthony
Michael D. Harvey
Jonathan B. Laronne
M. Paul Mosley

Water Resources Publications, LLC

For Information and Correspondence:

Water Resources Publications, LLC
P. O. Box 260026, Highlands Ranch, Colorado 80163-0026, USA

APPLYING GEOMORPHOLOGY TO ENVIRONMENTAL MANAGEMENT

Edited by

DEBORAH J. ANTHONY
MICHAEL D. HARVEY
JONATHAN B. LARONNE
M. PAUL MOSLEY

ISBN Number 1-887201-29-7

U.S. Library of Congress Catalog Card Number: 00-110430

This publication is printed and bound in Denver, Colorado, United States of America.

Tribute to

Stanley A. Schumm

The symposium and the resulting book are a tribute to a scientist whose long career has been characterized by a deep intellect, unswerving intellectual honesty, curiosity and integrity, total unselfishness and a willingness to provide ideas and assistance to all, and to a generous, honorable and genuinely modest man who has contributed significantly to the advancement of process geomorphology over a 45-year career.

The symposium and book are also a tribute to Stan's wife, Ethel, who has unfailingly supported him in his endeavors over the years, and who has always provided a warm heart, a sympathetic shoulder, and a well-stocked table for generations of hungry graduate students during Stan's weekly graduate student seminars at the Schumm home.

INTRODUCTION

This book is the outcome of a symposium held to honor Dr. Stanley A. Schumm, University Distinguished Professor at Colorado State University in Fort Collins, Colorado. When informed that the symposium was being planned and that his presence would be appreciated, his response was typical: "Why would they want to do this for me?"

Stan's research accomplishments have been recognized by many other organizations. In addition to the Horton Award from the AGU given in 1957, he was awarded the Kirk Bryan Award by the Geological Society of America in 1979. The British Geomorphological Research Group awarded him the David Linton Award in 1982, and the National Academy of Sciences awarded him the G. K. Warren Prize in 1986. He was selected as a Fellow by the American Association for Advancement of Science in 1989, and presented with the Distinguished Career Award by the Quaternary Geology and Geomorphology Division of the Geological Society of America in 1997. His accomplishments as a teacher at Colorado State University were recognized by his receipt of the Outstanding Educator of America Award in 1974, the L. W. Durrell Award in 1980, and his election as a University Distinguished Professor in 1986.

His overall contribution to the fields of geology, geomorphology and civil engineering has been extended by his service on journal editorial boards (e.g. GSA Bulletin, Water Resources Research, Journal of Geology), his service on panels and committees (e.g. NASA, ASCE, NSF, NRC), and by his international activities as a visiting professor and fellow at institutions in Australia, New Zealand, Israel, United Kingdom, Poland, Japan, South Africa, Taiwan, and Venezuela.

Stanley A. Schumm - November 2000

PREFACE

The papers presented by his former graduate students and colleagues during the two-day "Schummposium," in November 1996, reflect Stan's widespread interest in fundamental fluvial processes and applied geomorphology, and his fostering of a longstanding and beneficial relationship between the Departments of Earth Resources and Civil Engineering at Colorado State University. The number of his former students that returned to honor him demonstrated the high esteem and respect with which Stan is held: approximately 40 of 63 former students attended. Stan's national and international influence is apparent from the locations from which they came: the United States, United Kingdom, New Zealand, Australia, Israel, and Japan.

The papers in this volume address fluvial process (Section 1), extreme events (Section 2), anthropogenic effects on fluvial systems (Section 3), applied fluvial geomorphology (Section 4), and engineering geomorphology (Section 5), all areas in which Stan has had an interest during his career, and to which he has made many significant contributions. Stan began his geomorphic career at Columbia University by chance, since his initial interest lay in structural geology. His dissertation work conducted under Dr. Arthur Strahler in the Perth Amboy Badlands led him to resolve the long-standing apparent conflict between the Penck and Davisian models of slope development. Hired by the United States Geological Survey (USGS), he headed west in 1954 and with Dick Hadley and Walter Langbein tackled the effects of lithology and climate on sediment yield in the arid and semi-arid western United States. For this work, he and Langbein were awarded the 1957 Horton Award from the American Geophysical Union. Employed at the Water Resources Division of the USGS until 1967, Stan focused on the streams of the Great Plains, the role of sediment type on channel dimensions and characteristics, and channel and floodplain change over time. From this, he developed a sediment load-based classification of channel types that has since been much used, not only by fluvial geomorphologists, but also by stratigraphers and sedimentologists. Also during this period Stan addressed the philosophical issues of time and space in geomorphology in his frequently cited paper with Robert Lichty. Paleohydrological studies of the Murrumbidgee River system in Australia provided a quantitative analysis of the effects of climate change on river size and characteristics. In recent years he has revisited Australia and delved into the processes controlling the anastomosed channels of the Ovens and King Rivers.

In 1967 Stan left the USGS and joined the faculty of the Earth Resources Department at Colorado State University, where over the next 30 years he developed an internationally recognized graduate program in geomorphology. His work and that of his graduate students, whose contributions he has always acknowledged, has covered: channels on the Moon and Mars; flume studies of a range of channel types, including braided, meandering and incised systems; large river systems such as the Mississippi, Missouri, Sacramento, Nile, and Indus Rivers; the impacts of catastrophic sediment delivery on channel dynamics in steep mountainous terrain and recovery of impacted channels; incised channel dynamics in humid and arid environments; the influence of active tectonics on modern rivers; climate- and man-induced river metamorphosis of the Platte and Arkansas Rivers; among other topics. The Rainfall-Erosion Facility (REF) at Colorado State University's Engineering Research Center was central to the many experimental studies that were conducted by his graduate students. Resulting models of landscape development combined with his earlier work on sediment yield and climate have provided the conceptual basis for applied work such as land reclamation following mining and other man-induced landscape-scale disturbances. Experimental and field-based studies were at the core of his concepts of thresholds and complex response that provided intrinsic explanations for a range of fluvial system changes previously explained by extrinsic agents such as climate and tectonics. As a result of his research, and that of his students, Stan has authored and co-authored over 150 papers and written or edited books on general geomorphology, fluvial geomorphology, drainage basin and slope morphology, incised channels, experimental fluvial geomorphology, active tectonics, and potential pitfalls in geomorphology (To Interpret the Earth: Ten Ways to be Wrong).

Much of Stan's work has been not only scientifically innovative, but also has had practical application. Sustainable watershed and river management, mineral exploration, mined-land reclamation, mitigation of the environmental impacts due to development, and conflict resolution in courts of law are all areas in which his ideas have found application. It is not surprising that most of his students have found employment in the earth sciences, natural resource management, and the environmental consulting industry, as well as in academia.

Stan's research accomplishments have been recognized by many other organizations in addition to the Horton Award from AGU mentioned above. He was awarded the Kirk Bryan Award by the Geological Society of America in 1979. The British Geomorphological Research Group awarded him the David Linton Award in 1982, and the National Academy of Sciences awarded him the G. K. Warren Prize in 1986. He was selected as a Fellow by the American Association for Advancement of Science in 1989, and presented with the Distinguished Career Award by the Quaternary Geology and Geomorphology Division of the Geological Society of America in 1998. His accomplishments as a teacher were recognized at Colorado State University by his receipt of the L. W. Durrell Award in 1980, and his election as a University Distinguished Professor in 1986.

We, the editors, regard our involvement in the "Schummposium" and production of this volume as a genuine privilege. It is an opportunity to record our own depth of gratitude to Stan and Ethel, as teachers, colleagues in science and in life, and friends. Thank you, both.

Deborah J. Anthony
Michael D. Harvey,
Jonathan B. Laronne
M. Paul Mosley
July, 2001

TABLE OF CONTENTS

Section 1

FLUVIAL PROCESSES

BAR FORMING PROCESSES IN GRAVEL-BED BRAIDED RIVERS, WITH IMPLICATIONS FOR SMALL-SCALE GRAVEL MINING

Dru Germanoski[1]

ABSTRACT

Gravel-bed braided rivers are typically high energy, unstable, rivers that transport large quantities of bedload. As such they are often utilized as sources of gravel for construction, road aggregate, and other engineering needs. Braided rivers are common in Alaska because glaciers, which provide high annual discharge and high sediment loads, are common and also because Alaska is mountainous which provides high regional gradients.

This study of braiding dynamics in the Gerstle, Robertson, and Toklat gravel-bed braided rivers suggests that a dominant braiding and braid bar-forming process is the result of lobate bar dissection and accretion. The three-dimensional morphology of a migrating lobate bar plays an active role in the transformation of an actively migrating bedform into a stationary, subaerially exposed braid bar. Stalled lobate bars are typically dissected by flow concentration through the axial depression or along a bar-bounding scour channel, which in each case results in the subaerial emergence of a portion of the lobate bar, and transformation into a braid bar. Lobate bars stall and are dissected and converted into braid bars over periods of hours to several days.

Channel surveys also show that bar development and migration through a section of channel results in the development of new braid channels or reactivation of abandoned channels due to changes in channel hydraulic geometry. The development and migration of a lobate bar into a channel can force the channel to become shallower. This forces flow overbank into a low point in the adjacent gravel flat which produces a new channel through avulsion. Rehabilitation of gravel mining sites can be enhanced by considering these dynamics when locating extraction sites.

The results of preliminary studies suggest that gravel can be extracted successfully from these rivers while maintaining the natural appearance of the rivers, provided that: extraction is limited conservatively within the constraints of bedload transport rates; gravel extraction occurs with the geometry of a curved channel that mirrors natural channel segments in width, depth, length, slope, and plan-view geometry (Karle, 1990), and that extraction trenches be positioned to facilitate replenishment based on an understanding of braided river dynamics.

---

[1] Geology Department, Lafayette College, Easton, PA 18042.

KEYWORDS

Braided rivers, sediment transport, gravel mining, braid bar formation, hydraulic geometry, gravel-bed rivers

1. INTRODUCTION

Multiple channel braided rivers are high energy, high gradient rivers which transport large quantities of coarse sediment, primarily as bedload. Fahnestock (1963) lists the factors conducive to braiding as (1) erodible banks, (2) rapid variation in discharge, (3) steep valley slope, and (4) abundant sediment load. Therefore, braided rivers are typically gravel-bed or sand-bed rivers that drain mountainous terrain or extend from mountains across piedmonts or plains having steep regional gradients such as the western high-plains of the United States.

Because stationary, subaerially exposed mid-channel bars are a fundamental feature of the braided pattern, processes of bar development have been the focus of a great deal of research effort and theoretical consideration. In a paper that may be considered as the seminal paper on braid development, Leopold and Wolman (1957) described a process of in-channel longitudinal bar development, which results from the deposition of the coarse fraction of the load as a lag deposit in the center of the channel. The initial deposit directs flow into the banks and promotes further deposition on the bar surface and margins. Subaerial exposure of the mid-channel bar occurs in response to incision in the bar margin channels and flow shifting into and eroding the banks. This mechanism of bar development has been considered by many as the primary process by which braid bars develop in braided rivers– particularly in gravel-bed braided rivers.

Observations made in sand-bed braided rivers demonstrate that stationary, subaerially exposed braid bars typically develop through the dissection of formerly migrating subaqueous bedforms, often referred to as transverse bars because the bar crests are oriented at an angle transverse to flow (Brice, 1964; Smith, 1971; Cant, 1978). The plan-view geometry of actively migrating bedforms can vary considerably and ranges from straight-crested bedforms to lobate or tongue-shaped features (Smith, 1971). Because of the common tongue-shaped geometry of these bedforms, they are often called linguoid bars (Allen, 1968; Collinson, 1970; Blodgett and Stanley, 1980). Observations in small-scale laboratory braided rivers using both sand (Ashmore, 1982; 1991) and sand- and gravel-sized sediment (Germanoski, 1990; Germanoski and Schumm, 1993) along with field observations in gravel-bed braided rivers (Smith, 1974; Church and Jones, 1982) imply that 3-D/transverse bar dissection may be the dominant braid bar forming process in fully braided rivers regardless of bed-material size (Rundle, 1985a, 1985b; Germanoski, 1990, 1993). The term "fully braided" is used here to describe a river that has multiple braid channels (greater than 3 and typically greater than 5) to distinguish these rivers from channels that only locally bifurcate around a middle-channel bar or mid-channel bar complex.

Braiding can also result from flow avulsion across a braid bar or braidplain during high-flow events. Church (1972) referred to this process as "secondary anastomosis." Ashmore (1982) observed the same process in a flume study of braided river development.

The initial objective of this research was to observe and document braid-bar forming processes in three large gravel-bed braided rivers located in central Alaska to determine the relative significance of grain-by-grain accretion (Leopold and Wolman, 1957), lobate/transverse bar dissection (Cant, 1978; Ashmore, 1991), and secondary anastomosis (Church, 1972). This work was inspired by laboratory channel observations that suggested that the dissection of formerly migrating bedforms was a very significant, if not dominant, process of braid bar development in both sand and gravel-bed channels (Ashmore, 1982, 1991; Germanoski, 1989, 1993; Germanoski and Schumm, 1993).

As the field research progressed it became clear that the results could also have implications for small-scale gravel mining operations that are becoming increasingly common in gravel-bed rivers, particularly at high latitudes. The need for sand and gravel for road aggregate is particularly significant in high latitudes because roadbeds must be raised by fill above the natural landscape to insulate permafrost from the low-albedo road surface. Seasonal road damage due to intense freeze-thaw, frost-heave, solifluction, and snowmelt runoff is also significant for gravel-bed roads in these regions. One of the rivers selected for the study of braiding processes is the Toklat River located in Denali National Park, Alaska. Coincidentally, the Toklat River was being evaluated as a source of gravel for annual maintenance and re-surfacing of the main road in Denali Park. The major objectives for developing a source of gravel within the park are to select sites where the resource would be renewable, to select extraction sites where natural renewal and healing would occur rapidly, and to develop extraction techniques that facilitate natural rehabilitation (Karle, 1990).

2. STUDY AREA AND FIELD PROCEDURES

Braided rivers are ubiquitous in Alaska because mountain glaciers that produce high discharge annually and high sediment loads are common and the mountains provide high regional gradients. The three rivers included in this study are located on the north side of the Alaska Range in east-central Alaska (Fig. 2.1). The Gerstle and Robertson Rivers are located approximately 60 km apart, and both drain meltwater from glaciers into the Tanana and then Yukon Rivers to the north. The Toklat River is located farther west in Denali National Park and it also drains glaciers on the north side of the Alaska Range into the Yukon drainage. The physical dimensions of the study area segments of the three rivers are given in Table 2.1.

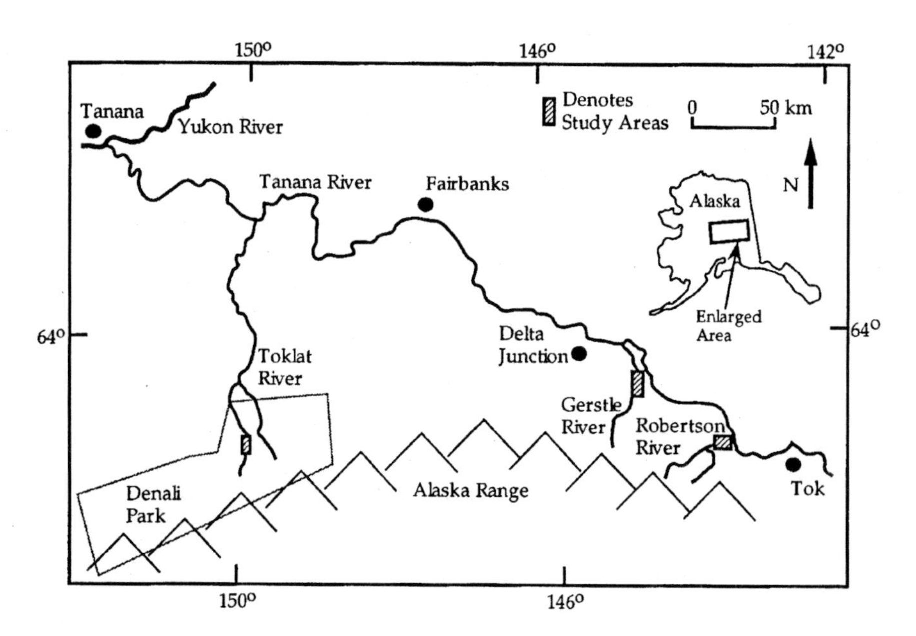

Fig. 2.1. Site location map showing the Gerstle, Robertson, and Toklat Rivers in east-central Alaska.

Table 2.1. Physical characteristics of the rivers at the study locations.

River	Drainage Area (km^2)	Braidplain Width (km)	Map Gradient (m/km)	Grain Median Diameter (mm)	Discharge* (m^3/s)
Robertson	1,485	0.8	6	10	130
Gerstle	565	0.6	10	19	70
Toklat	267	0.5	14	24	15

*Typical discharge measured during the time the river was being studied.

Each river is fully braided and had anywhere from 3 to 15 active channels at any given cross section during the field seasons. Flow depths ranged from centimeters to a maximum of 2 m in the Robertson River and channel widths ranged from meters to tens of meters. Field work was completed over a three year period from 1991 through 1993 during the months of June and July when glacial melt rates are high and river discharge and sediment transport rates are high (Lamke et al., 1990). Discharge and sediment transport rates were more than adequate to study bar forming processes and braiding dynamics in each instance because the river beds were active and bedforms were actively forming, migrating, and being transformed into braid bars in each river. Although flow depths and velocities were sufficiently high to permit active bedload transport and bar migration, conditions were still favorable for wading in active channels to measure flow velocities, bedload transport rates and to survey the channel bed.

Once a suitable study reach was located, the channel geometry was surveyed along multiple cross sections using total station surveying equipment. The channel margins, channel bed, bars, bedforms, and the water surface were repeatedly surveyed over several days to monitor channel change. In some cases, tag lines were stretched from steel reinforcing bar across active channels and measurements were taken relative to the taglines. Flow velocity was measured using an Ott rotating propeller flow meter, and bedload transport was measured using a standard 6-inch (152-mm) Helley-Smith bedload sampler. Flow and bedload data were collected along several transects across each bedform. The position of the flow meter or bedload sampler was determined by taglines or survey. The development of braid bars from the dissection of stalled lobate bars was documented with surveying data and photographs. Shear stress distributions over some submerged bars were determined from three-dimensional computer generated mesh maps based on surveys of the channel bed and water surface.

3. BAR TERMINOLOGY

The term "lobate bar" is used to denote migrating subaqueous bedforms that I formerly referred to as "linguoid bars" (Germanoski, 1987; 1989, 1990; 1991). Because the proliferation of bar terminology has proven to be an impediment to communicating our understanding of braided rivers (Smith, 1978), a simplified terminology scheme is adopted in this paper. Two types of channel-scale bedforms and bars are recognized here: (1) actively migrating lobate bars, and (2) stationary, subaerially-exposed braid bars.

"Lobate bars" are submerged, actively migrating, parabolic (or lobate) shaped bedforms bounded by an avalanche face along the downstream margin (Fig. 3.1). The lobate bars described here are equivalent to the linguoid bars

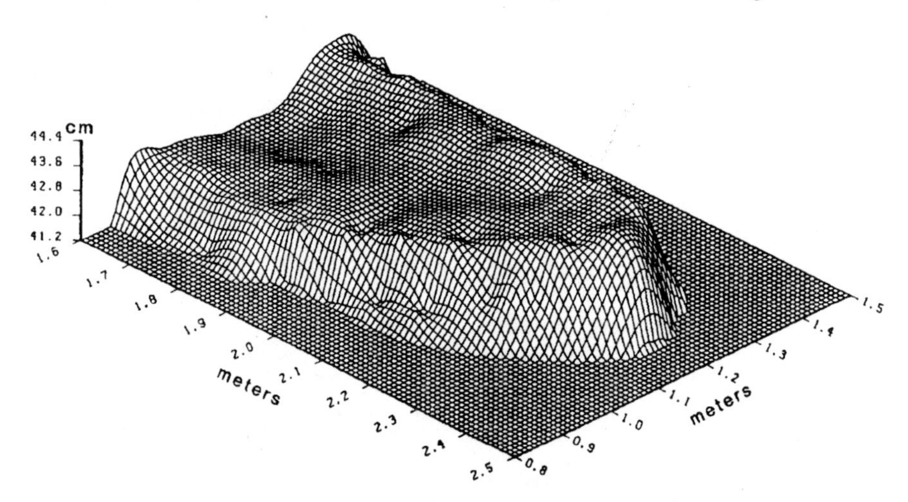

Fig. 3.1. **Three-dimensional morphology of an actively migrating lobate bar formed in a gravel-bed laboratory channel. Flow is from upper left to lower right. Note the parabolic geometry, axial depression on up-flow stoss side and high-angle avalanche surface along the downstream periphery. The vertical scale is exaggerated.**

described by Allen (1968), Collinson (1970), Boothroyd and Ashley (1975), and Blodgett and Stanley (1980), and they are essentially equivalent to the transverse bars described by Smith (1970, 1971, 1974), Church and Jones (1982) and others (see Miall, 1977 for a review). These bedforms exist in flume and natural river channels both as singular entities as well as periodically and semi-periodically spaced bedforms. Following the suggestions of an SEPM committee on bar and bedform nomenclature I believe these bedforms could be called "large, subaqueous 3-D dunes" (Ashley, 1990). However, the term "lobate bar" will be used throughout this paper following the lead of several authors in a recent Braided Rivers volume (e.g., Ferguson (1993), Leddy et al. (1993), in Best and Bristow, 1993).

"Braid bars" are stationary, subaerially exposed bars which are sites of sediment storage; they are elongate, oriented parallel or sub-parallel to streamflow and are referred to as braid bars because they separate flow into distinct channels. They are apparently equivalent to the mid-channel bars of Leopold and Wolman (1957), the braid bars of Smith (1974), the longitudinal bars of Williams and Rust (1969), Smith (1970), Boothroyd and Ashley (1975), the complex flats of Cant (1978) and Ashmore (1982), and the braid bar complexes of Ashley (1990). Braid bars can form or develop through a variety of processes, and therefore, as used here, the term braid bar is not meant to have any genetic connotation.

Subaerially exposed braid bars are frequently referred to as braid bar complexes because braid bars typically have complex histories and may be shaped and tangibly affected by a large number of erosional and depositional events (Ashley, 1990). Whether a simple braid bar formed by a single depositional event, or a more complex feature formed and was subsequently altered by a series of depositional and perhaps erosional events, a stationary, subaerially exposed bar separates flow into distinct channels and performs the same function in a braided river. Consequently, braid bars and braid bar complexes are viewed here as fundamentally the same feature with the exception that the term braid bar complex could be used to indicate that a particular braid bar bears unmistakable evidence of having experienced multiple depositional events.

4. SUBAQUEOUS LOBATE BARS AND BRAID BAR DEVELOPMENT

4.1. Lobate Bar Morphology

Observations of lobate bars in laboratory flumes and large, fully braided rivers revealed similar patterns of development, migration, and dissection. The parabolic-shaped lobate bars observed in both flume and natural braided rivers are dynamic features, which actively migrate through channels, although sometimes for only short distances. A well-developed lobate bar is characterized by an axial depression on the upstream surface, which forms in response to local flow convergence and scour (Fig. 3.1). Flow convergence in the axial depression causes scour and sustains transport of bed load that is transported onto the lobate bar from upstream. Flow diverges and fans out over the downstream portion of the lobate bar and bedload rolls down the downstream bar-bounding avalanche face. An apparent decrease in shear stress promotes deposition of a portion of the bedload on or near the crest of the avalanche face in the zone of flow divergence. The process of convergence and scour, and then divergence, deposition, and avalanche is well documented and has been described by Krigström (1962), Smith (1974), and more recently by Ashmore (1982, 1991; Ashmore and Parker, 1983), Rundle (1985a), Fujita (1989), Germanoski (1989, 1990, 1991), and Ferguson (1993).

Scour on the upstream portion and deposition on the downstream portion of the lobate bar allows the lobate bar to retain its overall shape while migrating down channel as a coherent entity or bedform. Bedload is continuously transported over the stoss (upstream) side of the bar, rolled down the avalanche face, and finally swept downstream by scour channels, which flank the bar-bounding slipface. Therefore, migrating lobate bars occur in zones of high bedload transport. The lee side avalanche faces are frequently steeper and have greater overall height on one side of the lobate bar than the other, which may in some instances grade imperceptibly into the channel bed. Differences in bedform height across a lobate bar were a common attribute of lobate bars observed in laboratory channels (Germanoski, 1989) as well as in natural braided rivers. Crowley (1983) described similar asymmetry in slipface elevation on dunes in the Platte River, Nebraska, which he referred to as major and minor slipfaces, respectively, and Gustavson (1978) described the same morphology on transverse bars in the gravel-bed Nueces River, Texas.

Migrating lobate bars have often been referred to as 'linguoid' bars because the parabolic or lobate shape is common in both laboratory and full-scale braided rivers (Fig. 4.1). However, many of the bars observed in all of these rivers have shapes, which are variations of the parabolic geometry such as nearly straight-crested transverse to flow, sinuous, or multi-lobed margins which form in response to local variations in the intensity and direction of the main flow paths. Although the lobate bar shown in Fig. 3.1 is a 3-dimensional bedform with a parabolic shape in planview, it is part of a more complex dual-lobed bedform composite.

Fig. 4.1. Lobate bar in the Robertson River, Alaska. The bar is approximately 6 m wide and grades into another adjacent bar whose avalanche face is visible on the right. Flow is toward the viewer.

The bars measured or surveyed in the Gerstle, Robertson, and Toklat Rivers are channel-scale features with widths on the order of channel widths and heights on the order of water depths (Figs. 4.2a and 4.2b). Because the lobate bars are subaqueous bedforms they are constrained by channel width and depth. The lobate bar widths are average widths based on several transects measured across each bar. Likewise, lobate bar heights and flow depths are average values based on several measurements of height along the lobate bar crests, and flow depths measured from the base of the avalanche face to the water surface. Whereas lobate bar widths are often closely constrained by channel boundaries, some bars are much narrower than the channels that

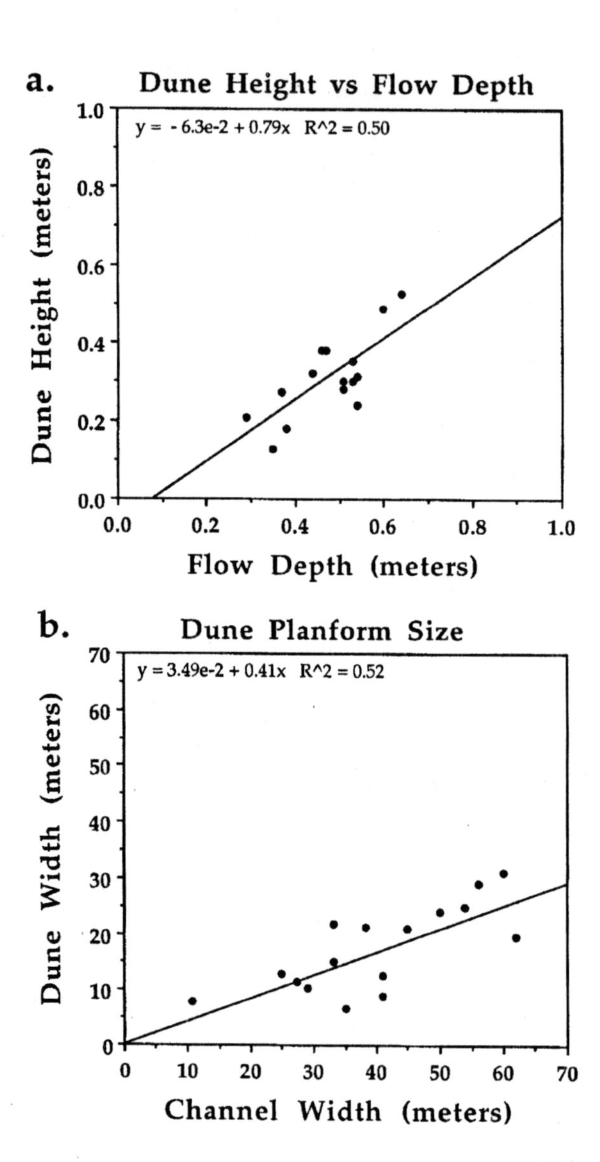

Fig. 4.2. (a) Relationship between lobate bar height and channel flow depth, and (b) Relationship between lobate bar width and channel width (all data are from the Gerstle, Robertson, and Toklat River, Alaska).

contain them (Fig. 4.2a). In either case, the bars are apparently related in scale to 3-dimensional flow structures that reflect either the interaction of flow and stationary braid bars bounding an individual channel, upstream channel boundaries or pools (Ferguson, 1993), or the interaction of flow structures with multiple, adjacent bedforms.

4.2. Lobate Bar Morphology and Flow

Lobate bar morphology, flow energy distribution, and bed-load transport appear to be closely related. Flow velocity and shear stress were consistent for all bars where measurements were taken as illustrated by the following examples. Average flow velocities and bed-load transport rates measured across an actively migrating lobate bar in the Robertson River are summarized in Fig. 4.3. Highest flow velocities were measured down the axis of the lobate

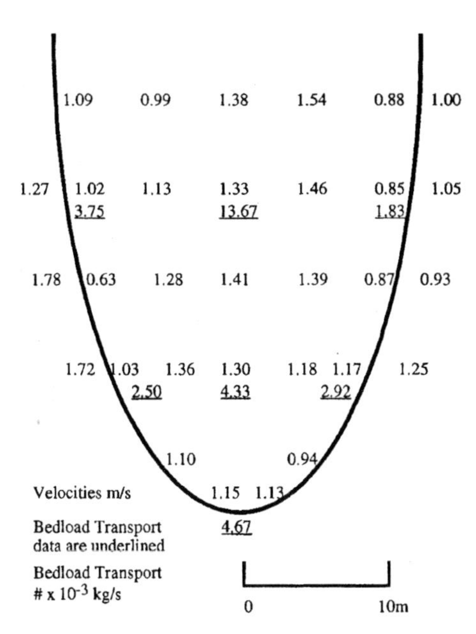

Fig. 4.3. Flow velocity and bed-load transport distribution over an active lobate bar in the Robertson River, Alaska. The bar is a schematic drawn to scale to show the relative distribution of flow intensity and bed-load transport. Bed-load transport rates are in kg/s entering the Helley-Smith.

bar in the zone of flow convergence. Bedload transport rates were highest over this portion of the lobate bar as might be expected. Flow velocities were also high in the bar-bounding scour channels, particularly where the lobate-bar-flanking scour channels converge at the downstream snout of the lobate bar (Fig. 4.3). High-energy flow in these scour channels sweeps bedload that avalanches down the slipface downstream. Lowest flow velocities were measured along the lobate bar crest, especially along the lateral margins of the top of the lobate bar (Fig. 4.3). This particular lobate bar was selected as an example because velocity and bedload data were collected over a short enough period of time (approximately an hour) that the lobate bar and associated flow structure maintained integrity. Although this was a particularly good example the results were consistent with data collected over several bars in the study rivers. Nonetheless, this data set is at best an approximate snapshot of a highly dynamic situation. Flow structure, intensity, and bedload transport rates would be expected to vary through time as evidenced by intensive measurements over a chute and lobe in a gravel-bed braided river by Ferguson and others (1992). Therefore, although actual values are time-varied, the relative distribution of flow velocity and bedload transport across the lobate bar appears representative.

Another example of the relationship between flow hydraulics and lobate bar morphology is illustrated by a large lobate bar surveyed in the Gerstle River. Figure 4.4a shows a 3-dimensional mesh map of this lobate bar while it was actively migrating. The similarity between this full scale lobate bar formed in a natural braided river and the lobate bar formed in a gravel-bed flume channel (Fig. 3.1) is obvious. The lobate bar has a parabolic shape in planview, an axial depression where flow is converging on the stoss side, lobate bar margin scour channels that converge at the downstream nose of the bar, and avalanche face crests that are the highest portions of the lobate bar (Fig. 4.4a).

The shear stress distribution over the lobate bar is also related to lobate bar morphology. Shear stress was calculated over the lobate bar using the Duboys equation:

$$\tau = \gamma d s \qquad (1)$$

where:

τ = shear stress
γ = specific weight of water
d = flow depth
s = local water-surface slope

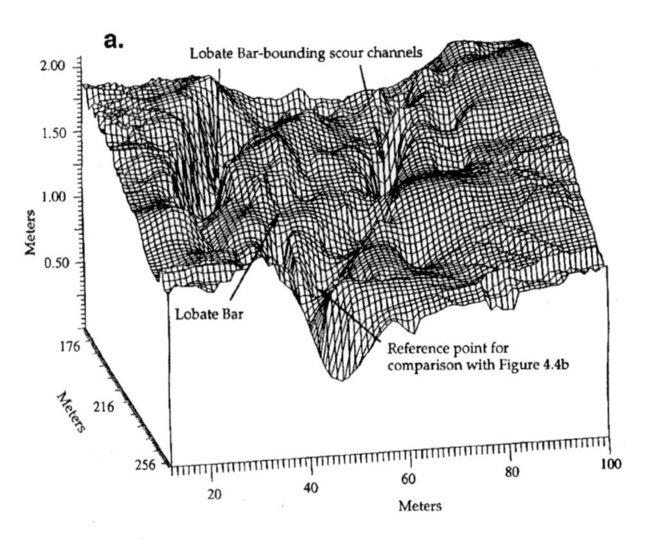

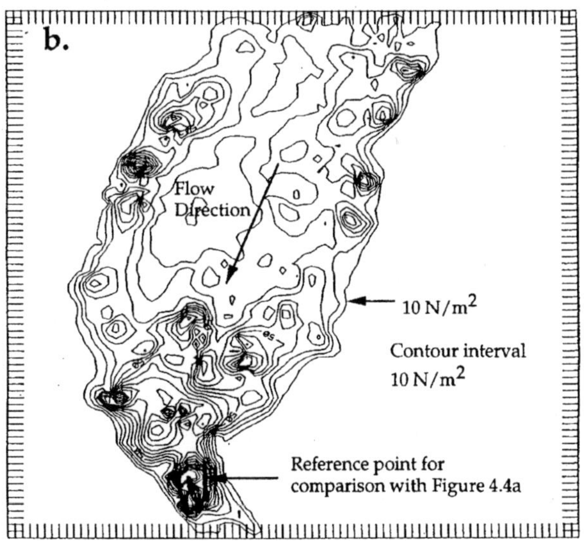

Fig. 4.4. **(a) Three-dimensional morphology of an actively migrating lobate bar formed in the Gerstle River. Note scales are distorted to present a view that shows the three-dimensional morphology to the best advantage;**
(b) relative shear stress distribution over the same bar in (a). Shear stress values are greatest down the axis of the bar and in the bar-bounding scour channels.

Specific weight of water was taken from tabled values for water at 10° C, depth was measured from a survey of the water surface and the bed of the channel, and local water-surface slopes were calculated from water-surface surveys. At each point, local surface slope was calculated by finding the direction of maximum slope through iteration and calculating local slope in that direction, parallel to the local flow line. This allowed slope and shear stress to be calculated around the lobate bar while taking into account a variable water surface with locally converging and diverging (or radial) flow over and around the bar.

The resultant shear stress distribution is shown in Fig. 4.4b. Although the actual shear stress values are unlikely to be absolutely correct due to the dynamic nature of the bar through the course of a topographic survey, the spatial distribution of shear stress is consistent with experiences wading over and around a large number of active bars in the Gerstle, Robertson, and Toklat Rivers. The highest shear stress values occur along the axis of the bar, and down the avalanche faces and lobate bar-bounding scour channels where water-surface slopes are steepest and flow depths are greatest (Fig. 4.4b). These are exactly the places where it is most difficult to wade and collect data as a result of high flow intensities.

Flow velocities over this particular lobate bar are also consistent with the shear stress distribution (Table 4.1). The highest flow velocities were measured in the lobate bar margin scour channels and along the axial trough on the stoss side of the lobate bar and the lowest velocities were measured at the crest of the avalanche face on the bar stoss surface (Table 4.1). These results are similar, but not as complete as those of Ferguson and others (1992) who as a large team made a successful effort to collect shear stress, velocity, and bedload transport data over a chute and lobe in the Sunwapta River.

Table 4.1. Average flow velocities across lobate bar #3.

	Average Velocity (m/s)	Velocity Range (m/s)
Bar Stoss Surface	1.19	0.94-1.34
Axial Trough	1.40	0.89-1.58
Flanking Scour Channels	1.65	1.08-2.13

4.3. Braid Bar Development

Direct observation of the development of stationary subaerially exposed braid bars in both small-scale laboratory braided rivers (Germanoski and Schumm, 1993) and in the Gerstle, Robertson, and Toklat Rivers indicates that the predominant mechanism of braid bar development was through subaqueous lobate bar dissection. This process has been described elsewhere as the dissection of "lobate bars" (Ashmore, 1982), the dissection of "tongue structures" (Rundle, 1985a), the "dissection and accretion of stalled linguoid bars" (Germanoski, 1989; 1990) and more recently as "transverse bar conversion" (Ashmore, 1991; Ferguson, 1993). In this paper, the process is called "lobate bar conversion" in order to be consistent with Ashmore's (1991) terminology ("transverse bar conversion") while incorporating the terminology "lobate bar" adopted by many authors in the recent *Braided Rivers* volume (Best and Bristow, 1993). The conceptual model of lobate bar dissection described here is based on observations of channel dynamics in sand- and gravel-bed braided rivers in a laboratory flume, sand-bed braided rivers (Germanoski, 1989, 1990), observations in the Gerstle, Robertson, and Toklat Rivers, and the observations and conclusions of Ashmore (1982, 1991) and others (Ferguson, 1993)

The development of braid bars by lobate bar dissection is a dynamic process resulting from the interaction between the fluid and the lobate bar topography, which continuously exert a mutual influence on each other. The 3-dimensional morphology (Fig. 3.1) of the bedform plays an important role in the interaction between the lobate bar and the flow. As the lobate bar migrates and sediment avalanches over the lee sides of the bedform, one or both flanks of the lobate bar may build in elevation, thereby shifting more flow away from the crests of the bar, and decreasing local shear stress. When local shear stress is reduced below the minimum necessary to sustain significant sediment transport and bedform migration, the lobate bar stalls. A portion of the lobate bar is left subaerially exposed above the water surface to form a braid bar, or nucleus of a larger braid bar complex (Fig. 4.5, Time 2). This process is favored by the fundamental geometry of the preexisting lobate bar because the bar margins are topographically higher than the rest of the bar (Figs. 3.1 and 4.4a). Although Fig. 4.5 Time 2 shows a braid bar nucleus forming from the flank margin of a dissected bar, both flank margins may be exposed when a lobate bar is dissected as the axial scour channel migrates downstream through the stalled lobate bar (Ashmore, 1982; Rundle, 1985a; Southard et al., 1984).

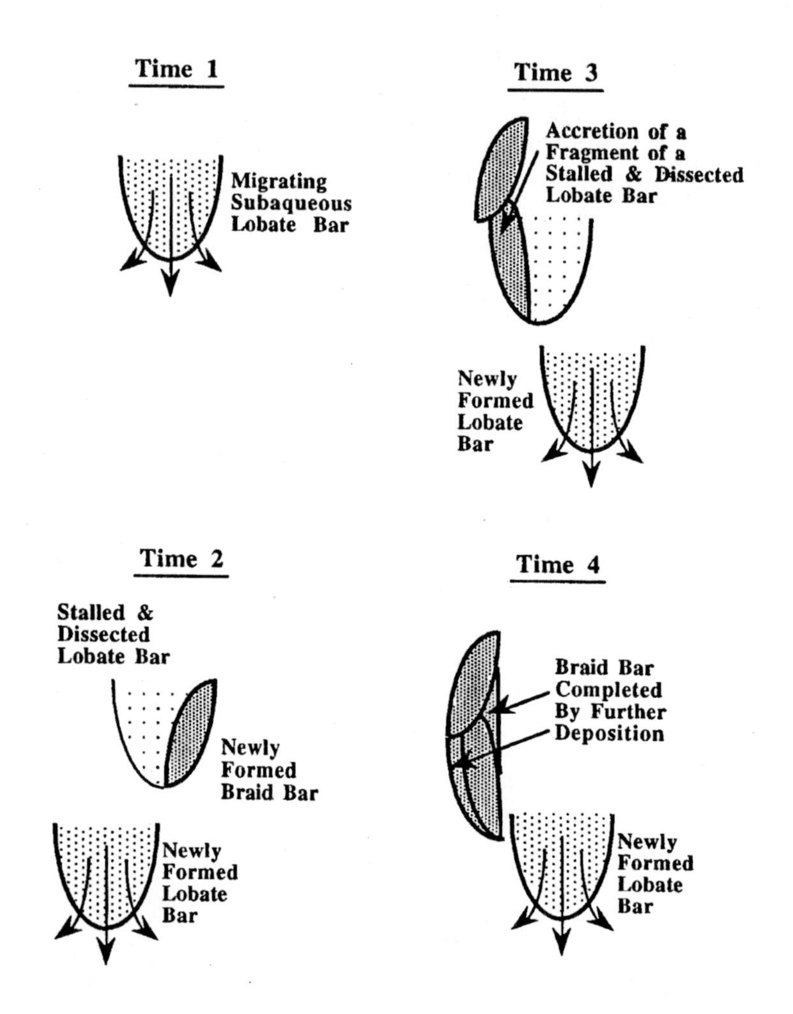

Fig. 4.5. **Process of braid bar development from the dissection and further accretion of portions of stalled lobate bars. See text for a description of the bar-forming process illustrated above.**

Once a braid bar forms through lobate bar dissection (or through any other process (Ferguson, 1993)), lobate bars migrating in channels adjacent to the newly formed braid bar may stall adjacent to the preexisting braid bar as a result of increased boundary effects created by the braid bar. The stalled lobate bar will then be dissected in a manner similar to its predecessor, and the stranded flank of the lobate bar becomes annealed onto the previously developed braid bar (Fig. 4.5, Time 3). Once a braid bar forms, it grows by: (1) further exposure of the initial lobate bar as flow shifts away from the newly formed exposed bar, (2) accretion of other passing lobate bar margins as flow shifts from the exposed braid bar, and (3) grain-by-grain accumulation of bed load on the braid bar margins, (Fig. 4.5, Time 4).

The development of braid bars from the dissection and further accretion of stalled lobate bars appears to be the same process observed in laboratory braided channels described by Ashmore (1982). On the basis of those observations, Ashmore argued that lobate bars are the "fundamental elements" of braided pattern development and the observations described in this paper support Ashmore's assertion. Rundle (1985a) has made similar observations in the gravel-bed Rakaia River in New Zealand, but he stressed the importance of receding discharge as a mechanism of lobate bar dissection (called "tongue structures" by Rundle, 1985a) whereas Ashmore (1982, 1991) and Germanoski (1989, 1990) demonstrated that this process could occur with constant discharge in laboratory braided rivers. Therefore, the development, cessation of migration, and dissection of an actively migrating lobate bar can occur as a result of bar and fluid dynamics under steady discharge. Careful examination of the interactions between the bedforms and the fluid illustrates the dynamic role played by both that causes the bedform to stall and be dissected under constant discharge.

There appears to be a negative feedback sequence that limits lobate bar migration in these gravel-bed braided rivers. Lobate bars develop in areas where flow convergence and high shear stress mobilize sufficient bedload to produce a lobate bar that migrates downchannel. As the slipface margin grows vertically from the deposition of a portion of the bedload that avalanches across the bar, flow depths and shear stress are progressively reduced along the downstream margins of the bar. Flow is simultaneously forced to converge by the bar's morphology into the axial depression in the bar surface and in the bar-bounding scour channels (Figure 3.1). Further scour in one or more of these channels either dissects the bar or leaves it stranded to form a braid bar.

4.4. Specific Field Examples of Lobate Bar Dissection and Braid Bar Development

Field observations suggest that lobate bar dissection can occur under steady or fluctuating discharge regimes. However, the processes are facilitated by a decrease in discharge over a bar. The decrease in flow over an individual lobate bar or complex of bars can occur either as a result of a decrease in discharge across the entire braidplain or as a result of a more localized reach-scale shift of flow away from a bar-bounding channel. Observations of braid bar development over three field seasons revealed that braid bars developed in the Gerstle, Robertson, and Toklat Rivers primarily through the process of lobate bar dissection and accretion illustrated by Fig. 4.5.

Subaqueous lobate bars occurred both as solitary bedforms (Fig. 4.6) as well as in groups or clusters stacked en échelon (Fig. 4.7a). Therefore, braid bars formed from solitary bars, the successive accretion of solitary bars, or from the dissection or emergence of clusters of bars. In some instances bars stalled and were converted into braid bars in the center of active channels following the sequence of events illustrated in Fig. 4.5, whereas in other instances, lobate bars stalled adjacent to unrelated, preexisting braid bars (Fig.

Fig. 4.6. Braid bar in the Gerstle River formed through the dissection and accretion of lobate bars. The lobate bar in the center of the photograph is being annexed onto the braid bar on the right as flow shifts to the left of the lobate bar.

a.

b.

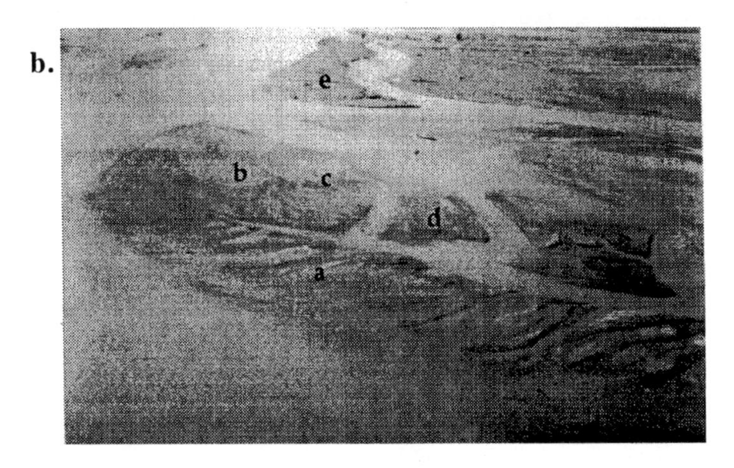

Fig. 4.7. **(a) Complex of stalled lobate bars in the Robertson River, Alaska;**
(b) one day later, the same lobate bars have been transformed into a braid bar as a result of a decrease in discharge. See text for discussion.

4.6). Figure 4.6 shows a large parabolic-shaped lobate bar that stalled and is emerging adjacent to a braid bar that had developed previously. As flow shifts away from the lobate bar and the preexisting braid bar, the lobate bar begins to emerge and form a braid bar complex with the pre-existing bar as a nucleus (Fig. 4.6; similar to conceptual scenario illustrated in Fig. 4.5 Time 3).

Figures 4.7a and 4.7b provide an example of braid bar development through lobate bar conversion that resulted from an obvious decrease in discharge over a two-day period in the Robertson River. Figure 4.7a shows the lateral slipface margin of a large active lobate bar (labeled a) adjacent to a group of low amplitude lobate bars labeled b, c, and d. The next day the lobate bar complex was fully emergent, and formed into a braid bar as a direct result of falling discharge. Figures 4.7a and 4.7b also show the development of another braid bar (labeled e) upstream and to the right of the lobate bar complex, also as a result of falling stage. In Fig. 4.7a, the stalled slipface of the lobate bar is visible as an incipient braid bar bounded by an active slipface-bounding scour channel, and then a day later fully emergent as a braid bar (Fig. 4.7b).

The lobate bar shown in Fig. 4.8a formed, stalled, was dissected, and became the nucleus for two braid bars that grew through accretion over a four-day period. After the lobate bar was fully developed and stalled, scour occurring in the axial channel dissected the lobate bar to form two braid bars, one on either side of the axial channel. Scour in the axial channel left the lobate bar margin avalanche crests above the surrounding water surface as the core of two braid bars (Fig. 4.8). The braid bars become larger as discharge dropped and flow shifted to other braid channels toward the center of the braidplain (to the left in Fig. 4.8). The braid bars shown in Fig. 4.8 were further modified by accretion of sediment along the margins and the emergence of the channel bed as flow shifted away from the bars and discharge decreased in this portion of the braidplain. The development and growth of these braid bars resulted from a combination of lobate bar and flow dynamics (lobate bar migration, growth, and dissection) and braid bar growth as a result of a local decrease in discharge as flow shifted away from the bar. Therefore, although the bar nuclei developed from the dissection of a bar, the braid bars developed and were modified by multiple events occurring over several days.

Figures 4.9a and 4.9b show another example of the dynamics of braid bar development and accretion from a migrating lobate bar over another four day period on the Gerstle River. From July 10 to early in the morning of July 12 a lobate bar formed and migrated several meters (Fig. 4.9a). Discharge and bedload transport through the reach increased through this period of three days (Fig. 4.9a). As discharge and bedload transport increased into the afternoon of July 12th the lobate bar migrated to the north (Fig. 4.9a). By July 14 the lobate bar had migrated more into the channel toward the east and ultimately stalled

Fig. 4.8. Braid bars formed from the dissection of a stalled lobate bar. The braid bars were formerly the avalanche face surface of the lobate bar shown in Fig. 4.4. As scour progressed down the axis of the lobate bar, the flanks of the bar emerged above the water surface. The braid bars were enhanced by a decrease in discharge through the reach and sediment deposition on the braid bar margins.

and was dissected as discharge decreased. The two main fragments of the lobate bar were attached to the pre-existing braid bar in the upper right-hand portion of Fig. 4.9b. In this case, the lobate bar that formed on July 10 formed in a channel between two pre-existing braid bars. It's further development and movement was influenced both by flow splitting downstream into two separate channels and also by the existence of the braid bar downstream to the right that blocked further downstream migration.

Whereas, laboratory results demonstrate that braid bars can form through the process of lobate bar conversion under constant discharge (Ashmore, 1982, 1991; Germanoski, 1989) the field results indicate that this process is facilitated by falling stage. This conclusion was also drawn by Rundle (1985a, 1985b) on the basis of his research on the Rakaia River in New Zealand. This is particularly true in glacial outwash rivers where discharge and

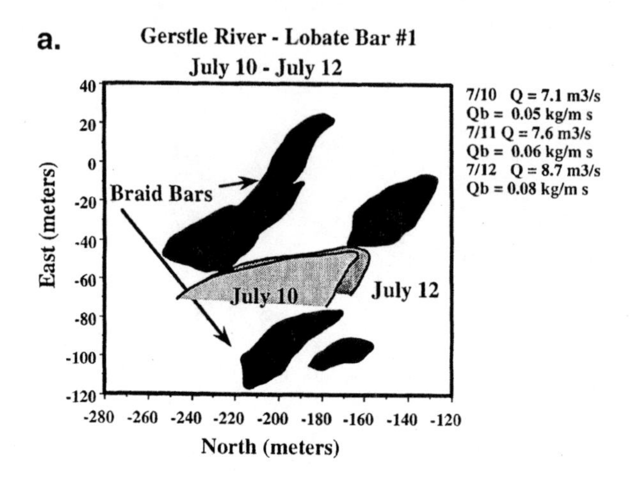

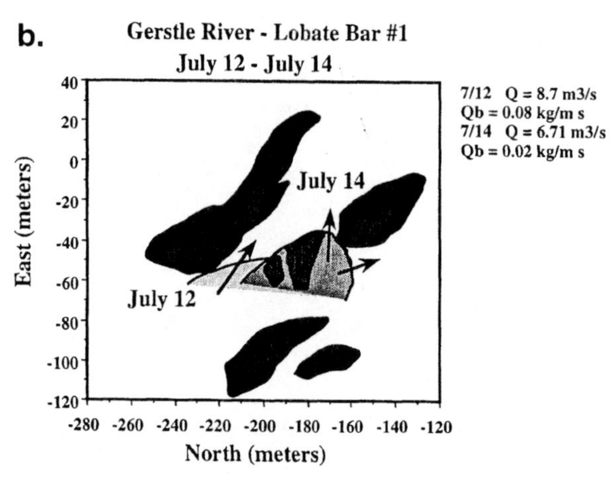

Fig. 4.9. **(a) Lobate bar migration over a two-day period in the Gerstle River, Alaska;**
(b) two days later, the bar had migrated toward the upper left hand portion of the map area, stalled and was dissected. The emergent fragments of this lobate bar merged with the pre-existing braid bar on the right.

local shear stress fluctuations may occur as a result of diurnal variation in glacial melting rates or as a result of several days of cloud cover alternating with several days of sunshine.

5. LOBATE BAR MIGRATION AND CHANNEL AVULSION

Weather-related increases in discharge also resulted in the re-occupation of abandoned anabranch channels as well as the development of new anabranch channels across braid bars and braid bar complexes on the Gerstle, Robertson, and Toklat River braidplains. The importance of re-occupation of abandoned channels during high discharge events has been emphasized by Church (1972) who referred to this process as "secondary anastomosis". The process has also been described by Krigström (1962) and Williams and Rust (1969) in outwash settings, and also by Ferguson and Werritty (1983) who called the process "channel switching". The channel switching described by Ferguson and Werritty (1983) frequently occurs as water is displaced from the main channel due to migration of an active bar which forces water over bank. This mechanism of avulsion has been termed "choking avulsion" by Leddy and others (1993) who examined processes of anabranch avulsion in a laboratory flume. Channels formed across the braidplain through "secondary anastomosis," "channel switching," or "choking avulsion," often exploited the topographic irregularities on the braidplain.

Observations in the Gerstle, Robertson, and Toklat Rivers show that avulsions are often driven by the development and migration of 3-D bedforms (the choking avulsion mechanism of Leddy et al., 1993). After witnessing this process in the Robertson and Gerstle Rivers, several reaches of the Toklat River were selected to monitor for a potential avulsion. Sites were selected where flow intensity was relatively high in channels located adjacent to abandoned braid bar bisecting channels, or where there were low areas on braid bars adjacent to channels where flow intensity was high. Figure 5.1 shows a reach where an avulsion occurred that we monitored over a period of five days. On June 9, 1993, the reach was characterized by high intensity flow and relatively high bedload transport. On June 10, discharge through the channel was 3.65 m^3/s and bedload transport was 2.44 kg/s. A pair of lobate bars developed and migrated into the reach on July 10 (Fig. 5.1a). When the bars migrated into the channel the channel aggraded an average of 28 cm and the average flow depth decreased in the channel from 26 centimeters to 8 cm. Discharge in the channel decreased from 3.7 to 1.5 m^3/s on July 11. Therefore,

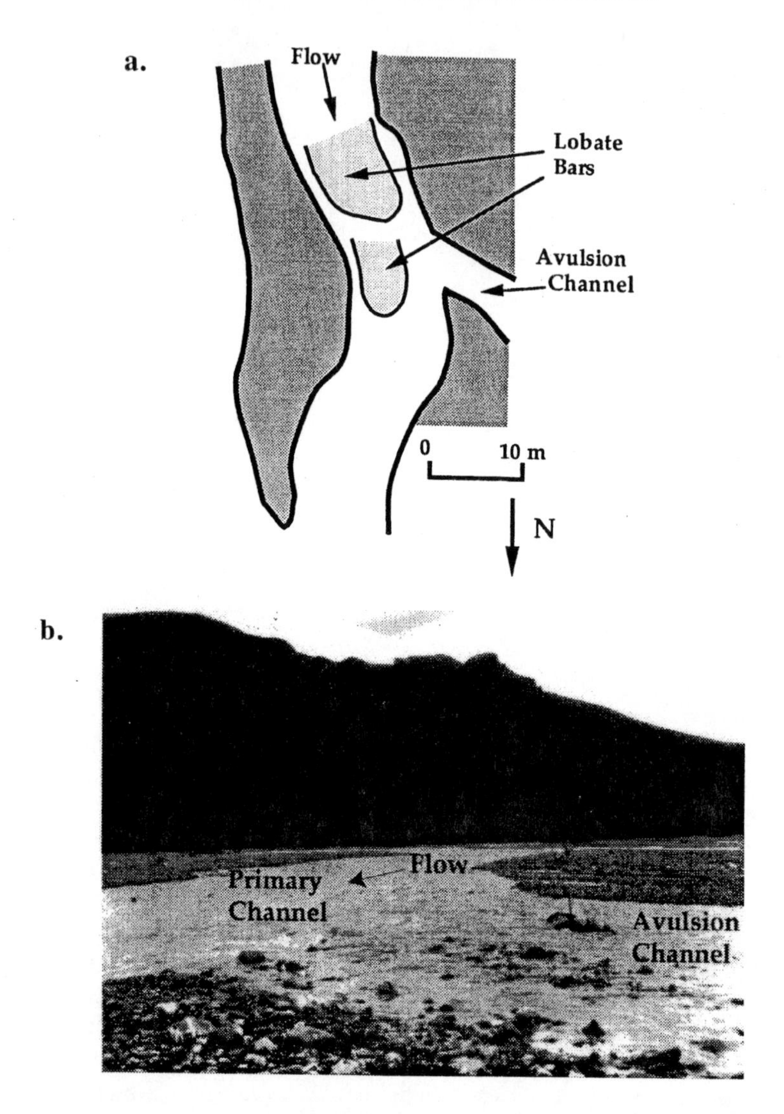

Fig. 5.1. (a) Map view of an avulsion channel in the Toklat River. Channel capacity was reduced in the primary channel when two lobate bars formed and migrated through the reach. As the lobate bars migrated into the channel, water was displaced into an abandoned braid channel on the right.
(b) View looking upstream; the avulsion channel is on the right and the primary channel is on the left.

the bars decreased the channel's capacity to convey water and forced the avulsion by displacing water from the active channel into an adjacent abandoned channel (Fig. 5.1b). Although avulsions are often driven by increasing discharge, this particular avulsion occurred during a period of decreasing discharge through the reach. Thus, the avulsion occurred primarily because the migrating bars filled the channel with sediment and displaced water from the primary channel.

6. IMPLICATIONS FOR SMALL SCALE GRAVEL MINING

The primary objective of mining gravel in Denali Park for road maintenance is to preserve the natural river equilibrium and natural river morphology while extracting gravel (Karle, 1990). At the time of the fieldwork for this study three test extraction trenches had been excavated on the Toklat River, one oriented parallel to flow, one oriented transverse to flow, and one designed to mimic the plan view and hydraulic geometry of natural channels in the reach (Karle, 1990). All three extraction pits were re-filled within two years by normal streamflows. However, the channel-like extraction pit was rehabilitated within 3 months, and in fact, because it mimicked a natural channel, it wasn't identifiable as a man-made feature to park visitors, even when newly created (Karle, 1990). The channel-like extraction pit filled in most rapidly apparently because it functioned as an active channel as a result of its geometry. The total extractable volume of material for this river to maintain equilibrium based on bedload transport measurements and integration of the flow duration curve was recommended conservatively at 5 percent of total annual bedload (Karle, unpublished report).

Understanding the dynamics of these gravel bed braided rivers can aid in developing a gravel mining procedure that will have a minimum impact on the systems and facilitate rehabilitation of the mine sites. The caveat is that mining has to be on a small scale in terms of both the physical dimensions of the extraction plot and the rate of extraction. Because lobate bars are channel-scale features they can effectively refill mined excavations, provided that the excavated channels are similar in size to natural channels and they are sited adjacent to, or downstream of, major channels with significant bedload transport. Observations in the Gerstle, Robertson, and Toklat Rivers support Ashmore's (1993) assertion that bars are very likely to develop downstream of the confluence of braid channels. Therefore, siting extraction channels downstream of channel confluences, which could be diverted into the excavation channels would more likely facilitate rapid sediment filling and channel rehabilitation through lobate bar migration. Natural avulsions can be used to maximize rehabilitation potential by siting the extraction channels downstream of pre-existing low points adjacent to major channels, which are located downstream of channel confluences.

7. CONCLUSIONS

Field observations and channel monitoring revealed that braid bars very typically form through the process of lobate bar dissection in the Gerstle, Robertson, and Toklat Rivers and lend further support to the hypothesis that lobate bar dissection is a primary (if not the primary) mechanism of braid bar development in most fully-braided sand- and gravel-bed braided rivers (Ashmore, 1991). Lobate bars may stall and become dissected as a result of the dynamic interaction between the developing bedform and the fluid under constant or transient discharge conditions. Once a braid bar forms, it may grow through the accretion of portions of additional lobate bars. However, braid bar development through this process can be aided significantly by a decrease in discharge. Given that many braided rivers are glacial outwash streams, which are inherently characterized by fluctuating discharge, it becomes clear that discharge reduction plays an active role in generating braid bars in many braided rivers. In these gravel-bed rivers the bars form and migrate short distances (meters to tens of meters) before stalling, becoming dissected, and being transformed into stationary, subaerially exposed braid bars.

Lobate bar development and migration through a section of channel also facilitated the development of new braid channels, or reactivation of abandoned channels, as a result of changes in channel hydraulic geometry. Ashmore (1982) observed the same process in his laboratory rivers and argued that while the process of channel switching occurs in braided systems, it is not the process by which the braided pattern develops initially.

Braided rivers can be used as gravel resources provided that gravel is mined within the constraints of annual bedload transport, and extraction pits mimic natural channel geometry. Extraction channel siting with consideration of river dynamics can facilitate channel rehabilitation.

8. ACKNOWLEDGEMENTS

Fieldwork in Alaska was supported by grants from the American Chemical Society Petroleum Research Fund grant PRF# 25016-B2 and Lafayette College. I am grateful to the US National Park Service for providing access to the Toklat River and lodging in Denali National Park, Alaska. Laboratory flume research was supported by the US Army Research Office Grant No. DAAG29-84-K-O189. Duncan Thomas, Geoffrey Hickey, and A. Randall Oser provided field assistance under harsh conditions and they helped process data. The manuscript was greatly improved by Michael D. Harvey's and Deborah Anthony's comments on an earlier draft.

9. REFERENCES

Allen, J.R. L. 1968. Current Ripples, Their Relation to Patterns of Water and Sediment Motion. Elsevier, Amsterdam.

Ashley, G.M. 1990. Classification of large-scale subaqueous bedforms: a new look at an old problem. Journal of Sedimentary Petrology, 60:160-172.

Ashmore, P.E. 1982. Laboratory modeling of gravel braided stream morphology. Earth Surface Processes and Landforms, 7: 201-225.

Ashmore, P.E. 1991. How do gravel-bed rivers braid? Canadian Journal of Earth Science, 28:326-341.

Ashmore, P.E. 1993. Anabranch confluence kinetics and sedimentation processes in gravel-bed braided streams. In Best, J.L. and Bristow, C.S. (eds), 1993, Braided Rivers. Geological Society Special Publication 75:129-146.

Ashmore, P.E. and Parker, G. 1983. Confluence scour in coarse-braided streams. Water Resources Research, 19:392-409.

Best, J.L. and Bristow, C.S. (eds), 1993, Braided Rivers. Geological Society Special Publication 75.

Blodgett, R.H. and Stanley, K.O. 1980. Stratification, bedforms, and discharge relations of the Platte River System, Nebraska. Journal of Sedimentary Petrology, 50:139-148.

Boothroyd, J.C., and Ashley, G.M. 1975. Processes, bar morphology and sedimentary structure on braided outwash fans, Northeastern Gulf of Alaska. Society of Economic Paleontologists and Mineralogists Special Publication, 23:193-222.

Brice, J.C. 1964. Channel patterns and terraces of the Loup River in Nebraska. U.S. Geological Survey Professional Paper 422-D.

Cant, D.J. 1978. Bedforms and bar types in the South Saskatchewan River. Journal of Sedimentary Petrology, 48:1321-1330.

Church, M.A. 1972. Baffin Island sandurs: a study of arctic fluvial processes. Canada Geological Survey Bulletin 216.

Church, M.A. and Jones, D. 1982. Channel bars in gravel-bed rivers. In R.D. Hey, J.C. Bathurst, and C.R. Thorne (eds.), Gravel Bed Rivers. John Wiley & Sons, New York 291-338.

Collinson, J.D. 1970. Bedforms of the Tana River, Norway. Geografiska Annaler, 52A:32-56.

Crowley, K.D. 1983. Large-scale bed configurations (macroforms), Platte River Basin, Colorado and Nebraska: primary structures and formative processes. Geological Society of America Bulletin, 94:117-133.

Fahnestock, R.K. 1963. Morphology and hydrology of a glacial stream - White River, Mount Rainier, Washington.U.S. Geological Survey Professional Paper, 422-A.

Ferguson, R.I. 1993. Understanding braiding processes in gravel-bed rivers: progress and unsolved problems. In Best, J.L. and Bristow, C.S. (eds), 1993, Braided Rivers, Geological Society Special Publication 75:73-87.

Ferguson, R.I., and Werritty, A. 1983. Bar development and channel changes in the gravelly River Feshie, Scotland. Special Publication of the International Association of Sedimentologists 6:181-193.

Ferguson, R.I., Ashmore, P.E., Ashworth, P.J., Paola, C., and Prestegaard, K.L. 1992. Measurements in a braided river chute and lobe 1. Flow pattern, sediment transport, and channel change. Water Resources Research, 28:1877-1886.

Fujita, Y. 1989. Bar and channel formation in braided streams. In S. Ikeda and G. Parker (eds.) River Meandering, AGU Water Resources Monograph 12:417-462.

Germanoski, D. 1987. Experimental study of braided river morphology. EOS, 68:44: 1297.

Germanoski, D. 1989. The effects of sediment load and gradient on braided river morphology. Unpublished Ph.D. Dissertation, Colorado State University, Fort Collins, Colorado.

Germanoski, D. 1990. Comparison of bar-forming processes and differences in morphology in sand- and gravel-bed braided rivers. Geological Society of America Abstracts with Programs, 22:6:A110.

Germanoski, D. 1991. Reassessment of the importance of linguoid bars in gravel-bed braided rivers. EOS, 72:17:138.

Germanoski, D., 1993. Braid bar development through the dissection of stalled linguoid dunes in gravel-bed braided rivers. Third International Geomorphology Conference, Programme with Abstracts, p. 145. Hamilton, Ontario Canada.

Germanoski, D., and Schumm, S.A., 1993. Changes in braided river morphology resulting from aggradation and degradation. The Journal of Geology, 101:451-466.

Gustavson, T.C. 1978. Bed forms and stratification types of modern gravel meander lobes, Nueces River, Texas. Sedimentology, 25:401-426.

Karle, K.F. 1990. Renewable gravel sources for road maintenance in Denali National Park and Preserve. Park Science, A Resource Management Bulletin, 10:5.

Krigstrom, A. 1962. Geomorphological studies of the sandur plains and their braided rivers in Iceland. Geografiska Annaler, 44:328-346.

Lamke, R.D., Bigelow, B.B., VanMaanen, J.L., Kemnitz, R.T., and Novcaski, K.M. 1990. Water Resources Data Alaska Water year 1990, U.S. Geological Survey Water-Data Report AK-90-1.

Leddy, J.O., Ashworth, P.J., and Best, J.L. 1993. Mechanisms of anabranch avulsion within gravel-bed braided rivers: observations from a scaled physical model. In Best, J.L. and Bristow, C.S. (eds), 1993, Braided Rivers, Geological Society Special Publication 75:119-127.

Leopold, L.B., and Wolman, M.B., 1957. River channel patterns: Braided, meandering and straight. U.S. Geological Survey Professional Paper 282-B:39-85.

Miall, A.D. 1977. A review of the braided-river depositional environment. Earth Science Review, 13:1-62.

Rundle, A.S. 1985a. The mechanism of braiding. Zeitschrift für Geomorphologie, Supplement Band, 55:1-13.

Rundle, A.S. 1985b. Braid morphology and the formation of multiple channels, the Rakaia, New Zealand. Zeitschrift für Geomorphologie, Supplement Band, 55:15-37.

Smith, N.D. 1970. The braided stream depositional environment: comparison of the Platte River with some Silurian clastic rocks, N. Central Appalachians. Geological Society of America Bulletin, 81:2993-3014.

Smith, N.D. 1971. Transverse bars and braiding in the lower Platte River, Nebraska. Geological Society of America Bulletin, 82:3407-3420.

Smith, N.D. 1974. Sedimentology and bar formation in the Upper Kicking Horse River, a braided outwash stream. Journal of Geology, 82:205-223.

Smith, N.D. 1978. Some comments on terminology for bars in shallow rivers. In Miall, A.D. (Ed), Fluvial Sedimentology, Canadian Society Petroleum Geologists, Mem. 5:85-88.

Southard, J.B., Smith, N.D., and Kuhnle, R.A. 1984. Chutes and Lobes: Newly identified elements of braiding in shallow gravelly streams. In: E.H. Koster and R.J. Steel (eds.), Sedimentology of Gravels and Conglomerates. Canadian Society Petroleum Geologists, Mem. 10:51-59.

Williams, P.F. and Rust B.R., 1969. The sedimentology of a braided river. Journal of Sedimentary Petrology, 39:649-79.

SHIFTING STAGE-VOLUME CURVES: PREDICTING EVENT SEDIMENTATION RATE BASED ON RESERVOIR STRATIGRAPHY

Jonathan B. Laronne[1] and Ralf Wilhelm[2]

ABSTRACT

Reservoirs undergo continuous decrease in their storage volume due to sedimentation. New stage-volume relations cannot always be determined by costly remapping. A methodology based on sedimentologic and geomorphic principles is established to evaluate this continuous shift. It not only affords a procedure to determine changing reservoir capacity, but it is also a key to evaluating historic event sediment yields. In the extreme arid environment of the Arava Rift Valley, these are shown to be low in comparison to sediment yields in the semiarid northern Negev Desert of Israel. The revisited Langbein-Schumm curve is demonstrated to apply even in this extremely harsh environment.

1. THE PREDICAMENT OF THE RESERVOIR HYDROLOGIST

A reservoir stage-volume curve is produced once dam and earth works have been completed. Such a curve is based on the original reservoir topography and should be continuously altered to account for sedimentation. Because reservoir topography commonly is not remapped regularly, flood volumes may be overestimated, particularly where sedimentation rates are high. Mapping of reservoir topography by soundings and land surveys has become cheaper and can be completed faster with total station, EDM theodolites (the latter where reservoirs are dry part of the year), but the remapping exercise is still too expensive to be frequently undertaken.

---

[1] Department of Geography & Environmental Development, Ben Gurion University of the Negev, P.O., Box 653, Beer Sheva 84105, Israel

[2] Institut fur Wasserbau, Ingineur Hydrologie und Wasserwitschaftung, Technische Hochschule Darmstadt, Darmstadt 64287, Germany.

Hence, to ascertain stored water volumes requires evaluation of the rate of reservoir sedimentation, utilizing it to periodically alter stage volume curves. Sediment cores of reservoir deposits may be measured to determine the average sedimentation rate. Where flow variability is considerable, it is inappropriate to modify the stage-volume curve by a steady, linearly decreasing capacity (which excludes the slow but measurable nonlinear decrease in trap efficiency). Utilizing a geomorphic-sedimentologic approach, we have attempted to identify event sediment layers in reservoirs, compute their volume and relate it to water inputs, for the purpose of establishing a continuously varying stage-volume curve based on historic event sediment yields.

2. THE GEOMORPHIC PROCESSES

Floodwater entering a reservoir or a lake transports both suspended sediment and bed load. The impingement of the flowing stream into the standing body of water involves a transfer of momentum, manifested by a decrease in flow. Measurements of the velocity field in a jet impinging into a standing body of water demonstrate that velocity gradients are very steep sideways, and they are gradual along the longitudinal jet axis (Albertson et al., 1950; Shteinman et al., 1993; Yu and Lee, 1993). Hence, bed-load deposition is gradual along the longitudinal jet axis because cessation of bed-load transport occurs under conditions similar to those of initiation of motion (Reid and Frostick, 1984). It is, however, abrupt cross-sectionally. Accordingly, the coarsest sediment can be transported only to fan-like depositional proximal areas, the upstream part of which has a longitudinal slope less steep than that of the original bed. Gravel will often be deposited in the proximal area, although in ephemeral reservoirs fine gravel and sand may, at times of a steep flood rise, be driven farther downstream while part of the reservoir is still dry. As the reservoir fills, consecutively more proximal areas are inundated. Hence, sand and fine gravel are deposited gradually upstream during reservoir filling. The depositional processes involving bed load are affected by pulses in the flood hydrograph (Laronne, 1988) and by moving bedforms (Allen, 1982).

Unlike bedload, suspended sediments are carried into, and distributed throughout, the entire reservoir. The suspended sediment may settle as soon as turbulence decreases distally and sideways from the turbidity current due to dampening and due to flood recession. The depth and the texture of the deposited suspended sediments depends not only on their location with respect the intruding turbidity current, but it also depends on the total depth of flow. A flood with several rises may lead to a temporary cessation of settling and resuspension (Bloesch, 1994), yet flood recession must ultimately bring about a progressive decrease in turbulence within the water body, leading to the settling of gradually finer sediment (Laronne, 1990).

3. THE STRATIGRAPHIC PRODUCT

The resultant stratification of flood sediment in a reservoir is a couplet, the lower singlet being the nongraded traction deposit, often sand, and the upper singlet being graded silt and clay (Laronne, 1987). Event stratification can be discerned in the field due to lamina (and bedding) surface partings (Friedman and Sanders, 1978). The partings are caused by high cohesion within the graded fines and low cohesion on the plane separating the clayey loam (settled at the end of an event) and the sand (settling at the onset of the following event). Stratification is also discerned due to the contrast between the lower and upper singlets, resulting from variations in moisture and organic contents and resistance to weathering.

Each couplet is deposited during one 'event' (Fig. 3.1). Couplets can be identified if most of the traction load has accumulated and most of the suspension has settled. Because the upper singlet is loamy, often a clay loam, incision into it is restricted to the incoming jet stream. Indeed, this is manifested by the entirety of upper singlets as viewed in trenches, be it in proximal or in distal areas (Fig. 3.1). A few unconformities have been observed, though only in the most proximal areas along the axis of the incoming streamflow. The determination of what is, and what is not, a hydrologic, geomorphologic and sedimentologic event is a bivariate question, depending on space and time. Cessation of motion by traction depends on the rate of hydrograph recession, but it can be almost instantaneous. Because the sedimentation rate of coarse sediment is almost immediate, we expect to identify those couplets derived from events large enough to generate transport of traction sediment with a sufficiently long duration to allow most of the fines to settle.

Event couplets are differentiated from varves and have been identified in a variety of settings (Wood, 1947; Murray-Rust, 1972; Christiansson, 1979; Lambert and Hsu, 1979). They have also been mapped, sampled and correlated (Hereford, 1987; Laronne, 1987, 1990). Couplets have been shown to be correlative throughout an entire reservoir provided that:

1. The reservoir is sufficiently small to allow incoming sediment to be deposited throughout its entire area; or
2. The event is sufficiently large so that sedimentation occurs throughout most of the reservoir. The sedimentologic character of the singlets is remarkably similar throughout the reservoir with the exception of coarse sediment lobes forming only at the proximal end (Laronne, 1991).

a.

b.

Fig. 3.1. Photograph of (a) couplet event layers at trench 11, En YahavReservoir. The bottom singlets are coarse bedload sediments deposited proximally.

b) Event couplet depositional sequence at Eshet Reservoir. The lower singlets are sand, the upper singlets silt grading to clay.

4. THE STUDY AREA

We worked in the Arava Rift Valley on the En Yahav and Eshet Reservoirs (Table 4.1), respectively draining the Neqarot and Hiyyon Wadis (Fig. 4.1). Mean annual rainfall in the Arava Valley decreases southwards from 75 mm/yr near the Dead Sea, to 55 mm/yr near En Yahav and 25 mm/yr at Eilat. Mean annual evaporation is 1,900 mm/yr at the Dead Sea, increasing to more than 2,700 mm/yr at Eilat. Mean potential evaporation in the Arava is more than 4,000 mm/yr, thus, the entire Arava Valley is classified as an extreme arid desert.

The main ephemeral wadi (Nahal in Hebrew) in the Arava Valley is the Nahal Arava. Fed by the large wadis draining the central Negev and Jordan, it flows northwards to the Dead Sea. Only two of its tributaries, Neqarot and Hiyyon, are reservoir controlled. Hence, most of its flood volume is not controlled. Three reservoirs on the lower Nahal Arava have recently been planned and constructed.

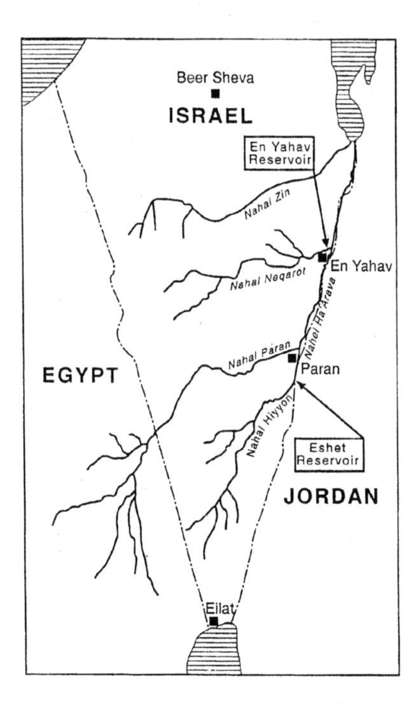

Fig. 4.1. The study area in the Arava Rift Valley.

Table 4.1. Characteristics of the Reservoirs.

Characteristic	Reservoir	
	En Yahav	Eshet
Nahal (= Wadi)	Neqarot	Hiyyon
Catchment area (km²)	984	1,256
Date of construction	1974/1975	July 1986
Volume of earth dam (m³)	650,000	85,000
Inundated area (km²)	0.8	0.8
Bulk density clay (t/m³)	1.14	1.06
Bulk density sand t/m³)	1.34	1.32

The Nahal Neqarot catchment extends from the eastern part of the central Negev highland, including the Makhtesh Ramon erosion cirque, to the north-south trending Arava Rift Valley. It has cut deep and steep-sided canyons into the highlands, widening into a broad, plainlike alluvial fan in the Arava valley. Large parts of the area are covered with hamadas, rock outcrops or desert lithosols (shallow rocky desert soils). Underlying these soils are mainly Upper Cretaceous limestones, except in the Makhtesh Ramon where sandstones, basalts and limestones outcrop.

The Nahal Hiyyon catchment is located south of the largest Arava Wadi, Nahal Paran, in the lower portion of the sedimentary Negev. The undulating plains are covered by barren regs, the regularity of which is broken by mesas and buttes. The Nahal Hiyyon has cut through this landscape, draining into the Arava Valley. It drains two Biq'at (basins): Biq'at Uvda and Biq'at Sayarim, valleys located between the Arava Valley and the Egyptian border.

5. METHODS

We dug 44 and 31 pits throughout En Yahav and Eshet reservoirs, respectively, using a backhoe. Pit depths varied according to the depth of the reservoir sediments, surpassing 5 m in several instances. The sedimentary sequence in these pits was described, paying particular attention to the thickness of event couplets. The location of the pits was mapped geodetically as was the entire area of the reservoirs. Because topographic maps were available only for the pre-dam topography (the post-dam including a borrow pit), calculated volumes of sediment accumulation were based solely on the stratigraphic sequences derived from the pits. Undisturbed bulk density of sediment (separately for the clay and sand, with no identifiable increase with depth) was determined using the sand cone method (ASTM D1556-64).

The date, number, and volume of flood inflows were derived from historic water level measurements, relying also on a digital logging stage recording station recently installed on the Eshet Reservoir (Rami Garti, pers. comm.). The flood volumes were calculated using the original stage-volume curves.

Stratigraphic sequences were compared in order to spatially correlate event couplets. Most of the correlations appear to be acceptably objective, as correlations were independently undertaken by two individuals. That the correlations are reasonable is evident in the homogeneity of couplet thickness and thickness sequence within the central and distal areas for all but the thinnest laminae (Figs. 5.1 and 5.2). The proximal facies were in places eroded and redeposited, bringing about less certainty in the legitimacy of their correlation. The correlation at En Yahav (Fig 5.1) is not as reproducible as the one at Eshet (Fig. 5.2), because the former has a very wide entrance, the depositional lobe having changed location between events. As demonstrated in Figs. 5.3 and 5.4, the number of identified event couplets decreases proximally, evidence that small events (those with smaller flow volumes) resulted in couplets covering only part of the reservoir. Utilizing Thiessen polygons and a variety of reservoir-boundary limitations, the thickness of event deposits throughout a reservoir was used to calculate the total volume of sediment deposited in it (Wilhelm, 1995). Because the thickness of gravel, sand, and graded silt-clay were measured for each couplet, we separately calculated the volume of each of these textural constituents for every depositional event.

In order to determine the true inflow volume of each historic flood, we recalculated the topography of the reservoir after each event, from which a stage-volume curve was established. The sediment inflow from the floods was regressed against each respective flood volume. Because we did not have at our disposal the shape of most of the hydrographs, we could not relate the incoming sediment volume to peak discharge, event power or other relevant hydraulic parameters. As such, we realized in advance that the sediment-water relations would contain a large degree of inherent variability.

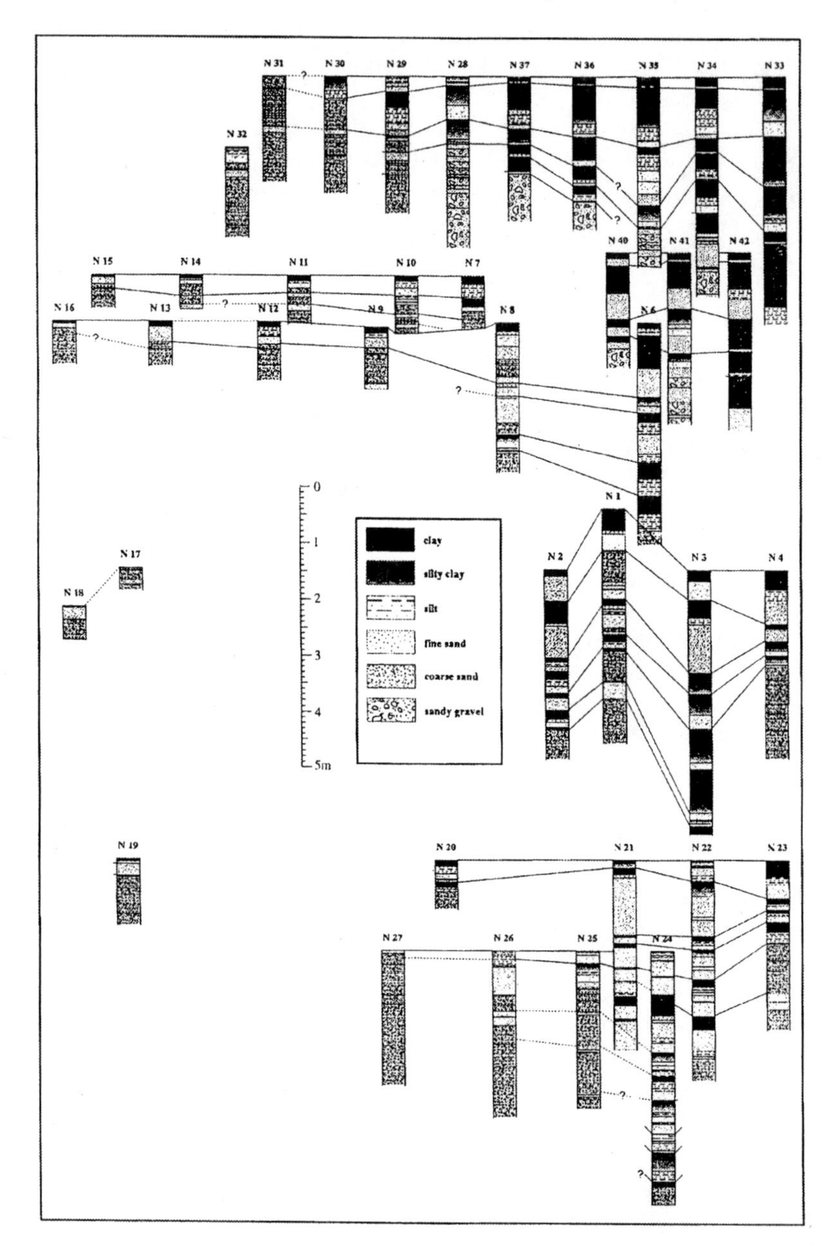

Fig. 5.1. Test pits of the En Yahav Reservoir displaying stratigraphic columns and spatial correlation between event layers.

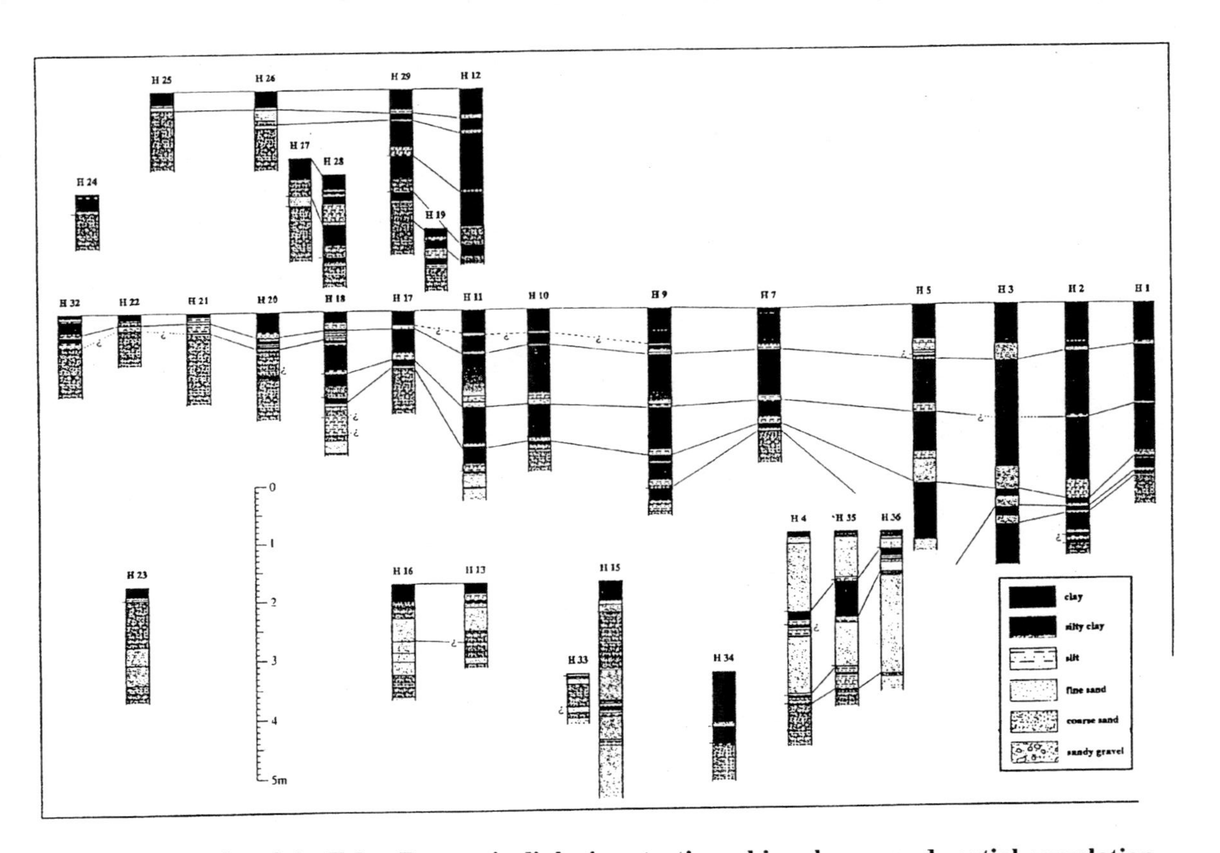

Fig. 5.2 Test pits of the Eshet Reservoir diplaying stratigraphic columns and spatial correlation between event layers.

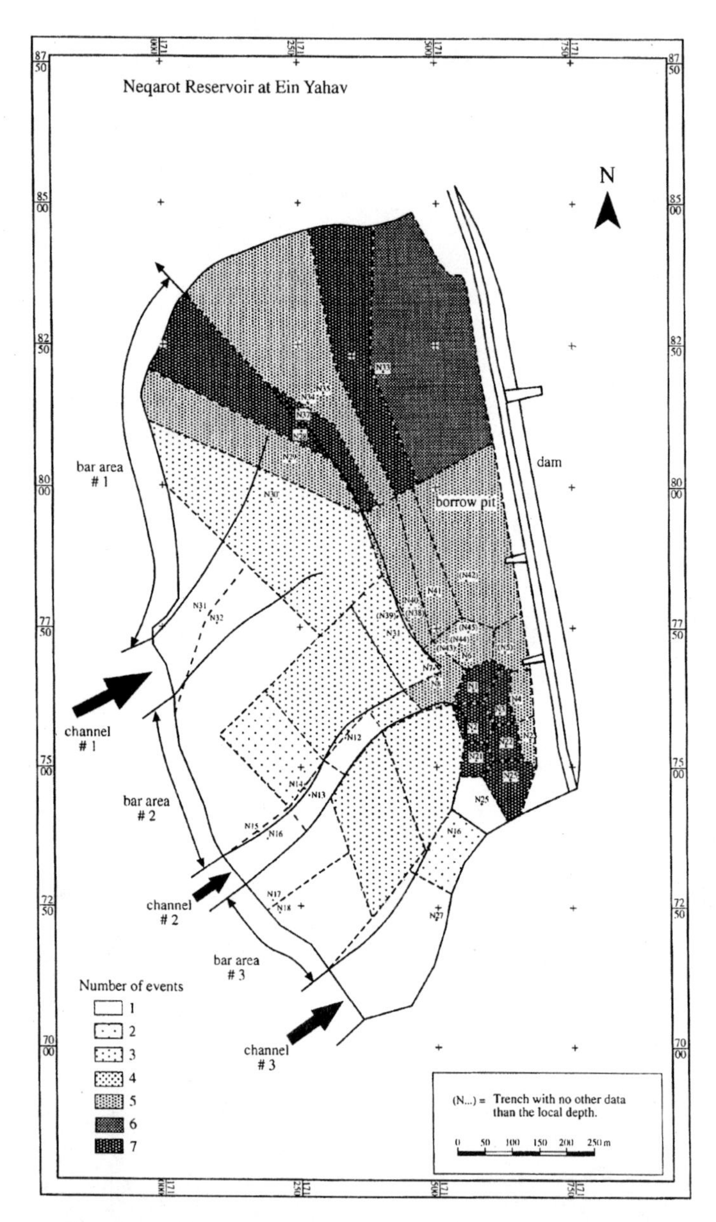

Fig. 5.3. Polygon network of the En Yahav Reservoir.

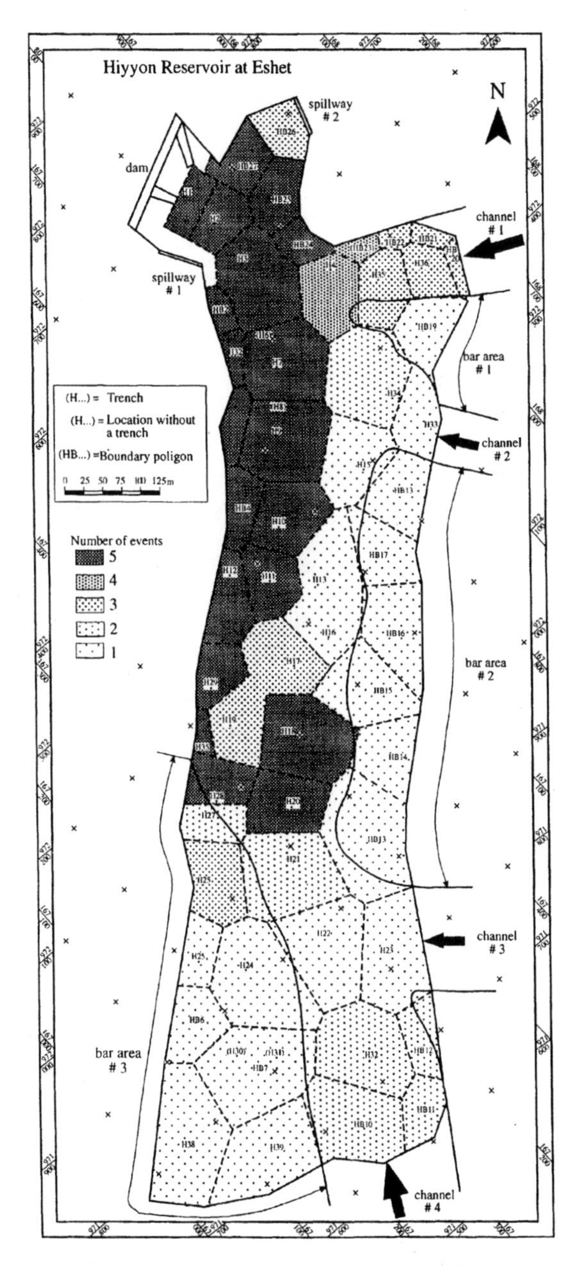

Fig. 5.4. Polygon network of the Eshet Reservoir.

6. RESULTS

Figures 5.3 and 5.4 show the location of pits and the number of identified event couplets at En Yahav and Eshet Reservoirs, respectively. The number of event layers increases distally. This is expected because the smaller events deposit fine sediment only in the vicinity of the dam. The proximal facies may be reworked and sediment may enter the reservoir from several locations, depending on the morphology of the braided wadi. This is particularly obvious at the En Yahav Reservoir (Fig. 5.3). Altogether 7 events were identified at En Yahav and 5 at Eshet, not all of which were correlative (Figs. 5.1 and 5.2; for details see Wilhelm, 1995). Sediment reworking is limited to a narrow area where flow was concentrated.

The volume of event couplets is summarized in Table 6.1. Based on our stratigraphic columns it appears that at least one, possibly two flow events were not recorded at Eshet. Referring to Table 6.1, it is apparent that a flow event in 1987/88 occurred but instrumentation was not yet available to record the maximum stage, so that the volume of the incoming flow is not known for this event. The stage sensor seems to have failed to register a sizeable event in the winter of 1990-1991, stratigraphic evidence of which is afforded in Fig. 5.1. Table 6.1 also shows that for En Yahav the event of December 23, 1993, and possibly the event of October 17, 1987, was large and some of the incoming water spilled. Hence, only the minimal incoming flow volume is known. We therefore conclude that the overall trapping efficiency of the reservoirs has hitherto been very high, almost 100 percent. As the reservoirs fill, the trapping efficiency must decrease.

The relationships between event sediment yield and flow volume are depicted in Fig 6.1. We used power functions for these regressions, since linear regressions yielded lower correlations. The functions are of the form $y = a\ x^b$, where y is the event sediment volume and x is the event floodwater volume. The dependence of sediment volume on water volume is, as predicted, low; R^2 being equal to 0.39 and 0.43, respectively. We assume that predictions of event sediment input would have been improved had we obtained higher quality flow volume data. The foregoing results indicate the extent to which the reservoirs have filled with sediment to date.

Assuming a continuous linear decrease in capacity, Fig. 6.2 depicts the predicted lifespan of the reservoirs. It indicates that within 11 years the remaining capacity will be smaller than the average flood event input at En Yahav (ca 50 years at Eshet), and within 33 years it will be less that the average annual flood input (considerably more at Eshet). Self evidently, such a prediction is a simplification of a process based on event variations, and it excludes not only the gradual decrease in trapping efficiency but also the potential large input of sizable, less frequent events. Indeed, that the trapping

Table 6.1. Hydrologic and event sedimentation reservoir input data.

2a: En Yahav Reservoir

Year	Date	Water Input [m³]	Real Water Input (Water Input - Sediment Volume) [m³]	Event Sediment Volume [m³]	Remaining Capacity [10⁶ m³]	Sediment Volume (sum) [m³]	Annual Sediment Load [t/yr]	Sediment Yield [t/km² – yr]
	01.06.1975				4.000	0		
1975/76	11.03.1976	700,000	700,000	27,944	3.930	27,944	34,650	35.2
1976/77							0	0
1977/78							0	0
1978/79							0	0
1979/80							0	0
1980/81	26.12.1980	800,000	772,056	111,319	3.611	139,263	138,036	140.3
1981/82							0	0
1982/83							0	0
1983/84							0	0
1984/85							0	0
1985/86							0	0
1986/87							0	0
1987/88	17.10.1987	3,300,000	3,160,737	101,382	3.509	240,645	125,714	127.8
1988/89	17.10.1988	1,930,000	1,689,355	158,179	3.351	398,824	196,142	199.3
1989/90							0	0
1990/91	event ?	?					0	0
1991/92	13.10.1991	1,470,000	1,071,176	210,223	3.141	609,047	260,677	264.9
1992/93							0	0
1993/94	23.12.1993	4,500,000[a]	3,890,953	296,712	2.844	905,759	367,923	373.9
	25.09.1994	350,000						
1994/95	06.11.1994	3,500,000	2,594,241	233,199	2.611	1,138,958	289,167	293.9
	Total	16550000	m³	1,138,958	m³			
Average	(event)	2068750	m³	162,708	m³		201,758	205.0
Average	(annual)	827500	m³	56,948	m³		70,615	71.8

[a] estimated value, reservoir spilled (B. Oshrowitz)

Table 6.1. Hydrologic and event sedimentation reservoir input data (continued).

2b: Eshet Reservoir.

Year	Date	Water Input [m^3]	Real Water Input (Water Input - Sediment Volume) [m^3]	Event Sediment Volume [m^3]	Remaining Capacity [10^6 m^3]	Sediment Volume (sum) [m^3]	Annual Sediment Load [t/yr]	Sediment Yield [t/km^2 – yr]
1986/87					1.700	0	0	0
1987/88	event ?	???		13,403	1.687	13,403	14,898	13.3
1988/89	25.12.1988	1,000,000	986,597	29,393	1.657	42,796	32,674	29.2
1989/90							0	0
1990/91	event ?	???		84,273	1.573	127,069	93,678	83.8
1991/92							0	0
1992/93							0	0
1993/94	23.12.1993	2,500,000	2,372,931	121,471	1.451	248,540	135,027	120.8
	24.09.1994							
	25.09.1994	400,000		-				
1994/95	05.10.1994							
	09.10.1994							
	10.10.1994	300,000		-				
	13.10.1994							
	02.11.1994							
	03.11.1994	1,820,000	1,571,460	268,201	1.183	516,741	298,132	266.7
	09.11.1994							
	02.12.1994							
	15.01.1995							
	22.01.1995	280,000		-				
	Total	**6,300,000**	m^3	**516,741**	m^3			
Average	(event)	**1,050,000**	m^3	**103,348**	m^3		**114,882**	**102.8**
Average	(annual)	**700,000**	m^3	**57,416**	m^3		**63,823**	**57.1**

[a] **estimated value, reservoir spilled (B. Oshrowitz)**

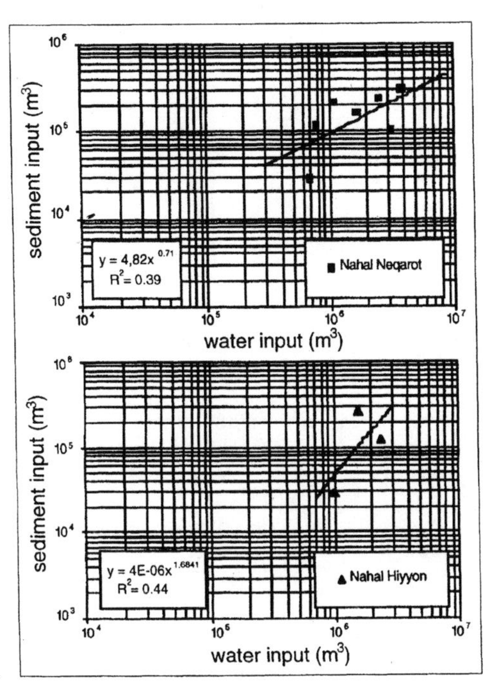

Fig. 6.1. Regressions of event sediment yield *vs* event flood volume at En Yahav and Eshet Reservoirs.

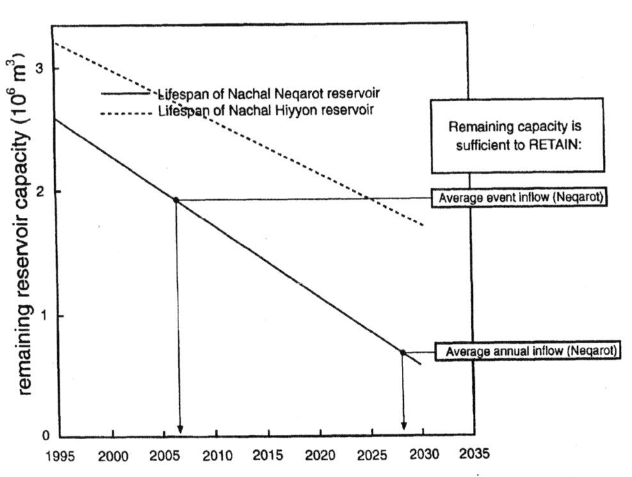

Fig. 6.2. Lifespan of the En Yahav and Eshet Reservoirs assuming a linear input of sediment.

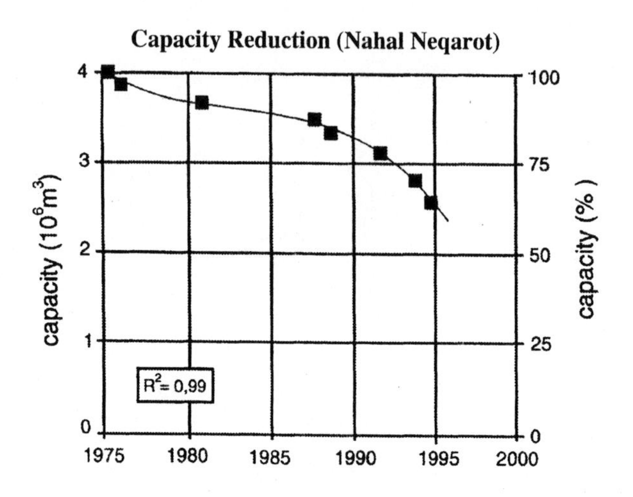

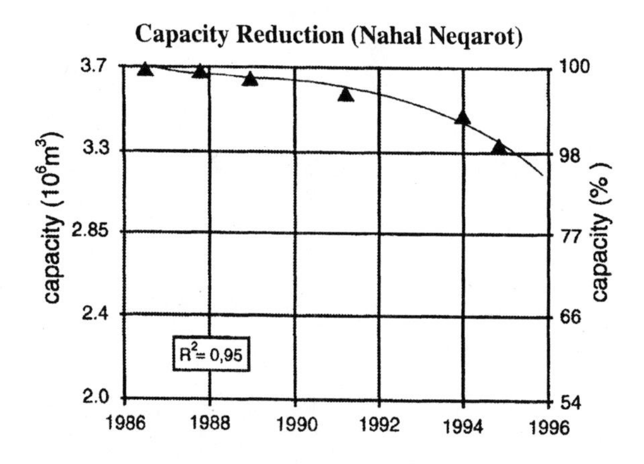

Fig. 6.3. Rates at which the En Yahav (a) and Eshet (b) Reservoirs have filled through summer 1995.

efficiency is not linear is manifested by the actual decrease shown in Fig. 6.3. The former effect may be evaluated using a Brune type trapping efficiency factor such as the capacity watershed area or the capacity inflow ratio (Table 6.2). The latter may be evaluated using individual future flood volume inputs and one of the regression equations (Fig. 6.1).

Table 6.2. Results of the reservoir sediment volume calculations.

Parameter	SI Units	En Yahav Reservoir	Eshet Reservoir
Total inflow volume	m^3/yr	16,150,000	5,900,000
Annual inflow volume	m^3/yr	807,500	655,556
Average event inflow volume	m^3	1,925,503	959,421
Total inflow volume: corrected	m^3/yr	13,478,518	4,797,105
Annual inflow volume: corrected	m^3/yr	673,926	533,012
Total sediment input	m^3/yr	1,158,163	390,802
Annual sediment input	m^3/yr	57,908	43,422
Annual sediment yield	m^3/km^2-yr	58.8	34.7
Average bulk density	t m^3	1.24	1.11
Sediment yield	t/km^2-yr	71.8	38.6
Original capacity		4,000,000	3,700,000
Total capacity loss: dam subsiding	m^3	250,000	a?
Drainage area	km^2	984	1,256
Period of record	Yr	20	9
Events per year	--	0.35	0.56
Capacity/watershed ratio C/W	m^3/km^3	4,065	2,946
Capacity inflow ratio C/I	m^3/m^3	5.9	3.2
Annual capacity loss[b]	Percent	1.8	1.2
Total capacity loss since construction[b]	Percent	35.2	10.6

[a]No data available

[b]Including dam subsidence

7. SEDIMENTOLOGIC/GEOMORPHOLOGIC CONTRIBUTION TO HYDROLOGY

The actual rate at which reservoir capacity has decreased is definitely non-linear. This may be deduced from an overview of the size and the timing of flow events (Table 6.1). It is best visualized in Fig. 6.3, which describes the rate at which the reservoirs have hitherto filled. Second order polynomial equations describe these rates rather well.

Our event sedimentation data may also be used to estimate the average sediment yield from this arid area, thereby comparing it to yields based on reservoir surveys draining similarly sized catchments in more humid regions throughout the United States (Langbein and Schumm, 1958). Apparently our data lie on the curve, thus strengthening the contention that low sediment yields typify deserts; whereas, highest yields are obtained in semiarid areas, such as the northern semiarid Negev Desert (Powell et al., 1996). In the extreme arid environment of the southern Arava, the sediment yield of the very small catchment of Nahal Yael (Schick, 1977) is considerably higher than our estimates (Fig. 7.1). This may derive from the locally steep schist and granite slopes and from the minute catchment area. It is interesting to observe that Nahal Yael is the first to have been observed to yield as much or more bedload than suspended load. Our calculations of the sand (or sand + gravel) fraction of the total event volume indicate a similar situation in the larger catchments. The event couplet technique cannot be used to identify the entire original coarse, proximal bedload deposits of reservoirs due to their remobilization. Given that

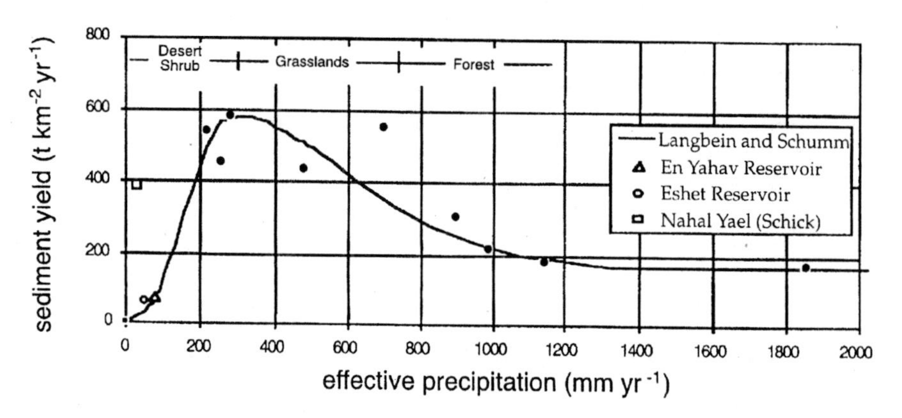

Fig. 7.1. The Langbein-Schumm (1958) curve revisited, adding data from an extreme arid environment.

our evaluation of the event bedload (= gravel + sand) fraction is an underestimate, we confirm that it is high in the arid region: it varies in the range of 17-65 percent at En Yahav and 16-31 percent at Eshet (Fig.7.2). These fractions of the traction load are considerably higher than the 5 percent often cited for humid regions.

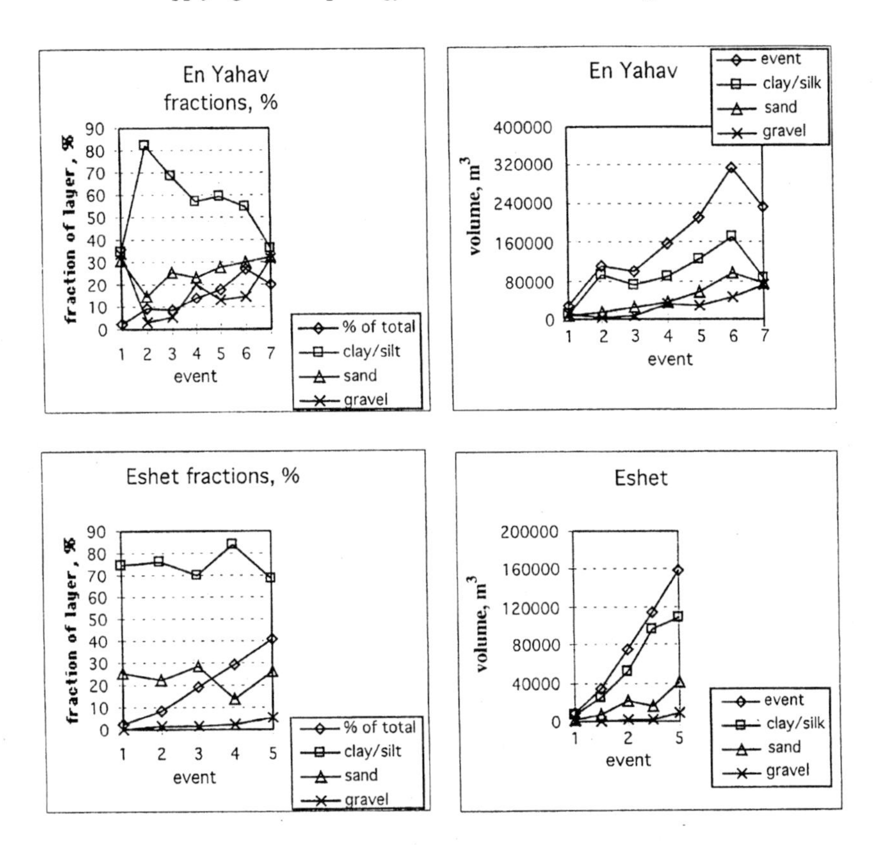

Fig. 7.2. **The percent and the actual volumes of gravel, sand and silt+clay (denoted 'clay' in legend) deposited in the En Yahav and Eshet Reservoirs. The graphs on the right depict total event volumes of deposited material as well as the volumes of gravel, sand, and silt+clay. The graphs on the left depict the fraction of deposited event volume relative to the entire sediment volume deposited until summer 1995, as well as the fraction of each of the 3 textural classes relative to the entire event volume.**

The sedimentologic/geomorphic approach we have presented is also useful for evaluating the past, shifting stage-volume relations for reservoirs. We hereby present such curves for the En Yahav Reservoir (Fig. 7.3). These may be useful not only for the more accurate determination of actual water inputs, but also for prediction of shifting stage-volume relations in these and similar environments.

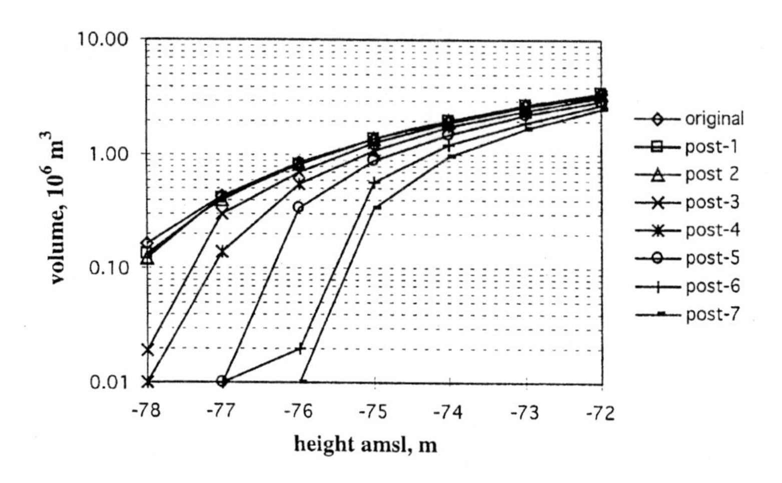

Fig. 7.3. Shifting stage-volume curves for the En Yahav Reservoir.

8. CONCLUSIONS

Slackwater deposits have opened the eyes of the engineering profession to the utility of this geomorphic-sedimentologic technique for the prediction of high magnitude floods (e.g., Kochel and Baker, 1988; Baker et al., 1993). Similarly, we foresee that this geomorphic-sedimentological technique may prove to be useful not only for the prediction of flow volume inputs into reservoirs and lakes, but also to postdict such paleohydrologic inputs. We realize that the quality of prediction is hampered by noise caused by the inter- and also intra-flood variability in sediment yield. A new approach to predict as yet unpredictable variations in sediment supply, particularly with respect to suspended sediment loads, awaits the next generation of engineers, working in unison with geomorphologists and sedimentologists.

9. ACKNOWLEDGEMENTS

This study is a joint venture of an engineer and a sedimentology-oriented geomorphologist. As such, it epitomizes one of several lessons that have emerged from the writings of Stan Schumm: the need for combined efforts to approach earth surface problems - be they academic or practical. As such, it is our acknowledgment of his imprint in this regard.

The project was funded by the Mekorot Water Supply Company - we acknowledge Alexander Dombe for his interest in this study. Tal Kedem, Yanai Shlomi, Oren Yekutieli and Jamil Akbar assisted in the field. We thank Frank Ethridge and Paul Mosley for their reviews.

10. REFERENCES

Albertson, M.L., Y.B. Dal, R.A. Jensen and H. Rouse. 1950. Diffusion of submerged jets. *Transactions American Society Civil Engineers* 115: 639-697.

Allen, J.R.L. 1982. *Sedimentary Structures - Their Character and Physical Basis.* Develop. in Sedimentol. 30A, Vol. II, Elsevier, Amsterdam, 663 pp.

Baker, V.R., G. Benito and A.N. Rudoy. 1993. Paleohydrology of Late Pleistocene superflooding, Altay Mountains, Siberia. *Science* 259: 348-350.

Bloesch, J. 1994. A review of methods used to measure sediment resuspension. *Hydrobiologia* 284: 13-18.

Christiansson, C. 1979. Imagi Dam - a case study of soil erosion, reservoir sedimentation and water supply at Dodoma, central Tanzania, *Geografisca Annaler* 61A(3-4): 113-145.

Friedman. G.M. and Sanders, J.E., 1978. Principles of Sedimentology. John Wiley and Sons, NY, 792pp.

Hereford, R. 1987. Sediment-yield history of a small basin in Southern Utah, 1937-1976: Implications for land management and geomorphology. *Geology* 15: 954-957.

Kochel, R.C. and V.R. Baker. 1988. Paleoflood analysis using slackwater deposits. Chapter 21, p. 357-376 in Baker, V.R., , R.C. Kochel and P.C. Patton (eds.): *Flood Geomorphology*. John Wiley

Lambert, A.M. and K.J. Hsu. 1979. Varve-like sediments of the Walensee, Switzerland. p. 287-294 in Schluchter, Ch., ed., *Moraines and Varves: Genesis, Classification:* Rotterdam, Balkema.

Langbein, W.B. and S.A. Schumm. 1958. Yield of sediment in relation to mean annual precipitation. *Am. Geophys. Union Trans.*, 1076-1084.

Laronne, J.B. 1987. Rhythmic couplets: sedimentology and prediction of reservoir design periods in semiarid areas. 229-244 in Berkovsky, Louis and G.W. Wurtele (eds.), *Progress in Desert Research*: Totowa, Rowman & Littlefield.

Laronne, J.B. 1988. Comment on 'Sediment-yield history of a small basin in Southern Utah, 1937-1976: Implications for land management and geomorphology'. *Geology*, 16: 956-7.

Laronne, J.B. 1990. Probability distribution of event sediment yields in the Northern Negev, Israel. p. 481-492 in Boardman, J., I.D.L. Foster and J.A. Dearing (eds.): *Soil Erosion on Agricultural Land*. Wiley, London.

Laronne, J.B. 1991. Sedimentology of reservoir sediments and its use for the determination of event sediment yields. *Israel Ministry of Energy and Infrastructure,* Publ. ES-47-91, 100 pp. (in Hebrew).

Murray-Rust, D.H. 1972. Soil erosion and reservoir sedimentation in a grazing area west of Arusha, northern Tanzania. *Geografisca Annaler*, 54A :3-4: 325-343.

Powell, M.D., I. Reid, J.B. Laronne and L.E. Frostick. 1996. Bedload as a component of sediment yield from a semiarid watershed of the northern Negev. *International Assoc. Hydrol Sciences,* 236:389-397.

Reid, I. and Frostick, L.E. 1984. Particle interaction and its effect on the thresholds of initial and final bedload motion in coarse alluvial channels. In Koster, E.H. and R.J. Steel (eds), Sedimentology of Gravels and Conglomerates. Canadian Society of Petroleum Geology, Mem. 10, 61-68.

Schick, A.P. 1977. A tentative sediment budget for an extremely arid watershed in the southern Negev. p. 139-163 in Donald O. Doehring (ed.): *Geomorphology in Arid Regions* , Publ. in Geomorphology.

Shteinman, B., A. Gutman and E. Mechrez. 1993. Laboratory study of the turbulent structure of a channel jet flowing into an open basin. *Boundary Layer Meteorology*, 62: 411-416.

Wood, A.E., 1947. Multiple banding of sediments deposited during a single season, *American Journal of Science* 245:5: 304-313.

Wilhelm, R. 1995. Event sediment yield and reservoir sedimentation in the Arava Rift Valley. Unpubl. Dipl. Arbeit, Ingineur Hydrologie und Wasserwitschaftung, Tech. Hoch. Darmstadt, 57 pp.

Yu, W-S and H-Y. Lee. 1993. Numerical simulation of turbidity current in reservoir.*International Journal of Sediment Research* 8:2: 43-65.

BEDLOAD SORTING IN RIVER BENDS

Deborah J. Anthony[1]

ABSTRACT

Transverse sorting of bedload in short channel reaches is accomplished by the association of transverse channel bed slopes with near-bed, up-slope flows. The resultant forces create the necessary hydrodynamic conditions which segregate coarser and finer portions of the bedload. In successive meander bends, the typical in-channel topography and associated helical flow patterns create the required forces that cause sediment sorting. The resulting sediment sorting patterns show that in successive bends, coarse bedload enters near the inside bank, crosses the channel at mid-bend, and exits the bend near the outer bank. Finer bedload displays the opposite pattern, entering the bend near the outer bank, crossing at mid-bend, and exiting near the inner bank. These sorting patterns for different size sediments do not support a model of cross-stream balance of forces. Only sediments whose size is close to the D_{50} of the bedload maintain a path about mid-channel. Without transverse bed slopes and the associated up-slope near-bed flow, sorting does not occur. Differential entrainment, coarse sediment depletion, armoring, and suspension are not required for bedload sorting. The different trajectories explain the observed patterns of downstream fining of point bars, and dramatic cross stream changes of bedload size in channel crossings.

KEYWORDS
Bedload sorting, river meanders, bedload transport

1. INTRODUCTION

The sorting of bedload in river channels is a process which transfers sediments of different sizes to different channel locations downstream. Sorting can be thought of as transverse or longitudinal. While longitudinal sorting (downstream fining) usually occurs (Iseya and Ikeda, 1987; Hoey and Ferguson, 1994), this paper focuses instead on transverse sorting - those processes which separate sediment of different sizes from one another in a

---

[1] Assistant Professor, Colorado State University, Fort Collins, CO 80523.

small channel reach. This phenomenon has been particularly noted in meander bends (Miall, 1978, Bridge and Jarvis, 1982), and it is demonstrated in this study.

The interactions between flowing water, channel topography, and moving sediment govern the sorting process. The process is conceptually simple, but the application in river channels with a variety of topographic zones and secondary flows can be complex. As bedload moves downstream, the individual particles are acted on by drag-and-lift forces created by the flow, gravity, and impacts with other moving sediment. An understanding of the net effect of these forces, and the feedback between bedload movement and flowing water is necessary to fully understand the trajectory of different size bedload particles, and to also explain adjustments that occur in rivers in response to changing hydrologic conditions.

2. SEDIMENTS

2.1. Bedload Movement

Traditionally, bedload has been described as the sediment moving on the channel bed by rolling, sliding and saltation. Different definitions for bedload (Bagnold, 1966, 1977; Bridge and Dominic, 1984) emphasize both the role of the fluid and the bed interactions in sediment movement. An analysis of these processes usually begins with an examination of the forces acting on a stationary particle (Kirchner et al., 1990; Wiberg and Smith, 1987; Komar and Li, 1986; Miller et al., 1977; Simons and Senturk, 1976; White, 1940; Shields, 1936). These are illustrated in Fig. 2.1, and include the forces of gravity (F_G), buoyancy (F_B), lift (F_L), and drag (F_D).

The gravitational and buoyant forces are simply a function of the volume and density of the individual particle. The gravitational force acts vertically downward, while the buoyant force (assuming hydrostatic conditions) acts vertically upward, and is a function of the volume of water displaced by the particle. Most analyses of critical conditions for entrainment only consider the net gravitational force ($F_{G'}$) acting on a bedload particle, the gravitational force minus the buoyant force.

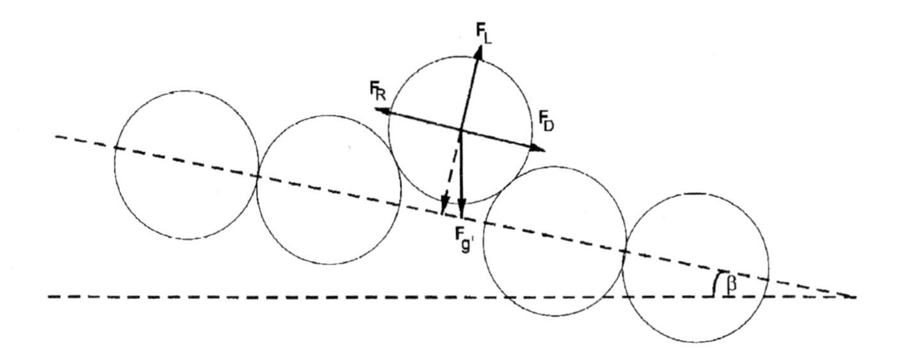

Fig. 2.1. Forces acting on a stationary particle resting on the bed of a river. $F_{G'}$ is the net gravitational force (gravity minus buoyancy), F_L is the lift force due to the Bernoulli effect, and F_D is the drag force created by the flowing water. F_R is the resisting "force."

The forces of lift and drag act on the surface area of the exposed bedload particle, and they are created by the flow of water around the particle. The drag force (F_D) is usually considered a function of the square of the average velocity acting on the exposed area perpendicular to flow of the particle, and it acts in the direction of the bottom velocity vector. The lift force (F_L) is created by the existence of a velocity gradient across the grain (and thus a pressure differential between the bottom and top of the grain) and acts perpendicular to the free surface. Estimation of these effects are dependent on assumptions about the velocity gradient near the bed, and on coefficients of lift and drag.

A resisting force (F_R) is implicit but not defined in some studies, while other researchers (Wiberg and Smith, 1987; Komar and Li, 1986) have defined it as a combination of the gravitational force acting normal to the bed surface and the angle of repose of the particle. This approach allows for calculation of the rotation that a particle must go through as it starts to move. Various studies have shown that in stream beds composed of heterogeneous sizes, smaller particles may be shielded between larger ones, whereas larger clasts project higher into the flow and so have greater forces of lift-and-drag acting on them. Analyses of these differences have lead to a realization that in streams with a range of bed material sizes, the critical conditions for motion (shear, unit discharge, velocity, etc,) for a wide range of grain sizes within the bed material matrix are similar, and these different sizes move together.

This simple "balance of forces" approach is only a first step in understanding the processes of sediment entrainment and subsequent sediment trajectories through river channels of complex topography. Missing are factors which address actual bedload characteristics, with variations in packing, sorting, size, density, shape, and relative exposure (Einstein, 1950; Komar and Li, 1986; Wiberg and Smith, 1987; Richards, 1990; Petit, 1994). In addition, local flow conditions show a great deal of variability at many scales. Large-scale turbulence, caused by planform roughness, includes fluctuating bed- and bank-cells that are shed off roughness elements at the outer banks (Anthony and Harvey, 1991). Turbulent flow over bedforms at intermediate scales (Mendoza and Shen, 1990; Robert, 1993), and smaller scale, random fluctuations at the bed, called ejections and sweeps (Lapointe, 1992; Nakagawa and Nezu, 1981; Grass, 1970) may be associated with bedform and grain roughness.

2.2. Bedload Sorting

When investigators have attempted to model bedload sorting, these complexities are addressed using a variety of assumptions and simplifications (Allen, 1970; Bridge, 1977; Diegaard, 1980; Odgaard, 1982; Parker and Andrews, 1985; Ikeda, 1989). An understanding of the processes of sorting (and their end product, sorted bed sediment) requires an analysis of the forces discussed above, and a determination of how the net effect of these forces varies for grains of different sizes to create different trajectories through individual channel reaches.

In the simplest force balance approach, all forces act in a plane (Fig. 2.1). But in reality, these forces are three dimensional vectors, which act at different angles from one another throughout a river channel. River bends are characterized by transverse slopes and helical flows (Leopold and Wolman, 1960; Rozovskii, 1961; Smith and McLean, 1984; Thorne et al., 1985; Odgaard and Berg, 1988; Anthony and Harvey, 1991). Thus gravity and drag are not acting only in the plane parallel to the primary flow direction, which may explain the patterns of grain size sorting through meander bends that have been described in the literature (Jackson, 1975; Dietrich 1982; Bridge and Jarvis, 1982).

The difference in vector direction and magnitude is not the only way in which the sorting models can differ from a simple force balance. Sorting implies transport, and hence, the problem is one of dynamics, not statics. Lift-and-drag are proportional not to the velocity of the flow near the particle but to the difference in the velocity between the particle and the flow (Parker and Andrews, 1985; Ikeda, 1989). This changes both the magnitude and direction of the lift-and-drag vectors, especially for larger particles. Also, the lift force is significantly diminished when the particle is in motion, because the velocity gradient over the particle is reduced (Middleton and Southard, 1984).

Several models have been developed to address the process of bedload sorting in meander bends. Allen (1970) was the first to look at differential bedload transport and subsequent sorting and then use it to explain the sedimentary deposits created by meandering rivers. Allen's model assumed a transverse force balance between the drag (from helical flow) and gravity, omitted the lift force, and assumed a constant discharge and lateral slope. For this constant lateral slope, transverse forces are just balanced. This predicts a constant lateral distribution of grain sizes around a bend, a situation which does not represent the conditions in rivers (Bridge, 1977; Bridge and Jarvis, 1982; Anthony and Harvey, 1991), or flumes (Hooke, 1975).

Engelund (1974) also considered a transverse balance of forces to predict large-scale bottom geometry of a channel bed and overall hydraulics. He realized that channel form was created by an interaction between the flow and sediment movement, including sorting (although his model had fixed, vertical side walls). Using a more complex model than Allen (1970), Engelund (1974) considered both transverse and longitudinal forces, and included the lift force. Coupled with a flow model similar to Rozovskii (1961), this model was used to predict Hooke's (1975) flume results, with some success. Sorting was thus only an unanalyzed byproduct of his model, and it was not directly addressed. This type of implicit sorting analysis is also included in other models of meander hydraulics and topography (Ikeda, Parker, and Sawai, 1981).

Bridge (1977) developed a model which also included longitudinal and transverse force balance equations, with gravitational, drag, and lift forces combining to produce a net cross-stream force vector of zero. In addition, a component of friction for the cross-stream force balance was included, but only applied to grain size predictions for unsteady flow. Bridge (1977) stated that, since "sediment is not traveling parallel to the mean downstream direction, a component of the frictional force must be present in the transverse direction", which he assumed only occurred during rising or falling stages. However, detailed sampling of moving bedload (Anthony and Harvey, 1991; Bridge and Jarvis, 1982) shows that sorting occurs for much of the bend length even during steady, equilibrium flows.

Diegaard (1980) constructed a model in which sediment continuity was maintained, in contrast to earlier attempts. This model focused not on a transverse force balance, but on the differences between suspended sediment and bedload transport. Coarser material moves downslope as it passes through the bend, and finer material is swept upslope in suspension. Diegaard (1980) noted that no transverse sorting would occur in the absence of significant amounts of suspended sediment.

Using a very different approach, Odgaard (1982) developed a model to predict cross-stream variations in depth, depth-averaged velocity, and grain size distribution, based on the principle of conservation of moment-of-momentum flux and an analysis of the critical conditions for entrainment. Parker and Andrews (1985) noted that this static type of model is complimentary to the mobile models discussed above. However, it is difficult to see how this model, which assumes critical shear stress conditions at every point in the channel, would apply to rivers with significant sediment transport. Even so, the model was applied to the Sacramento River with success (Odgaard, 1982).

Parker and Andrews (1985) used the mobile sediment model preferred by Bridge (1977), but also included a condition of continuity for each grain size and assumed nonsuspendable bed material. Their model, which is applicable only for gently curving channels, also focused on the cross-stream and downstream force balance. A more sophisticated approach to the drag force was used, with the net force a function of the difference between the flow and sediment vectors. However, this model is limited to the case of weak sorting, due to the gentle curvature assumption. Sorting is accomplished by the preferential downslope movement of coarse material, leaving the bedload near the inner bank depleted of the coarser fraction and, therefore, finer in size.

Ikeda (1989) attempted to include, with the transverse balance of forces due to flow and gravity, the transfer of sediment due to suspension and vertical sorting due to armoring. His approach assumes an equilibrium position for individual grain sizes, as in many other models (Allen, 1970; Bridge, 1977; Engelund, 1974), and is only applicable in the "fully developed" downstream portion of a bend.

Kovacs and Parker (1994) approached the problem of bedload motion using a vector analysis technique. Their approach might be used successfully for analysis of bedload sorting, but currently they have applied it only to straight channels with uniform grain sizes.

In summary, most models of bedload sorting assume a cross-stream balance of forces, where the downslope pull of gravity equals the upslope drag of the flow. This implies no cross-stream movement of a particular size sediment, but rather a trajectory which is parallel to the channel centerline. This is contrary to observations of sediment movement through a bend

(Jackson, 1975; Bridge and Jarvis, 1982; Dietrich, 1982). Even more questionable is the assumption of a longitudinal balance of forces (Bridge, 1977; Engelund, 1974), which implies no sediment movement (and thus no sorting) whatsoever. Other than by a force balance, sorting has been explained through the processes of suspension, armoring, or grain size depletion through downslope movement of coarse particles.

In addition, models have assumed a channel topography dominated by a transverse slope and the accompanying helical flow cell. However, this set of conditions is limited to the point-bar slope, and this sometimes represents only about one third of the channel width (Anthony, 1987; Anthony and Harvey, 1991). Outward flow over the point-bar platform, where there is almost no cross-stream slope, and the turbulent flow in the thalweg, with mild up- and downstream slopes, do not fit the assumptions used in the reviewed models.

Therefore, the objectives of this study were to:

- Determine patterns of bedload movement (including the trajectories of individual grain sizes) with respect to the overall patterns of topography and flow through meander bends
- Test the hypothesis that bedload sorting occurs when a large range of grain sizes are in motion because the cross-stream forces acting on each grain size are unbalanced, thus creating different net trajectories
- Test the hypothesis that sorting can be accomplished without significant suspension, armoring, or differential entrainment.

3. STUDY AREA

Fall River is a tributary of the Big Thompson River in the Front Range of northern Colorado (Fig 3.1). This sinuous, low gradient, snowmelt-fed river contains significant amounts of mobile bedload ranging in size from pebbles to fine sand. The source of most of the mobile sediment is an alluvial fan created by the July, 1982 failure of an earthfill dam located in the tributary headwaters of Roaring River valley (Jarrett and Costa, 1985; Blair, 1987; Pitlick, 1993). Additional information on this site can be found in Anthony (1992), Anthony and Harvey (1991), and Pitlick (1993).

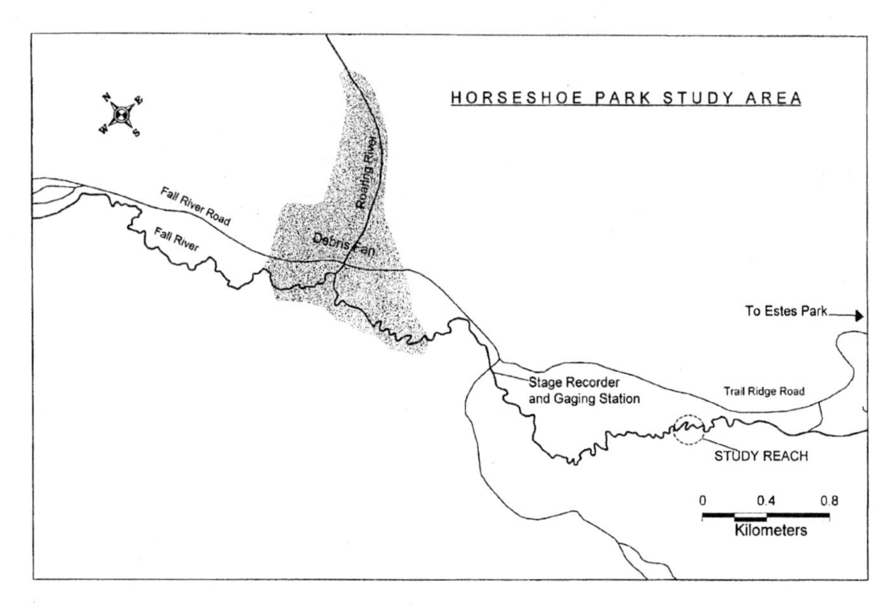

Fig. 3.1. Location map for the Fall River study site. Note the alluvial fan which supplied significant amounts of sand and gravel size material to the downstream reaches after failure of the Lawn Lake dam in July 1982.

The 90 km^2 drainage basin above the study reach provides discharges ranging from over bank flows of 9.0 m^3/s to less than 0.5 m^3/s (Pitlick, 1985). The yearly hydrograph is simple, with a peak discharge during the spring snowmelt, decreasing flows through the summer and fall, and lowest discharges in the winter. Because this basin lies above 2300 m, it is not subjected to the massive summer convective thunderstorms (Jarrett, 1987) that create significant hydrograph peaks, although smaller thunderstorms do occur.

The bedload transport study site is a two-bend reach located about four kilometers downstream from the alluvial fan (Fig. 3.1). This sinuous (P = 2.3), low gradient (S = 0.00132) reach of Fall River has a bed composed of a wide range of grain sizes (32 mm pebbles to 0.06 mm fine sand) which are mobile in significant quantities at virtually all flows above 0.5 m^3/s. Thus the river, which has stable root-reinforced banks and a bed with large amounts of mobile sediment, also has a relatively uncomplicated hydrograph, making this an ideal study site.

4. DATA COLLECTION AND ANALYSIS

Data were collected from twenty-two monumented cross-sections established on the two-bend reach. Measurements made at each cross section included bed- and water-surface topography, velocity profiles at one meter intervals (both longitudinal and transverse velocity vectors), and bedload transport. All data were collected from mobile bridges that spanned the channel.

During the first field season, one complete set of measurements (for all 22 cross sections) was made for discharge levels at representative portions of the hydrograph (rising, bankfull, falling, and low stages). Due to the high variability of velocity and bedload transport measurements (Carey, 1985, Dietrich and Smith, 1984), data collection the following year focused on reproducibility. Those cross sections that had undergone the most change during the annual hydrograph of the preceding year were measured repeatedly.

4.1. Bedload Measurements and Analysis

A Helley-Smith sampler was used to measure bedload transport rates at one meter intervals across the channel. Two one-minute samples were taken at each location, and were weighed separately so that point values of bedload transport could be determined. Samples of bedload material were retained for grain-size analysis.

Uncertainty in the bedload transport measurements occurs because of the suction created by the Helley-Smith sampler (Pitlick and Harvey, 1986). However, the sampler used in this study was a "Denver" type bedload sampler which is more accurate than the traditional "Berkeley" sampler for small grain sizes because it creates less suction (Pitlick and Harvey, 1986). To further protect from this type of error, no sampling was done in shallow water where only medium to fine sand occurred, and where it could be seen that the bedforms were stationary.

Further uncertainly is created by the inherent variability of bedload transport rates when bedforms are present (Carey, 1985). Sampling on a dune crest will provide dramatically greater transport rates than that done in the lee of a dune, where flow separation around the bedform causes velocities that can be very low or even negative, and where transport is effectively zero. Therefore, the point value of bedload transport depends on where sampling takes place on the dune. To overcome this problem, six to eight sampling traverses were made at the cross-sections during each flow stage during the second field season. In spite of the variability, overall sediment transport patterns showed clear trends, and cross-stream averages were consistent.

The wet bedload weights measured in the field were converted to dry weights, and unit bedload transport rates were calculated in units of kg/m-s. Dry bedload was sieved into even phi intervals, from +5 to -5, and the standard calculations for grain size distribution were done for each sampling site (i.e., the D_{95}, D_{84}, D_{50}, D_{16}, D_{05}, Folk's graphic mean and inclusive graphic standard deviation). The bedload moving past a cross section was divided into percent of total bedload by size and position within the cross section. This information was plotted on study reach maps, by percent of a given size class at each position (see Figs.5.4 and 5.5 below). These are the first maps of bedload trajectory by grain size known to the author. Other maps (Jackson, 1975; Dietrich, 1982; Bridge and Jarvis, 1982) show grain size (or D_{50}) patterns for a channel reach, but do not show the movement by individual size fractions through a channel bend.

5. RESULTS

5.1. Primary Patterns

The large-scale pattern which emerged for the entire study reach was one of significant in-channel topographic adjustment to changing discharge, which can be expected to have a significant effect on the sorting of bedload sediments. The details of this metamorphosis, especially with regards to the first field season, have been discussed in detail elsewhere (Anthony, 1987; Anthony and Harvey, 1991), and will only be summarized here.

In early spring, the channel topography inherited from the previous winter low flows was very symmetric, with rectangular cross-sections even through the bends. Increasing discharge due to spring snowmelt caused helical flow to become stronger. As the depth of flow increased, the stronger helical cell and stronger bed shear stress eroded the sediment deposited at low flow in the thalweg. The outward movement of sediment near the inner bank at the bend entrance, due to outward flow over the point-bar platform, coupled with downstream decreasing bed shear stress along the inner bank and small negative slopes built up the point-bar platform. This build up, and the thalweg excavation, created the point-bar slope. The meeting point of the outward velocity over the point-bar platform and the helical cell defines the sharp break between the point-bar platform and slope (Fig. 5.1a).

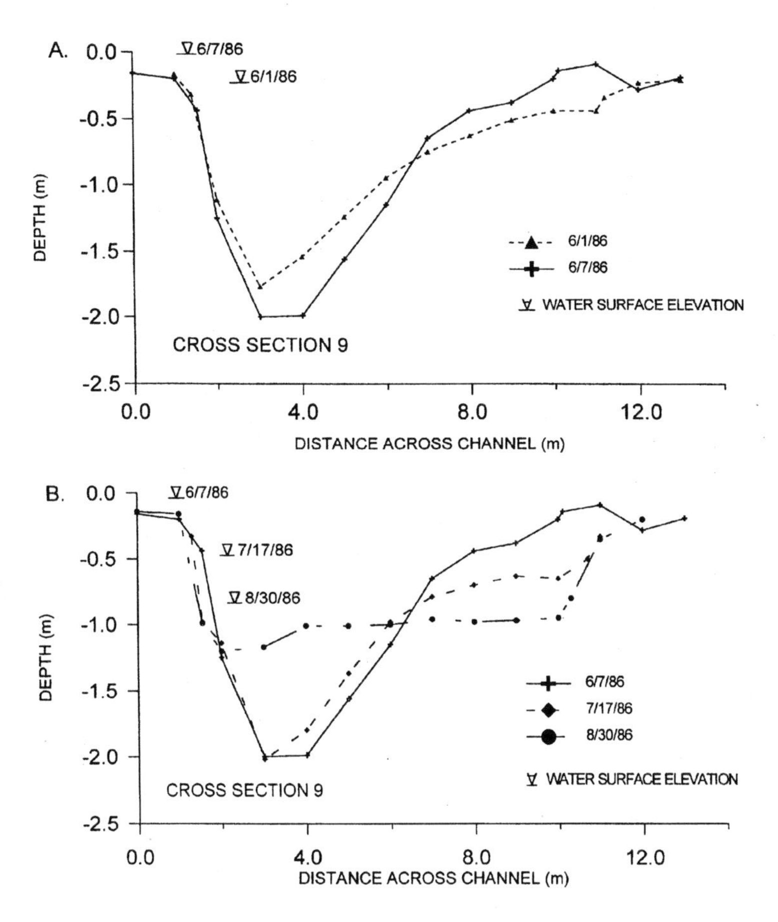

Fig. 5.1. Changes in channel cross-section topography due to changes in discharge: (a) Erosion of the thalweg and deposition on the point bar due to spring rising stage; (b) Deposition in the thalweg and erosion of the point bar during the long summer falling stage, which created a more symmetric cross section.

As discharge increased, the point bar built up closer to the floodplain level (Fig. 5.1a). Bed- and bank-cells caused by the large outer bank roughness elements (slump blocks) formed in the thalweg. These high velocity, high turbulence cells, in conjunction with the stronger helical cell and greater shear stress, keep the thalweg sediment free. This was accomplished in the upstream portion of the bend by rapidly moving the finer sediment upward onto the point-bar slope. The coarse sediment did not move into the thalweg until well after the bend apex, so that maximum sediment transport rates were confined to the point-bar slope through much of the channel. Cross-section asymmetry was at a maximum at bankfull stage, due to the buildup of the point-bar platform. At slightly lower stages, thalweg excavation was just as deep, but active point-bar platforms were not as close to the floodplain level.

Falling stage reversed the sequence (Fig. 5.1b). High velocity flow eroded into the subaerially exposed platform, causing the inner edge of the channel to retreat while new point bars were formed upstream and downstream of the bend apex. The bed- and bank-cells weakened, and the helical flow cell extended into the thalweg. The highest bedload transport rates moved closer to the concave bank, and the thalweg finally became a region of sediment transport.

As stage continued to fall, both the strength of the helical cells and their size in each bend decreased. The thalweg began to aggrade, since lower shear stress could no longer move the coarse material up out of the thalweg against negative slopes. Aggradation of the thalweg decreased the depth of flow, further weakening the helical cell, and causing positive feed-back that induced more thalweg aggradation as stage dropped (Fig. 5.1b).

The most important variation in bedload transport patterns through this transformation were in the changing distribution of bedload transport rates across the channel with changing stage. At high discharges, almost all bedload moving through the bends was concentrated on the point-bar slope. Large dunes, with heights up to 50 cm, traveled across the point-bar slope, while a thin layer of moving sediment (of all grain sizes) with no discernable bedforms was transported across the point-bar platform. Almost no sediment transport was measured in the thalweg region, except near the bend exit. This is in contrast to the findings of other studies (Hooke, 1975; Dietrich and Smith, 1984) where the zone of maximum bedload transport coincided with the thalweg at high flows. When moving sediment was captured in the upstream portions of the thalweg, it consisted of gravel or small pebble clasts and a small amount of very fine, dark sediment of local origin. Thus at bankfull discharges, the thalweg region was sediment starved. Suspended sediment concentration, even at bankfull discharge when the water was cloudy, was minimal (Pitlick, 1985).

At intermediate flows, higher transport rates were measured in the thalweg. The zone of maximum bedload transport shifted toward the outer bank, but it was still located on the point-bar slope, which also shifted. Transport rates were lower during rising stage than during falling stage. This is in contrast to the opposite pattern noted by Kuhnle (1992) for short term events such a floods in small streams. Very little suspended sediment was transported, and clear water made the bottom quite visible.

During low flows, there was no clear preferential location for maximum bedload transport. Transport was generally highest near the channel center, and decreased gradually to zero at the banks. There was almost no suspended sediment. Finally, no armoring was seen, even at these low flows. As with all higher flows, the channel was covered by mobile sediment. As discharge fell to its lowest level, the dunes simply became stationary.

5.2. Bedload Sorting

The processes of differential bedload movement (or trajectories which varied by size fraction) which cause bedload to become sorted were due to the topographic variations and flow patterns discussed above. Between high and intermediate discharges, little variation in sorting pattern occurred. At these discharges, both cross-section asymmetry and strong helical flow develop and sediment sorting was continuous. At each bend entrance, the coarse sediment was located near the inner bank, and the finer near the outer bank (Fig.5.2a). As the sediment moved through the bend, the coarser fraction moved preferentially down-slope toward the outer bank, while the finer fraction was swept inward. At the bend apex, some of the coarsest sediment was located near the outer bank, but the individual grain size distributions at all cross-section locations overlapped (Fig. 5.2b). However, the sorting process continued through the remainder of the bend, so at the bend exit the grain sizes were segregated, and thus more "well sorted" in terms of the grain-size distribution at each location, (Fig. 5.2c), with the finest sediment near the inner bank and the coarsest sediment in the outer part of the channel.

At low discharges this continuous process did not occur. Cross-sectional asymmetry was confined to the pools in each bend, as was any helical flow. At each bend entrance, the general bedload size distribution was the same as at higher discharges. However, as the sediment moved through the bend, no cross-stream movement occurred until a pool was reached (Fig. 5.3). Then abrupt changes in the grain size distribution were measured between adjacent cross sections. Downstream from each pool, little change occurred until the next pool was reached. Thus the sorting process was gradual and continuous at high and intermediate flows, and abrupt and discontinuous at low flows.

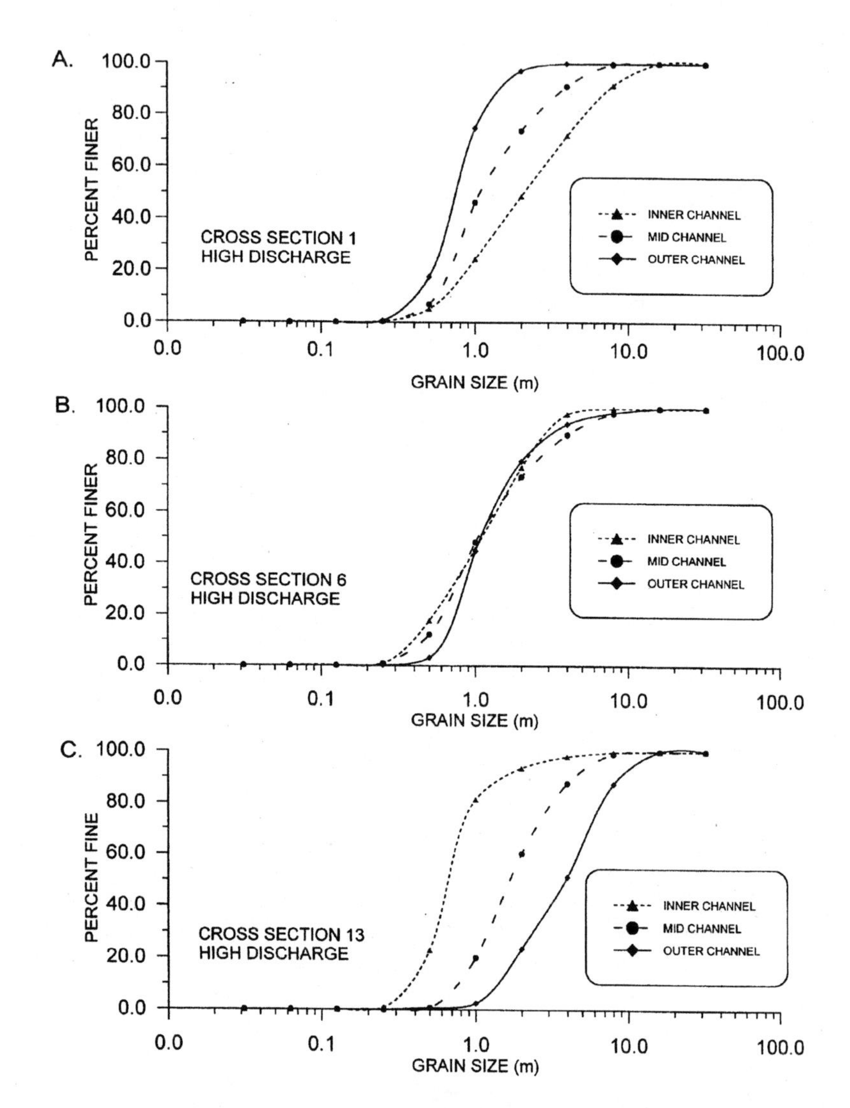

Fig. 5.2. **Sorting in the upstream bend of the two bend study reach: (a) size distribution at the bend entrance; (b) size distribution near the bend apex (note the overlapping grain-size distributions); (c) size distribution near the bend exit (note that the pattern in (a) has been reversed).**

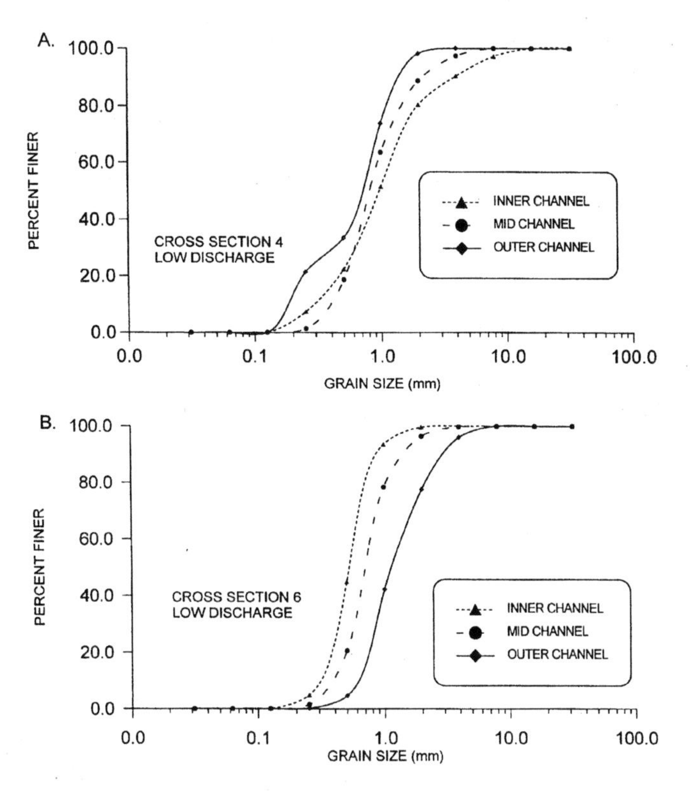

Fig. 5.3. **Sorting at low flow: (a) grain-size distribution in the cross section upstream of a pool; (b) grain-size distribution in the cross section downstream from that same pool (note the sorting which has occurred and that the spread of the three grain-size curves at each cross section are less than at higher discharges before and after sorting).**

At bankfull and overbank discharges, the sorting process was well displayed by the 8 mm bedload fraction (Fig. 5.4a) (about the top 7 percent by weight for this stage). Sorting of this fraction was continuous through the bends, with the band of moving material narrow in the crossings, and widest through the center of the bends. In the upstream bend, it appears that this size class has two separate trajectories, one over the top of the point-bar platform, and one down the center of the point-bar slope. This may be a sampling problem or may be due to a real difference in trajectory controlled by topography.

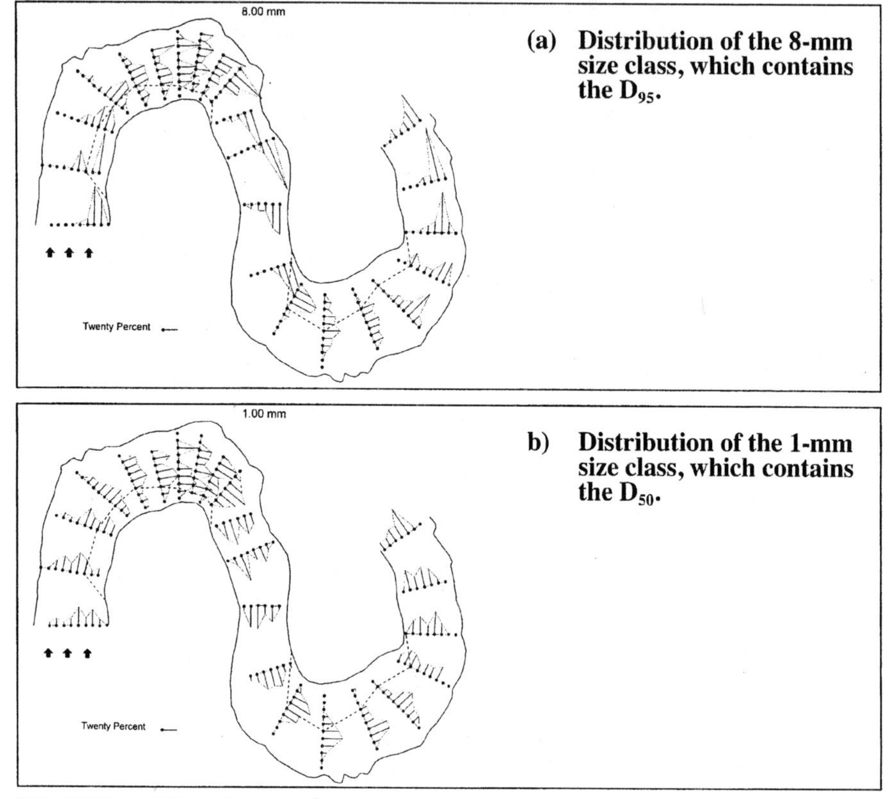

(a) Distribution of the 8-mm size class, which contains the D_{95}.

b) Distribution of the 1-mm size class, which contains the D_{50}.

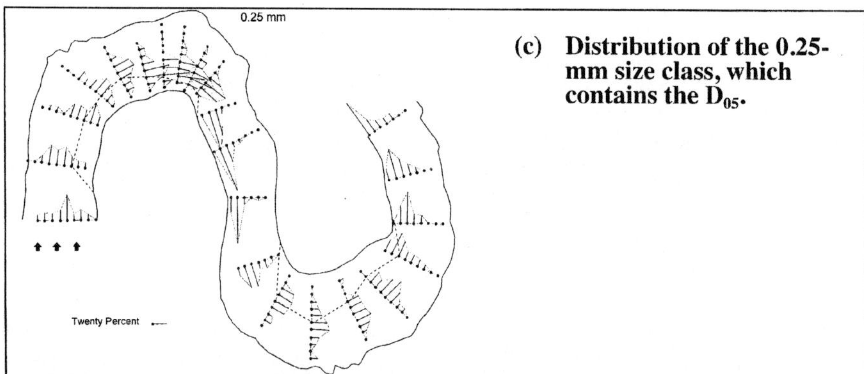

(c) Distribution of the 0.25-mm size class, which contains the D_{05}.

Fig. 5.4. **The trajectory of representative size fractions in moving bedload at high discharge. For all maps, the dotted line represents the edge of the point bar platform, and the heavy arrows show the direction of flow.**

The size class centered on 1 mm particles (Fig. 5.4b) contains the D_{50} of the high discharge bedload. The trajectory here shows a path of sediment that is uneven, showing movement first toward one bank, then another through the bends. This could be due to sampling problems, or it could be because this fraction of the bedload is most sensitive to the relative strength of cross-streamflow and cross-stream topography.

The finer fractions of bedload at high discharge (Fig. 5.4c) show the reverse pattern to the coarse fraction, being most abundant near the outer bank at the bend entrance, and moving toward the inner bank through the bend. Most noticeable is the absence of the fine material from the top of the point-bar platform. The finest 5 percent of the bedload (the 0.25-mm size class) is not transported to the inner bank until downstream of the point-bar platform. This is because the helical flow cell that "sweeps" the finer fraction inward does not extend up onto the point-bar platform. This size fraction was composed of material caught in the Helley-Smith, but also includes the "unmeasured" load, some of which may be suspended.

During falling discharges, coarse sediment in the 8- and 4-mm size classes showed patterns similar to high flows. However, the 8-mm size class moved in a tighter trajectory, less dispersed than at higher discharges. The 1-mm fraction, which contained the D_{50} and the median grain size for this flow stage, stayed more centrally located than in the higher discharges, concentrated roughly at the center of each cross section. The finest fractions showed an interesting pattern, due to the point bar split into two smaller subaqueous forms at this lower discharge level. The upstream point-bar platform showed essentially the same exclusion of the fine material that occurred during high discharge. However, the lower downstream point-bar platform was composed of finer sediment. By that point in the bend, the coarse sediment was almost all in the thalweg region, and flow was all inward between the two point bars. This produced not only a lower point-bar platform, but one with different attendant bedforms. In the lee of this secondary point bar, ripples formed, which were unusual for this river. They occurred because the sediment was finer than the 0.6 mm maximum necessary for ripple formation (Simons and Senturk, 1976; Simons and Richardson, 1966), and was also well sorted (since ripples do not form in poorly sorted sediments, Chiew, 1991). The finer sediment was also not as well segregated from the other size classes in the crossings as it was at higher discharges. This may point to weaker helical flow strength at this stage.

The sorting process during low flows was much more difficult to distinguish (Fig. 5.5). The coarsest portion of the bedload (the 4-mm fraction,

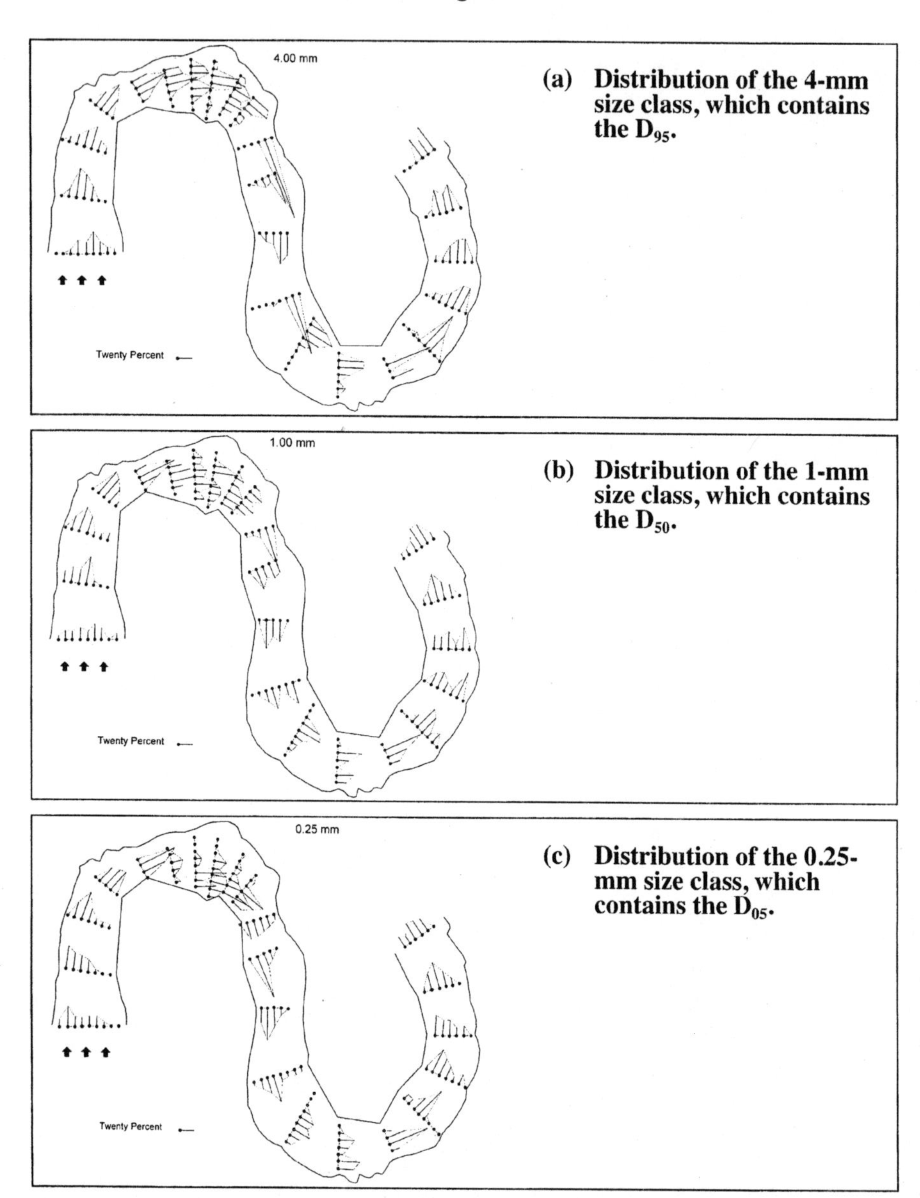

Fig. 5.5. **The trajectory of representative size fractions in moving bedload at low discharge. For all maps, the heavy arrows show the direction of flow.**

Fig. 5.5a) showed a cross-stream movement similar to that of higher discharges, but with most changes occurring near the pools. The 1-mm fraction (Fig. 5.5b) showed little pattern and was present at most sampling localities. The most obvious differential transport appeared in the finest fraction (the 0.25-mm fraction, Fig. 5.5c), where sorting occurred preferentially near the pools.

6. DISCUSSION AND CONCLUSIONS

The patterns of channel adjustment and the trajectories of individual size fractions discussed above show that published models of sediment transport do not adequately address the processes occurring during transverse sediment sorting. The fact that only the D_{50} has a trajectory parallel to the channel centerline demonstrates that for most size fractions a balance of cross-stream forces is not occurring. Rather, these forces are unbalanced, since cross-stream movement is observed. In addition, suspension and/or armoring are not necessary for this sorting to be accomplished. While at bankfull flows the water was turbid, and some material was in suspension (and measured as part of the "unmeasured" suspended load by the Helley-Smith), sorting was accomplished in almost exactly the same pattern during intermediate flows, where the water was clear and the bottom of the channel was quite visible. Channel-bed armoring was never observed. Finally, sorting is not simply accomplished by grain-size depletion of coarse particles from the inner channel, and movement toward the outer bank. Rather, all particles are in motion, with trajectories determined by their grain size.

It appears that this is essentially a cube-square problem. The forces acting on a particle can be divided into those caused by gravity and those caused by the flowing water. Gravity acts on a particle in proportion to its weight, hence its diameter cubed. Flowing water acts on the surface area of the particle perpendicular to the flow and so this force is proportional to its diameter squared. As a particle's size increases, the effects of gravity become more important, whereas smaller particles are more influenced by flow direction near the bed. The net effect is that, as particle size increases, it moves preferentially downslope through a bend, toward the outside bank, while finer material is swept toward the inner bank by helical flow. The emergent pattern shows that at the bend entrance, the coarsest material is near the inner bank, while the finest is near the outer, concave bank. As the sediment moves downstream, the finer material moves inward, and the coarser material moves outward, so that at the bend exit, the pattern is reversed.

6.1. Dynamics of the Sediment-Channel System

Fall River can be viewed as a system dominated by both positive and negative feed-back. Between its outer, rigid banks, changing discharge levels influence the strength of the helical flow cell and the amount of sediment in transport. The helical flow cell and the bed- and bank-cells that form at higher discharges control the position of the zone of maximum sediment transport. The negative feed-back between the patterns of sediment movement, bed slope and shear stress in the downstream direction control and stabilize bed topography at each discharge level, while the positive feed-back between them creates the point bars and excavates the thalweg during changing stage. This influences both sorting and sediment transport. When discharge changes, the new flow and sediment transport patterns are influenced by the preexisting channel topography, but eventually modify it.

Surprisingly, differential cross-stream bedload movement remains relatively unchanged during these dramatic topographic adjustments. This is because the sorting process is a balance between local bed slope and local grain shear stress. As the channel topography becomes more symmetric, the local slope from the thalweg to the point-bar platform becomes flatter. But due to decreasing depth, the magnitude of the grain shear stress vector, as well as the angle that it makes with the downstream axis decreases. Since both factors decrease and increase together, the patterns of bedload sorting remain relatively constant for rising, high, and falling stages. However, weak helical flow and cross-section asymmetry only occur sporadically at the lowest flows. The new pools, or areas of cross- section asymmetry, were associated with narrow zones in the channel, downstream from sharp curvature changes. This tends to confirm the importance of the depth to radius of curvature ratio in determining the strength of the helical cell. Since the process of sorting is sporadic at these low flows, it also confirms the importance of topography and helical flow on sorting in a bend.

6.2. Conclusions

This investigation addressed transverse sorting processes and differential bedload trajectories in a channel undergoing dramatic topographic adjustments. It also identified aspects of channel bend dynamics that are not incorporated in many of the published bedload sorting models. In summary:

1. Transverse sorting of moving bedload in a meander bend occurs when particles of different sizes have different trajectories through each bend. Since cross-stream movement is characteristic of all these trajectories (except the D_{50}), then cross-stream forces are not balanced.

2. The unbalanced forces (drag, lift, gravity) change with particle size, as seen in the different trajectories for each size fraction. Since drag-and-lift forces are proportional to the diameter of the particle squared, and gravitational forces are proportional to the diameter cubed, this would explain the preferential downslope movement of coarser material, and the sweeping of fines toward the channel center.

3. Suspension, armoring, or depletion of coarser bedload are not required for bedload sorting.

4. Through a broad range of discharges, bedload sorting patterns remain stable, due to the fact that helical flow strength, bed shear stress, and transverse channel slopes increase and decrease together.

5. While helical flow and transverse slope explain the primary patterns of each trajectory, the topographic zones of the channel where these do not occur (point bar platform, areas near the outer bank) can also affect each grain size trajectory at some flow levels.

7. REFERENCES

Allen, J.R.L. 1970. A quantitative model of grain size and sedimentary structures in lateral deposits. Jour. Geol.,v. 7, p. 129-146.

Anthony, D.J. 1987. Stage dependent channel adjustments in a meandering river, Fall River, Colorado. Unpublished M.S. thesis, Colorado State University, Ft. Collins, CO, 80523. 180 p.

Anthony, D.J. and M.D. Harvey. 1991. Stage-dependent cross-section adjustments in a meandering reach of Fall River, Colorado. Geomorphology, v. 4, p. 187-203.

Bagnold, R.A. 1966. An approach to the sediment transport problem from general physics. U.S. Geol. Surv. Prof. Paper 422I, p. 1-37.

Bagnold, R.A. 1977. Bedload transport by natural rivers. Water Resources Research, v. 13, no.2, p. 303-312.

Blair, T.C. 1987. Sedimentary processes, vertical stratification sequences, and geomorphology of the Roaring River alluvial fan, Rocky Mountain National Park, Colorado. Jour. Sed. Petrology, v. 57, p. 1-18.

Bridge, J.S. 1977. Flow, bed topography, grain size, and sedimentary structure in open-channel bends: a three-dimensional model. Earth Surface Processes & Landforms, v. 2, p. 401-416.

Bridge, J.S. and D.F. Dominic. 1984. Bed load grain velocities and sediment transport rates. Water Resources Research, v. 20, p. 476-490.

Bridge, J.S. and J. Jarvis. 1982. The dynamics of a river bend: a study in flow and sedimentary processes. Sedimentology, Vol. 29, p. 499-541.

Carey, W.P. 1985. Variability in measured bedload transport rates. Water Resour. Bull., v. 21, p. 39-48.

Chiew, Y.M. 1991. Bed features in nonuniform sediments. Jour. Hydraul. Div., Am. Soc. Civ. Eng., v. 117, p. 116-120.

Diegaard, R. 1980. Longitudinal and transverse sorting of grain sizes in alluvial rivers. Ser. Pap. 26, Tech. Univ. of Denmark, Lyngby. 108 p.

Dietrich, W.E. 1982. Flow, boundary shear stress, and sediment transport in a river meander. Unpublished Ph. D. dissertation, University of Washington, Seattle, WA, 261 p.

Dietrich, W.E. and J.D. Smith. 1984. Bed load transport in a river meander. Water Resources Research, v. 20, p. 1355-1380.

Einstein, H.A. 1950. The bedload function for sediment transportation in open channels. U.S. Dept. of Agriculture, Soil Conservation Serv., Tech. Bull. 1026. 71 p.

Engelund, F. 1974. Flow and bed topography in channel bends. Jour. Hydraul. Div., Am. Soc. Civ. Eng., v. 100, p. 1631-1648.

Grass, A.J. 1970. Instability of fine sand bed. Jour. Hydraul. Div., Am. Soc. Civ. Eng., v. 96, p. 619-631.

Hoey, T.B. and R. Ferguson. 1994. Numerical simulation of downstream fining by selective transport in gravel bed rivers: model development and illustration. Water Resources Research, v. 30, n. 7, p. 2251-2260.

Hooke, R.L. 1975. Distribution of sediment transport and sheer stress in a meander bend. Jour. Hydraul. Div., Am. Soc. Civ. Eng., v. 83, p. 543-565.

Hooke, R.L. and C.E. Chase. 1978. Flow, bed topography, grain size and sedimentary structure in open channel bends: A three dimensional model: A discussion, Earth Surf. Processes and Landforms, v.3, p. 421-422.

Ikeda, S. 1989. Sediment transport and sorting at bends. In River Meandering, S. Ikeda and G. Parker, eds., Am. Geophy. Union, Washington, D.C., p. 103-126.

Ikeda, S., G. Parker, and K. Sawai. 1981. Bend theory of river meanders, Part I: Linear development. Jour. Fluid Mechanics, v. 112, p. 363-377.

Iseya, F. and H. Ikeda. 1987. Pulsations in bedload transport rates inducted by a longitudinal sediment sorting: a flume study using sand and gravel mixtures. Geografiska Annaler, Series A, v. 69A, n.1, p. 15-27.

Jackson, R.G., II. 1975. Velocity-bedform-texture patterns of meander bends in the lower Wabash River of Illinois and Indiana. Geol. Soc. Am. Bull., v. 86, p. 1511-1522.

Jarrett, R.D. 1987. Flood hydrology of foothills and mountain streams in Colorado. Ph.D. thesis, Colorado State University, Ft. Collins, CO. 239 p.

Jarrett, R.D., and J.E. Costa. 1985. Hydrology, geomorphology and dam-break modeling of the July 15, 1982 Lawn Lake Dam and Cascade Lake Dam failures, Larimer County, Colorado. U.S. Geol. Surv. Open File Report, 84-612, 109 p.

Kirchner, J.W., W.E. Dietrich, F. Iseya, and H. Ikeda. 1990. The variability of critical shear stress, friction angle, and grain protrusion in water-worked sediment. Sedimentology, v. 37, p. 647-672.

Komar, P.D. and Z. Li. 1986. Pivoting analyses of the selective entrainment of sediment by shape and size with application to gravel threshold. Sedimentology, v. 33, p. 425-436.

Kovacs, A. and G. Parker. 1994. A new vectorial bedload formulation and its application to the time evolution of straight river channels. J. Fluid Mech., v. 267, p. 153-183.

Kuhnle, R.A. 1992. Bedload transport during rising and falling stages on two small streams. Earth Surface Processes and Landforms, v. 17, p. 191-197.

Lapointe, M.L. 1992. Burst-like sediment suspension events in a sand bed river. Earth Surface Processes and Landforms, v. 17, p. 253-270.

Leopold, L. B. and M.G. Wolman. 1960. River meanders. Geol. Soc. Amer. Bull., v. 71, p. 769-794.

Mendoza, C. and J.W. Shen. 1990. Investigation of turbulent flow over dunes. Jour. Hydraul. Div., Am. Soc. Civ. Eng., v. 116, p. 459-477.

Miall, A.D. 1978. Fluvial sedimentology: an historical review. Progress in Physical Geography. v. 2, p. 1-45.

Middleton, G.V. and J.B. Southard. 1984. Mechanics of Sediment Movement. Soc. Econ. Paleontologists and Minerologists Short Course No. 3., 2nd Edition, Providence. 371 p.

Miller, M.C., I.N. McCave, and P.D. Komar. 1977. Threshold of sediment motion under unidirectional currents. Sedimentology, v. 24, p. 507-527.

Nakagawa, H. and I. Nezu. 1981. Structure of space-time correlations of bursting phenomena in an open channel flow. Jour. Fluid Mechanics, v. 104, p. 1-43.

Odgaard, A.J. 1982. Bed characteristics in alluvial channel bends. Jour. Hydraul. Div., Am. Soc. Civ. Eng., v. 108, p. 1268-1281.

Odgaard, A.J., and M.A. Bergs. 1988. Flow processes in a curved alluvial channel. Water Resources Research, v. 24, p. 45-56.

Parker, G. and E.D. Andrews. 1985. Sorting of bedload sediments by flow in meander bends. Water Resources Research, v. 21, p. 1361-1373.

Petit, F. 1994. Dimensionless critical shear stress evaluation from flume experiments using different gravel beds. Earth Surface Processes and Landforms, v. 19, n. 6, p. 565-576.

Pitlick, J.C. 1985. The effect of a major sediment influx on Fall River, Colorado, unpublished M.S. Thesis, Colorado State University, Ft. Collins, CO, 80523, 127 p.

Pitlick, J.C. 1993. Response and recovery of a subalpine stream following a catastrphic flood. Geol. Soc. of Amer. Bull., v. 105, n.5, p. 657-670.

Pitlick, J.C., and M.D. Harvey. 1986. A summary of 1985 channel changes and sediment transport on Fall River, Rocky Mountain National Park, CO. Unpublished manuscript, Dept. of Earth Resources, Colorado State Univ., 38 p.

Richards, K. 1990. Fluvial Geomorphology: initial motion of bed material in gravel-bed rivers. Progress in Physical Geography, v. 14, p. 395-415.

Robert, A. 1993. Bed configuration and microscale processes in alluvial channels. Progress in Physical Geography, v. 17, n. 2, p. 123-136.

Rozovskii, I.L. 1961. "Flow of Water in Bends of Open Channels". Academy of Science of the Ukrainian SSR, Kiev, Translated by Y. Prushansky, Israel Program for Scientific Translations, Jerusalem, PST Cat. No. 363.

Shields, A. 1936. Anwendung der Ahnlichkeitsmechanik und der Turbulenzforschung auf die Geschiebebewegung. Mitt. Preuss. Vers. Anst. Wasserb. Schiffb. v. 36, 26 pp.

Simons, D.B. and F. Senturk. 1977. Sediment Transport Technology. Water Resources Publications, Ft. Collins, CO, 807 p.

Simons, D.B. and E.V. Richardson. 1966. Resistance to flow in alluvial channels. U.S. Geol. Surv. Prof. Paper 422-J. 61 p.

Smith, J.D. and S.R. McLean. 1984. A model for flow in meandering streams. Water Resources Research, v. 20, p. 1301-1315.

Thorne, C.R., L.W. Zevenbergen, J.C. Pitlick, S. Rais, J.B. Bradley, and P.Y. Julien. 1985. Direct measurements of secondary currents in a meandering sand - bed river. Nature, v. 316, p. 746-747.

Thorne, C.R. and M.C. Tovey. 1981. Stability of composite river banks, Earth Surface Processes, v. 6, p. 469-484.

White, C.M. 1940. The equilibrium of grains on the bed of a stream. Proc. Roy. Soc., Ser. A 174, pp. 332-338.

Wiberg, P.L. and J.D. Smith. 1987. Calculations of the critical shear stress for motion of uniform and heterogeneous sediments. Water Resources Research, v. 23, p. 1471-1480.

DEVELOPMENT OF MINIATURE EROSION LANDFORMS IN A SMALL RAINFALL-EROSION FACILITY

Shunji Ouchi[1]

ABSTRACT

Miniature erosional landforms were created by applying artificial rainfall on a square mound made of a mixture of fine sand and kaolinite, and their surface topography was analyzed as a self-affine fractal surface. Two runs were performed with different ratios of mixed material (the ratios of sand and clay are 15:1 for Run 1 and 30:1 for Run 2), which gave the mounds different permeabilities. Erosion started as rapid valley incision and then widening of valleys accompanied by slow degradation of interfluves and ridges followed. In Run 1, valley sideslopes tended to decline as the degradation proceeded; whereas, in Run 2, in which the material to be eroded was much more permeable, parallel retreat of steep valley sideslopes seemed dominant. Low and gentle topography appeared at the end of Run 1 and the surface became mostly flat in Run 2. The general trend of surface lowering is reflected in the changes of the average heights (*zmean*) and the standard deviation of heights in a 10x10 cm square area (*Zi*). The former showed an exponential decrease, and the latter increased rapidly in the early stages and then decreased gradually in both runs. The increase of *Zi* was smaller and less rapid but the decrease was larger and more rapid in Run 2. The value of *H'*, which indicates a fractal characteristic of self-affine curves and surfaces, decreased rapidly to about 0.7~0.8 in the early stage of Run 1, and then gradually decreased towards 0.5, which is the value of random Brownian motion. In Run 2, however, the value of *H'* decreased to about 0.5 at first and then gradually increased towards 1.0, which represents a smooth and planar surface. This difference apparently reflects the difference in the manner of slope retreat, declining (Run 1) and parallel (Run 2). Declining slopes probably have a limit, at which point erosion does not proceed any more; whereas, parallel retreat of steep slopes can produce planar surfaces ("pediplains"). The parameter *H'* measured on a land surface may well indicate the processes of erosion that have been dominant in the area.

---

[1] College of Science and Engineering, Chuo University, 1=13=27 Kasuga, Bunkyo, Tokyo, 112 Japan.

KEYWORDS

erosion, rainfall, landforms, degradation

1. INTRODUCTION

Experimentally developed miniature erosional topography was analyzed as a self-affine surface for the purpose of understanding landform development within the concept of fractal geometry. Landforms have been considered a good example of fractal geometry found in nature, especially after Mandelbrot (1967) pointed out the statistical self-similarity of coastlines. The concept of fractal geometry, which points out the existence of certain characteristics in complex forms regardless of scale, may provide a tool to quantitatively analyze landform development by erosion. However, a standard method for expressing landform morphology with fractal geometry has not been established yet, despite many attempts (e.g., Xu et al., 1993). This is due at least to the self-affine nature that landforms show when the distribution of elevation is considered. The self-similar fractal curves such as the well-known triadic Koch (snowflake) curve and coastlines can be scaled isotopically to keep the similarity in different scales. The self-affine fractal curve, such as the time record of a one-dimensional random walk (Brownian motion) and the skyline of mountains, on the other hand, cannot be scaled isotopically. The skyline of mountains, for example, flattens if the viewing point is moved farther away. For this type of self-affine fractal curve, the vertical variation must be scaled differently from the horizontal, because the vertical and horizontal directions have different meaning. Matsushita and Ouchi (1989) developed a method to analyze the self-affinity of various fractal curves, including topographic transects, with the parameter H', which is equivalent to the scaling parameter of fractional Brownian motion, H (Hurst exponent). This method was extended to analyze surfaces developing in three-dimensional space, and H' values of landforms measured on 1:25,000 scale topographical maps indicated that high mountain areas have larger values of H' and well dissected and lowered areas have values close to 0.5 (Ouchi and Matsushita, 1992). The parameter H' may represent a certain characteristic of topography, and its scale-free nature suggests that H' may be a useful tool to study landform evolution.

The significance of H' in the formation of landforms is not clear yet, because the actual course of long-term landform evolution by erosion is largely unknown. The development of erosional surfaces generated experimentally by artificial rainfall can be observed and measured, however, and may provide some insights into the meaning of H' in landform evolution.

2. EQUIPMENT AND PROCEDURE

A square mound (about 100 x 100 x 20 cm) was made of a mixture of fine sand (D_{50} about 0.12 mm) and kaolinite in a small rainfall-erosion facility. The ratios of sand and kaolinite were 15:1 by weight for Run 1 and 30:1 for Run 2. The coefficients of permeability were $k=2.8\text{x}10^{-4}$cm/s and $k=5.3\text{x}10^{-3}$cm/s, respectively. Artificial, very fine drop rainfall was applied to this mound from sprinkler tubes, four (Run 1) and eight (Run 2) strips (1 m long each), which were set around the mound 1.0~1.2 m above the mound surface. The number of sprinkler tubes (i.e., precipitation) was doubled in Run 2 to increase the rate of erosion by generating surface runoff, because much slower erosion rates on more permeable material had been expected from preliminary experiments. Generating uniform rainfall over the mound surface was technically very difficult, and precipitation varied widely, but average precipitation rates of about 38 mm/hr in Run 1 and about 77 mm/hr in Run 2 were achieved. Precipitation tended to be highest in the central area, and this led to the development of a large, main valley at the middle and residual hills at the four corners of the mound. This condition determined the broad outline of the miniature landform, but apparently had no significant effect on the characteristics of landform evolution. The sand mound was mainly eroded by surface runoff. Creep-like movements of surface material and occasional slope failures were observed in the early stages of experiments. No erosion by rainsplash was observed. The surface topography was measured by a point gage, which automatically reads and stores data (x, y and z) in memory when the rainfall is discontinued. Erosion was more rapid immediately after the rainfall resumed, but this did not result in any change in the characteristics of the topography. Measurements were made along 77 cross-section lines spaced at 1 cm intervals on the inner part of the mound surface. Every break point along the cross section was measured, and the data were later converted to 76 x 76 cm gridded data for the analysis.

3. MEASUREMENT OF SELF-AFFINITY WITH THE PARAMETER H'

Self-similar fractal curves, which have an infinite number of overlaps with various scales, can be quantitatively characterized by a fractal dimension, D. The length of the curve (number of unit length), N, is scaled with the unit length of measurement, a, as $N \sim a^{-D}$. Consider the case of the triadic Koch curve, which is generated by an infinitely recursive procedure of dividing a simple line into thirds and replacing the middle segment by two equal segments forming part of an equilateral triangle, $a=(1/3)^n$ and $N=4^n$, where n is the stage number of the recursive procedure. Then, D is log4/log3=1.26··· for the triadic Koch curve. The fractal dimension, D, is practically invalid on self-affine curves, which do not have an infinite number of overlaps with various scales. Matsushita and Ouchi (1989) developed a method, which treats x and y coordinates separately, to analyze the fractal characteristics of self-affine curves as well as self-similar curves. The curve is divided into sections of equal length by the yardstick (or walking divider) method, and values for curve length N, and standard deviations of x and y coordinates of all measured points on the curve (X and Y) are obtained for each section. The respective average values of X, Y and N are regarded as the representative values for this yardstick length. X and Y are related to N as:

$$X \sim N^{\nu x} \tag{3.1}$$

$$Y \sim N^{\nu y} \tag{3.2}$$

X and Y are scaled with each other as:

$$Y \sim X^{H'} \tag{3.3}$$

$$H' = \nu_y \,/\, \nu_x \tag{3.4}$$

$H' = \nu_y$, because $\nu_x = 1$ for the curves without overlaps in the x direction, such as topographic transects. The variance of increments of the time record of one-dimensional random walk (Brownian motion), X_t^2, is known to be related to the corresponding time interval, T, as $X_t^2 \sim T^{2H}$ (H=0.5, and 0< H<1 for the fractional Brownian motion, Mandelbrot and Van Ness, 1968). The parameter H' is apparently equivalent to this scaling exponent H (Hurst exponent) of fractional Brownian motion. If $v_y=v_x=v$ (or $H'=1$), then the curve is defined as self-similar, with a fractal dimension $D=1/v$. Ouchi and Matsushita (1992) extended this "line scaling method" to self-affine surfaces ("area scaling method"). The surface is divided into squares of nearly equal surface area, applying a unit surface area as a scale. The elevation variance Z^2, basal area A and surface area S are obtained for each square unit. The respective average values are regarded as representative values for this scaling unit. Z^2 and A are related to S as:

$$Z^2 \sim S^{vz} \tag{3.5}$$

$$A \sim S^{vA}. \tag{3.6}$$

Z^2 and A are scaled with each other as:

$$Z^2 \sim A^{H'} \tag{3.7}$$

$$H'=v_z/v_A. \tag{3.8}$$

$H'=v_z$, because $v_A=1$ for surfaces without overlaps. The measured values of H' by this "area scaling method" on fractional Brownian surfaces, any transect of which has characteristics of fractional Brownian motion, reproduced well the values of H with which these surfaces were generated (Ouchi and Matsushita, 1992). This method appeared practical to obtain the value of H', and was exclusively used in this study. The correlation in the equation $Z^2 \sim A^{H'}$ must be strong if the surface is to be analyzed as a self-affine surface. In this study, relations with R^2 values less than 0.93 were rejected.

4. DEVELOPMENT OF MINIATURE EROSIONAL TOPOGRAPHY

Run 1: Erosion started at the edges of the mound right after rainfall began. After 1 hour, some valleys a little more than 10 cm deep had developed on the surface (Fig. 4.1). The largest valley appeared on the southern edge of the mound, draining from north to south along the subtle initial relief. The positions of the major valleys did not change throughout the experiment. The incision of these valleys and their tributary valleys was rapid, with development of the valley system in the first several hours, and was then followed by slower change. The valleys increased in width with declining

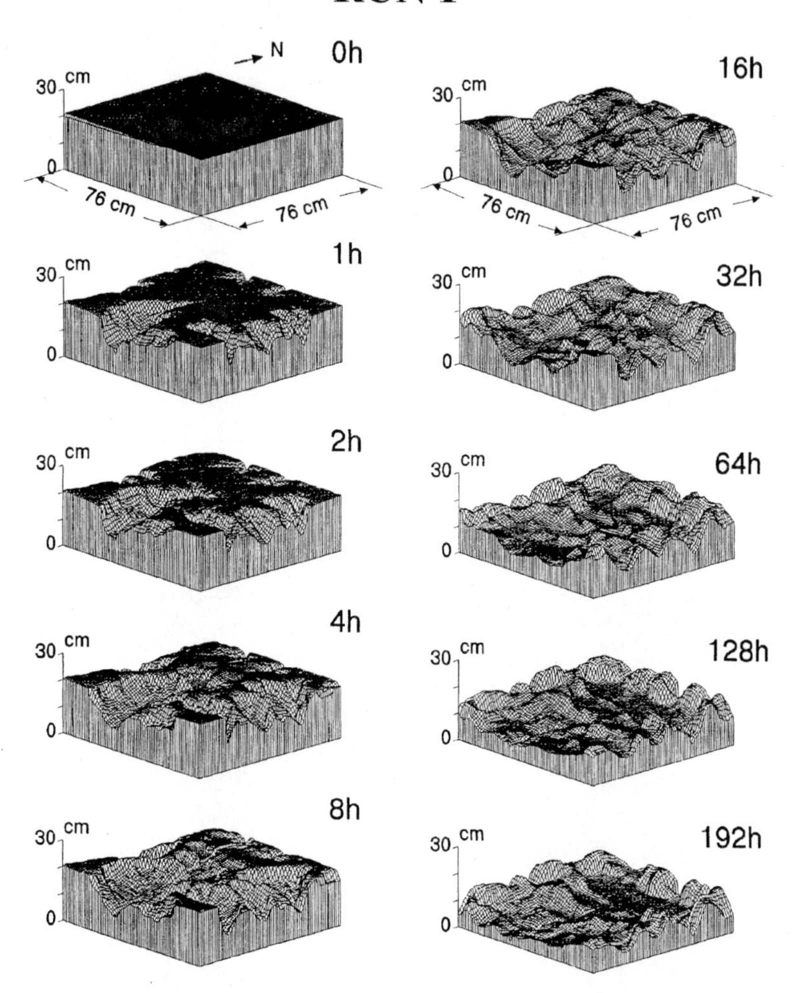

Fig. 4.1. Block diagrams showing the surface topography of the mound in Run 1. All the blocks are viewed from the southeast.

valley sideslopes accompanied by slow degradation of interfluves and ridges. The remnants of the original flat surface could be observed clearly for the first 8 hours of rainfall, but they were obscured after 16 hours when the entire valley system was well developed (Fig. 4.1). At the end of the experiment (192 hours of rainfall), the surface had become very gentle.

The value of *Zi*, which is the average standard deviation of elevation in a 10 x 10 cm square area, increased rapidly in the first several hours (14.4 mm at 8 hours) and then decreased gradually (Table 4.1, Fig. 4.2). This reflects the rapid valley incision in the early stages and the later domination of slow degradation of interfluves and ridges with declining valley sideslopes. The average height from the arbitrary base level, *zmean*, shows an exponential

Table 4.1. Values of average height (*zmean*), average standard deviation of elevation in a 10 x 10 cm square (*Zi*), and *H'*.

Time of Rainfall (hours)	*zmean* (mm)	*Zi* (mm)	*H'*
Run 1			
0	269	0.3	1.0
1	257	11.1	0.76
2	253	11.8	0.79
4	241	13.6	0.74
8	234	14.4	0.72
16	218	13.5	0.75
32	197	13.6	0.71
64	181	13.2	0.66
128	166	11.1	0.61
192	160	11.6	0.59
Run 2			
0	203	0.3	0.91
1	193	8.2	*
3	184	8.5	0.54
7	171	9.9	0.54
15	156	10.6	0.57
31	140	10.6	0.62
63	120	8.7	0.59
127	98	5.2	0.72
191	83	3.4	0.81
255	74	2.4	0.97

*The correlation between Z^2 and *A* was not sufficient to calculate *H'*.

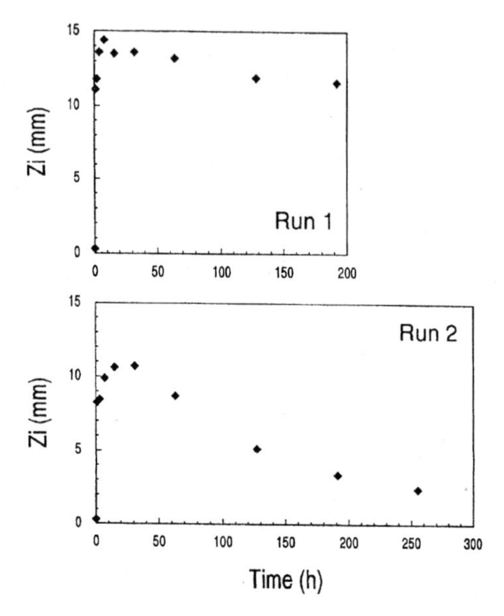

Fig. 4.2. The average standard deviation of elevation in a 10 x 10 cm square *Zi*, Run 1 and Run 2.

decrease (Table 4.1, Fig. 4.3) with an overall average erosion depth of 109 mm (about 63,000 cm^3 of material was eroded from the measured area). The value of H' decreased rapidly from about 1.0 to 0.7~0.8 (0.79 at 2 hours), in a somewhat unstable manner in the early stages. It then decreased gradually to 0.59 by the end of experiment (Table 4.1, Fig. 4.4). The tendency of H' shown in Fig. 4.4 indicates that H' would decrease gradually towards 0.5, which represents the random Brownian motion surface, if the experiment continued.

Run 2: The higher permeability of the mound (one order of magnitude larger than Run 1) required more intense rainfall to generate surface flows and this caused more slope failures in the early stages. The surface topography after 1 hour of rainfall, which did not show enough correlation between Z^2 and A to calculate H' (R^2=0.86), seems to reflect the influence of slope failures. Rapid incision occurred in the main valley that developed south to north at the middle of the mound (Fig. 4.5), and this incision reached a certain base level of erosion within less than 1 hour of rainfall. The erosion proceeded with lateral valley widening and parallel retreat of steep valley walls, on which many small rills developed. This manner of erosion produced a wide, flat-bottomed valley from the relatively early stages of the experiment. The surface became very flat

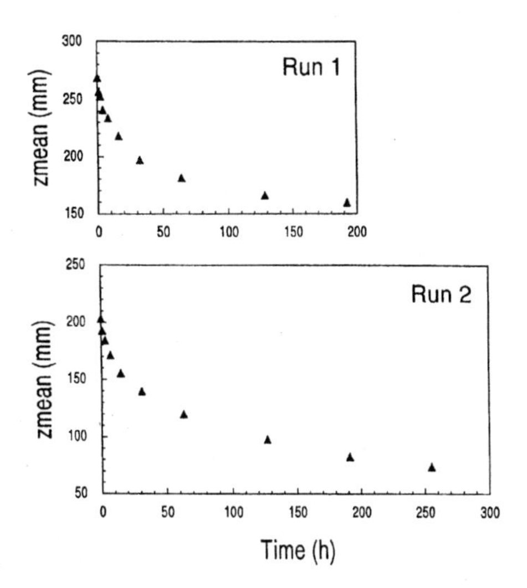

Fig. 4.3. The overall average height from the arbitrary base level, *zmean*, Run 1 and Run 2.

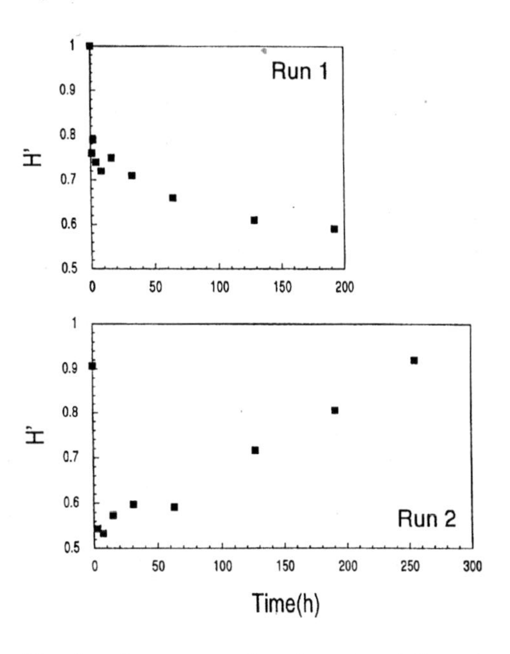

Fig. 4.4. Parameter *H '*, Run 1 and Run 2.

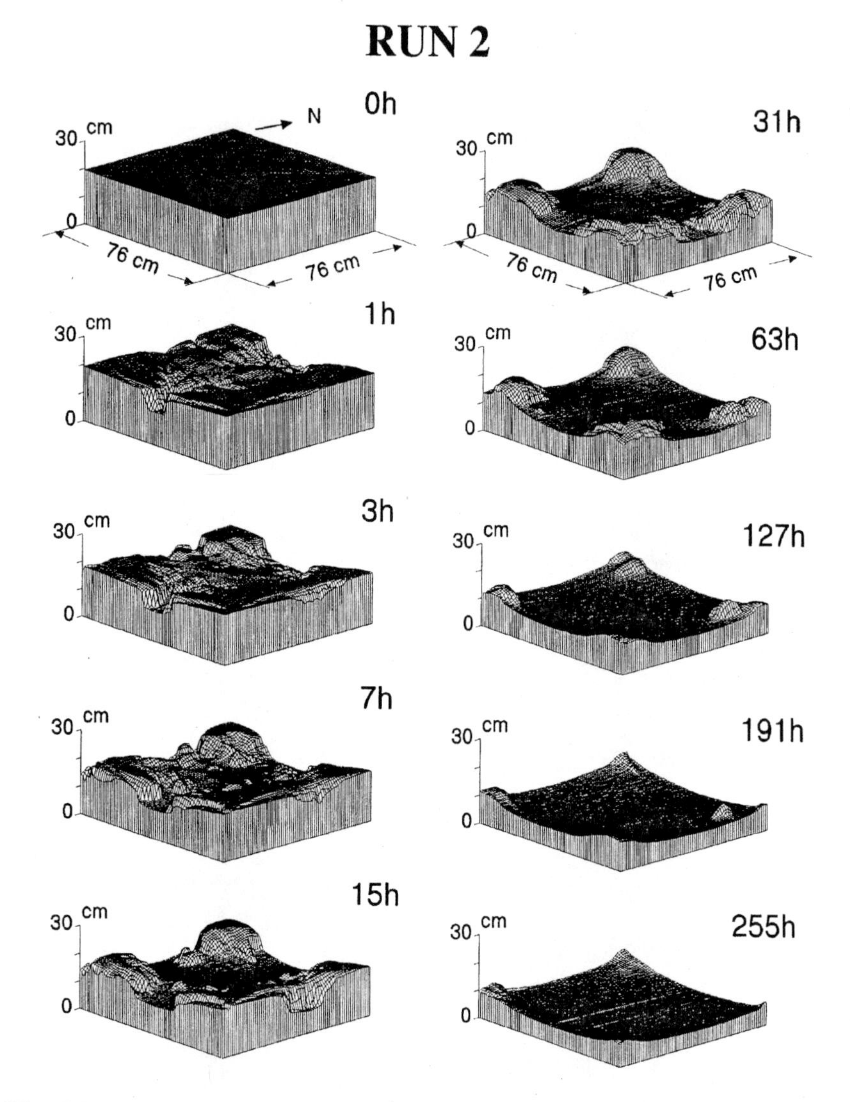

Fig. 4.5. Block diagrams showing the surface topography of the mound in Run 2. All the blocks are viewed from the southeast.

after 255 hours of rainfall, except for the four corners of the square mound, where precipitation was not sufficient and water flow concentration hardly occurred.

The value of *Zi* increased in the early stages (10.6 mm at 15 hours, smaller and slower than in Run 1), and then decreased more rapidly (Table 4.1, Fig. 4.2). The average height, *zmean*, shows an exponential decrease (Table 4.1, Fig. 4.3) with an overall average erosion depth of 129 mm (about 74,510 cm^3 of material was eroded from the measured area). The value of *H'* decreased rapidly from 0.91 to 0.54 in the early stage (at 3 hours), then increased gradually to 0.97 at the end of the experiment (Table 4.1, Fig. 4.4). Figure 4.4 shows that *H'* in Run 2 increased towards 1.0, the number which represents a smooth and planar surface. This direction of change in *H'* is almost opposite to the change in Run 1, and this difference apparently results from the difference in the dominant process of erosion.

In both runs, surface runoff induced gully-type erosion from the edge of the mound, and this incision reached a certain base level in the early stages. The erosion rate, however, was much more rapid in Run 2. The initial surface was extensively modified after 3 hours of rainfall and a relatively flat area developed in the middle from the early stages in Run 2. The slope failures may have played an important role. In Run 1, on the other hand, gullies did not cut all the way through the initial surface even after four hours of rainfall. The smaller increase and more rapid decrease of *Zi* in Run 2 indicates the relative dominance of planation-type erosion in Run 2. The most significant difference in erosional processes between Runs 1 and 2 seems to be the method of slope retreat. While valley slope decline and development of a valley system seemed dominant in Run 1, valley widening with parallel slope retreat was conspicuous in Run 2. This difference is apparently reflected in the change of *H'* towards 0.5 in Run 1 and 1.0 in Run 2.

5. DISCUSSION

Czirók and Somfai (1993) made miniature landforms by spraying water on a smooth bank-like mound made of a mixture of silica sand and earthy soil with organic matter. They took a photo of the "profile" when a feature like a real mountain ridge had developed, and analyzed the digitized profile to get an α value, which is equivalent to our *H'*. They pointed out that the value of α they got, 0.78 ± 0.05, agreed well with those obtained for pictures taken from ridges in the Dolomites, Italy (0.8 ± 0.1). These values also agree with the values of exponent *H*, which is equivalent to *H'* and α, in the relationship that Hurst (1951) pointed out for the rescaled range R / σ of some geophysical time series (where *R* is the cumulative sample range and σ is the standard deviation of the data, versus the period of record *n* as $R / \sigma \sim n^H$). Hurst (1951) showed that *H* is about 0.7~0.8, instead of 0.5, which is predicted from statistical theories, and this has been called the Hurst Phenomenon (Klemeš, 1974). The experiment conducted by Czirók and Somfai (1993), however, was relatively short, as it ended when they saw the mountain-like feature. If their experiment had continued for longer, the α value likely would have decreased towards 0.5.

In Run 1 of this study, the larger value of H', 0.7~0.8 appeared in the early stages and then decreased towards 0.5, which is the value for a random Brownian surface. A value of H' smaller than 0.5 can hardly be imagined for a surface formed by erosion. The surface would become flatter (H' increases) later in a longer time period with the continuation of erosion. However, the surface would not be a flat plain with the value of H' equal to 1.0, which indicates a smooth and planar surface in the case of landforms (a completely horizontal and flat surface has H' of 0.0, but this is very improbable). Declining slopes will finally reach a certain low but non-zero gradient, on which erosion will not proceed any more. This gradient is determined by the balance between eroding force and erodibility of the slope. A value of 0.5 may be the ultimate value of H' for a surface that develops with this type of erosional process. In Run 2, however, an H' value close to 0.5 appeared in the early stages and H' gradually increased towards 1.0 as the surface flattened. Valley widening with parallel retreat of steep valley sideslopes caused extension of flat surfaces from the early stages of the experiment, and this process apparently caused the increase of H' towards 1.0. Under the conditions of these experiments, the H' value of the surface degraded towards 0.5 as a result of valley incision and slope decline. When the initial surface was eroded by parallel retreat of valley walls to produce a flat residual surface, the value of H' increased toward 1.0, during the long(er) period of erosion. Using the values of H' and Zi, which indicate the magnitude of relief, one may be able to infer the process of erosion by which the surface has been developing. Permeability of material and rainfall rate are probably the main controlling factors over the erosional processes in the experiments. The effects of these factors, however, cannot be determined because both permeability and precipitation were changed from Run 1 to Run 2. Here, only differences in the morphological development of surface topography were considered, and what causes the difference in slope retreat is left for future examination.

Although it is too early to compare the results of these experiments with real landforms, I would like to outline some possible directions for future study. The results derived from Run 1 suggest that erosional landforms developed by valley incision and slope decline over a long time period, which seem common in many mountain areas, tend to have a value of H' that decreases towards 0.5 with the advance of degradation. In order for a high value of H' to remain for a long time in this condition of erosion, some continued input of renewing energy, namely uplift, would be necessary. The relatively large H' value of 0.72 for the Yarigatake area (Ouchi and Matsushita, 1992), which is characterized by high mountain ridges and deep valleys, is consistent with this interpretation, and the small H' value for the well dissected Yamizosan area, 0.48, seems to reflect the long-term tectonic stability of this area. Valley widening with the parallel retreat of steep valley walls resulted in the increase of H' value towards 1.0 in the experiments. This implies that the low relief plain (small Zi) with a large value of H' can be considered as a "pediplain," and in the same manner the low relief plain with a value of H' close to 0.5 may be regarded as a "peneplain". Schumm (1956a) stated that parallel and decline slope retreat in the evolution of erosional topography occurs depending on the available relief and on infiltration rates of soil. In badlands at Perth Amboy, New Jersey, he observed parallel slope retreat proceeding with channel downcutting and with lateral planation after the cessation of downcutting, until basal deposition occurs. In the experiments (Run 2), parallel slope retreat with lateral planation occurred after rapid valley incision in the early stages. The coefficient of permeability of the material forming the badlands at Perth Amboy, about 4.0×10^{-3} cm/s, is similar to that of Run 2, and this implies that permeability of material may have some controlling effects on the method of slope retreat. Schumm (1956b) also pointed out the difference in slope retreat between less permeable Brule and more permeable Chadron formations in Badlands National Monument, South Dakota. The Brule slopes retreat parallel, as in the badlands at Perth Amboy, under the action of surface runoff, while the Chadron slopes tend to decline by the action of creep. This conclusion seems completely inconsistent with the results of the experiments, in which slope decline occurred on less permeable material and parallel slope retreat on more permeable material. In Run 2, however, precipitation was increased to generate surface runoff and this, apparently, caused the parallel slope retreat on more permeable material. In Run 1, on the other hand, the development of valley system and slope decline were the characteristic processes, but creep, which is considered to cause slope decline, seemed not to play an important role in the slope retreat. The rainfall probably was not enough to generate sufficient lateral erosion to keep parallel slope retreat in Run 1. Permeability and erodibility of material, precipitation, and available relief that relates to the rate of uplift should be considered as important controlling factors in the course of erosional landform evolution, which may be reflected in the value of H'.

6. ACKNOWLEDGMENTS

This study was supported by the Institute of Science and Engineering, Chuo University, which is gratefully acknowledged. I also thank Yuko Nakano, Kenji Kawana, Minoru Kawazu and Kei Suwa, students of Chuo University, for their assistance in the experiment.
I would like to dedicate this short article to Professor Dr. Stanley A. Schumm who kindly instructed a foreign student throughout his graduate course in Colorado State University from 1979 to 1983. I will never forget those bright days I spent at Colorado State University (CSU).

7. REFERENCES

Czirók, A. and E. Somfai. 1993. Experimental evidence for self-affine roughening in a micromodel of geomorphological evolution. *Physical Review Letters*, 71, 2154-2157.

Hurst, H.E. 1951. Long-term storage capacity of reservoirs. *Trans. Am. Soc. Civil Eng.*, 116, 770-808.

Klemeš, V. 1974. The Hurst phenomenon: A puzzle?. *Water Resources Research*, 10, 675-688.

Mandelbrot, B.B. 1967. How long is the coast of Britain? Statistical self-similarity and fractional dimension. *Science*, 165, 636-638.

Mandelbrot, B.B. and J.W. Van Ness. 1968. Fractional Brownian motions, fractional noises and applications. *SIAM Rev.*, 10, 422-437.

Matsushita, M. and S. Ouchi. 1989. On the self-affinity of various curves. *Physica D*, 38, 246-251.

Ouchi, S. and M. Matsushita. 1992. Measurement of self-affinity of surface as a trial application of fractal geometry to landform analysis. *Geomorphology*, 5, 115-130.

Schumm, S.A. 1956a. Evolution of drainage systems and slopes in badlands at Perth Amboy, New Jersey. *Bull. Geol. Soc. Am.*, 67, 597-646.

Schumm, S.A. 1956b. The role of creep and rainwash on the retreat of badland slopes. *American Journal of Science*, 254, 693-706.

Xu, T., I.D. Moore, and J.C. Gallant. 1993. Fractals, fractal dimensions and landscapes - a review. *Geomorphology*, 8, 245-262.

CHANNEL MORPHOLOGY ON THE EASTERN SLOPE OF THE COASTAL RANGE IN TAIWAN[1]

Jui-Chin Chang[2]

ABSTRACT

The Coastal Range is the most active tectonic area in Taiwan. Subject to continual stress coming from the southeast, the Coastal Range has been deformed by several faults and folds. Uplift has led to the emergence of marine terraces in the coastal area. Most rivers have a straight course with gorges and entrenched valleys in the steep slopeland, but ingrown meanders and entrenched meanders in the gentle slopeland and basin areas. On alluvial fans and in the coastal area, channels exhibit straight courses or meanders, depending on channel slope. In response to baselevel lowering, a river adjusts by changing its channel form, depending on the magnitude and rate of baselevel change, valley slope, and drainage area. It is concluded that the channel morphology is controlled by the resistance of the rocks, topography, geological structure and tectonic movement.

KEYWORDS

Channel morphology, sinuosity, coastal range (eastern Taiwan)

1. INTRODUCTION

The Coastal Range is the most active tectonic area in Taiwan (Hsu, 1956; Wang, 1976; Teng and Wang, 1981; Chen, 1988; Yao et al., 1989; Chen et al., 1991). It is 140 km long and 10 km wide with a maximum elevation of 1600 m above sea level. Several high gradient, consequent rivers dissect the eastern slope. Their channel patterns are the response to lithologic, structural and

---

[1] The author's manuscript has been edited by M P Mosley for inclusion in this volume.

[2] Department of Geography, National Taiwan Normal University, Taipei, Taiwan, ROC.

tectonic controls. However, there have been few studies on marine terraces and river terraces (Wang, 1960; Shih et al., 1988; Liew et al., 1990; Teng and Shen, 1990; Chang et al, 1991; Hsieh et al., 1994) and fewer still have addressed channel morphology. This paper considers how the rivers adjust under the multiple effects of the heterogeneous rock, various geological structures and rapid tectonic uplifting.

2. REGIONAL SETTING

2.1. Stratigraphy and Uplift

The collision of the Asian Continental Plate and Philippine Sea Plate has caused the uplift of a variety of different sedimentary and igneous rocks (Hsu, 1956). Recently, a comprehensive stratigraphic system was proposed by Teng et al. (1988), in which the rock formations are divided chronologically in ascending order as: (1)Tuluanshan Formation, andesitic volcanic and volcastic sedimentaries; (2) Lichi Formation, a melange including ophiolitic, andesitic and other sediments; (3) Fanshuliao Formation, with turbidities composed of alternating sandstones and shales dominated by volcanic detritus; (4) Paliwan Formation, with turbidities consisting of intercalated conglomerates, sandstone, and shales which are dominated by metasedimentary rock fragments. It is divided into Shuilien, Taiyuan and Futien Members, which are conglomerate, sandstone and shale dominated facies respectively; and (5) Peinanshan Conglomerate, which consists of alluvial to shallow marine conglomerates dominated by metamorphic rock fragments. (Fig. 2.1). Except for the Peinanshan Conglomerate, the formations appear on the eastern slope on the Coastal Range. Although there are few data on rock mechanics, field observation indicates that the Tuluan Formation and Shuilien Conglomerate are more resistant to erosion than the Fangsuliao, Paliwan and Lichi Formations.

Due to continual stress from the southeast, the Formations have been deformed by several faults and folds. In general they strike northeast to southwest. The Chimei Fault divides the Coastal Range into northern and southern blocks, and extends from Fengping to the Longitudinal Valley (Chen et al., 1991). The southern part has a higher uplift than the northern part (Hsu, 1954). Recent studies of marine terrace and river terrace correlations support this finding (Shih et al., 1988; Liew et al., 1990; and Chang et al., 1991).

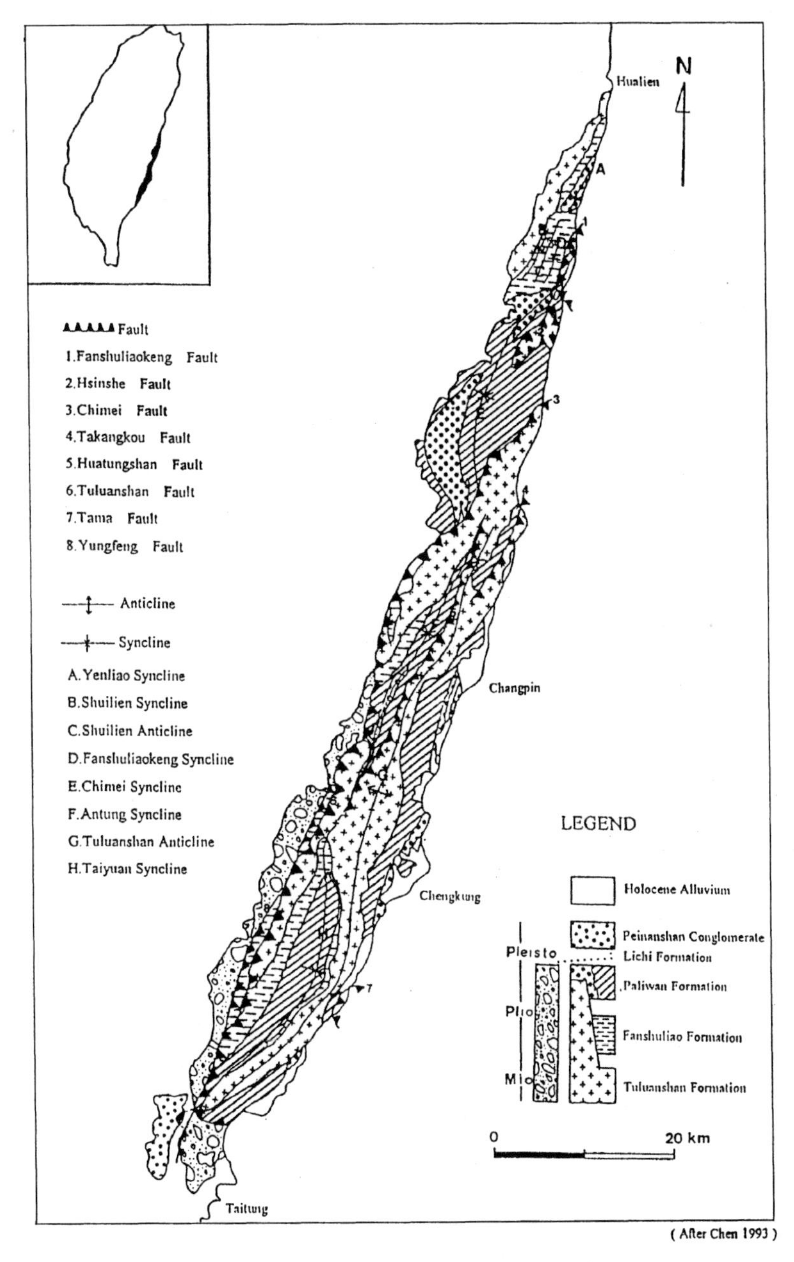

Fig. 2.1. Geological map of the Coastal Range (after Chen, 1993).

2.2. Geomorphic Classification

The geomorphology of the eastern slope of the Coastal Range reflects lithology and structure. Based on the slope and morphology of the terrain, the area can be divided into steep slopeland, gentle slopeland, basin, alluvial fan and marine terraces (Fig. 2.2). The steep slopeland, associated with the Tuluan Formation,

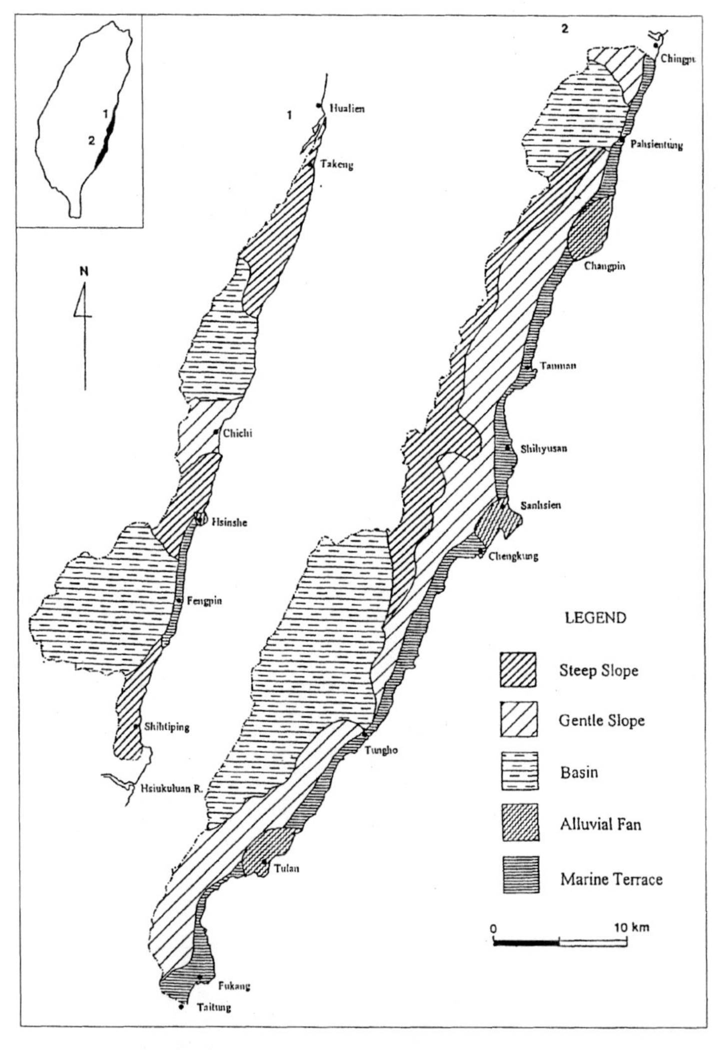

Fig. 2.2. Geomorphological classification of the Coastal Range.

is located in the northern part of the Coastal Range and on the eastern ridge in the middle part. Gentle slopeland occupies much of the southern part of the Range. Basins, including Shuilien, Fansuliao, Fengpin, Schuimuting and Mawuku, are surrounded by steep slopeland or gentle slopeland, and are well developed in soft rock. Alluvial fans and marine terraces lie along the coast, and are capped with a layer of volcanic and sedimentary clasts.

3. CHANNEL MORPHOLOGY

In general, river morphology is controlled by variations in discharge, sediment load and the slope of the valley floor. Channel pattern may change at locations where critical values of stream power, gradient and sediment load are encountered (Schumm and Khan, 1972). Rivers also may respond to climatic change, baselevel change and human activities.

The rivers on the eastern slope of the Coastal Range are relatively short and small, with high gradients. Except for the dendritic drainage developing in their watersheds, most rivers are consequent and flow into the sea individually (Fig. 3.1). The morphology of the channels in plan view can be classified as straight and meandering. Depending on the degree of incision, valley cross sections can be classified as shallow ditch, wide valley and gorge (Figure 3.2). In steep slopeland areas, valleys are deeply incised gorges where they have a straight course, and feature entrenched meanders and ingrown meanders where incision of meandering courses has occurred. On floodplains, rivers exhibit free meandering.

Channel morphology varies in response to changes in rock resistance, structure and tectonic movement along a river. Measurements of drainage area and bifurcation ratio by Shih (1977) and of channel length, gradient and sinuosity by the author indicate that each has a characteristic value in each geomorphic region (Table 3.1). The variables are correlated; for example, a river developing on a steep slope is relatively short with a high gradient, a small drainage area, low sinuosity and low bifurcation ratio. Most rivers have a straight course with a gorge and entrenched valley on steep slopes, but ingrown meanders and entrenched meanders in the gentle slopeland and basins. On alluvial fans and marine terraces, rivers are straight or meander, depending on the overall topographic gradient.

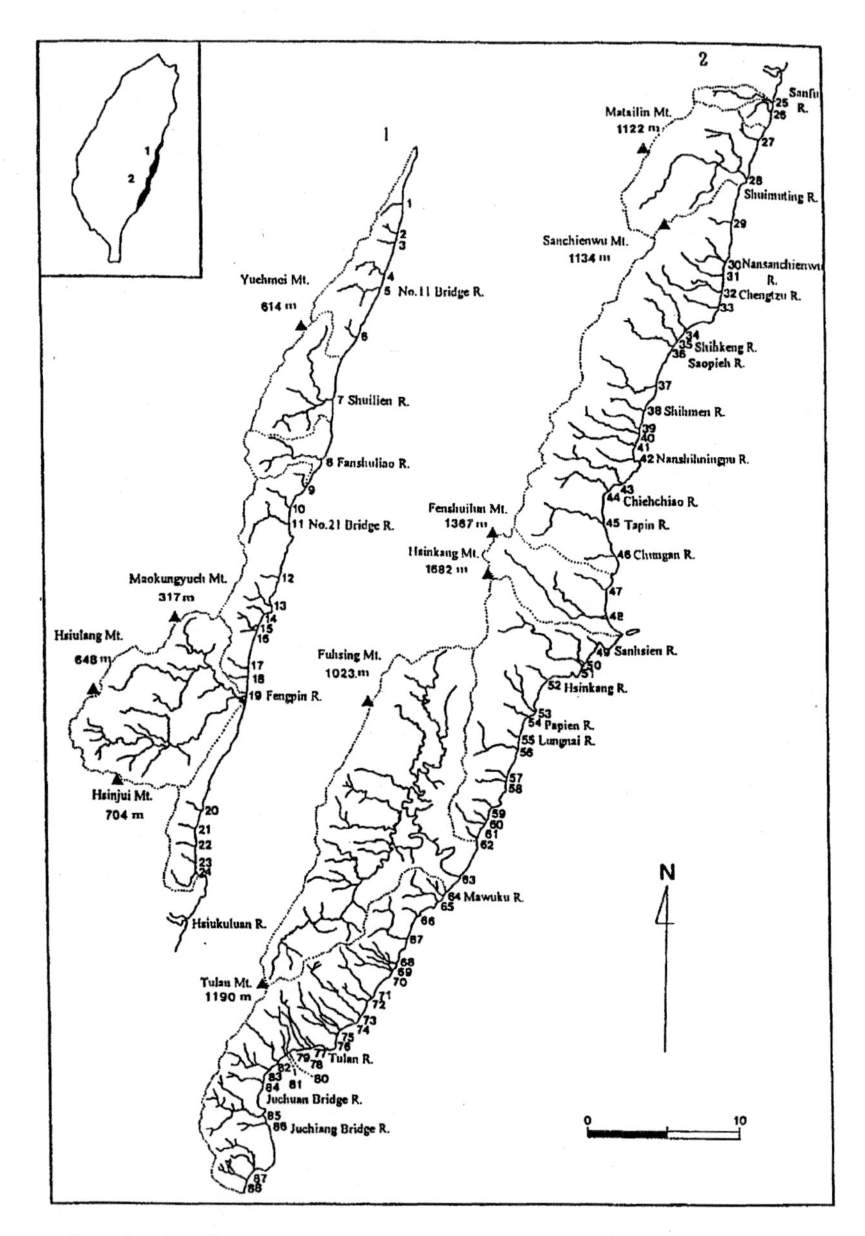

Fig. 3.1. Drainage system of the eastern slope of the Coastal Range.

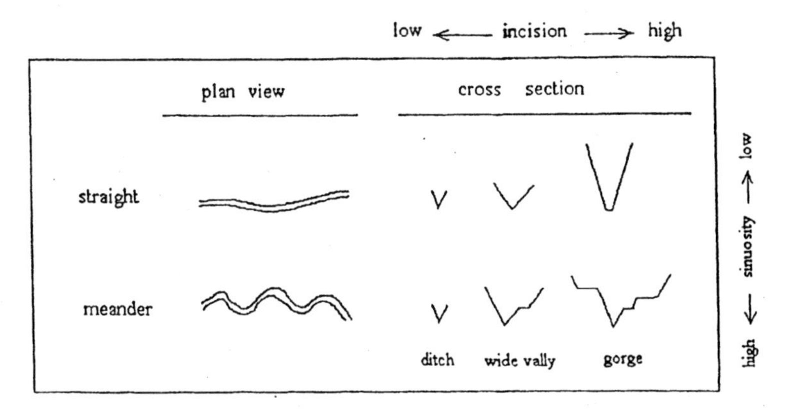

Fig. 3.2. Idealized sketch of the morphology of channels: straight and meandering.

Table 3.1. Characteristics of the drainage network in different geomorphic regions.

	Area	Length	Gradient	Sinuosity	Bifurcation Ratio	Channel Form
Steep slopeland	Small	Short	High	Low	Low	Straight
Gentle slopeland	Medium	Medium	Medium	Medium	Medium	Meander
Basin	Large	Long	Low	High	High	Meander
Alluvial fan	Medium	Medium	Medium	Low	Low	Straight
Marine Terrace	Small	Short	Low	Medium	Low	Straight or meander

4. FACTORS CONTROLLING CHANNEL MORPHOLOGY

Schumm and Lichty (1965) proposed a set of controlling variables for the morphology of the fluvial system, including time, initial relief, geology, climate, and so on. In this paper, the focus is on the rivers' long term response to lithology, geological structure and tectonic movement (baselevel change).

4.1. Lithologic Control

In response to the heterogeneous resistance of rock, valleys clearly show different characteristics in each segment of a river. For example, the Sanfu River (Hsi) flows in a gorge in the hard rocks of the Tuluan Formation, exhibits ingrown meandering with unpaired terraces in the soft rocks of the

Paliwan Formation, and also has a characteristic alluvial reach. Some nickpoints are concordant with discontinuities in lithology (Fig. 4.1). On the other hand, rivers have a regular form in regions with homogeneous lithology. For example, the Juchiang River flows through the soft rock known as the Lichi Formation, and has a relatively wide valley and a graded longitudinal profile (Fig. 4.2). Heterogeneous resistance can be found in the Fanshuliao River. The river winds freely, with four step terraces, through soft rock but where it cuts into hard rock there is a 30-m deep gorge (Fig. 4.3). Hard rock clearly confines lateral erosion and retards incision.

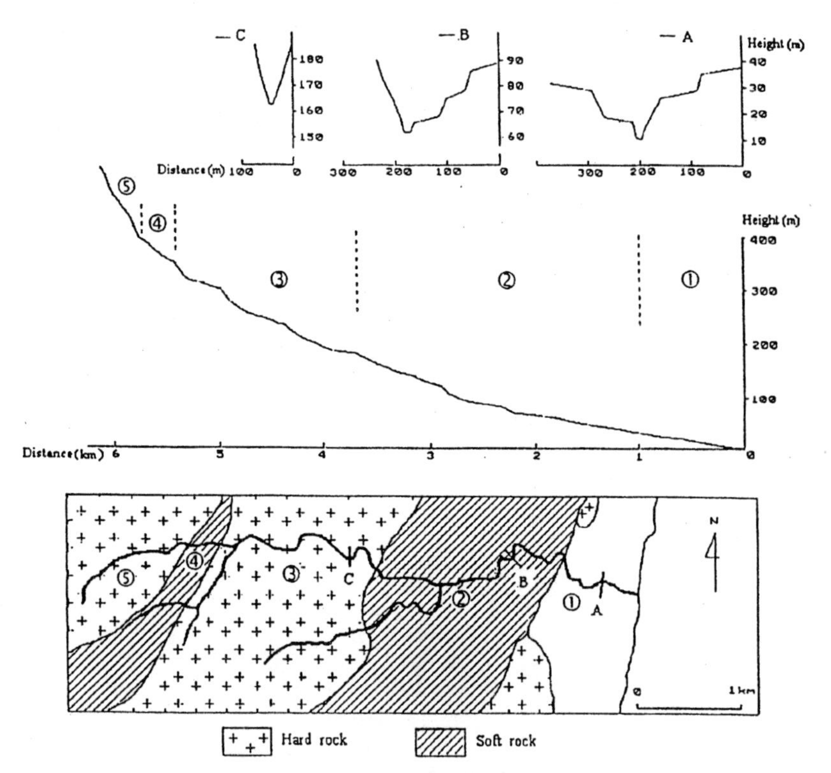

Fig. 4.1. Rock control of the Sanfu River. A and B cross profiles indicate the river developing in the alluvium and Paliwan Formation (soft rock) are ingrown with meanders. Profile C shows the river developing in the Tuluan Formation (hard rock) is a gorge with a narrow and deep valley. The longitudinal profile shows the river has a high gradient and several nick points in the hard rock segments.

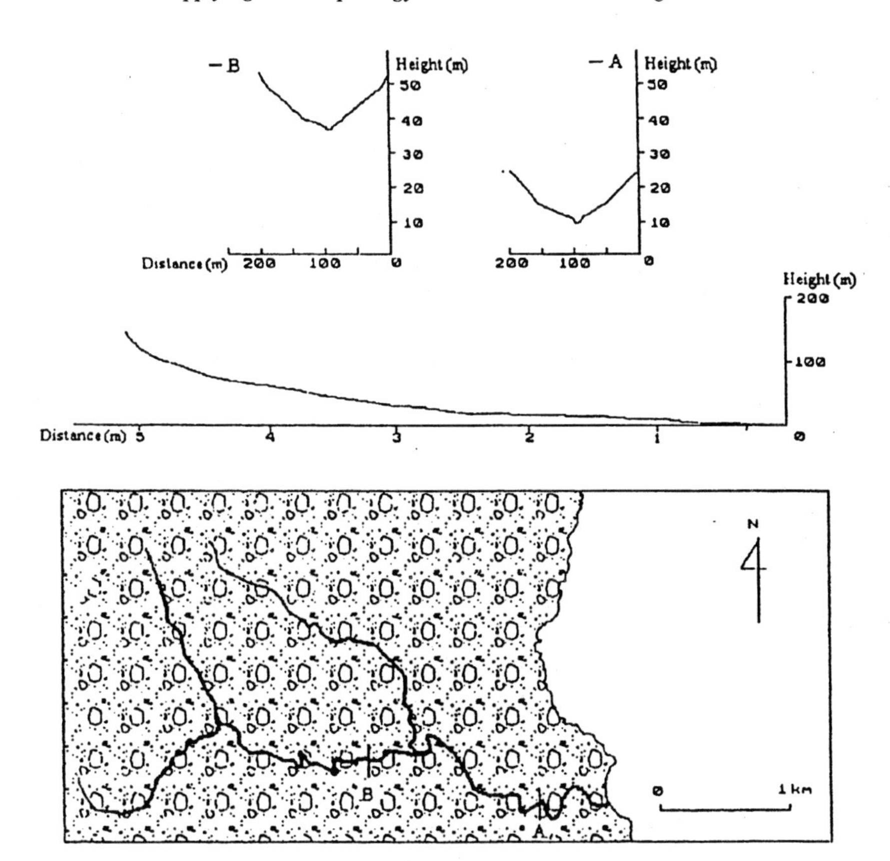

Fig. 4.2. **A wide and regular valley in homogeneous rock (Lichi Formation), in the case of the Juchiang River. Cross profiles A and B show the Juchiang River has the same valley form in the homogeneous rock. The longitudinal profile indicates the river tends to be graded without significant nickpoint.**

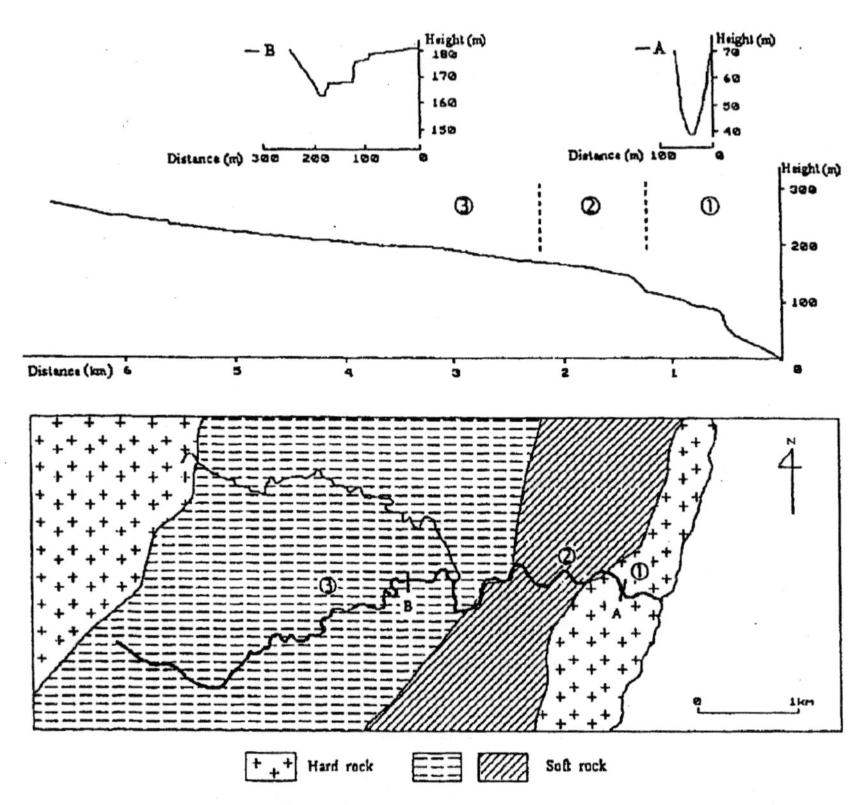

Fig. 4.3. Rock control of Fanshuliao River with a deep valley in hard rock (Tuluan Formation) and ingrown meanders in soft rock (Paliwan Formation and Fanshuliao Formation).

4.2. Structural Control

In general, consequent rivers develop in accordance with the initial slope, but when geological structures such as synclines and faulting affect the surface, rivers develop along the structural lines (Fig. 4.4). The branches of the Shuilien River flow through small parallel synclines, the Shuilien syncline and Fanshuliaokeng syncline. The Maoping River flows along a fault line and the boundary line of the Tuluanshan Formation (hard rock) and Paliwan Formation (soft rock) (Fig. 4.4). Both exhibit a relatively straight form, even though they have developed on soft rock. The Mawuku River is similar; its northern and southern branches wind along the Taiyuan syncline.

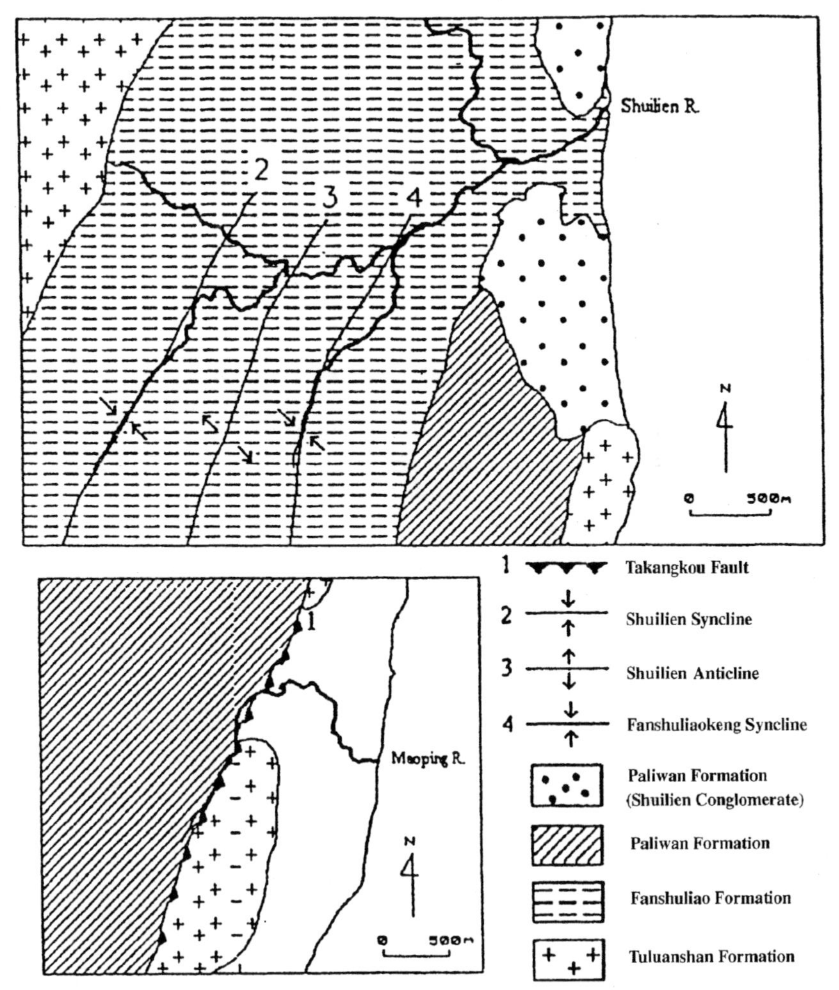

Fig. 4.4. Structural controls (synclines and fault line) on rivers. Map A shows the tributaries of the Shuilien River flow along the Shuilien and Fanshuliaokeng synclines; map B shows the Maoping River flows along the Takangkou Fault.

4.3. Tectonic Control

The Coastal Range is an active tectonic area, with differential uplift. Shih et al (1988) and Hsu (1988) have identified three Holocene marine terraces. Hsieh (1990) conducted a detailed study of the late Quaternary deposits and marine terraces. Liew et al.(1990) estimated average uplift rates of 9-10 mm/yr to 2-3 mm/yr from Chengkung to the north of Shuilien. Wang and Burnett (1990) estimated that the mean uplift was 4.7-5.3 mm/yr for the eastern Coastal Range. Based on the correlation of marine terraces and the observation of marine deposits in the field, the present author found Terrace I had a wider surface and more significant scarp than Terraces II and III. This could imply that the land remained relatively stable relative to sea level during the period of Terrace I. The alluvial fans consequently could have prograded more during that time. The existing scarp could have been formed by rapid uplift and wave erosion. If wave erosion is assumed to have been the same during the three different stages, the larger scarp could have been formed by a rapid uplift. Because Terrace II and Terrace III have a narrower surface and lower scarp than Terrace I, the land may have been stable for a short period, but then uplifted rapidly. Below Terrace III, there are 2 or 3 terraces, which could be good evidence for active intermittent tectonic uplift (Fig. 4.5).

The southern part of the Coastal Range, south of Hsiukuluan River, has a higher uplift than the northern part (Takeng to Shihtiping, Sec A.). This is evident from the height of the marine terrace: Terrace I is 20-40 m in Section A, but 50-60 m, 40-50 m, and 50-70 m in Sections B (Pahsientung to Shihyusan), C (Chengkung to Tuli) and D (Chili to Tulan), respectively (Table 4.1). This implies that Sections B, C and D are located at the crest of up-warping.

Although there is not much dating in the area, and some dating is contradictory, there is a tendency to show greater uplift in Changpin and Tulan. For example, two samples of driftwood are dated 5070 and 5230 B.P. at 40 m elevation in Changpin; whereas, samples with almost the same date are located below 20 m in Shihtiping and Chengkung (Fig. 4.6). This also indicates differential uplift in the Coastal Range.

How does the river adjust under the effect of baselevel change? Ouchi (1985) explained that alluvial rivers respond to valley-slope deformation caused by active tectonics in various ways, depending on the rate and amount of surficial deformation and the type of river. Miller (1988) demonstrated that gradient has an influence on the channel pattern. Gomez and Marron (1991) pointed to sinuosity as an indicator of the effects of neotetonic activity on alluvial channels. Schumm (1993) mentioned that Ware's experimental studies show that baselevel lowering can be compensated for by an increase in sinuosity, a decrease in depth (hydraulic radius) and an increase in roughness. He also supported the notion that increasing valley slope causes river patterns to change from low to higher sinuosity, which maintains the channel gradient on a range of valley slopes. If the change is too great, the channel pattern may

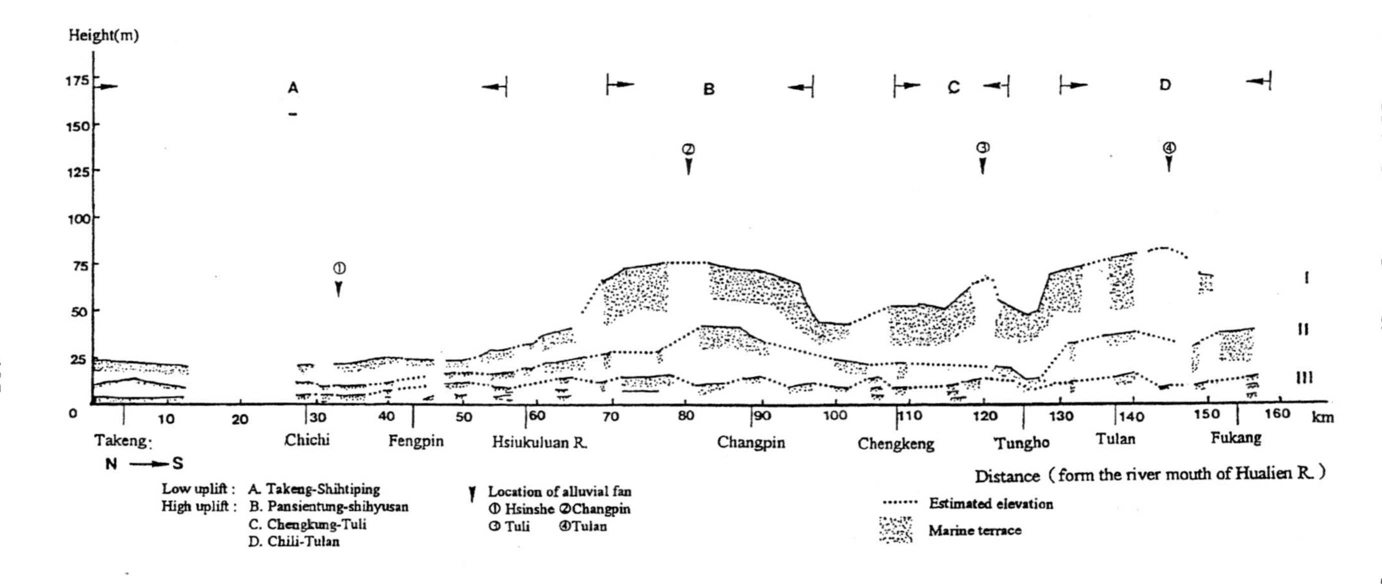

Fig. 4.5. Three steps of marine terrace along the Huatung Coast (Hualien to Taitung).

Table 4.1. Height of marine terraces in low uplift and high uplift areas.

Marine Terrace		I	II	III
Northern section (low uplift)	A. Takeng-Shihtiping	20-40 m	15-20 m	5-10 m
Southern section (high uplift)	B. Pansientung-Shihyusan	50-60	20-40	10-15
	C. Chengkung-Tuli	40-50	20-25	10-15
	D. Chili-Tulan	50-70	30-40	10-15

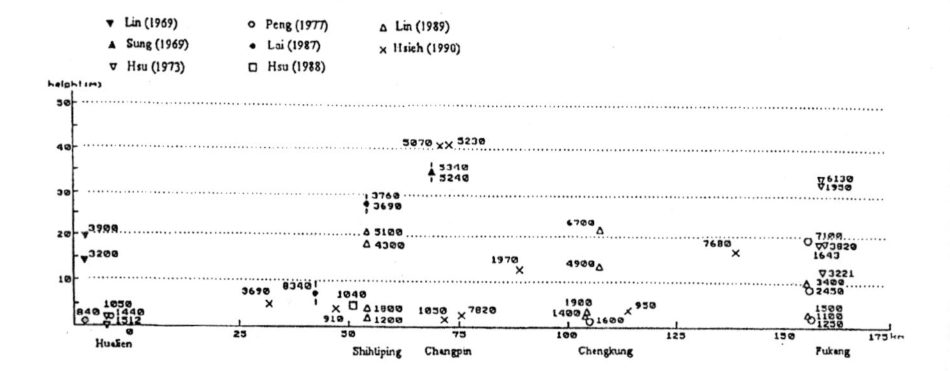

Fig. 4.6. Dating of various samples along the Huatung Coast.

change, such as from meandering to braided. Figure 4.7 attempts to illustrate this concept. A-D represents the channel profile of the Shihmen River, while A-G represents the valley profile. Points E and B, and F and C, are at the same points in the valley, and G and D are at the river mouth. When the baselevel is lowered from E to F, the gradient A-F is larger than A-E. The sinuosity increases from 1.23 (A-B) to 1.35 (B-C), in order to absorb the increase in the gradient. In contrast, when the gradient A-G is smaller than A-E, sinuosity decreases to 1.06.

Due to the baselevel change, a larger influence is produced on the lower part of the river than on the upper course. Channel sinuosity in the low uplift area is smaller than that in the high uplift area (Fig. 4.8). Further, the Changping River, flowing on the alluvial fan, has tended to become relatively straight in response to the rapid uplift , and has a steeper fan slope.

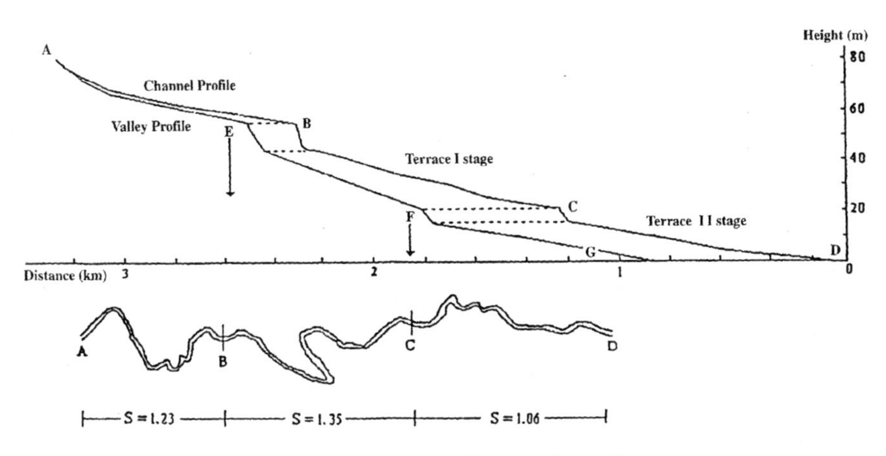

Fig. 4.7. Sinuosity responding to slope change.

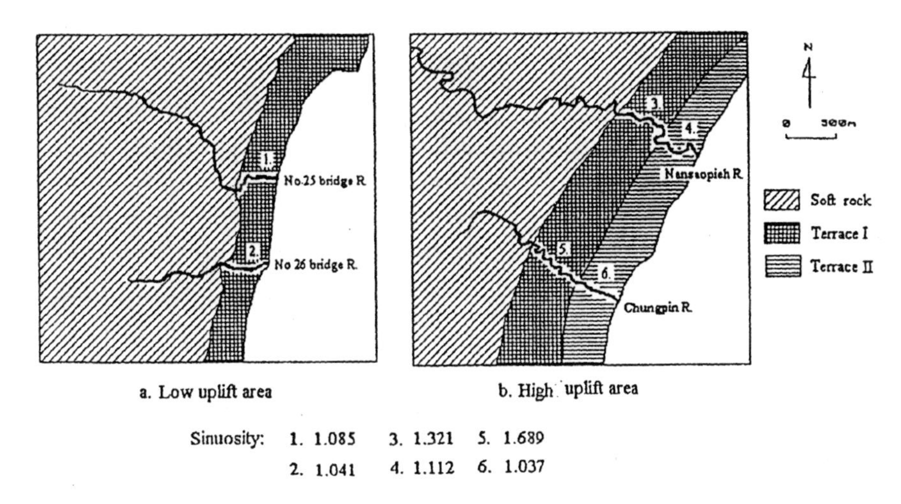

Fig. 4.8. Sinuosity in the low uplift area and the high uplift area.

How much increase in the gradient is required to cause channel pattern to change? Schumm and Khan (1972) conducted an experimental study of the relation between slope and sinuosity. They found that when the valley floor slope increases, sinuosity increases in order to maintain a constant channel gradient. There were two thresholds for the channel pattern change. In the first, the slope reached about 0.2 percent and the channel changed from straight to meandering. In the second, the slope was about 1.4 percent and the channel changed from meandering to braided (Fig. 4.9). This means if the slope increases, the energy of the stream is concentrated on incision rather than lateral erosion, which makes the river channel tend to become or remain straight.

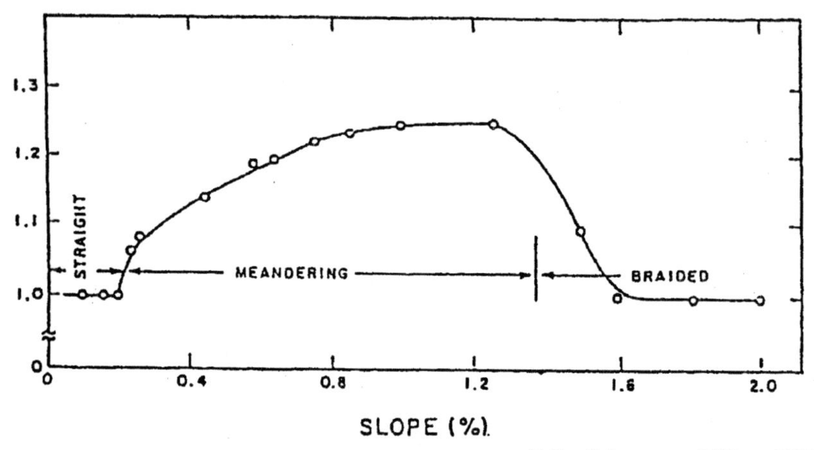

(After Schumm and Khan, 1972)

Fig. 4.9. Relation between flume slope and sinuosity (after Schumm and Khan, 1972).

Measurements of the sinuosity and gradient (slope) of 88 rivers developed on the eastern slope of the Coastal Range do not show a relationship, which is exactly similar to the experimental study. However, three groups can be identified. The first exhibits high sinuosity (1.25-1.50) and low gradient (5-10 percent); the second has sinuosity of 1-1.25 and medium gradient (10-15 percent), and the third has sinuosity of 1-1.20 and steep gradient (17-35 percent). It seems that sinuosity has a slightly inverse relation with gradient, but it is difficult to determine the exact threshold value at which sinuosity decreases when the slope is increasing (Fig. 4.10).

Many geomorphologists support the idea that baselevel lowering will rejuvenate fluvial systems, eroding and transporting sediments downstream to restore a new equilibrium profile. Schumm (1993) stated that Leopold and Bull had pointed out that a baselevel change affects not only stream gradient but also channel bed roughness. He emphasized that the impact of baselevel change on the fluvial system depends on the magnitude of baselevel lowering.

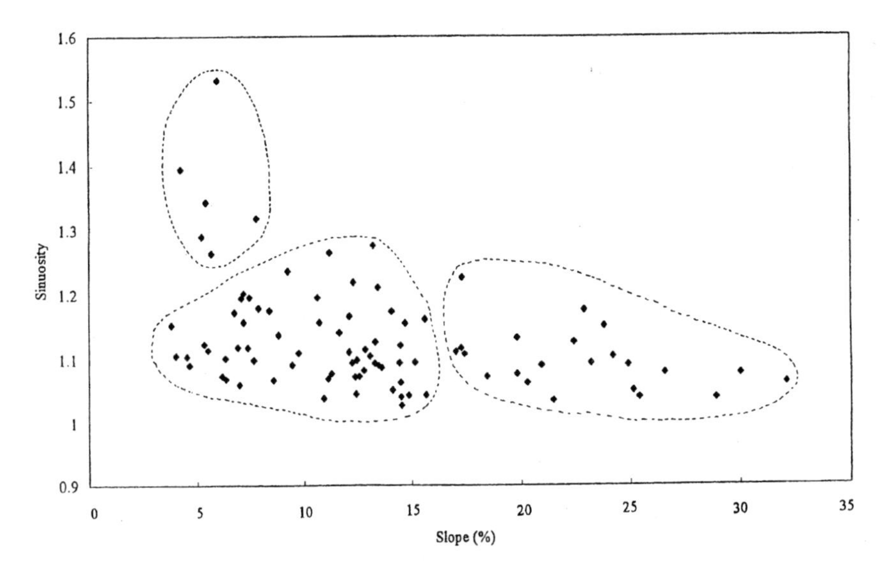

Fig. 4.10. Relation between the sinuosity and gradient of the rivers of the eastern slope, Coastal Range.

If baselevel lowering is small, a river can adjust by changing its pattern, by increasing bed roughness, or by changing shape. If the change is large, river incision is likely, and if the change is very large, rejuvenation may extend throughout the entire drainage basin.

The rate of baselevel change and drainage area are important factors. If baselevel changes rapidly, a stream may incise vertically, whereas a slow change may be accommodated by lateral migration. If drainage area is small, rejuvenation could extend to the entire drainage, whereas if it is large, the impact is likely to be restricted to lower reaches. South of Changpin, a series of rivers are incised into a marine terrace, and their valleys are wider upstream of the scarp at the seaward edge of Terrace I, where there has been much greater incision than in Terrace II and Terrace III (Fig. 4.11). There are two nickpoints corresponding to the baselevel lowering. When a large river is rejuvenated, adjustment can be achieved by increasing sinuosity, width or bed roughness. This is the case with Shihmen River, and with the rivers in the area between Changpin and Pahsientung, and near Tulan.

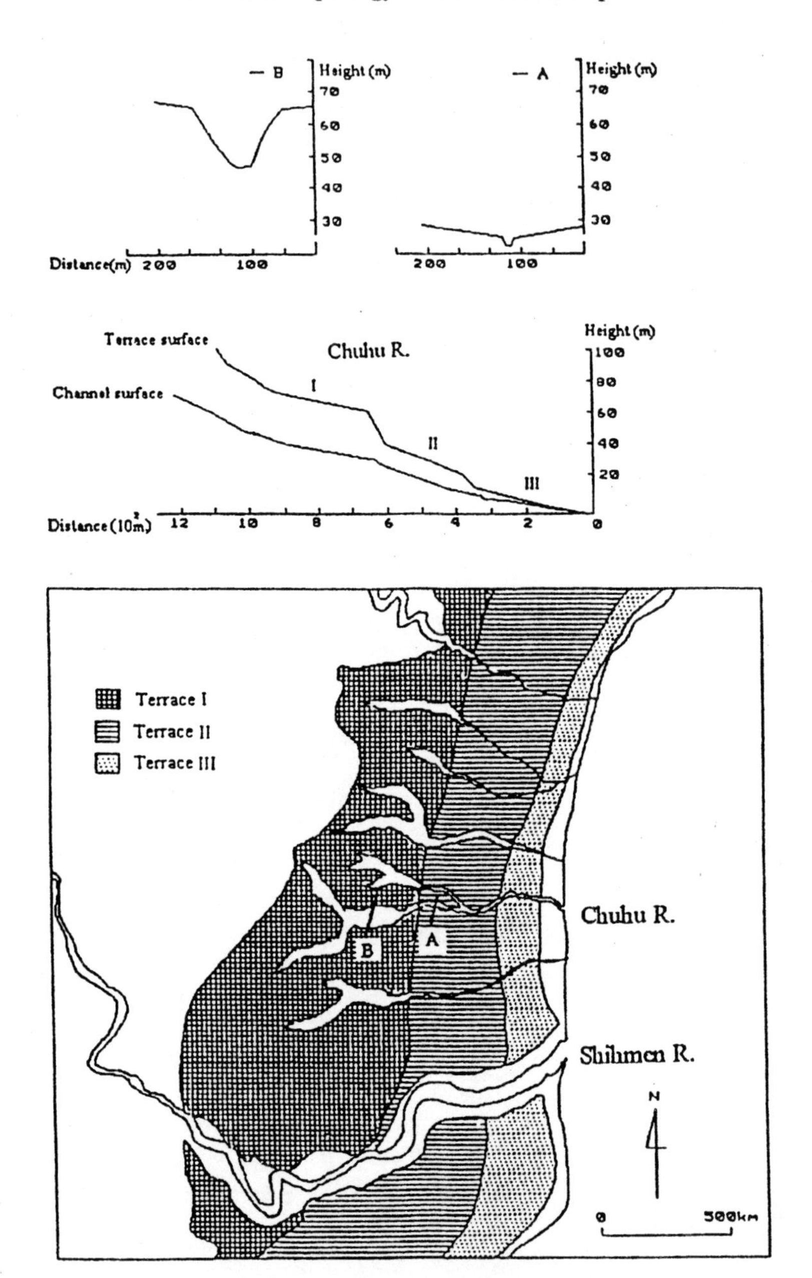

Fig. 4.11. River rejuvenated upstream of the scarp of marine Terrace I.

Differential uplift may also lead to a shift in the channel course. For example, the Sanhsien River has shifted from south (high uplift) to north (low uplift), leaving a series of eight terraces on the southern bank (Fig. 4.12).

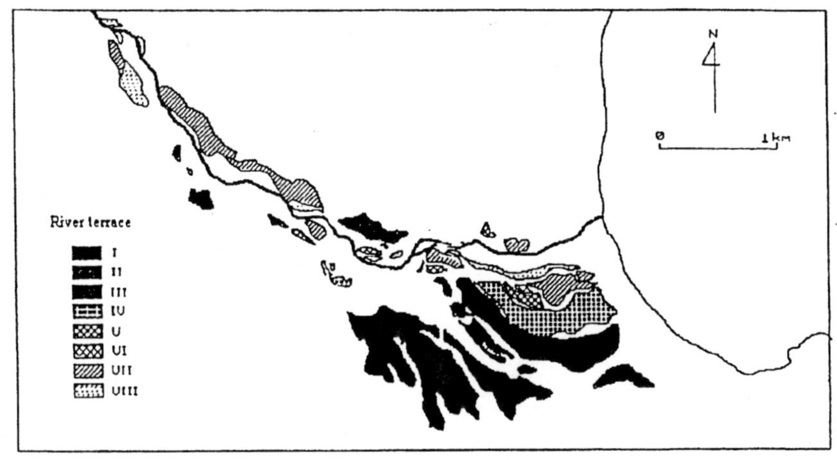

(After Chang et al. 1991)

Fig. 4.12. River terraces of the Sanhsien River.

Overall, there may be a complex response to the effects of lithology, structure, and tectonic uplift. Mawuku river is a good example. In its upper course, the mainstem flows along the synclinal valley with a high sinuosity and eight terraces in the soft rock (Paliwan Formation). It tends to have entrenched meanders through the hard rock (Tulwan Formation) with a deep valley, and is incised into marine terraces with a straight and shallow channel farther downstream (Fig. 4.13).

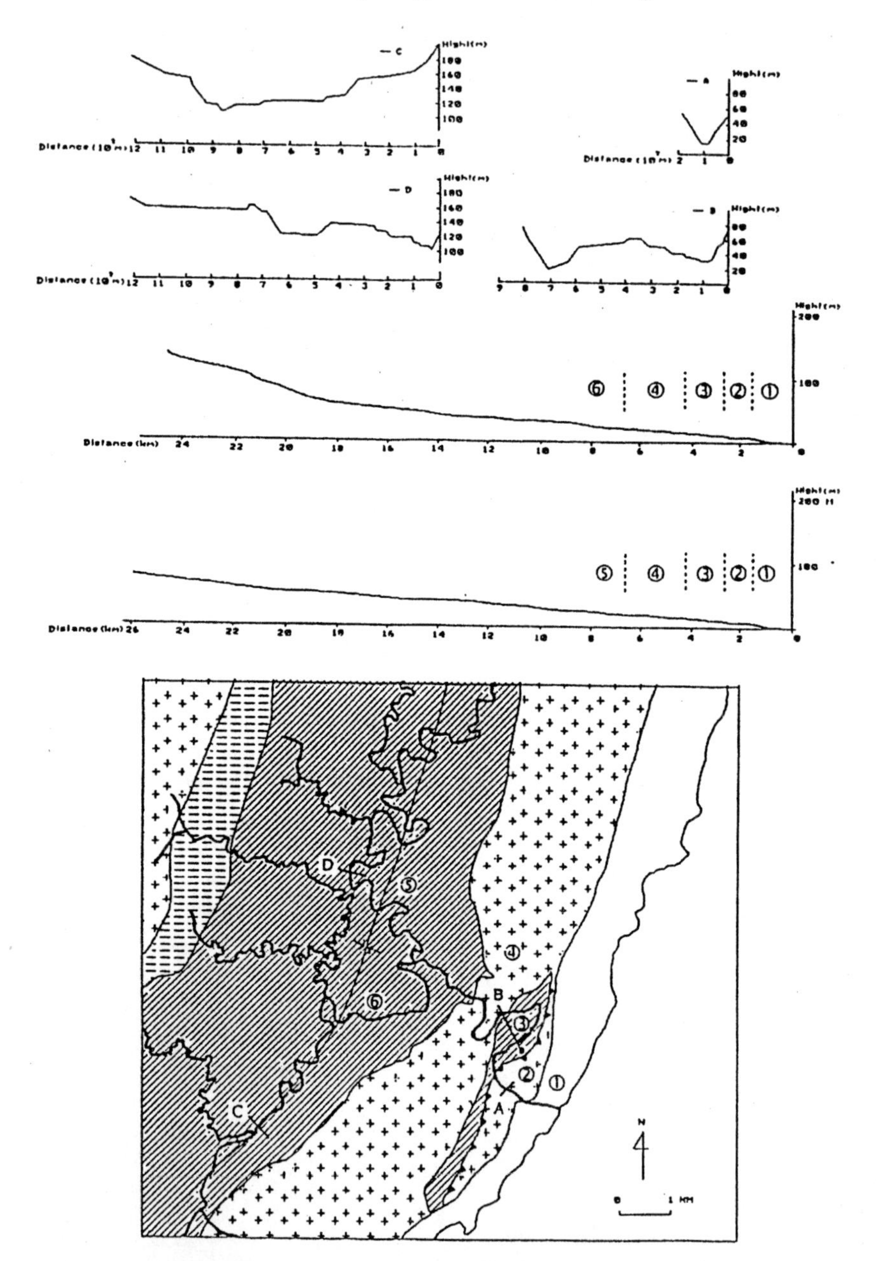

Fig. 4.13. Complex response of rivers to multiple effects of lithology, structure, and tectonics, the case of Mawuku River.

5. CONCLUSIONS

In response to the effects of heterogeneous resistance of rock, geological structure, and differential uplift, rivers developing on the eastern slope of the Coastal Range have varied morphology. River channels are straight, with gorges and entrenched valleys, in steep slopeland (hard rock), but exhibit ingrown and entrenched meanders in gentle slopeland and basins (soft rock). On alluvial fans and marine terraces, channels are straight or meandering, depending on the slope. Locally, rivers develop along structural lines such as synclines and fault lines. In the coastal area, rivers respond to baselevel lowering by changing channel morphology. The nature of the change depends on the magnitude and rate of baselevel change, valley slope, and the size of the river. When the magnitude and rate of baselevel lowering are small, the change of valley slope is small, or the river is large, a river adjusts by increasing sinuosity, width and/or bed roughness. When they are large, a river adjusts by rapid incision, and extends rejuvenation upstream throughout the entire drainage system.

6. ACKNOWLEDGEMENTS

The author wishes to thank Prof. T.T. Shih for his kind encouragement and Dr. C.L. Chang for his great help and discussion on the paper. The author also appreciates the constructive suggestions of Prof. K.H. Teng and S.M. Shen during the field observation. Gratitude is also extended to Prof. C.Y. Wong for his correcting of this manuscript and to M.C. Liu, Y.S. Liu and H.M. Kao for their help in preparing the diagrams and data processing.

7. REFERENCES

Chang, J. C., Shih, T. T., Tseng, C. H., Kao, P. F., Chen, M. L. (1991). A geomorphological study of the river terraces in Eastern Coastal Range: Geographical Research 17, 99-146.

Chen, W. S. (1988). Tectonic evolution of sedimentary basins in the Coastal Ranges, Taiwan: unpub. Ph.D. thesis, Institute of Geology, National Taiwan University, 304 pp.

Chen, W. S., Huang, M. T. and Liu, T. K. (1991). Neotectonic significance of the Chimei Fault in the Coastal Range, eastern Taiwan: Proc. Geol. Soc. China, 34, 43-56.

Gomez, B., Marron, D. C., (1991). Neotectonic effects on sinuosity and channel migration, Belle Fourche River, Western South Dakota. Earth Surface Processes and Landforms, 16, 227-235.

Hsieh, M. L., Lai, T. H., Liew, P. M. (1994). Holocene climatic river terrace in an active tectonic uplifting area, middle part of the Coastal Range, eastern Taiwan, J. Geol. Soc. China, 37, 97-113.

Hsieh, M. L. (1990). The late Quaternary deposits, marine terraces and neotectonism of the coastal area between Hualien and Taitung, eastern Taiwan: unpub. M.S. thesis, Institute of Geology, National Taiwan University, 168 pp.

Hsu, T. L. (1954). On the geomorphic features and the recent uplift movement of the Coastal Range, eastern Taiwan: Bull. Geol. Surv. Taiwan, 7, 51-59.

Hsu, T. L. (1956). Geology of the Coastal Range, eastern Taiwan: Bull. Geol. Surv. Taiwan, 8, 39-63.

Hsu, T. L. (1962). A study on the coastal geomorphology of Taiwan. Proc. Soc. China, 5, 29-46.

Hsu, M. Y. (1988). A geomorphological study of marine terraces in Taiwan. unpub. Ph.D. thesis, Institute of Geography, Chinese Culture University, 178 pp.

Liew, P.M., Hsieh, M. L., and Lai, C. K. (1990). Tectonic significance of Holocene marine terraces in the Coastal Range, eastern Taiwan: Tectonophysics, 183, 121-127.

Miller, T. K. (1988). Stream channel pattern: a threshold model, Physical Geography, 9, 373 -384.

Ouchi, S. (1985). Response of alluvial rivers to slow active tectonic movement, Geol. Soc. Am. Bull., 96, 504-515.

Schumm, S. A. (1993). River response to baselevel change: Implications for sequence stratigraphy: J. Geol., 101, 279-294.

Schumm, S. A., and Khan, H. R. (1972). Experimental study of channel patterns: Geol. Soc. Am. Bull., 83, 1755-1770.

Schumm, S. A., and Lichty, R. W., (1965). Time, space, and causality in geomorphology: Am. J. Sci., 263, 110-119.

Schumm, S. A., Mosley, M. P., and Weaver, W. E. (1987). Experimental Fluvial Geomorphology: Wiley, New York, 413p.

Shih, T. T. (1977). A geomorphological quantitative study on the Huatung coast of eastern Taiwan: Geographical Research, 3, 143-170.

Shih, T. T., Teng, K. H., Hsu, M. Y. and Yang, G. S. (1988). A geomorphological study of marine terraces: Geographical Research, 14, 1-50.

Teng, C. S., Wang, Y. (1981). Island arc system of the Coastal Range, eastern Taiwan, Proc. Geol. Soc. China, 24, 99-112.

Teng, K. H. and Shen, S. M. (1990). The study on the stream landform of the Taiyuan Basin of the Coastal Range in eastern Taiwan: Quaternary Research, 33, 73-85.

Teng, L. S., Chen, W. S., Wang, Y. (1988). Toward a comprehensive stratigraphic system of the Coastal Range, eastern Taiwan, ACTA Geological Taiwanica, 26, 19-35.

Wang, C. H. and Burnett, W. C. (1990). Holocene mean uplift rates across an active plate-collision boundary in Taiwan: Science, 248, 204-206.

Wang, C. S. (1960). Coastal terraces and relative sea-level change, Proc. Geol. Soc. China, 3, 113-115.

Yao, T. M., Tien, P. L., Wang-Lee. C. (1989). Stratigraphy and depositional environments of the Taiyuan Basin, eastern Taiwan, Ti-Chih, 9, 95-112.

Section 2

EXTREME EVENTS

EARTHQUAKES, GROUNDWATER, AND FLUVIAL SYSTEMS -- SOME SPECULATIONS

John Adams[1]

ABSTRACT

Water plays an interesting role in the geology of earthquakes, and it has some unique possibilities for increasing our understanding of the earthquake process. Fluvial and lacustrine sequences can record earthquake deformation or shaking effects, and recognition of these paleoearthquakes can improve hazard estimates otherwise only based on short historical records. Human activity sometimes induces earthquakes through the injection of water or the filling of reservoirs. Furthermore, the post-earthquake response of groundwater provides insight into deep crustal processes that may bear on earthquake prediction. Hence earthquakes and their effects should be on the minds of all who work with fluvial systems.

KEYWORDS

Groundwater, geology, earthquakes, fluvial systems, neotectonic activity

1. ON THE FLUVIAL SEDIMENTARY EVIDENCE FOR EARTHQUAKES

The ground deformation due to the cumulative displacement of many large earthquakes is an obvious manifestation of neotectonics. In many geologically-young regions, neotectonic activity is the dominant control on river morphology and behavior (e.g., Schumm, 1986), both directly, as in rising mountain ranges diverting river drainage, and indirectly through uplifted mountains and high sediment loads requiring braided rivers. The New Zealand landscape has been built up by earthquakes and cut down by erosion (Adams, 1985). The dynamic balance between the two processes has had a major influence on the country's history and human geography, through alluvial gold

[1] National Earthquake Hazards Program, Geological Survey of Canada, Ottawa, Canada.

deposits, abundant hydroelectric power, communication routes between coastal plains, and so forth. More recently, interest has turned to the tectonic deformation from individual earthquakes (both historic and prehistoric) and the information they contain for seismic hazard assessment.

An example from New Zealand's North Island illustrates some of the consequences through a hypothetical analysis. The next large earthquake continuing the neotectonic deformation that raised the Tararua/Ruahine ranges across the Manawatu River — requiring the cutting of the Manawatu Gorge — will probably cause ponding of the river upstream of the gorge, with some logistical complications! A recent analogy is the 1980 El Asnam earthquake in Algeria where a lake was formed by warping across the course of the Chelif River (Vita-Finzi, 1986, p. 82). In Algeria, each past large earthquake had left its geological mark, so at some point upstream of the Manawatu Gorge there is probably a record of relative subsidence during the most recent (prehistoric, or possibly the 1855) earthquake and its predecessors. A record of sudden, incremental subsidence upstream of the gorge might be recorded in a floodplain site away from the river channel but still affected by overbank deposition. It might comprise thick silts deposited soon after the earthquake while the plain is relatively low and thin peaty soil layers representing most of the time between earthquakes (the bedrock gorge will act as a base level in the short term). This sedimentary sequence could be interpreted as a paleoseismic record — the amount of subsidence indicates the relative magnitudes of the earthquakes, and radiocarbon dating of the silts can determine how often the earthquakes recur. Thus the geological record could provide a much longer history of large earthquakes than we have from New Zealand's short (circa 150-year) historical record, and a more reliable estimate of the seismic hazard for the city of Palmerston North, just downstream of the gorge.

Similar sedimentary sequences, relatively thick (though perhaps fine-grained) clastics inferred to be rapidly deposited, capped by thin, often organic-rich layers representing slow deposition, have been interpreted as recording earthquake events in varied environments. Example are: subsidence of Oregon and Washington coastal marshes during great subduction earthquakes (Atwater et al., 1995); turbidite deposition off the Oregon and Washington coast related to the same earthquakes (Adams, 1990); a sequence of silt and peat layers in a drained lake in central China (Adams, see below); and sag pond and colluvial wedge stratigraphy used to date the adjacent fault's movements (e.g., Yeats et al., 1997).

1.1 Paleoearthquakes Near Xiji, China

The rounded, loess-covered hillsides of Gansu and Ningxia, China are disfigured by massive landslide scars from the magnitude 8.5 Haiyuan earthquake of 1920. The hilltops are at an elevation of about 2100 m, and the valleys about 350 m lower. Near Xiji, Ningxia Province, (about 40 km from the fault rupture) some of the landslides formed 41 lakes, like Dangjiacha (State Seismological Bureau, 1980; Zhu, 1982), which at the present time contains clear water and has muddy shores.

The road from Xiji to Mung Xie village (sited by Dangjiacha Lake) crosses the Maojiaping River at approximately 35.90°N, 105.56°E. On the right hand bank of the river, 200 m downstream of the bridge built in 1964, a tributary gulley has cut down 6.5 m into the bedded sediments of a drained lake. The sequence consists of alternating massive chocolate-brown muddy gyttja and faint-bedded buff-colored silt (Fig. 1.1). The silt grades upwards into the gyttja, but the top of the gyttja is sharp, and commonly deformed. The rapid rate of silt accumulation is clear from the absence of organic layers in the silt and from load deformation features at its base.

Given the experience of widespread landsliding and newly-formed landslide dammed lakes in 1920, the paleo-lake, which from my cursory visit seems to have been as large or larger than any of the current lakes, is likely to have been formed by a landslide triggered by a prehistoric or early-historic Haiyuan-sized earthquake. During an inter-seismic period like the present, there was relatively little silt washed into the lake, and muddy organic material accumulated steadily. When a major earthquake occurs, it would be expected to cause shaking, liquefaction and deformation of the topmost gyttja (although no unequivocal evidence for this, as distinct from load features, was discerned, the lowermost gyttja contains a deformed layer covered by bedded gyttja and so at least some of the deformation seems unrelated to loading by the overlying silt). At the same time, the earthquake would cause massive landslides in the loess hillsides of the catchment, and the silt eroded from these would be rapidly washed into the lake. The period of rapid silt deposition is perhaps a few years to a few decades before the landslide scars revegetate or are terraced and cultivated, then the erosion rate slows. In the absence of the rapid silt deposition, muddy gyttja accumulates slowly until the next major earthquake, when the cycle repeats. Nine cycles are visible in the described section (Fig. 1.1), and it is noted that both the gyttja and the silt layers decrease in thickness upwards, perhaps due to the lake shallowing as it was filled in or as the outlet downcut.

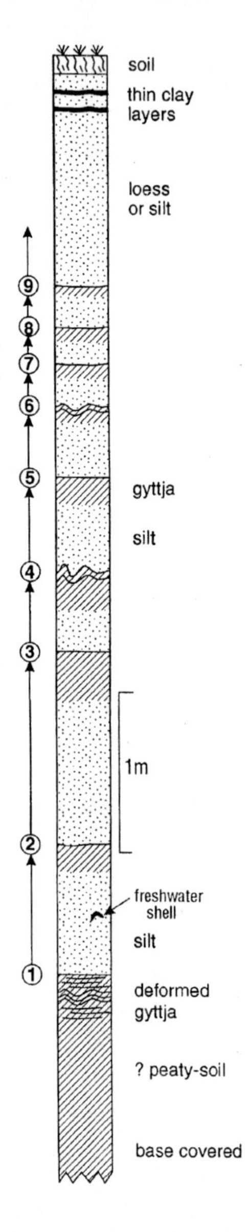

Fig. 1.1. **Exposed section of the bed of a drained lake near Xiji, China, showing the repetitive cycles of slowly-deposited gyttja and rapidly-deposited silt inferred to represent earthquake shaking events in the loess catchment.**

According to Chinese Working Group of the Project 206 (1989, p. 51) three colluvial wedges were formed by pre-1920 vertical movements on the Haiyuan fault near Xiji in the last 2,000-3,000 years, suggesting great earthquakes about every 1,000 years on average. Written records indicate six magnitude 7, landslide-causing earthquakes (A.D. 734, 1125, 1352, 1622, 1654, and 1718) have occurred within 200 km of Xiji, for a 200-year return period. The accumulation rate of the gyttja is not known, but if 0.5 mm/yr (an approximate rate for gyttja in temperate climates, see Doig, 1991), its cumulative thickness (1.5 m) represents a duration of about 3000 years for the lake. In the middle of the section the average gyttja thickness is 150 mm, or about 300 years at the hypothetical rate. These estimates are consistent with great (lake-forming) earthquakes every few millennia and major (landslide-producing) earthquakes every few centuries, and in fair agreement with the earthquake history. Confirmation of this interpretation requires radiocarbon or archaeological dating of the gyttja sediments, their correlation with equivalent sections in the same and nearby lakes, and a detailed examination of the sediment deformation features.

2. WATER-INDUCED EARTHQUAKES

Human activity induces earthquakes in two main ways, both a consequence of manipulating water. In declining oilfields it is now common to enhance recovery of the oil by injecting fluids at high pressure to push the oil towards recovery wells. The fluids are often oil- or salt-contaminated water produced with the oil, so this injection procedure removes them from the surface and solves an environmental problem. Given increasing concerns with surface environmental pollution, such disposal is being more widely used for general waste disposal. Unfortunately, in a fraction of cases such injection can cause earthquakes, because the high pressure fluids migrate, reduce the stability of existing faults, and trigger the release of their stored strain energy. While most of the earthquakes attributed to fluid injection are not very large (e.g., magnitude 4 or less at Ashtabula, Ohio, but magnitude 5.5 at Rocky Mountain Arsenal, Colorado), some large earthquakes considered by many to be natural might just have been triggered by injection wells (e.g., Cleveland, Ohio, 1986, magnitude 5; see Ahmad and Smith, 1988, for a pro-triggering view). Davis and Frohlich (1993) provide a discussion of several cases and some criteria for assessing whether injection has, or will, cause earthquakes.

The filling of reservoirs will also induce earthquakes under some circumstances (Yeats et al., 1997, p. 467-471, provide a modern review). In particular, the filling of deep and large reservoirs seems more likely to induce earthquakes. In most cases the earthquakes are small, less than magnitude 4, and tend to be shallow, in the top 5 km of the crust. However, the Koyna Reservoir in western India is notable for having induced a magnitude 6.3 earthquake. Dam construction is continuing in India, and the devastating Killari (Latur) magnitude 6.1 earthquake of 1993, with over 8,000 killed, occurred near a new reservoir and might have been related to its load or pore-pressure effects on the basement fault that ruptured (Seeber et al., 1996). It seems unlikely that either the injection wells or the reservoirs truly *cause* the earthquakes, rather they are more likely to have induced or triggered movement on faults that were already close to failure. Reservoir-induced earthquakes are more significant in regions where earthquakes are rare and seismic design provisions are often scant. That much larger earthquakes have occurred even in the relatively aseismic parts of the continents suggests that some day a devastating earthquake might be accidentally induced. At present, however, we have no way of anticipating problem areas or testing the stability of large faults at depth.

The speculative concept of "hydroseismicity" (e.g., Bollinger and Costain, 1988), not unrelated to the above, has been aired in the last decade — that rivers by providing ample water can lubricate nearby faults and cause or induce earthquakes. Various papers discussing possible correlations between river flow (or annual rainfall) and earthquake activity have been published, but appear unconvincing because the magnitude of the effect is so small. There is a general spatial association between rivers and earthquakes in eastern North America (examples being seismicity near the St. Lawrence, Ottawa, and Mississippi Rivers), but this seems better attributed to the rivers eroding along ancient fault-weakened zones in the bedrock, zones that also tend to localize contemporary earthquake strain accumulation and release. It is ironic that the very geographical features that led to the founding of cities such as Montreal, New York and Memphis may have guaranteed that these cities lie in regions of above average seismic hazard.

3. EARTHQUAKE-INDUCED WATER

Many early reports of earthquakes are tainted by fanciful notions because there was no conceptual framework (i.e., sudden movement on faults) to explain them. However, some reports include descriptions by astute observers and by those affected economically after the earthquake. It is common to find references to water gushing from the ground, for springs or wells to have run dry (and the consequent hardships), for other wells to have increased their yield, or for new springs to have started. Clearly something was happening in the earth, but what?

Most obvious is that shaking during an earthquake can cause saturated sandy alluvium to liquefy and compact, expelling sediment-laden water. This earthquake-induced liquefaction occurs up to a hundred kilometers from large earthquakes and the ejected water deposits large sand lenses on the surface, termed sand volcanoes. The 1811/12 sequence of large earthquakes in the Mississippi valley near New Madrid caused one of the most extensive fields of liquefaction features known (Obermeier et al, 1990; Johnston and Schweig, 1996), because these were very large earthquakes in a region where much of the surficial material was saturated alluvium (the ideal material is clean channel sand covered by a muddy overbank deposit — the sands liquefy easily and the overbank deposits confine the liquefied sand until it bursts forth).

Similar liquefaction phenomena occur near any large earthquake where the sediment conditions are right — thus they were formed in abundance during the 1886 earthquake near Charleston on the U.S. Atlantic coastal plain, but are of minor importance in drier environments such as southern California (e.g., 1994 Northridge earthquake). The investigation of prehistoric sand volcanoes and their feeder dykes near New Madrid, Charleston and elsewhere has provided important information about the age and extent of paleo-earthquakes, allowing a better assessment of the interval between large earthquakes than possible from the short historical record in North America (e.g., Tuttle and Schweig, 1996; Adams, 1996; and Yeats and Prentice, 1996, and other papers included in that issue).

Liquefaction phenomena have two important characteristics — they occur only in sediment-covered areas, and the expulsion of sandy water slows and stops soon after the shaking is over. A longer-lived phenomena involves large-scale changes in groundwater production which occur over a period of months to years and which affect bedrock groundwater at a considerable distance from the earthquake rupture. The key paper is Muir-Wood and King (1993), from which the following examples are taken. The 1983 Borah Peak earthquake in Idaho (magnitude 7.0) initiated their investigation of the phenomena. Near the northern end of the fault rupture, flow was increased by almost an order of magnitude from Ingram's Warm Springs into Warm Springs Creek. The earthquake happened in October, a time of the year when surface runoff is low and declining. Thus stream flow increases like this were easily discerned on daily hydrograph readings from nearby river basins (Fig. 3.1). The flow increases were of the order of 1 to 10 cubic meters per second, depending on the basin, and represent flow increases of +20 to +200% above the expected average flows at this time of the year. A detectable effect was still present after six months. For the entire region, the excess water amounted to 0.3 km^3. A similar analysis for the August 1959 Hegben Lake earthquake, Montana (magnitude 7.3), indicated that the excess water amounted to 0.5 km^3.

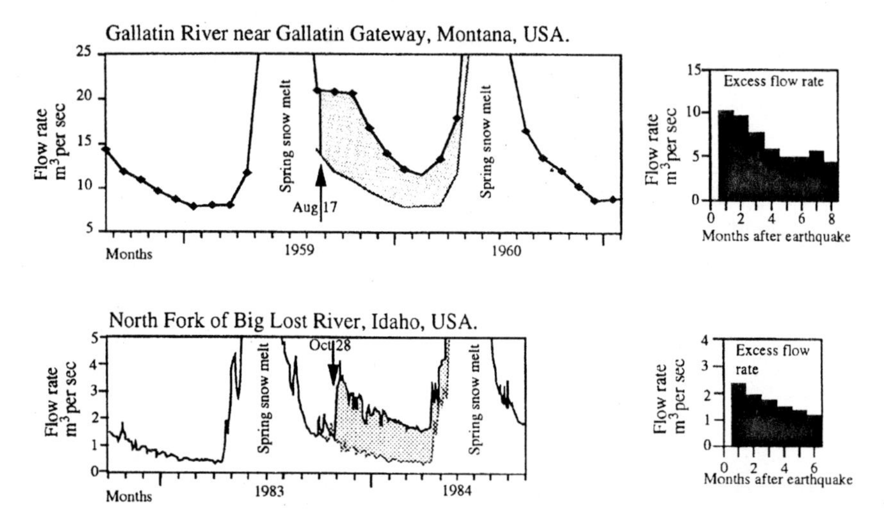

Fig. 3.1. Examples of excess river flow (shaded) following the Hegben Lake (top) and Borah Peak (bottom) earthquakes [part of Figure 1 of Muir-Wood and King, Journal of Geophysical Research, v. 98 p. 22,037, copyright by the American Geophysical Union, 1993].

The data from Borah Peak and Hegben earthquakes have the following common features: (1) a significant increase in the groundwater discharge following a normal faulting (extensional) earthquake, (2) a smaller effect at larger distances from the fault rupture, and (3) a slow decline in the effect over subsequent months.

Muir-Wood and King (1993) attribute these observations to the strain effects near an earthquake rupture. Before a normal-faulting earthquake, the ground is stretched slightly over a volume corresponding in depth to the depth of incipient faulting and in diameter to 2 or 3 times the fault length. The stretching slightly opens vertical cracks in the upper crust, sucking in ground water. For the Borah Peak and Hegben earthquakes, the volume subsequently released corresponded to the order of 50 mm of rainfall over the catchments. This water had accumulated in the ground in the years before the earthquake, instead of running off. During the earthquake, the fault ruptures, the strain is released, and the previously extended crust contracts suddenly. Water which filled the widened cracks is now under pressure and flows towards the surface, emerging as springs or through a rise in water table, and adding to the streamflow where the excess flow is recorded by hydrographs. Muir-Wood and King (1993) suggest from their physical modeling of the process that cracks to depths of 2 to 5 km are affected, and that it is the slow passage of water from these depths along tortuous cracks that causes the extended period of extra flow. Their model for normal-faulting earthquakes works in complementary fashion for reverse-faulting earthquakes: during compressional strain build-up water is slowly expelled from the ground before the earthquake, and after the earthquake groundwater is sucked-in (causing, e.g., a drop in well levels). As might be expected, the effects of strike-slip earthquakes are generally smaller in scale and more varied in style (Muir-Wood and King, 1993, p. 22,049). With this conceptual model, it is possible to infer the mechanism of historical earthquakes from their reported hydrological effects, as done by Muir-Wood and King (1993) for some earthquakes in the Apennines of Italy.

Other authors have challenged the Muir-Wood and King (1993) interpretation and argued instead that the excess water flow represents rapid drainage of near surface groundwater through a zone of near-surface fractured rock whose permeability has been increased an order of magnitude by the earthquake shaking (Rojstaczer et al., 1995). Arguments for this interpretation come from the rather constant geochemistry of the excess waters from the 1989 Loma Prieta earthquake, their young age as inferred from their tritium content, and the observations that well levels dropped in some areas of excess flow. However, the counter-example is complicated by the fact that the 1989 Loma Prieta is a dominantly strike-slip earthquake, and can be expected to generate a complex response, and that the samples studied were taken in 1992, arguably after most of the excess water had already passed from the system.

Recent observations of pockmarks may also suggest that such subaerial permeability changes do not provide the whole answer. Pockmarks are large conical depressions in lake or sea-bottoms, caused by water or gas escaping through underwater sediments; the fine sediment is disturbed and then dispersed by bottom currents. In certain parts of coastal Maine pockmarks exist in sediments lacking appreciable gas, and may be formed by water escape alone (Kelley et al., 1994). Furthermore, some of these pockmarks occur in linear trends which might mark bedrock fractures that have concentrated the sub-sediment water flow. While most pockmarks do not seem to vent continuously, most have a fresh morphology and they are inferred to be long-lived features that are excavated episodically. A pockmark field in the Gulf of Patras, Greece, became active at the time of a magnitude 5 earthquake

(Hasiotis et al., 1996). During the 24 hours prior to the earthquake the water bottom temperature anomalously increased by 5° C, while for a few days after the earthquake the majority of the pockmarks vented gas. Perhaps these changes might be explained by the expulsion of warm groundwater from the deeper sediments and upper crust and by the flushing out of gas. An earthquake, either causing sediment liquefaction and water expulsion at the time of the shaking, or more speculatively, the prior or post-earthquake slow expulsion of deep groundwater, might provide an episodic explanation for the Maine pockmarks and possibly other enigmatic bottom features in the southern Great Lakes (e.g., as cited by Adams, 1996).

4. CONCLUSIONS

In many fluvial environments it is not the catastrophic exogenic event that causes cyclic sedimentation, but the normal flood cycle or changes in the river channel position. Thus those that wish to interpret unusual geological sequences in terms of neotectonics need to reject the normal sedimentary processes before invoking the abnormal. One hopes, too that those working on the engineering, geographic or sedimentological aspects of rivers will keep an eye open for the truly abnormal and alert a collaborator interested in earthquake potential to investigate!

An understanding of water-induced earthquakes will help to minimize the chance that future injection projects or reservoirs will trigger a damaging earthquake. Furthermore, this understanding and the study of water induced by earthquakes together provide a unique opportunity for increasing our knowledge of the natural earthquake processes. In this way, simple hydrological observations, made at the right time and place, can check sophisticated physical models for strain changes near large earthquakes, and so give insight into pre- and post-earthquake processes.

Most seismologists are skeptical that earthquake prediction will attain what the public might regard as suitable ("a magnitude 6.5 earthquake, within the next 24 hours, near Parkfield"), let alone the level of prediction suggested by Clarke and McQuag (1996) in their science fiction novel ("a magnitude 8.3 earthquake, between 5 and 6 pm on Saturday the 25th February next year, 50 km north of Memphis"). Nevertheless it is tantalizing to think that certain hydrological signatures might indicate those regions of the crust that are being rapidly strained in preparation for rupture. This hope may seem unrealistic given today's knowledge and monitoring abilities, but it is clear that without sound observations and a conceptual framework to interpret the results there can be no significant progress.

5. ACKNOWLEDGMENTS

This paper was prepared for the Symposium "Applying Geomorphology to Environmental Management" Colorado State University, Fort Collins, Colorado, November 1-2, 1996. (A Reunion-Symposium of the "Schumm School"). I appreciate Stan and Ethel Schumm's help in learning to ride a bicycle and adjusting to life in Colorado during two stays at Colorado State University in 1976 and 1978, and incidentally getting me to think about what rivers were *really* doing when we weren't watching. Frank Ethridge (CSU) and Harold Wellman (Victoria University of Wellington, New Zealand) provided additional confidence that astute observations often lead to new insights about the earth. I thank Zhu Haizhi for organizing my 1985 visit to the Xiji area. GSC contribution number 1996462.

6. REFERENCES

Adams, J. 1985. Large-scale tectonic geomorphology of the Southern Alps, New Zealand. p. 105-128, in M. Morisawa and J. Hack (eds) *Tectonic Geomorphology* Proceedings of the 15th Annual Binghamton Geomorphology Symposium, George Allen and Unwin, 390 pp.

Adams, J. 1990. Paleoseismicity of the Cascadia subduction zone: evidence from turbidites off the Oregon-Washington margin. *Tectonics*, 9:569-583.

Adams, J. 1996. Paleoseismology in Canada — A dozen years of progress. *Journal of Geophysical Research,* 101:6193-6207.

Ahmad M. U. and J. A. Smith. 1988. Earthquakes, injection wells, and the Perry nuclear power plant, Cleveland, Ohio. *Geology,* 16:739-742.

Atwater, B. F., A. R. Nelson, J. J. Clague, and 13 others. 1995, Summary of coastal geologic evidence for past earthquakes at the Cascadia subduction zone. *Earthquake Spectra,* 11:1-18.

Bollinger, G. A. and J. K. Costain. 1988. Long-term cyclicities in earthquake energy release and major river flow volumes in Virginia and Missouri seismic zones. *Seismological Research Letters,* 59:279-283.

Chinese Working Group of the Project 206. 1989. *Atlas of active faults in China.* Seismological Press, Xian Cartographic Publishing House, 121 pp.

Clarke, A. C. and M. McQuay. 1996. *Richter 10.* Bantam Books, New York, 373 pp.

Davis, S. D. and C. Frohlich. 1993. Did (or will) fluid injection cause earthquakes? - Criteria for a rational assessment. *Seismological Research Letters,* 64:207-224.

Doig, R. 1991. Effects of strong seismic shaking in lake sediments, and earthquake recurrence interval, Témiscaming, Quebec. *Canadian Journal of Earth Sciences,* 28:1349-1352.

Hasiotis, T., G. Papatheodorou, N. Kastanos and G. Ferentinos. 1996. A pockmark field in the Patras Gulf (Greece) and its activation during the 14/7/93 seismic event. *Marine Geology,* 130:333-344.

Johnston, A.C. and E.S. Schweig. 1996. The enigma of the New Madrid earthquakes of 1811-1812. *Annual Reviews of Earth and Planetary Science,* 24:339-384.

Kelley, J. T., S.M. Dickson, D. F. Belknap, W. A. Barnhardt and M. Henderson. 1994. Giant sea-bed pockmarks: Evidence for gas escape from Belfast Bay, Maine. *Geology,* 22:59-62.

Muir-Wood, R. and G. C. P. King. 1993. Hydrological signatures of earthquake strain. *Journal of Geophysical Research,* 98:22,035-22,068.

Obermeier, S. F., R. B. Jacobson, J. P. Smoot, and 4 others. 1990. Earthquake-induced liquefaction features in the coastal setting of South Carolina and in the fluvial setting of the New Madrid seismic zone. *U.S. Geological Survey Prof. Paper* 1504, 44 pp.

Rojstaczer, S., S. Wolf and R. Michel. 1995. Permeability enhancement in the shallow crust as a cause of earthquake-induced hydrological changes. *Nature,* 373:237-239.

Schumm, S. A. 1986. Alluvial river response to active tectonics. p. 80-94 in Wallace, R.E. (chairman) *Active Tectonics* National Academy Press, Washington, D.C.

Seeber, L., G. Ekstrom, S. K. Jain, and 3 others. 1996. The 1993 Killari earthquake in central India: A new fault in Mesozoic basalt flows? *Journal of Geophysical Research,* 101:8543-8560.

State Seismological Bureau. 1980. *The 1920 Haiyuan great earthquake.* (in Chinese) Edited by the State Seismological Bureau of Lanzhou and the Seismological team of Ningxia. Beijing, 134 pp.

Tuttle, M. P. and E. S. Schweig. 1996. Recognizing and dating prehistoric liquefaction features: Lessons learned in the New Madrid seismic zone, central United States. *Journal of Geophysical Research,* 101:6171-6178.

Vita-Finzi, C. 1986. *Recent Earth Movements.* Academic Press, London, 226 pp.

Yeats, R. S. and C. S. Prentice. 1996. Introduction to special session: Paleoseismology. *Journal of Geophysical Research,* 101:5847-5853.

Yeats, R. S., K. Sieh and C. R. Allen. 1997. *The Geology of Earthquakes.* Oxford University Press, New York, 568 pp.

Zhu, H., 1982. Earthquake hazards, prehistoric seismicity and seismic microzonation. *Proceedings, Third international earthquake microzonation conference,* June 28 - July 1, Seattle, III:1379-1392.

APPLICATION OF QUANTITATIVE MORPHOLOGIC DATING OF PALEO-SEISMICITY OF THE NORTHWESTERN NEGEV, ISRAEL

Zeev Binyamin Begin[1]

ABSTRACT

Paleo-seismicity can be studied through the geometry of alluvial fault scarps associated with earthquakes. The erosion of alluvial fault scarps was modeled in recent years through a diffusion equation and the rate of the process is characterized by the diffusion coefficient k. If a minimum value of k can be established, then the mathematical model enables one to estimate the maximum height of an alluvial scarp, which is obliterated after a given period of time. The statistical relationship between fault scarp height and earthquake magnitude allows transformation of maximum scarp height to maximum magnitude of an earthquake, which leaves no trace in alluvial surfaces after a given period of time.

In this study, values of k were estimated for alluvial slopes in the Negev, Israel, by measuring the geometry of slopes of known ages. These data permit the estimation of maximum non-detectable earthquake magnitude for the Negev.

KEYWORDS

Morphology, paleo-seismicity, earthquakes, alluvial fault scarps

1. INTRODUCTION

The available record of past major earthquakes may sometimes by augmented through the study of fault scarp morphology, which can be used to date faults either qualitatively (Wallace, 1977), semi-quantitatively (Bucknam and Anderson, 1979; Mayer, 1984, p. 304), or through quantitative modeling. The

---

[1] Geological Survey of Israel, 30 Halkhe Yisrael Street, Jerusalem, 95501, Israel.

latter avenue was explored by several authors, assuming that the governing equation of scarp erosion is a two-dimensional diffusion equation.

$$\frac{\partial y}{\partial t} = k \frac{\partial^2 y}{\partial x^2} \tag{1.1}$$

where y is elevation, x is horizontal distance measured from the midpoint of the scarp profile, t is time, and k is the diffusion-erosion coefficient. This coefficient is a measure of the rate of the erosion process, and it is assumed---as a first approximation---that it is independent of time. If so, k can be used as a geomorphological clock (Kirkby and Kirkby, 1976; Nash, 1980a, 1980b, 1984; Colman and Watson, 1984; Hanks et al., 1984; Mayer, 1984; Andrews and Hanks, 1985; Pierce and Colman, 1986). These studies show that, at least in some circumstances, Eq. 1.1 can be regarded as a fair approximation of the process of degradation of scarps in alluvial materials. This specific field is relatively young and there has not yet been much experience with it. It is noted however, that in addition to its advantages, some difficulties in the application of the diffusion model to scarp dating have already been encountered.

The aim of this study is to apply this method in an attempt to define parameters to paleo-seismicity in the desert of Negev area in Israel. It was originally carried out following a request by the Licensing Division of the Israeli Atomic Energy Commission (Begin, 1987). From the point of view of licensing a nuclear power plant, an earthquake risk assessment is required, for which the following questions are pertinent: (1) Given an undated fault scarp, what is it reasonable maximum age? (2) Given an undisturbed alluvial surface, what is the maximum initial height of a fault scarp (caused by an earthquake of a certain magnitude) that is eroded beyond recognition within a given period of time? In other words, what is the highest magnitude of an earthquake, which occurred in the past, but is not detectable through fault scarps?

2. SOLUTIONS FOR EQUATION 1.1

Solutions for Eq. 1.1 were suggested for two different initial conditions: (1) the initial scarp morphology is that of a vertical step (Colman and Watson, 1983; Hanks et al., 1984), and (2) the initial morphology is that of a planar scarp whose slope is equal to the angle of repose, characteristic of the materials underlying the scarp.

For an initial vertical scarp, Colman and Watson's (1983) Eq. 7 can be slightly rearranged as:

$$t = \frac{1}{\pi k} \left(\frac{d}{2\,(\tan\theta - \tan\alpha)} \right)^2 \qquad (2.1)$$

where d is the fault offset; θ is the measured maximum scarp angle; α is the slope of the surface up and down the fault scarp (this is the original pre-faulting surface slope, see Fig. 2.1). Equation 2.1 is equivalent to Hanks et al. (1984, Eq. 8).

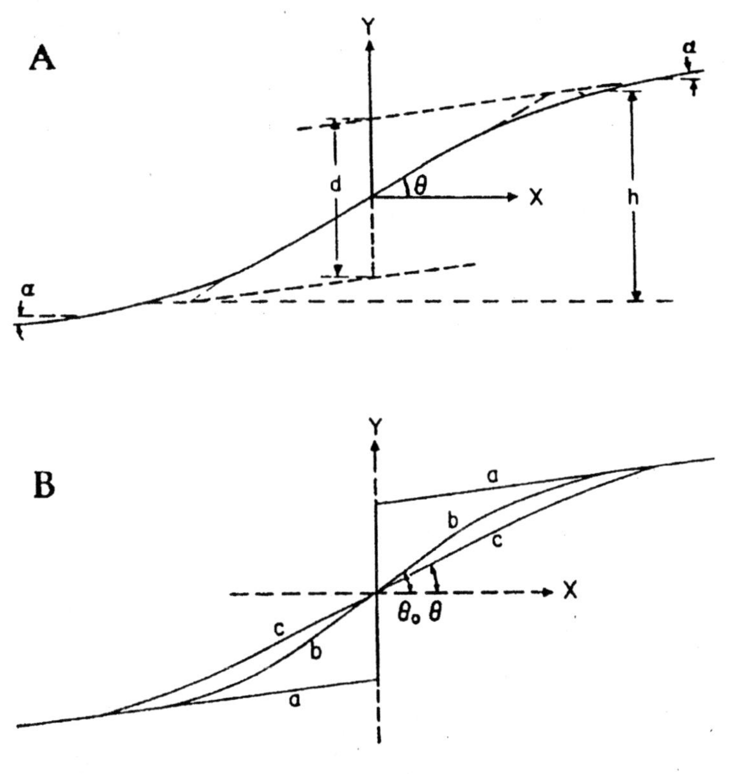

Fig. 2.1. The geometry of a fault scarp and definitions of parameters (from Colman and Watson, 1984).

For an initial sloping planar scarp, two solutions were presented: Colman and Watson (1983, note 11) suggested:

$$\tan\theta = \tan\alpha + (\tan\theta_0 - \tan\alpha)\, erf\left(\frac{d}{2\tan\theta_0\,(4kt)^{1/2}}\right) \qquad (2.2)$$

where θ_0 is the initial scarp slope, *erf* is the error function, and the other notations as above.

Equation 2.2 was rearranged by Pierce and Colman (1986, Eq. 3), assuming a pre-faulting horizontal surface ($\alpha = 0$):

$$k = \left(\frac{d}{4t^{1/2}\,\tan\theta_0\; erf^{-1}(\tan\theta\;/\;\tan\theta_0)}\right)^2 \qquad (2.3)$$

here erf^{-1} is the inverse error function. Equation 2.3 can be solved for *t*:

$$t = \left(\frac{d}{4k^{1/2}\,\tan\theta_0\;[erf^{-1}(\tan\theta\;/\;\tan\theta_0)]}\right)^2 \qquad (2.4)$$

and the error function can be approximated as a series of expansion of the form:

$$eft(z) = 2/\sqrt{\pi}\,(z - z^3/1!\,3 + z^5/2!\,5 - z^7/3!7 \pm \ldots) \qquad (2.5)$$

(Mayer, 1984, p 311).

Nash (1984, Fig. 12) suggested a numerical solution for this case, and he presented a curve relating

$$[\tan(\theta - \alpha)/\tan(\theta_0 - \alpha)]\ to\ [(kt/d^2)\tan^2(\theta_0 - \alpha)] \qquad (2.6)$$

Note that his curve applies only for $\alpha < 10°$. According to Pierce and Colman (1986), Eq. 2.3 yields results which are identical to Nash's numerical solution.

3. LIMITATIONS ON THE APPLICABILITY OF THE DIFFUSION-EROSION MODEL

Theoretically, the application of the above solutions of Eq. 1.1 is straightforward, because both Eqs. 2.1 and 2.4 show that the product of the diffusion-erosion coefficient *k* and time *t* can be calculated from the initial and final configurations of the scarp: First *k* is calibrated through the application of Eq. 1.1 and its solutions to scarps of known ages. Then, for a scarp of an unknown age, the geometry of the scarp profile is measured and d, θ, and α are determined. With the known value of k, t can be determined. The limitations of this procedure stem from several sources, as outlined below.

3.1 Applicability of Equation 1.1

The implicit assumption concerning Eq. 1.1 is that the rate of erosion (that is, the volume per unit length of scarp eroded at any time from any point on the scarp profile) is directly proportional to the scarp slope (that is, the tangent of its slope angle). Pierce and Colman (1986) summarized those slope erosion processes, which may be considered as complying with this assumption ("Group A"). For other processes, which are known to operate on certain slopes ("Group B"), this assumption is not valid and hence, Eq. 1.1 and its solutions are not applicable. Scarp profiles formed by Group A processes are different from profiles formed by Group B processes, and therefore they may be distinguished (Pierce and Colman, 1986).

A different distinction was suggested by Nash (1984), according to whom the diffusion-erosion model is applicable only to transport-limited hillslopes (slopes on which more loosened debris is available for transport than the transportational processes are capable of removing).

3.2. Applicability of the Diffusion Model at the Initial and Late Stages of Degradation

3.2.1. Initial Stage of Degradation

"The processes that remove the free face from a scarp and lead to a slope at the angle of repose do not satisfy the assumptions used in Eq. 2.1. Therefore, the diffusion equation model applies only after the scarp has reached its angle of repose and is controlled by creep, raindrop impact, and slope wash" (Colman and Watson, 1984). For fault scarps in the western United States, the time necessary to remove the free face appears to vary from a few tens of years to a few hundred years (Colman and Watson, 1984) or even several years only (Pierce and Colman, 1986). This time is the starting time for the diffusion equation model. Because the diffusion equation does not apply to the processes that remove the free face from a scarp, Colman and Watson (1984) suggested that the age of the scarp (t) be divided into t_0, the time required to remove the free face, and t_1, the time for which the diffusion equation applies to scarp degradation. Thus, t_0 is the time required to go from profile a to profile b in Fig. 2.1b, and t_1 is the time required to go from profile b to profile c. To calculate kt, one first obtains $kt_1 = k(t_0+t_1)$ and then subtracts kt_0, both calculated from Eq. 2.1 by using a vertical initial profile. These calculations are combined in Eq. 3.1:

$$kt_1 = \frac{d}{4\pi}\left[\frac{1}{(\tan\theta - \tan\alpha)^2} - \frac{1}{(\tan\theta_0 - \tan\alpha)^2}\right] \qquad (3.1)$$

and the true age of the scarp is t_1 plus the time required to remove the free face (Colman and Watson, 1984).

3.2.2. The problem of the Initial Scarp Slope

"Most scarps, especially fault scarps, are formed with free faces, steep segments having angles commonly more than 50°. These free faces are modified by slumping, spalling, and other processes until the scarp slope reaches approximately the angle of repose of the material in which the scarp is developed. The angle of repose of unconsolidated deposits depends on climate and materials, but angles of 25° to 35° are common." (Colman and Watson, 1984).

Even if a vertical initial scarp is assumed, one is soon faced with the necessity to define the angle of repose, which is a prerequisite for the application of Eq. 2.3.

Colman and Watson (1984) estimated θ_0 by trial-and-error calculations to determine the initial angle that results in the lowest correlation coefficient between kt and surface offset. Estimates of θ_0 for actual data sets are a few degrees lower than what one might expect for these materials. Therefore, θ_0 is more accurately described as the angle at which the diffusion equation begins to apply to the degradation of the scarp, rather than as the angle of repose.

Pierce and Colman (1986) suggested that the "starting angle" θ_0 is related to the angle of repose. They recommended that it be determined from consideration of the materials involved and field measurements of scarps at or near the angle of repose (as suggested by Nash, 1984), rather than by selection of a starting angle that minimizes the dependence of k on scarp height.

At any rate, for scarps, which degrade to angles which are much lower than the angle of repose, the calculated kt values are not very sensitive to the choice of θ_0 (Fig. 3.1).

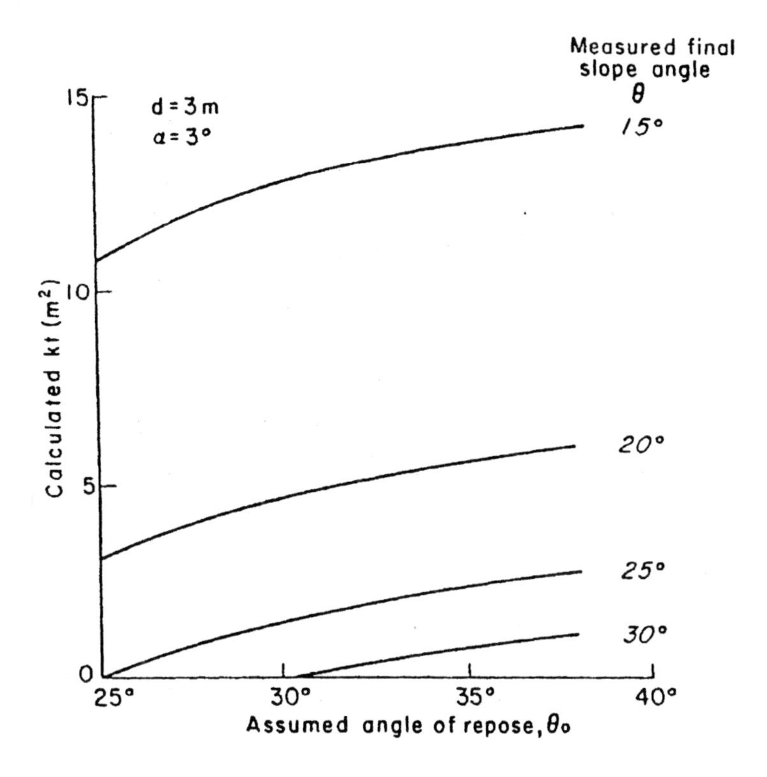

Fig. 3.1. **The relationship between kt and the angle of repose θ_0, calculated through Eq. 2.6, for d = 3 m and α = 3°, and different values of final slope angle, θ.**

3.2.3. Late Stage of Degradation

The possibility may be raised that at a certain low slope angle degradation actually comes to a halt, and that such slopes may be stable and unchanged for very long periods of time. In that case, the diffusion model is not valid and calculations based on it might lead to age estimates which are too young.

There is no explicit reference in the literature to the question of the lowest slope angle for which the diffusion model is still valid, but slope angles as low as 4° were shown to comply with it (Hanks et al., 1984, Fig. 14). In the Elat area, Israel, pebble-slopes higher than 1.5 m are thought to have been stabilized at angles of 5° to 10° (R. Gerson, Hebrew University, Jerusalem, pers. comm., 1987). Hanks et al. (1984, p 5788) addressed this issue indirectly.

They assumed that the minimum observable scarp slope (relative to the fan slope on which such a scarp might be placed) is 3°, noting that there is some uncertainty as to what is observable and what is not. It should be noted that Hanks et al. refer to "observable" topographic differences and not "detectable" ones, but even the "observable" angle that they assume is very low.

4. FACTORS AFFECTING THE VARIABILITY OF THE DIFFUSION-EROSION COEFFICIENT

4.1. Variability of *k* Independent of Time

4.1.1. Scarps in Different Materials

Mayer (1984) demonstrated variability in scarp morphology as a function of materials. "*Analysis of covariance of the Pitaycachi fault scarp data indicate clear and significant differences in morphology between the groups of scarp segments formed in different materials. The morphologic data suggest either that there has been a more rapid degradation of the scarps formed in the slightly cohesive materials, or that the initial scarp angle was smaller in the slightly cohesive materials, compared to the more cohesive materials.*" Hence, he concluded that "*the constraint on morphological dating imposed by variability in scarp materials is that different k values should be used for different material categories.*"

4.1.2. The Effect of Climate

The diffusion coefficient *k* must depend to a great extent on the amount of water available for erosion. Variability of *k* due to climate is clearly noted in Table 5.1. Hanks et al. (1984) related the 15-fold difference in *k* between the Raymond fault, California and Lake Bonneville shoreline in Utah (both cut in weakly consolidated material) to differences in rainfall.

The variability of *k* due to microclimate was explored by Pierce and Colman (1986) who compared *k* values for alluvial scarps facing different directions. Especially in a semi-arid area, differences in *k* values are related to differences in potential solar-beam radiation, which is a strong determinant of microclimate for the different orientations and inclinations of scarps. Solar radiation differences determine differences in soil moisture and type and density of vegetation, especially in a semi-arid climate. These in turn determine erosion processes and rates, and the result is a strong dependence of k on scarp orientation. For semi-arid central Idaho, south-facing scarps have degraded to slopes 65 to 75 percent as steep as north-facing scarps, and the calculated k values have a range of nearly two orders of magnitude. In Israel, several archeological mounds (tells) were found to be much more degraded in their southeast slopes than in their northwest slopes (Rosen, 1986, Table 2).

4.1.3. The Effect of Scarp Height

Hanks et al. (1984) suggested that observations of a scarp of a certain age can be plotted on a diagram relating $\tan\theta$ to scarp offset, d. If it can be assumed that k is constant along the scarp, then for a scarp of a certain age t the product kt is a constant. Therefore, it follows from Eq. 2.1 that the regression of $\tan\theta$ on d should be a straight line with an intercept $\tan\theta$ and a slope of $1/2(kt)^{1/2}$. This permits calculation of a kt value which averages all the observations.

In several studies of scarps, however, considerable scatter is observed on such diagrams, and apart from the naturally occurring random scatter, some scatter indicates departures from the basic model. It was noted by Hanks et al. (1984) that the linearity between $\tan\theta$ and d, as predicted by Eq. 2.1, is not always retained for higher scarps and they found a correlation between k values and scarp height.

Despite the fact that Colman and Watson (1984) (see also Hanks et al., 1984) succeeded in formulating a procedure that defines θ_0 in a manner which complies with the diffusion model, it must be noted that in those instances in which d is correlated with kt, violation of the basic assumptions is indicated. In other words, if one adheres to the procedure recommended by Pierce and Colman (1986), then θ_0 is not artificially determined (usually lower than the expected angle of repose) and the dependence of k on scarp height is not eliminated. This in itself indicates that in such cases the diffusion-erosion model does not strictly apply. Pierce and Colman (1986) explored that correlation and suggested that on higher scarps the assumption of linearity between scarp and slope and sediment discharge-as assumed for Eq. 1.1 - is not valid, and that different processes occur on these high scarps. To Pierce and Colman (1986, p. 877) this suggested that Group B processes are involved in the scarp erosion, together with Group A processes. These authors, following Hanks et al. (1984), showed a prominent dependence of k on scarp height for a large sample of scarps in Idaho and Nevada, with higher scarps being associated with larger k values. For scarp heights ranging between 1 and 12 m, *k* increases 20-fold! This dependence was found to be much smaller for north-facing scarps than for south-facing scarps.

4.2. Changes of *k* with Time

The dependence of *k* on climate, as explained above, must have led to changes in *k* over time, especially in the Quaternary, in which climate changes occurred on a scale of thousands or tens of thousands of years. Over a period of time what one actually measures is the effect of an average *k* value. Comparisons of *k* values from distant sites should therefore reflect their climate history.

Also, during the process of scarp degradation, the materials exposed to erosion may change due to two factors: (1) lower layers may be exposed, which may be of different *k* values, and (2) sorting may change the characteristics of the scarp sediments (Mayer, 1984). These also lead to changes in *k* with time.

5. CHOICE OF A K VALUE FOR THE NORTHERN NEGEV

5.1. Literature Survey

Applying the diffusion model, the actual parameter resulting from measurements of fault scarp geometry is the product *kt* and the practical problem (from the point of view of a nuclear licensing agency) is to evaluate how large *t* really is, once *kt* is calculated. In other words, what is needed is an assessment of the smallest reasonable value of *k*.

The few studies published demonstrate a very wide range of *k* values (Table 5.1), reflecting not only measurement errors or limited applicability of the diffusion model, but also a wide range of scarp materials, height, microclimate, and climatic history. Table 5.1 shows that the minimum measured value of *k* is *k*(min) = 0.0001 m^2/yr.

Table 5.1. Value of *k* as Determined in Previous Studies.

Site	k(m^2/yr)	Age(yrs BP)	Reference	Remarks
Raymond Fault, California	0.0160	230,000	1	d =30 m
Lake Algonquin, wave cut bluffs, Michigan	0.0120	10,500	2	d = 20 m
Nippising Great Lakes, wave cut bluffs, Michigan	0.0120	4,000	2	d = 20 m
Santa Cruz sea cliffs, California	0.0110	230,000	1	d = 35 m
Alluvial scarps, central Idaho	0.0100-0.0001	15,000±4,000	3	k dependent on d and on orientation
River terrace scarps, West Yellowstone, Montana	<0.0020	>7,100	4	d = 2.5-3.1 m
Lake Bonneville shore terraces, Utah	0.0002-0.0023	15,000	3	k dependent on d d = 1-12 m
Fish Spring Range fault scarps, Utah	>0.0003	<12,000	1	d = 0.5-3.2 m
Oqirrh Mountains fault scarps, Utah	<0.0023	>15,000	1	d = 2-6 m
Drum Mountains fault scarps, Utah	>0.0005	<10,000	3	d = 2 m
Lake Lisan shore terraces, Dead Sea, Israel	0.0004	12,000	5	d = 1-7 m(at one site d = 16m)

References: (1) Hanks et al., 1984; (2) Nash, 1980b; (3) Pierce and Colman, 1986; (4) Nash, 1984; (5) Bowman and Gerson, 1986

5.2. Some Preliminary Measurements of *k* in the Northern Negev

5.2.1. Geomorphological Setting

Several measurements of *k* in the northern Negev were carried out in the course of this study in order to determine a preliminary minimum value of *k*. The geometry of some stream terraces was measured in Nahal Besor, some 6 km northwest of Sede Boqer, and in Nahal Be'er Sheva, east and west of Ze'elim (Table 5.2). The geomorphologic setting of these terraces is schematically presented in Fig. 5.1: The upper terrace in Najal Besor (H in Fig. 5.1) contains a paleosol which was dated at about 30,000 years old (Zilberman, 1990). The

Table 5.2. Values of *kt*, calculated by Eq. 2.6 for some scarps in the Northern Negev.

Number	Scarp Parameters						Calculated kt_l (mi²) (θ_0=31°)
	Topographic setting		Material	Geometry			
	Coords.	Azimuth		Scarp offset (m)	θ	α	
H-1	12280354	060	coarse colluvium	4.8	26	1.0	2.90
H-2	12260355	060	"	4.4	22	2.5	6.90
H-3	12430355	200	"	5.4	27	1.5	2.90
H-4	12430355	200	"	4.1	21	3.0	7.73
H-5	12530343	320	"	5.9	28	3.0	2.85
H-6	12630342	050	"	4.6	25	4.0	4.74
H-7	12650346	270	silt	3.4	27	3.0	1.34
HAZ-1	11870683	320	"	0.9	13	5.0	2.89
HAZ-2	11870683	320	"	1.3	13	5.0	6.03
HAZ-3	11870683	320	"	5.4	22	3.0	11.05
HAZ-4	11870683	320	"	3.2	20	3.5	6.09
ZEL-3	10250742	270	"	3.1	30	5.5	0.30
ZEL-4	10250742	270	"	5.0	30	4.0	0.67
					mean = 4.34 std. dev. = 3.12		
L-1	66424211	180	"	1.3	22.5	2.5	0.55
L-2	66424211	180	"	1.8	28	3.0	0.27

youngest age of this terrace is epi-paleolithic, and it can be as young as 10,000 years BP (E. Zilberman,P. Goldberg, pers. Comm., 1987). Therefore, the erosion of this terrace and the formation of its scarps is presumable younger than 10,000 years. The lower terrace (L in Fig. 5.1) is replete with dams and other agricultural structures which are of Byzantine age, 4th-7th centuries AD (P. Goldberg, Hebrew University, Jerusalem, pers. Comm., 1987).

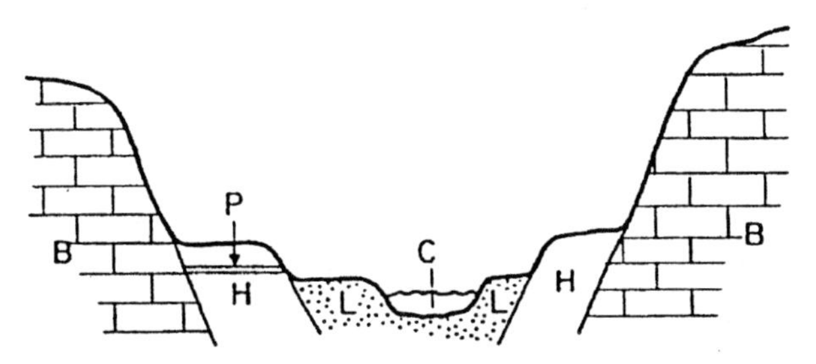

Fig. 5.1. Schematic representation of stream terraces in the northern Negev. B — bedrock; H — higher terrace; P — paleosol, about 30,000 years BP; L — lower terrace of the Byzantine period; C — present channel fill.

Defining *t* as the time that elapsed since a scarp was left to erode without the influence of a stream undercutting it, the following constraints can be defined for scarps in Nahal Besor and Nahal Be'er Sheva: For scarps formed in the upper terrace, t <10,000 years and for scarps formed in the lower terrace, t <1,400 years.

5.2.2. Measurements

The geometry of 13 scarps of the upper terrace and two scarps of the lower terrace was measured. The results are presented in Figs. 5.2 and 5.3, and are summarized in Table 5.2. Using d, θ, and θ_0, kt_1 was calculated for the surveyed scarps. The value of $\theta_0 = 31°$ was chosen because in two scarps $\theta = 30°$ was measured. This value is close to the angle of internal friction φ which was measured in loess in the Be'er Sheva region (Michaeli and Morell, 1986) resulting in a mean value of 29.4° for 15 samples, with a standard deviation of 3.2°. It is concluded from Table 5.2 that for scarps of the higher terrace, the mean value of kt_1 is 4.34 m^2 with a standard deviation of 3.12 m^2, and for scarps of the lower terrace, the mean value of kt_1 is 0.41 m^2. Note that kt_1 is not correlated to scarp offset.

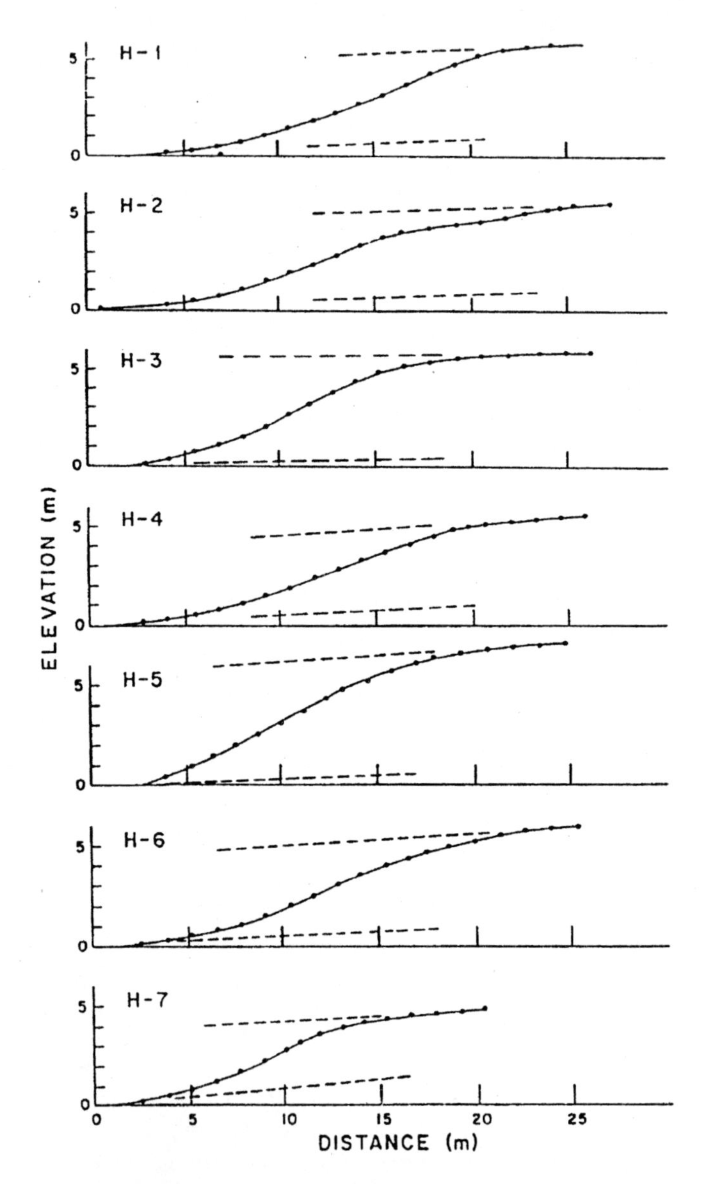

Fig. 5.2. **Profiles of scarps in Nahal Besor west of Sede Boqer. For location, see Table 5.2.**

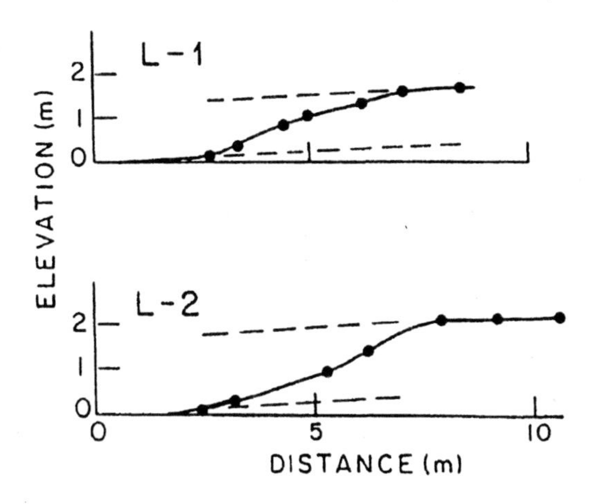

Fig. 5.3. **Profiles of scarps in Nahal Be'er Sheva west of Be'er Sheva. For location, see Table 5.2.**

Table 5.3. **Calculation of k for the Northern Negev.**

	Upper Terrace Scarps kt_1(min) = 1.22 sq m $t \leq 10{,}000$ y	Lower Terrace Scarps kt_1(min) = 0.27 sq m $t \leq 1{,}400$ y
t_0 = 100 y	$t_1 \leq 9.900$ yr $k \geq 0.00012$ sq m/y	$t_1 \leq 1{,}300$ y $k \geq 0.00019$ sq m/y
t_0 = 1000 y	$t_1 \leq 9.000$ y $k \geq 0.00013$ sq m/y	$t_1 \leq 400$ y $k \geq 0.00068$ sq m/y

We may now proceed to evaluate k from kt_1 by using the above-mentioned constraints on the time t that elapsed since the scarps were abandoned to erode without the influence of the stream. Since $t_1 = t - t_0$, and since the time (t_0) that is necessary to erode a vertical scarp to its angle of repose is not known, we use two estimates, namely t_0 = 100 years and t_0 = 1,000 years. A summary of the relevant considerations is presented in Table 5.3. Since we are interested in the minimum value of k (see discussion below), a value of mean kt_1 minus 1 standard deviation is used, based on Table 5.2. It is concluded that a reasonable minimum value of k in the northern Negev, derived from this limited sample is 0.0001 m^2/yr.

6. SIGNIFICANCE FOR NORTHERN NEGEV PALEO-SEISMICITY

6.1. Application to a Fault Scarp at Sede Boqer

A 1.5-m scarp formed in silty alluvium was detected near the Ben-Gurion College at Sede Boqer (coordinates 6712/4147) by Tahal (Ltd) geologists, while studying the area as part of a feasibility study for the proposed nuclear power plant near Shivta. The scarp is located near the Zin Fault, and it was trenched, revealing an offset in the underlying alluvium. The age of the disturbed alluvium is Early to Middle Pleistocene (Issar et al., 1984), and an attempt is made here to determine the age of scarp formation through the diffusion model.

The geometry of the scarp was measured (Fig. 6.1), and the resulting parameters are: $\theta = 17°$, $\alpha = 1.5°$, $d = 1.60$ m. Calculating kt by Eq. 2.6 (the results are insensitive to θ_0), kt_1 is estimated to be 3.0 m^2. If it is accepted that $k \geq 0.0001$ m^2/yr, then t_1 is estimated as $t_1 \leq 30{,}000$ years. For $t_0 = 100$ years, the age t of the scarp is $t < 30{,}100$ years, while for $t_0 = 1{,}000$ years, $t < 31{,}000$ years. If the value of k is assumed to be higher than 0.0001, the assumed time of scarp formation will be accordingly lower.

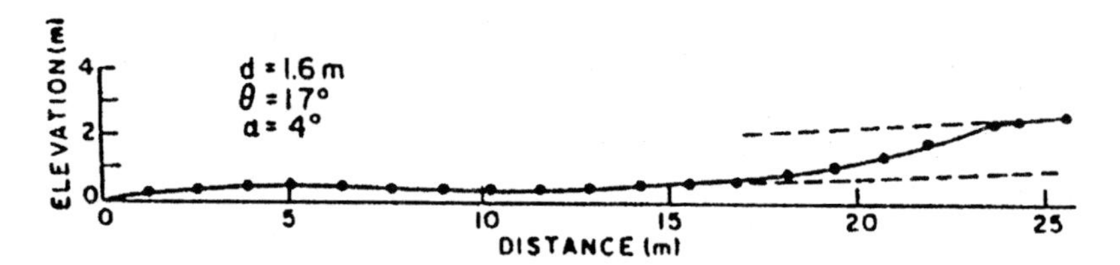

Fig. 6.1. Profile of a fault scarp at Ben-Gurion College, Sede Boqer.

6.2. Limits of Detectibility of Paleo-Seismicity in the Negev

With the above limits of k, it is possible to determine some constraints on morphologic dating of alluvial fault scarps in the northern Negev and on the level of detection of paleo-seismic activity. Based on a conservative approach, we seek the maximum earthquake magnitude which occurred t years before present, but has left no noticeable fault scarp, Defined otherwise, we seek the minimum earthquake magnitude which is still detectable after it formed a fault offset d, at time t before the present. We are, therefore, interested in the minimum offset of a scarp that for a period t is just obliterated, that is, eroded until its maximum slope angle, θ, is very close to the original slope, α. Using Eq. 6.1:

$$d(\min) = (2\sqrt{\pi k t}) \tan(\theta - \alpha)_{\min} \qquad (6.1)$$

Adopting the criterion proposed by Hanks et al. (1984), the limit of scarp recognition is assumed to be a slope segment 3° steeper than the segments below and above it, that is: $(\theta - \alpha)_{\min} = 3°$. The minimum detectable fault offset can thus be determined for different time periods with minimum k value estimated as above (in Eq. 6.1, minimum value of k leads to minimum value of d). The results are shown in Fig.6.2, which is interpreted as follows: assuming, for example, $k \geq 0.0001$ m²/yr within 10,000 years, a fault scarp with $d \leq 0.2$ m would be eroded "beyond recognition" and will not be detectable.

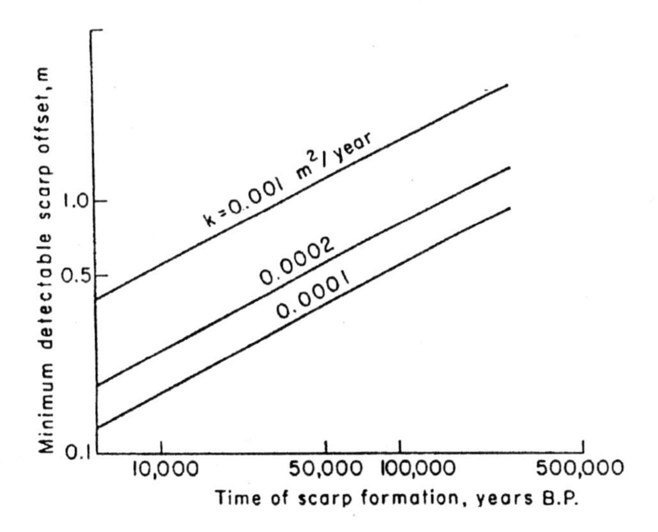

Fig. 6.2. Minimum-detectable fault offset in alluvial material for different time periods. Curves are calculated by Eq. 6.1. *k* is in m²/year, and the limit of morphological detectibility is assumed to be a slope steepened by 3°.

Fault offsets can be related to earthquake magnitude through the relationship proposed by Slemmons (1977, Table 13) for normal faults:

$$M = 6.83 + 1.05 \log(d) \qquad (\text{standard deviation} = 0.45) \qquad (6.2)$$

where M is earthquake magnitude and d is fault displacement in meters. Combining Eqs. 6.1 and 6.2, we get:

$$M(\min) = 6.06 + 0.53 \log(kt) \qquad (6.3)$$

from which it is possible to construct curves depicting minimum-detectable earthquake magnitudes for different periods of time, as in Fig 6.3, which is interpreted as follows: assuming $k \geq 0.0001$ m^2/yr, it is argued that an earthquake of magnitude $M \leq 6.1$ which occurred 10,000 years BP may leave no recognizable traces in alluvial material. Note that since M is proportional to log(d) (Eq. 6.2) and d is proportional to the square root of k (Eq. 6.1), a 10-fold difference in k leads to a difference in $M(min)$ of 0.5. Therefore, even if one assumes an extremely low $k = 0.00001$ m^2/yr, the minimum recognizable magnitude for 10,000 years would be 5.6, which is close to the earthquake magnitude below which surface rupture does not occur.

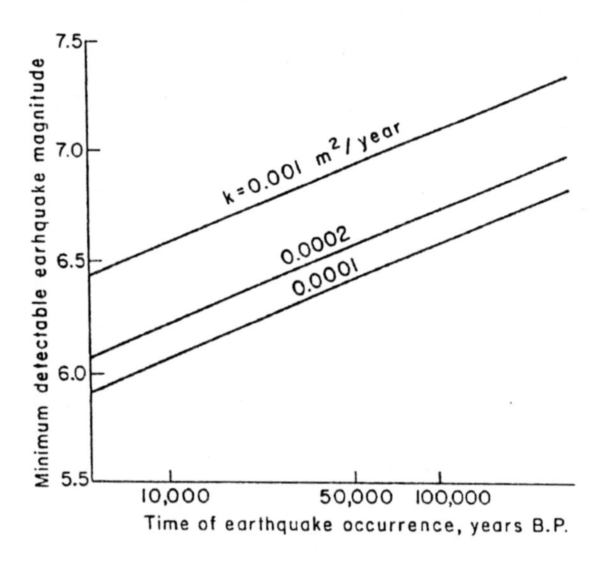

Fig. 6.3. Minimum detectable earthquake magnitude for different time periods. Curves are calculated by Eq. 6.3.

7. ACKNOWLEDGMENTS

I first met Ran Gerson in 1964 when he took some courses in geology with my class. Having concluded his B.A. studies in geography, it was typical of him to expand his knowledge in geology, and since then I watched him expanding evermore. Some years ago, Ran and I participated in two conferences on tectonic geomorphology in Nevada and New Mexico, and there was not doubt that he dominated those scenes, giving several lectures and actively participating in most discussions. His contributions in this field are reflected in this study, which he critically reviewed, and to which he added so much through many discussions of both concepts and details.

Thanks are due to Dr. Y. Weiler, Dr. A. Sneh, Mr. E. Zilberman, and Mrs. B. Katz for their critical reading of the manuscript, to Mrs. V. Arieh for typing it, to Mr. S. Levy for drafting, and to Mr. E. Ram for field assistance.

8. REFERENCES

Andrews, D.J., T.C. Hanks. 1985. Scarps degraded by linear diffusion: Inverse solution for age. Journal of Geophysical Research 90, 10193-10208.

Begin, Z.B. 1987. Quantitative morphologic dating of fault scarps: a review, with application to the northwestern Negev, Israel, Geological Survey of Israel, Report GSI/25/87, 37p.

Bowman, D., R. Gerson. 1986. Morphology of the latest Quaternary surface faulting in the Gulf of Elat region, Eastern Sinai. Tectonophysics 128:97-119.

Bucknam, R.L., R.E. Anderson. 1979. Estimation of Fault-scarp ages from a scarp-height-slope-angle relationship. Geology 7: 11-14.

Colman, S.M., K. Watson. 1984. Ages estimated from a diffusion equation model for scarp degradation. Science 221: 263-265.

Hanks, T.C., R.C. Bucknam, K.L. Lajoie, R.E. Wallace. 1984. Modification of wave-cut and faulting-controlled landforms. Journal of Geophysical Research 89: 5771-5790.

Issar, A., A. Karnieli, H.J. Bruins, I. Gilead. 1984. The Quaternary geology and hydrogeology of Sede Zin, Negev, Israel. Israel Journal of Earth Sciences 33: 34-42.

Kirkby, A., M.J. Kirkby. 1976. Geomorphic processes and the surface survey of archaeological sites in semi-arid areas. In: Davidson, D.A., Shackley, M.L. eds., Geoarchaelogy: Earth Science and the Past. London: Duckworth, p 229-253.

Mayer, L. 1984. Quaternary fault scarps formed in alluvium using morphologic parameters. Quaternary Research 22: 300-313.

Michaeli, L. and W. Morell. 1986. The influence of hydrochemical conditions of loess on slope stability. Geological survey of Israel, Report GSI/50/86, 73 p (in Hebrew).

Nash, D.B. 1980a. Morphologic dating of degraded normal fault scarps. Journal of Geology 88: 353-360.

Nash, D.B. 1980b. Forms of bluffs degraded for different lengths of time in Emmet County, Michigan, USA. Earth Surface Processes 5, p. 331-345.

Nash, D.B. 1984. Morphologic dating of fluvial terrace scarps and fault scarps near West Yellowstone, Montana. Geological Survey of America Bulletin 95: 1413-1424.

Pierce, K.L. and S.M. Colman. 1986. Effect of height and orientation (microclimate) on geomorphic degradation rates and processes: late glacial terrace scarps in central Idaho. Geological Survey of American Bulletin 97: 869-885.

Rosen, A. 1986. Cities of clay, the geoarcheology of tells. Chicago and London: University of Chicago Press, 167 p.

Slemmons, D.B. 1977. State-of-the-art for assessing earthquake hazards in the United States. Report 6, Faults and earthquake magnitude, U.S. Army Corps of Engineers Waterways Experiment Station, 129 p.

Wallace, R.E. 1977. Profiles and ages of young fault scarps, north-central Nevada. Geological Society of America Bulletin 88: 1267-1281.

Zilberman, E. 1990. The development of the landscape in the central and northwestern Negev in the Neogene and the Quaternary, Ph.D., Thesis, Hebrew University, Jerusalem, 108 p. (in Hebrew, English abstract).

CHANNEL CHANGE FROM EXTREME FLOODS IN CANYON RIVERS

Ellen Wohl[1], Daniel Cenderelli[2], and Mario Mejia-Navarro[3]

ABSTRACT

Geomorphic hazards associated with flooding along canyon rivers take the form of erosional and depositional modification of the pre-flood channel and valley geometry. Because of the complexity of the variables controlling flood modifications, it is difficult to adequately predict the distribution of channel change in terms of both location and magnitude of erosion and deposition likely to result from a specified flood. However, flood hazard studies become increasingly important with growing human use and occupation of canyon-river environments.

There are two basic components to understanding flood hazards along canyon rivers: predicting flood occurrence, and determining the capacity of the flood to modify the existing channel configuration. At the most fundamental level, the latter factor will depend on the ratio of driving forces to resisting forces. Most of the variables constituting the driving and resisting forces at work during floods along canyon rivers remain poorly quantified or completely unquantified. This is a result, in part, of the high spatial and temporal variability of driving and resisting forces during a flood. In a brief review of the existing literature on channel modifications associated with extreme floods along canyon rivers, we demonstrate that values of hydraulic variables versus channel geometry cannot be used to consistently separate predominantly erosional reaches from predominantly depositional reaches in a data set of several rivers. However, this data set can be used to define a lower threshold value for substantial flood modification of channel boundaries. The threshold is based on a ratio of stream power/drainage area, and takes the form of $y = 21x^{0.36}$, where y is stream power per unit area and x is drainage area. Focusing on two recent glacier-lake outburst floods in Nepal, we demonstrate that channel modification correlates well with the combination of channel-boundary composition and channel geometry. This facilitates a qualitative prediction of potential channel modification by a possible third outburst flood in the region.

---

[1] Department of Earth Resources, Colorado State University, Fort Collins, Colorado 80523.

[2] Assistant Professor, Department of Geology, University of Alabama, Tuscaloosa, Alabama 35487-0338.

[3] Department of Earth Resources and Integrated Decision Support Group, Colorado State University, Fort Collins, Colorado 80523.

KEYWORDS

Geomorphic hazards, floods, canyon rivers, channel change, erosion

1. INTRODUCTION

Geomorphic hazards associated with flooding along canyon rivers take the form of erosional and depositional modification of the pre-flood channel and valley geometry. The magnitude and distribution of such modifications are controlled by numerous variables, including flood magnitude, duration, and rate of rise and recession (Costa and O'Connor, 1995); the shear stress and stream power per unit area relative to the resistance of the channel to erosion (Baker and Costa, 1987); the flow history of the channel; sediment supply, grain size, and resistance to abrasion; and channel geometry. Because of the complexity of these controlling variables, it has been difficult to adequately predict the distribution of geomorphic hazards in terms of both location and magnitude of erosion and deposition likely to result from a specified flood. However, increasing human use and occupation of canyon-river environments creates a need for flood hazard studies.

There are two basic components to quantifying flood hazards along canyon rivers. The first component is predicting flood occurrence. Floods may be generated by snowmelt, various types of rainfall, rain-on-snow, failure of natural and artificial dams, or ice jams. Recurrence intervals for these differing flood types may be a function of meteorological conditions, hillslope or glacial processes within the drainage basin, or dam maintenance. The second component is to determine the capacity of a given flood to modify the existing channel. At the simplest level, this capacity will depend on the ratio of driving forces to resisting forces. The driving forces are the hydraulic and abrasive forces of water and sediment acting on the channel. These forces are governed both by the supply of water and sediment, and by the reach-scale and cross-sectional distribution of energy as a function of channel gradient and geometry. The resisting forces, or the ability of the channel boundaries to remain unchanged by the passage of water and sediment, will be a function of rock-mass strength of bedrock, and of inertial resistance of unconsolidated sediments. The relatively immovable boundaries of rivers located in bedrock canyons cause flow depth, flow velocity, and bed shear stress to increase more rapidly with discharge than in alluvial rivers (Baker, 1984). Resistant boundaries also promote such distinctive hydraulic phenomena as sudden transitions between subcritical and supercritical flow, hydraulic jumps, large-scale turbulence, extreme downstream variability in tractive force (Scott and Gravlee, 1968), and strong flow separation zones (Kieffer et al., 1989).

Most of the variables that control driving and resisting forces operating during floods along canyon rivers remain poorly quantified or completely unquantified. This results in part from the high spatial and temporal variability of these forces. The lack of knowledge regarding site-specific flood processes has lead us to take another approach. We use a literature review on channel changes during large floods along canyon rivers to search for patterns that can be used to predict the potential for flood modification as a function of flood hydraulics in relation to channel characteristics. In this paper, we will briefly review several case studies of channel modifications associated with extreme floods along canyon rivers. The examples include single, extreme floods, and channels dominated by recurring large floods. These examples are used to illustrate general relationships that may aid in predicting patterns of erosion or deposition among various canyon rivers. We then focus on a case study of two recent glacier-lake outburst floods in Nepal. Using a one-dimensional flow simulation model, and GIS (Geographic Information Systems) techniques to analyze the relations between hydraulic variables and mapped patterns of erosion and deposition, we extrapolate from these past floods to predict probable future patterns of erosion and deposition along a neighboring channel that presently has a moraine-dammed lake with a dangerously high water level.

2. EXAMPLES OF FLOOD MODIFICATION ALONG CANYON RIVERS

2.1. Big Thompson River, Colorado, USA

For approximately 30 km of its length, the Big Thompson River of Colorado flows through a bedrock canyon incised into the Precambrian crystalline rocks of the Colorado Front Range. Valley morphology varies greatly downstream (Table 2.1). Surficial deposits of coarse-grained alluvium and colluvium mantle bedrock fairly continuously throughout the canyon. The course of the river is bordered by discontinuous cobble and boulder terraces, and by occasional tributary debris fans or colluvial talus slopes. The debris fans and terraces are widely used for building sites.

An extreme flash flood occurred in Big Thompson Canyon during the night of July 31-August 1, 1976. Up to 30 cm of rainfall fell in the upper

Table 2.1. Summary of selected characteristics of the various floods.

River	Drainage Area (km^2)	Peak Flood Discharge (m^3/s)	"Normal" Flood Discharge (m^3/s)	Average Gradient	Flood Width (m)	Average Unit Stream Power Erosional, Depositional (W/m^2)	Rock Type	Types of channel modification
Big Thompson River, USA	150	880	100	0.010-0.040	30-180	4900 2200	Precambrian crystalline rocks	scour along steep, narrow reaches; deposition in wider reaches
South Branch Potomac River, USA	478	1250	20	0.004-0.006	100-200	300 284	Silurian shales, limestones	erosion as longitudinal grooves, channel widening
Snake River,USA	—[a]	570000	2000	0.0006-0.002	50-150	20000 6000	Permian/Triassic volcanic rocks	slackwater deposits, depositional bars, scabland topography
Burdekin River, Australia	30000	15-30000	1260	0.002	200-700	5400 1700	Carboniferous welded rhyolitic tuffs	slackwater deposits, boulder bars, incised bedrock
Herbert River, Australia	8800	12-15000	1000	0.004	100-200	8000 5400	Permian/Triassic granite	slackwater deposits, boulder bars
Katherine River, Australia	6390	6000	1000	0.0016	15-150	3240 980	Precambrian sandstones	boulder bars
Piccaninny River, Australia	61	100	0	0.004	15-40	284 110	Devonian sandstones	incised bedrock, cobble deposition
Langmoche Khola, Nepal [b]	—[a]	1375-2400	35-205	0.011-0.231	15-220	12500 7900	Precambrian crystalline rocks	depositional bars in wide reaches; erosion of terraces in narrow reaches

a. Drainage area not applicable; damburst flood.

b. All values listed are ranges for the various reaches studied along the 1985 outburst-flood route. The average unit stream power values are for reach 1 (L1), which is approximately 8 km downstream from the outburst-flood source.

portion of the canyon, producing a peak discharge of 880 m^3/s at the canyon mouth (the previous maximum discharge for 88 years of gaging records was 200 m^3/s) (McCain et al., 1979). As a result, many tributary channels scoured to bedrock through thick surficial fills. Heavy rains and undercutting of side slopes by fluvial erosion triggered numerous slope failures that also introduced sediment to the channels. In the main channel, limited scour and widespread deposition occurred along channel reaches more than 80 m wide and having gradients less than 0.02. Reaches 40 to 80 m wide with gradients of 0.02 to 0.04 had extensive scour and deposition, whereas reaches less than 40 m wide and with gradients of 0.02 to 0.04 had intensive scour and limited deposition (Shroba et al., 1979). Scour was most intense on the outside of bends, in steeper reaches, and where the channel was constricted. In such reaches, the outer margins of debris fans from side slopes and tributaries were eroded. Deposition was greatest upstream from constrictions, at gentler or wider reaches, at the insides of bends, and where the flow was impeded by bridges, buildings, or dense vegetation.

The erosional and depositional patterns observed along Big Thompson Canyon demonstrate trends common to many canyon rivers; erosion along steep and narrow channel reaches where a flood may generate high values of stream power, and deposition where a drop in stream power occurs because of channel geometry. The widespread existence of these trends suggests the possibility of developing a threshold to differentiate between predominantly erosional and predominantly depositional reaches along canyon rivers during floods.

2.2. Bedrock Canyons in the Monsoonal Regions of Australia

A series of four bedrock canyons located in the northern, monsoonal region of Australia illustrates various forms of flood modification. The Burdekin River of northeastern Australia (Table 2.1) flows through an 18-km-long canyon. Most cross-sections within the canyon are characterized by a three-part channel morphology: a narrow inner channel flanked by either bedrock benches and fine-grained flood levees, or by gravel or boulder bars and flood levees. The maximum monsoonal floods of 15,000 to 30,000 m^3/s are competent to transport the boulders forming the flood bars, and it is these flows, which probably control the location and morphology of coarse-clast deposits within the gorge (Wohl, 1992a).

The neighboring Herbert River (Table 2.1) flows through a 70-km-long canyon. Channel morphology alternates between wider reaches with extensive boulder deposition, and narrow gorges that are relatively sediment-free except for fine-grained slackwater deposits along the upper channel margins. As on the Burdekin River, only the extreme monsoonal floods of 12,000 to 15,000 m^3/s are competent to fully mobilize the clasts forming the boulder bars (Wohl, 1992b).

The Katherine River (Table 2.1) flows through a 20-km-long canyon in north-central Australia. Similar to the Burdekin Gorge, the Katherine Gorge has a pool-and-boulder-rapid morphology along an inner channel flanked by bedrock benches. Flow competence calculations suggest that the boulder bars are stable except during the most extreme floods, which are on the order of 6,000 m^3/s (Baker and Pickup, 1987).

Piccaninny Creek (Table 2.1) flows through a 19-km-long canyon in northwestern Australia. Channel morphology alternates downstream between broader, lower gradient reaches of sand and gravel deposition, and reaches of exposed and incised bedrock (Wohl, 1993). Bedrock incision occurs as parallel longitudinal grooves, offset potholes, and inner channels with sinuous walls. Piccaninny Creek is an ephemeral channel with periods of short, intense floods of 50 to 100 m^3/s caused by rainfall from dissipating tropical cyclones, interspersed with long periods of no flow. Floods are thus responsible for both sediment transport, and for the bedrock incisional features along the channel.

As in the case of the Big Thompson River, stream power per unit area is substantially higher along erosional reaches of the four Australian channels than along depositional reaches. Deposition occurs in the form of boulder bars composed of clasts that are only mobilized during the largest floods, whereas erosion occurs as removal of alluvial veneer, and incision into bedrock.

2.3. South Branch Potomac River, West Virginia, USA

The South Branch of the Potomac River (Table 2.1) flows through a 20-km canyon before its junction with the North Fork of the South Branch. Widespread rainfall on November 4-5, 1985, produced the largest recorded flood in the South Branch Potomac River basin. A peak discharge of 1,250 m^3/s caused substantial erosion and deposition. Erosional features included longitudinal grooves, scour marks, widened channels, stripped flood plains, chutes, and anastomosing erosion channels (Miller and Parkinson, 1993). Depositional features included channel gravel bars, gravel splays and sheets, and backwater deposits. Erosion was classified at four levels, from minimal to catastrophic (Miller and Parkinson, 1993). Catastrophic erosion occurred primarily downstream from valley constrictions or at the mouths of bedrock canyons. However, Miller (1990) and Miller and Parkinson (1993) concluded that the severity of erosion along a channel is heavily influenced by local variations in channel gradient, valley width, discharge, and erosional resistance of the valley floor, and is not readily explained by a single parameter such as unit stream power. The trends of higher stream power along erosional reaches, noted in the preceding flood examples, seem to be complicated along the South Branch Potomac River by the influence of local variations. This suggests that the rate of downstream change of hydraulic and channel morphologic variables may be as important as absolute values of these variables for explaining flood modification along some canyon rivers.

2.4. The Bonneville Flood, Idaho, USA

Approximately 14,500 years ago, Pleistocene Lake Bonneville flooded the divide between the closed Bonneville Basin and the watershed of the Snake River, producing a flood with a peak discharge of approximately 0.6 million m^3/s through Hells Canyon (O'Connor, 1993) (Table 2.1). The canyon consists of a series of straight, fault-controlled segments separated by abrupt bends. Major accumulations of flood deposits along the canyon bottom are restricted to wider locations; these include bar complexes, angular and poorly sorted gravel at the flow margins, and suspended sediments deposited in back-flooded tributaries. Erosional features, including scabland topography and cataracts, were associated with narrow reaches of higher stream power per unit area. The Bonneville Flood clearly had a major impact on channel and valley morphology along the Snake River, for features produced by the flood remain clearly discernible several millennia later.

The Bonneville Flood features along the Snake River canyon seem to display a clearer distinction between predominantly erosional or depositional reaches on the basis of unit stream power than do the South Branch Potomac River flood features. In contrast to the South Branch Potomac River, the Snake River, Big Thompson River, and Australian river canyons have resistant bedrock, relatively low width/depth ratios, climates (semiarid or seasonal tropical) not conducive to chemical weathering or the production of thick soils and abundant fine-grained alluvium, and large ratios of extreme flood discharges to mean annual flood discharge. These factors may produce more easily discerned reach-scale variations between erosional and depositional reaches associated with fluctuations in unit stream power.

2.5. Glacier-lake Outburst Floods in the Dudh Kosi Drainage, Nepal

Two recent glacier-lake outburst floods (GLOFs) have occurred along upper tributary channels of the Dudh Kosi drainage basin in central northern Nepal (Fig. 2.1). In 1977 an ice-cored moraine dam at 5120 m elevation on the Nare

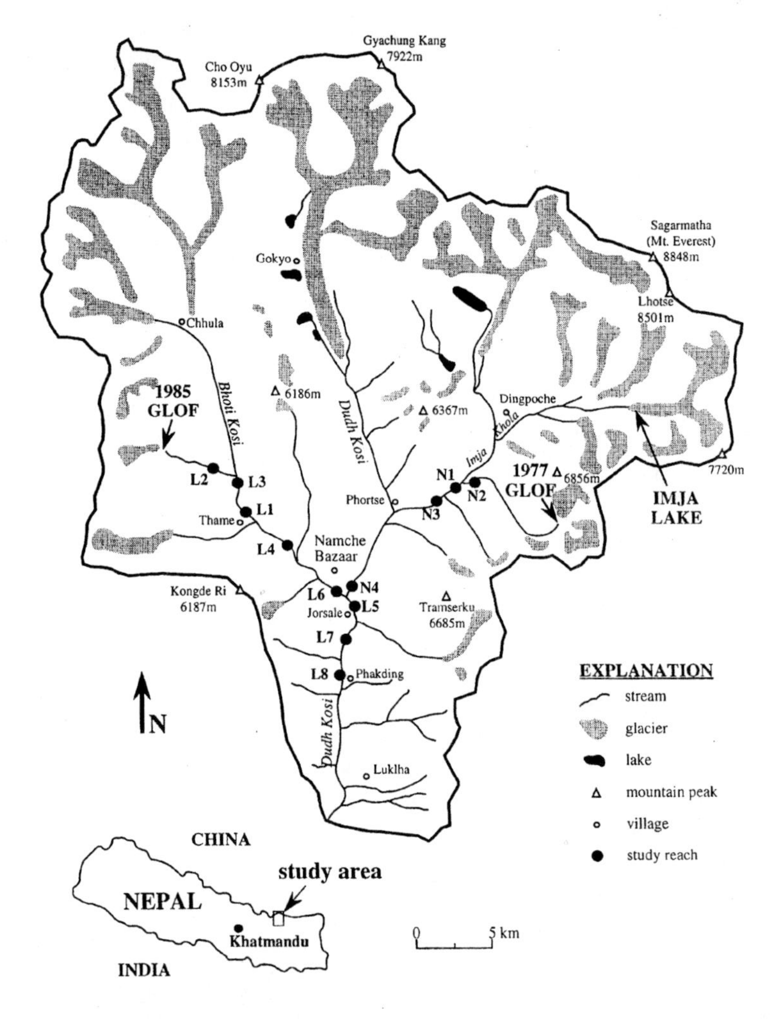

Fig. 2.1. **Location map of the Nepalese glacier-lake outburst floods (GLOFs). The 1977 GLOF (reaches designated by N) occurred on the Nare Khola; the 1985 GLOF (reaches designated by L) occurred on Langmoche Khola.**

Khola failed, draining a lake that contained approximately 500,000 m^3 of water (Fushimi et al., 1985). The resulting flood peaked at approximately 1,850 m^3/s. Peak discharge lasted for about an hour, and the flood lasted 6-8 hours (Fushimi et al., 1985; Zimmermann et al., 1986). In 1985 a moraine-dammed lake holding 5,000,000 m^3 at 4400 m elevation failed on the Langmoche Khola (Vuichard and Zimmermann, 1987). The resulting flood had a peak discharge of 2,250 to 2,400 m^3/s along the upper 10 km of the flood route. The peak discharge lasted about an hour and the duration of the entire flood was 6 to 8 hours. The channels above 3,400 m elevation in the Dudh Kosi drainage basin are incised into glaciofluvial terraces.

Both channels are characterized by downstream alternations between narrow gneissic bedrock gorges (15 to 60 m wide, gradients > 0.07) with minimal deposition, and broader (150 to 200 m wide, average gradients 0.02-0.06) reaches with extensive boulder and cobble deposits from these large floods. Flows subsequent to the glacier-outburst floods have been contained within a relatively small (10 to 15 m wide, 1 m deep) channel incised into the flood deposits in the depositional reaches, and it is apparent that normal annual high flows, on the order of 35 to 205 m^3/s, are incapable of reworking the flood depositional features. Surveys of flood stage limits during a 1995 field season in the region allowed us to model flood discharge and hydraulics along the channel. Erosional reaches tended to have higher values of velocity, shear stress, and unit stream power, as a result of the confined cross sections and high bed gradients. Depositional reaches had lower values of these hydraulic variables. Estimates of hydraulic variables were obtained by routing the GLOF discharges along multiple surveyed reaches using the one-dimensional step-backwater model HEC-RAS (Hydrologic Engineering Center, 1995). In this paper we will focus on one reach near the outburst source for each of the two floods; estimates of hydraulic variables are presented in Table 2.2. Although average hydraulic variables differ between the two types of reaches, and threshold values may be defined for each hydraulic variable, maximum hydraulic values for depositional reaches overlap with values for erosional reaches (Fig. 2.2).

Erosion primarily occurred as valley widening, with previously stable glaciofluvial terraces being partially removed during the flood flows. Deposition consisted mostly of boulder bars up to 2.5 m thick. At reach 1 of the 1985 GLOF, large bar complexes were deposited in wider reaches where, in general, values of unit stream power were less than 10,000 W/m^2 (Table 2.2). The bar complexes consist of multiple, linear and curvilinear longitudinal

Table 2.2. Summary of hydraulic data for predominantly erosional and depositional reaches along Reach 1 of the 1985 outburst-flood routes.

Process -Variable	Shear Stress (N/m^2)	Stream Power per Unit Area (W/m^2)	Velocity (m/s)
	EROSIONAL		
-Threshold	> 2,500	> 10,000	> 4.0
-Average	2,700	12,500	4.5
-Maximum	3,500	17,100	5.2
	DEPOSITIONAL		
-Threshold	< 2,500	< 10,000	< 4.0
-Average	2,100	9,900	3.8
-Maximum	2,500	14,000	4.1

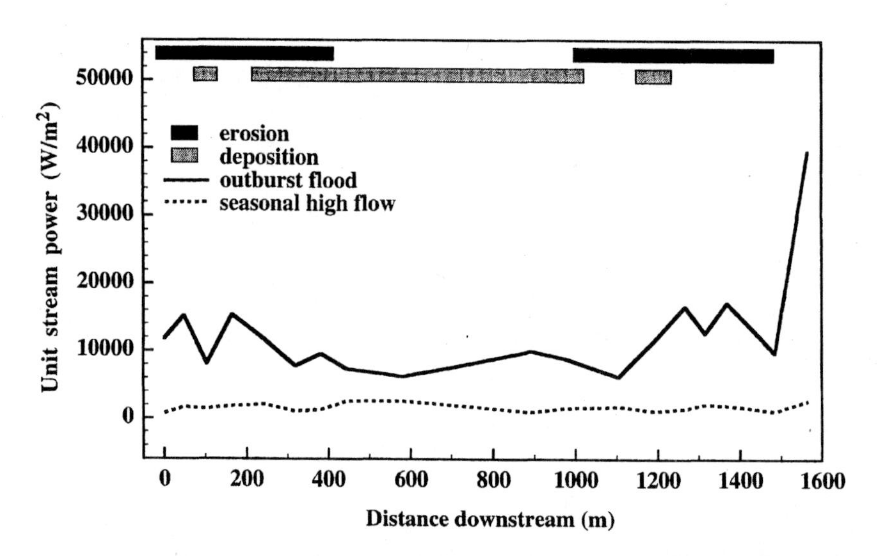

Fig. 2.2. Spatial variation in stream power per unit area along reach 1 for the 1985 outburst flood and seasonal high flow. The distribution of erosional and depositional features from the 1985 outburst flood is also delineated.

bars that are parallel or subparallel to valley alignment. The bar complexes are clast-supported, poorly to moderately imbricated, moderately to very poorly sorted, and composed primarily of cobbles and boulders.

The geomorphic effects of the GLOFs decrease downstream from the outburst point (Fig. 2.3). We attribute this decrease to three factors: (1) Downstream attenuation of the peak flood discharge, and thus less flood power: the 1985 flood had attenuated by approximately 43 percent 27 km downstream from the lake, peak discharge at Phakding (Fig. 2.1, site location L 8) was 1,375 m^3/s; (2) a lower ratio between the GLOF discharges and "normal" monsoonal flood discharges as a result of larger monsoonal floods at lower elevations; and (3) a greater frequency of GLOFs in the lower channel reaches, which have more tributary channels with headwater glaciers.

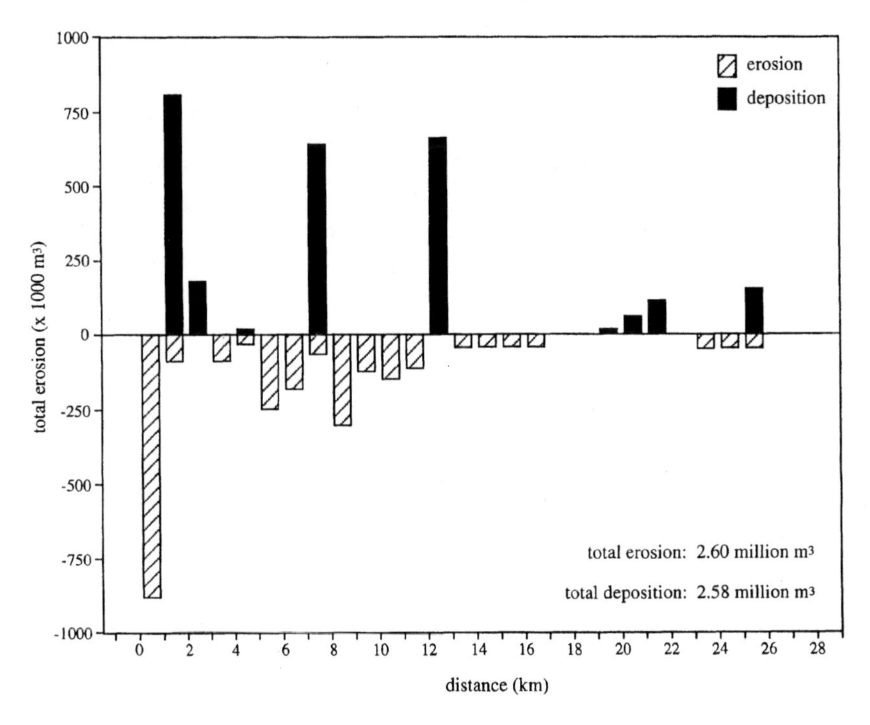

Fig. 2.3. Distribution of eroded and deposited sediment along the route of the 1985 Langmoche Khola outburst-flood, Nepal (modified from Vuichard and Zimmermann, 1987).

3. PATTERNS OF FLOOD MODIFICATION ALONG CANYON RIVERS: SUMMARY OF CASE STUDIES

Bedrock canyons seem to be generally characterized by downstream alternations between relatively wide and low gradient reaches, and relatively narrow and steeper reaches. These variations may be controlled by bedrock lithology or structure, or they may be seemingly unrelated to substrate changes. A given discharge flowing downstream between such varied sections will have higher values of velocity, shear stress, and stream power per unit area in the confined, steeper reaches than in the broader reaches, and will thus be more competent to transport sediment and erode channel boundaries in the confined reaches. Numerous studies have demonstrated a strong correlation between values of stream power and erosional/ depositional modification of channel boundaries (e.g., Baker, 1973; O'Connor et al., 1986; Baker and Pickup, 1987; Wohl, 1992a, 1993; O'Connor, 1993). Most of these studies have focused on depositional features of coarse sediments because boulder size and distribution may be quantitatively related to critical tractive force to predict the magnitude of flow necessary to transport boulders (e.g., O'Connor, 1993). The specific morphologies of depositional features will depend on cross-sectional to reach-scale distribution of flow power, as well as on sources (Scott and Gravlee, 1968) and grain-size of coarse sediment. Erosional features have been more qualitatively related to the potential for cavitation (e.g., Baker and Pickup, 1987; O'Connor, 1993), but quantification of total erosion of cohesive channel boundaries during a single extreme flood has not yet been attempted. Much of the difficulty results from uncertainty about the erosional threshold of bedrock-lined channels in narrow gorges. Erosion of bedrock may occur via corrosion (chemical weathering), abrasion, cavitation, or plucking. Relatively little is known of how these processes operate along natural channels, and few event-based rates of erosion have been published.

Description of the erosional resistance of unconsolidated alluvium forming channel boundaries is equally challenging. Alluvial boundaries will be greatly influenced by vertical and horizontal gradations in grain size; although the meter-diameter boulders forming part of a channel bank may have a high entrainment threshold, the erosion of the 10-cm-diameter cobble layer underlying them may cause bank erosion well below that entrainment threshold. Erosional resistance of alluvial boundaries thus includes a substantial stochastic component that will make precise quantification difficult.

Until we understand more about the mechanics of rock erosion by flowing water, our predictions of flood erosional/depositional patterns along canyon rivers will be limited to extrapolations from comparisons of existing data. As an example of this approach, Fig. 3.1 shows three representative plots of the flood data discussed in this paper. We plotted a variety of potential relations: width/depth ratio vs. unit stream power/discharge; w/d ratio vs. drainage area; w/d ratio vs. discharge; w/d ratio vs. gradient; unit stream power vs. gradient; unit stream power vs. discharge; discharge/unit stream power vs. gradient; and unit stream power/discharge vs. gradient. When these data are plotted, there is a clear distinction between erosional and depositional reaches for individual channels, but there is no threshold value of stream power, gradient, or width/depth ratio that separates erosional and depositional reaches on all the channels (Fig. 3.1). These results suggest the importance of the **ratio** of driving forces to resisting forces, and the rate of downstream change in channel morphology and flow hydraulic variables, in controlling channel modification during a flood along a canyon river. Until we can better quantify resisting forces, in particular, statistical analyses of the type of data examined here are likely to be inconclusive. Field data quantifying with any degree of accuracy relationships between total erosion (or deposition) and a combination of geometric variables such as width/depth ratio and gradient are also lacking.

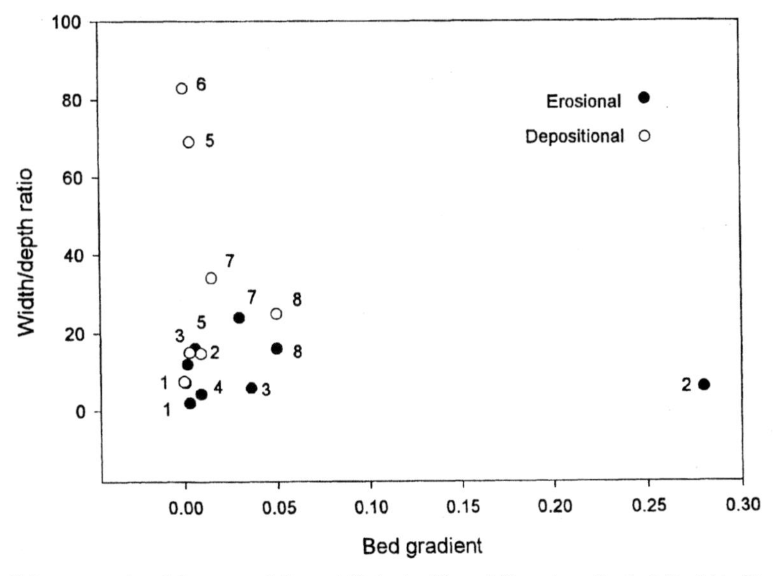

For all figures, numbered data are as follows: 1 Katherine River, 2 Piccaninny Creek, 3 Burdekin River, 4 Herbert River, 5 South Branch Potomac River, 6 Snake River (Bonneville Flood), 7 Big Thompson River, 8 Langmoche Khola.

Fig. 3.1(a). Width/depth ratio vs. bed gradient for the flood case studies summarized in Table 2.1.

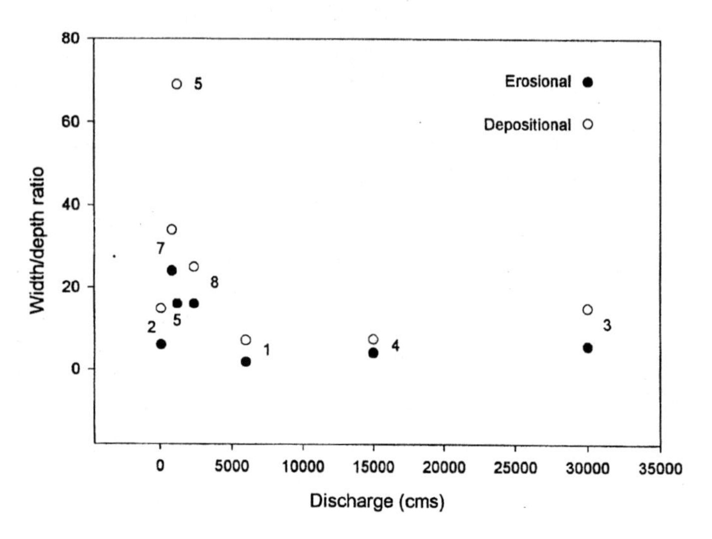

Fig. 3.1(b). Width/depth ratio vs. discharge. (The Bonneville Flood was not plotted because the discharge is so much larger than those of the other floods.).

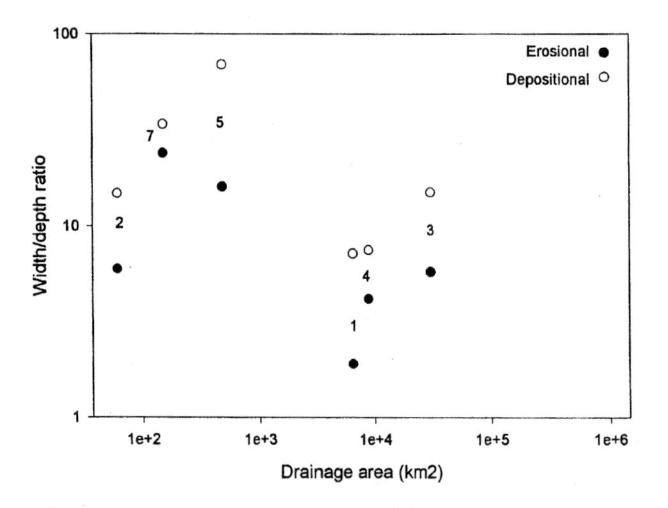

Fig. 3.1(c). Width/depth ratio vs. drainage area for the rainfall floods summarized in Table 2.1.

In a similar exercise focused on low gradient, alluvial channels, Magilligan (1992) used data from five floods to plot stream power per unit area against drainage area. He identified a minimum threshold of 300 W/m^2 for substantial erosional modification. A similar plot using the canyon-river data suggests a threshold that is a function of drainage area (Fig. 3.2). Although it might initially appear that this plot of a dependent variable (unit stream power) against an independent variable (drainage area) would necessarily produce a threshold, the lack of linear correlation between unit stream power and either drainage area or discharge argues against an inevitable threshold. Unit stream power is strongly controlled by both discharge and channel geometry, and some of the world's largest drainages have quite low values of unit stream power (Rathburn, 1993). In addition, floods along canyon rivers are characterized by extreme downstream variability of unit stream power (e.g., Wohl, 1992a, b; Wohl et al., 1994) (Fig. 3.3).

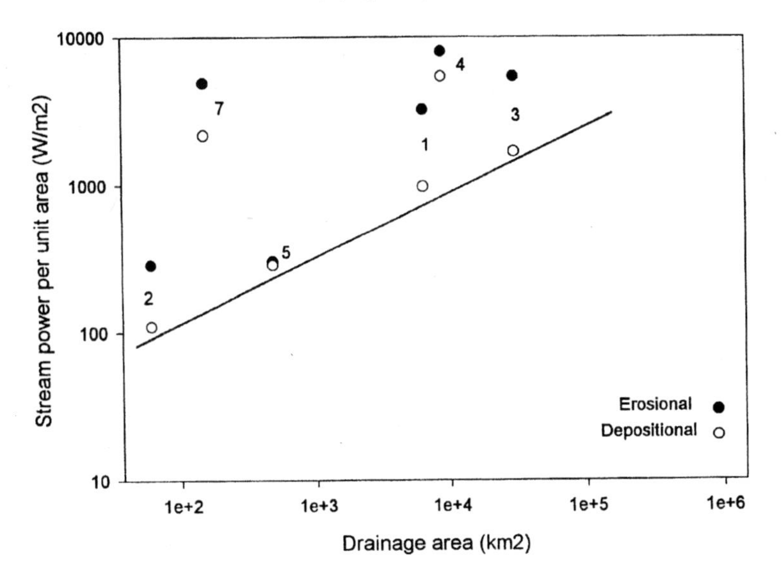

Fig. 3.2. Relation between stream power per unit area vs. drainage area for the rainfall-flood case studies summarized in Table 2.1. Numbered data are as listed in Fig. 3.1. The limiting threshold for substantial geomorphic modification of the channel boundaries is shown as a solid line at the base of the plotted data. Equation for the line is $y = 21\ x^{0.36}$.

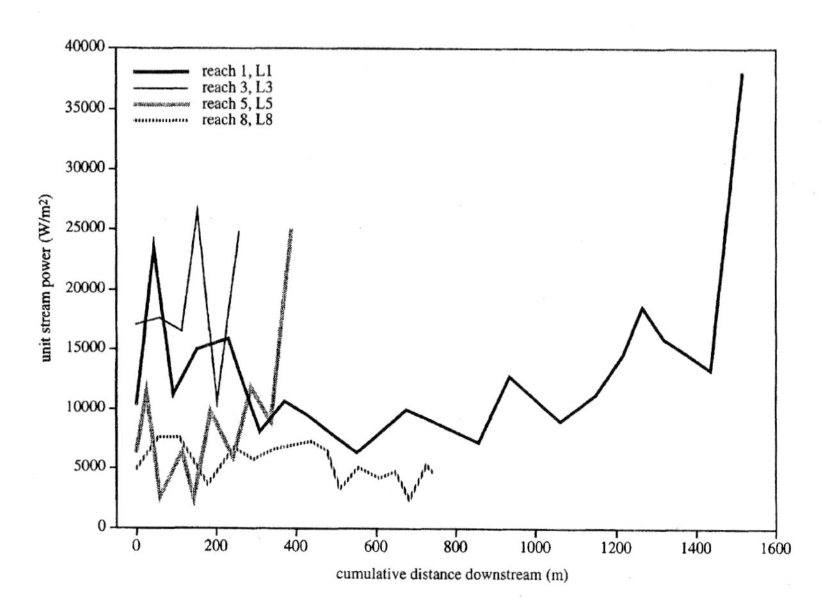

Fig. 3.3. Stream power per unit area vs. distance downstream for four reaches along the 1985 Langmoche outburst-flood in Nepal.

The data in Fig. 3.2 define a threshold above which substantial channel-boundary modification is likely to occur during rainfall floods. This threshold is only moderately useful for hazard prediction, however, because it does not distinguish between erosional and depositional modification. Another approach is to compare the average unit stream power in depositional reaches to the stream power in erosional reaches for the channels summarized in Table 2.1 (Fig. 3.4). Such a plot indicates that the ratio of erosional to depositional stream power is consistently between 1.5 and 3.5. (Erosional threshold may be similar to depositional threshold when valley trend is oblique to dominant flow direction (Miller, 1995; Cenderelli and Cluer, in press)). The variability of this ratio may reflect the numerous factors, which influence erosional thresholds (such as packing, cohesion, and imbrication for unconsolidated boundaries, and joints, weathering, and bedding for bedrock), whereas depositional thresholds are mainly a function of clast size. In any case, the generalized relation of 1.5 to 3.5 times higher erosional stream power may be useful in predicting the magnitude of erosional and depositional channel modification resulting from a flood. However, we examined another approach to flood-hazard prediction in an attempt to obtain more site-specific predictive capabilities. This approach is illustrated with the data from the two Nepalese GLOFs.

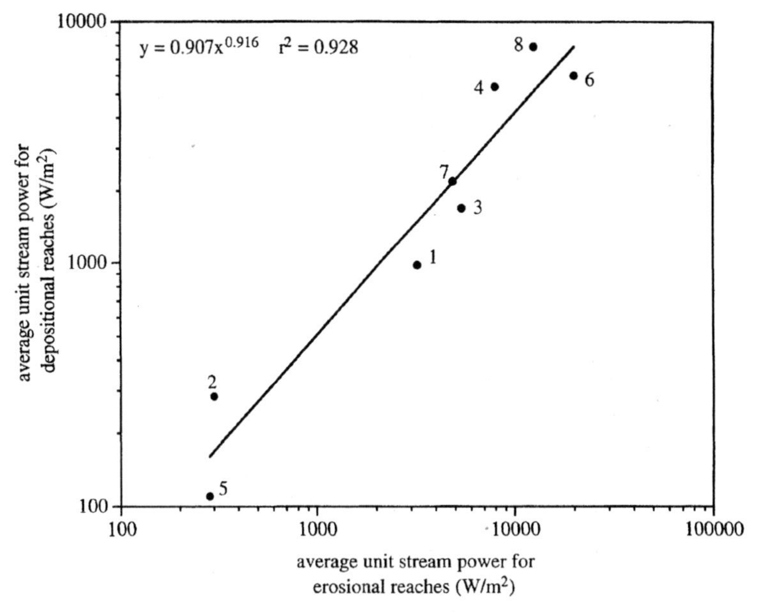

Fig. 3.4. Average values of stream power per unit area for depositional and erosional reaches along the channels summarized in Table 2.1. Numbered data are as listed in Fig. 3.1.

4. PREDICTION OF FLOOD HAZARDS

The analysis summarized above suggests that one means of predicting channel modification during an extreme flood is to use data from past floods along very similar channels, where such data are available, to delineate zones of probable flood erosion and deposition. As an illustration of such an approach, we examined the spatial distribution of erosional, depositional, and minimally modified channel reaches along the 1977 and 1985 GLOF routes in Nepal, to predict potential erosional and depositional patterns from a possible future outburst flood from Imja Lake. This glacial lake lies along a third major tributary in the upper Dudh Kosi basin (Fig. 2.1). The lake has been monitored by the Nepalese government and foreign scientists for several years, and artificial drainage of the lake has been proposed to alleviate the hazard of an outburst flood (Watanabe et al., 1994).

We created one data set by dividing the 1977 and 1985 outburst-flood routes into channel reaches that had substantial deposition during the flood, reaches that had substantial erosion, and primarily bedrock reaches that showed minimal flood modification. We created a second data set by subdividing channel longitudinal profiles into reaches of consistent gradient. The third data set consisted of a division between channel reaches formed predominantly in bedrock, and those formed predominantly in Quaternary alluvium. All three data sets are based on 1:50,000 scale topographic maps with a contour interval of 40 m. We then analyzed these three data sets using both statistical analyses and the GIS software ARC/INFO (ESRI, 1995) and GRASS (USACERL, 1993).

Spatial comparisons of the characteristics of the 1977 and 1985 outburst-flood routes indicate that predominantly erosional or depositional reaches correlate strongly with alluvial channel boundaries, whereas minimally modified reaches are found primarily in bedrock (Fig. 4.1, Table 4.1).

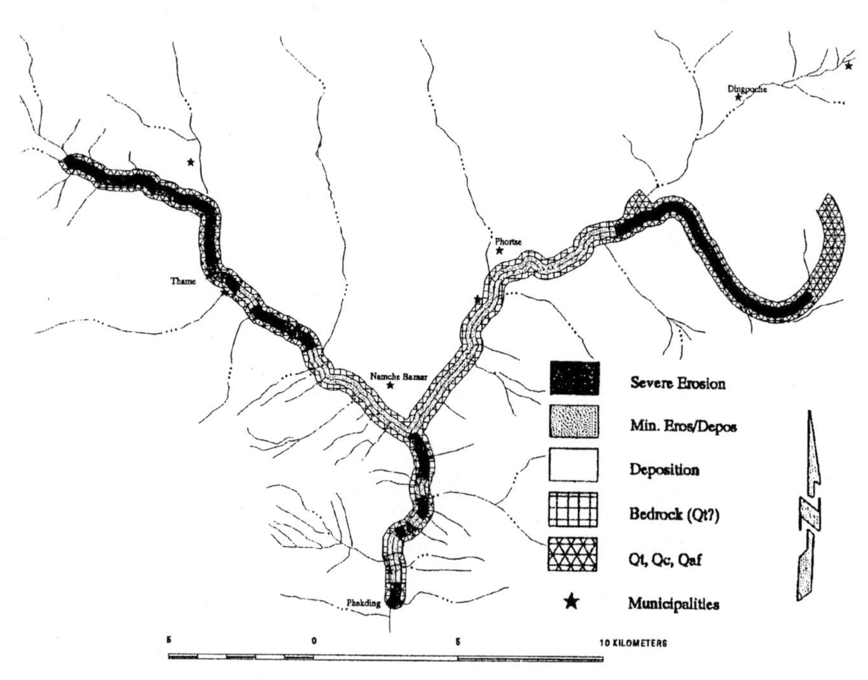

Fig. 4.1. Map combining channel-boundary composition and predominant form of channel modification during the 1977 and 1985 glacier-lake outburst-floods in Nepal.

Table 4.1. Summary of reach characteristics for the 1977 and 1985 outburst floods.

							Boundary Conditions	
Outburst flood -Geomophic Process	No. of Reaches	Total Reach Length (m)	Average Gradient	Gradient std. dev.	Maximum Gradient	Minimum Gradient	Percent Sediment	Percent Bedrock
1977								
-Depositional	5	3836	0.0644	0.0246	0.0970	0.0390	100	0
-Erosional	10	5428	0.1912	0.0565	0.2630	0.1000	100	0
-Transport	15	11,395	0.0887	0.0369	0.1500	0.0310	0	100
1985								
-Depositional	16	8,884	0.0458	0.0315	0.1105	0.0108	95	5
-Erosional	18	10,224	0.0740	0.0465	0.1942	0.0221	87	13
-Transport	16	7,988	0.1054	0.0609	0.2312	0.0362	0	100

An analysis of variance indicates that the mean gradients of depositional, erosional, and minimally modified reaches along the outburst flood routes are statistically different ($\alpha = 0.05$). A multiple comparison of means using the Student-Newman-Keuls (SNK) method indicates that the mean gradients of the three reaches along the 1985 flood route are statistically different ($\alpha = 0.10$) from one another (Fig. 4.2a). For the 1977 flood route, the

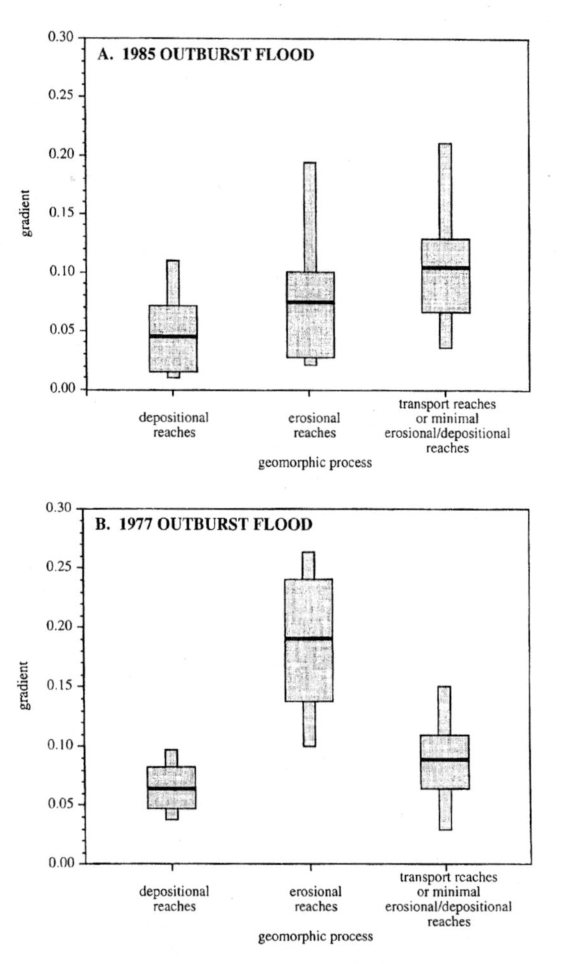

Fig. 4.2. Box plots comparing gradient distributions between the three geomorphic process reaches identified along (a) the 1985 outburst-flood route (Langmoche Khola), and (b) the 1977 outburst-flood route (Nare Khola).

SNK results indicate that the mean gradient of erosional reaches is statistically different from the mean gradient for depositional and transport reaches (α = 0.05 or α = 0.10), but the mean gradients of depositional and transport reaches are not statistically different (Fig. 4.2b).

In summary, our analyses of the 1977 and 1985 outburst-floods indicate that: (1) channel reaches that were not substantially modified by the flood had bedrock boundaries which limited erosion, and steep, narrow geometries that limited deposition; (2) predominantly erosional reaches were steep and narrow, and had unconsolidated channel boundaries from which sediment was readily entrained; and (3) depositional reaches tended to be less steep and wider than both erosional and transport reaches. The hydraulic thresholds of Table 2.2, when combined with channel-boundary composition, could be used to route a flood discharge along the Imja Khola and to predict channel reaches likely to be substantially modified as the result of a flood.

The valley morphology of the Imja Khola is very similar to that of the Langmoche Khola (Fig. 4.3). We, therefore, expect the erosional and

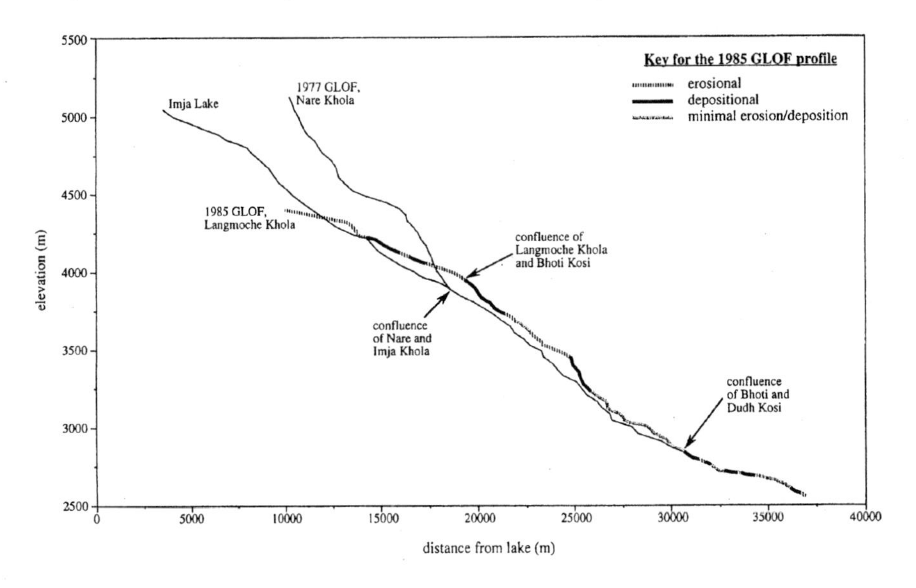

Fig. 4.3. Longitudinal profiles of Imja Khola, Nare Khola, and Langmoche Khola. The type of channel modification that occurred along the 1985 outburst flood route is illustrated along the Langmoche Khola longitudinal profile.

depositional patterns along the Imja outburst-flood route to mimic those along the 1985 Langmoche flood route. Along the upper 10 to 15 km of the river, substantial change may occur where the valley is steep, narrow, and bounded by unconsolidated sediment, and where the valley is wide and less steep. The lower Dudh Kosi drainage (below Namche Bazaar) has experienced two outburst-floods during the past twenty years. Thus, the geomorphic effects from a potential Imja Lake outburst-flood along this section of the river would be minimal because, (1) the channel boundaries have already adjusted to the hydraulics of an extreme flood, (2) attenuation of an Imja Lake outburst-flood would presumably be similar to that of the 1985 Langmoche flood, further reducing the flood's geomorphic effectiveness in the lower basin, and (3) monsoonal precipitation more strongly influences flood hydrology at lower elevations, so that the ratio of outburst-flood magnitude to seasonal high-flow magnitude would be lower in the lower basin than in the upper basin. These results suggest that hazard mitigation strategies for a potential Imja Khola flood should focus on the upper 10 to 15 km of the channel.

5. CONCLUSIONS

We have focused on geomorphic flood hazards in the form of erosional or depositional modification of channel boundaries during a flood. Steep, narrow channel reaches with resistant bedrock boundaries may have high values of flow velocity or shear stress during a flood, and thus also be quite hazardous, without undergoing substantial channel-boundary modification. With the exception of bridges and dams, most structures built along canyon rivers are likely to be along wider channel reaches with some alluvial deposition, and we have, therefore, focused on these reaches.

As demonstrated in the review of case studies (Table 2.1), hydraulic variables and channel geometry are likely to vary substantially between predominantly erosional and predominantly depositional reaches along a single channel. However, threshold values that consistently separate erosional from depositional reaches for a variety of canyon rivers are not yet discernible (Fig. 3.1). Although more sophisticated statistical techniques may reveal such thresholds, the use of these techniques will only be justified when more precisely quantified, reach-scale field data are available. We attribute the present absence of a consistent erosional/depositional threshold to the role of several other factors in controlling channel modification during a flood along a canyon river. These factors include: (1) the importance of the ratio of driving to resisting forces; (2) the rate of downstream change in channel morphology and flow hydraulic variables; (3) the history of recent large floods; (4) the ratio of extreme floods to more frequent floods; and (5) the high spatial and temporal variability of flood hydraulics and channel-boundary resistance. We can define a lower threshold value for substantial modification of channel boundaries during rainfall floods; this value is a function of stream power per unit area and drainage area and takes the form of $y=21x^{0.36}$, where y is stream power per unit area and x is drainage area (Fig. 3.2).

For more detailed prediction of spatial patterns of erosion and deposition during a flood, it is probably best to use similar channels and types of floods (i.e., outburst versus rainfall) as analogs. Analyzing two recent glacier-lake outburst floods in Nepal, we found that erosional and depositional modification correlated well with the combination of channel-boundary composition and channel geometry. This allowed us to qualitatively predict potential channel modification resulting from another outburst-flood that may occur in the region. Quantitative prediction of channel modification would require a detailed survey of the predicted outburst-flood route.

Our results suggest that future studies of floods along canyon rivers should devote more attention to quantifying the ratio of driving to resisting forces, and particularly to quantifying channel-boundary resistance to erosion. Also, the potential importance of the rates of downstream change in hydraulic and morphologic variables, as these influence channel modification, has seldom been addressed in previous studies. Although at present we can make only general, qualitative predictions regarding channel modification during a flood, the use of geomorphic records of past flood modifications within a given drainage region may be most helpful in predicting future flood modifications in that region.

6. ACKNOWLEDGMENTS

Ian Rutherfurd and Deborah Anthony provided helpful reviews of this manuscript.

7. REFERENCES

Baker, V.R. 1973. Paleohydrology and sedimentology of Lake Missoula flooding in eastern Washington. Geological Society of America Special Paper 144, 79 pp.

Baker, V.R. and Costa, J.E. 1987. Flood power. In, L. Mayer and D. Nash, eds., Catastrophic flooding. Allen and Unwin, Boston, p. 1-21.

Baker, V.R. and Pickup, G. 1987. Flood geomorphology of the Katherine Gorge, Northern Territory, Australia. Geological Society of America Bulletin 98: 635-646.

Costa, J.E. and O'Connor, J.E. 1995. Geomorphically effective floods. In, J.E. Costa, A.J. Miller, K.W. Potter, and P.R. Wilcock, eds., Natural and anthropogenic influences in fluvial geomorphology. American Geophysical Union, Washington, D.C., p. 45-56.

ESRI (Environmental Systems Research Institute, Inc.). 1995. ARC/INFO software, version 6.1. Redland, CA.

Fushimi, H., Ikegami, K. and Higuchi, K. 1985. Nepal case study: catastrophic floods. In, G.J. Young, ed., Techniques for prediction of runoff from glacierized areas. IAHS Publication no. 149, p. 125-130.

Hydrologic Engineering Center. 1995. HEC-RAS, River analysis system, version 1.1. U.S. Army Corps of Engineers, Davis, CA.

Kieffer, S.W., Graf, J.B. and Schmidt, J.C. 1989. Hydraulics and sediment transport of the Colorado River. In, D.P. Elston, G.H. Billingsley, and R.A. Young, eds., Geology of Grand Canyon, northern Arizona. American Geophysical Union, Washington, D.C., p. 48-66.

Magilligan, F.J. 1992. Thresholds and the spatial variability of flood power during extreme floods. Geomorphology 5: 373-390.

McCain, J.F., Hoxit, L.R., Maddox, R.A., Chappell, C.F., and Caracena, F. 1979. Storm and flood of July 31-August 1, 1976, in the Big Thompson River and Cache la Poudre River basins, Larimer and Weld Counties, Colorado. Part A. Meteorology and hydrology in the Big Thompson River and Cache la Poudre River basins. U.S. Geological Survey Professional Paper 1115, p. 1-85.

Miller, A.J. 1990. Flood hydrology and geomorphic effectiveness in the central Appalachians. Earth Surface Processes and Landforms 15: 119-134.

Miller, A.J. and Parkinson, D.J. 1993. Flood hydrology and geomorphic effects on river channels and floodplains: The flood of November 4-5, 1985, in the South Branch Potomac River basin of West Virginia. In, R.B. Jacobson, ed., Geomorphic studies of the storm and flood of November 3-5, 1985, in the upper Potomac and Cheat River basins in West Virginia and Virginia. U.S. Geological Survey Bulletin 1981, chapter E, 96 pp.

O'Connor, J.E. 1993. Hydrology, hydraulics, and geomorphology of the Bonneville Flood. Geological Society of America Special Paper 274, 83 pp.

O'Connor, J.E., Webb, R.H., and Baker, V.R. 1986. Paleohydrology of pool-and-riffle pattern development: Boulder Creek, Utah. Geological Society of America Bulletin 97: 410-420.

Rathburn, S.L. 1993. Pleistocene cataclysmic flooding along the Big Lost River, east central Idaho. Geomorphology 8: 305-319.

Scott, K.M. and Gravlee, G.C., Jr. 1968. Flood surge on the Rubicon River, California - hydrology, hydraulics and boulder transport. U.S. Geological Survey Professional Paper 422-M, 38 pp.

Shroba, R.R., Schmidt, P.W., Crosby, E.J., Hansen, W.R. and Soule, J.M. 1979. Storm and flood of July 31-August 1, 1976, in the Big Thompson River and Cache la Poudre River basins, Larimer and Weld Counties, Colorado. Part B. Geologic and geomorphic effects in the Big Thompson canyon area, Larimer County. U.S. Geological Survey Professional Paper 1115, p. 87-152.

USACERL (U.S. Army Construction Engineering Reseach Laboratory). 1993. Geographic Resources Analysis Support System (GRASS) software, version 4.1. Champaign, Illinois.

Vuichard, D. and Zimmermann, M. 1987. The 1985 catastrophic drainage of a moraine-dammed lake, Khumbu Himal: Cause and consequences. Mountain Research and Development 7: 91-110.

Watanabe, T., Ives, J.D. and Hammond, J.E. 1994. Rapid growth of a glacial lake in Khumbu Himal, Himalaya: Prospects for a catastrophic flood. Mountain Research and Development 14: 329-340.

Wohl, E.E. 1992a. Bedrock benches and boulder bars: Floods in the Burdekin Gorge of Australia. Geological Society of America Bulletin 104: 770-778.

Wohl, E.E. 1992b. Gradient irregularity in the Herbert Gorge of northeastern Australia. Earth Surface Processes and Landforms 17: 69-84.

Wohl, E.E. 1993. Bedrock channel incision along Piccaninny Creek, Australia. Journal of Geology 101: 749-761.

Wohl, E.E., Greenbaum, N., Schick, A.P. and Baker, V.R. 1994. Controls on bedrock channel incision along Nahal Paran, Israel. Earth Surface Processes and Landforms 19: 1-13.

Zimmermann, M., Bichsel, M. and Kienholz, H. 1986. Mountain hazards mapping in the Khumbu Himal, Nepal, with prototype map, scale 1:50,000. Mountain Research and Development 6: 29-40.

Section 3

ANTHROPOGENIC EFFECTS ON FLUVIAL SYSTEMS

THE EFFECTS OF FLOW AUGMENTATION ON CHANNEL GEOMETRY OF THE UNCOMPAHGRE RIVER

R. A. Mussetter and M. D. Harvey[1]

ABSTRACT

Geomorphic, hydraulic and channel stability conditions in the approximately 43-km-long reach of the Uncompahgre River between Montrose and Delta, Colorado, were analyzed to evaluate the potential physical effects of increased flows associated with the proposed AB Lateral Hydropower Project. Incipient motion analyses based on one-dimensional (1-D) hydraulic modeling and sampled bed material-size gradations indicate that the channel bed will armor under the higher flows, and is, therefore, unlikely to degrade significantly. The channel will respond to the higher flows by increasing width by approximately 26 percent (8 m). Evaluation of an approximately 3-km-long reach of the river located 8.3 km upstream from Montrose, that has been subjected to a range of flows similar to the proposed project flows, provided verification for the changes that were predicted for the project reach. The width adjustment will occur rapidly during the initial years of operation, reducing asymptotically with time. The bank erosion rate is expected to decrease to less than 25 percent of the initial rate after about 5 years of operation and to about 11 percent of the initial rate after 10 years of operation. Erosion associated with the widening will increase the existing sediment load in the Gunnison River by less than 3.5 percent during the initial adjustment period. The elevated sediment load is very small in comparison with the historic loads in the Gunnison River prior to construction of upstream dams (Aspenall project).

KEYWORDS

Channel stability, geomorphology, hydropower, erosion, flow augmentation.

---

[1] Principal Engineer and Principal Geomorphologist, Mussetter Engineering, Inc., P.O. Box 270785, Fort Collins, Colorado 80527.

1. INTRODUCTION

Transbasin diversion of flow into an alluvial channel can cause significant morphological changes, including an increase in channel dimensions, modification of the planform and gradient and changes to the gradation of the boundary materials (Kellerhals et al., 1979). The magnitude of these changes, and their specific nature, is dependent on the amount and timing of the imported flows in relation to the natural flows, the amount of sediment carried by the imported flows, and the geomorphic characteristics of the receiving stream. Lane (1955) and Schumm (1977), for example, predict that bed degradation is likely to occur if relatively sediment-free flows are imported to a channel whose bed material is mobilized. In coarse-grained channels, relatively sediment-free, imported flows that are less than the bed-mobilizing flows are unlikely to cause bed degradation, even if those flows are of long duration. If the banks of the coarse-grained channel are composed of erodible materials, however, sustained flows of moderately high magnitude may cause channel widening (Thorne and Tovey, 1981). Reliable estimates of the rate and magnitude of the expected changes can be made by identifying the geomorphic characteristics that control the dynamics of the receiving stream, and then applying computational techniques that mathematically describe the hydraulic and sediment transport processes that are related to those characteristics (Mussetter Engineering, Inc., 1995).

The proposed AB Lateral Hydropower project will divert water from the Gunnison River into the Uncompahgre River through the Gunnison Tunnel, which has been in place since about 1913. Historically, the Gunnison Tunnel has been operated during the irrigation season (typically April through October), delivering up to 14 m^3/s to the Uncompahgre River (HDR, 1995). The imported flows constitute from 35 percent to over 70 percent of the total flow in the river on an average monthly basis. The existing point of discharge is at the South Canal outfall, which is located approximately 8 km upstream from Montrose, Colorado (Fig. 1.1). The proposed hydropower outfall will be located just downstream from Montrose, thus any effects of flow augmentation that are associated with the project will occur in the approximately 43-km reach of the river between Montrose and the confluence with the Gunnison River at Delta, Colorado. The reach between the South Canal outfall and Montrose and Delta (M&D) Diversion (South Canal reach) provides an excellent prototype for evaluating the expected changes downstream from Montrose because it has been subjected to flow augmentation that is similar in magnitude to the proposed project flows for several decades.

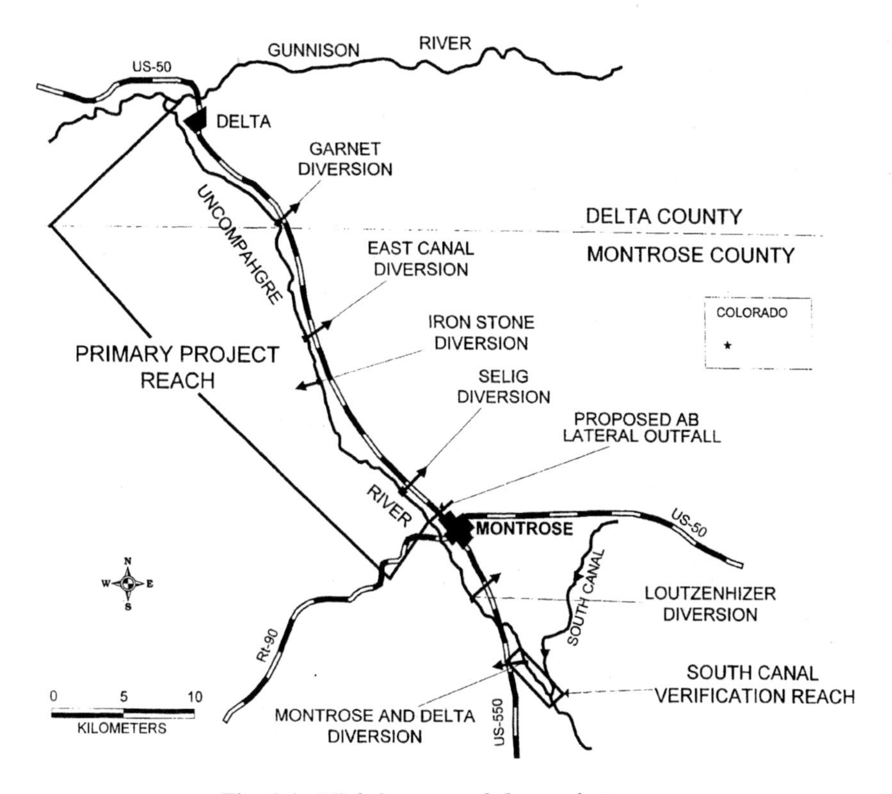

Fig. 1.1. Vicinity map of the project area.

A study was performed to predict the likely response of the channel to the imported flows, and the results were compared with channel changes that have occurred in the upstream South Canal reach.

2. FLOW REGIME MODIFICATIONS

Natural flows in the river follow a typical snowmelt pattern, with high flows during the spring runoff period, which typically occurs in May through July (Fig. 2.1). Under existing conditions, the average annual flow increases from about 6.8 to 15.6 m^3/s at the South Canal outfall due to the imported flows. Six major irrigation diversions extract water from the river during the irrigation season, significantly depleting the flows in the river (Fig. 2.1). The average

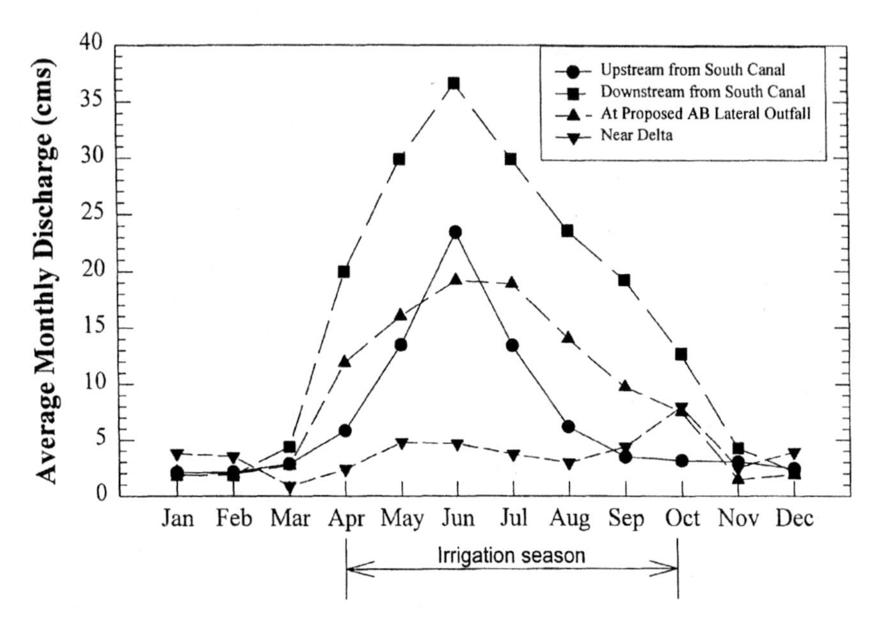

Fig. 2.1. Average monthly flows in the Uncompahgre River under no-project conditions.

annual flow decreases in the downstream direction from 15.6 m^3/s at the South Canal outfall to about 8.8 m^3/s at Montrose and 3.3 m^3/s at Delta due to the diversions. Under the proposed project condition, with a 24-m^3/s plant capacity, the average annual flow in the river immediately upstream from the AB Lateral outfall will decrease to about 2 m^3/s, and it will increase to about 19 m^3/s downstream from the outfall. During the irrigation season, the flows will decrease significantly in the downstream direction due to the diversions, but during the non-irrigation season, sustained flows that are similar to the historic flows that occurred during the runoff season would be present throughout the reach (Figs. 2.2 and 2.3).

The analyses that were performed for this study, and that are described in this paper, were based on the average monthly flows. Instantaneous and mean daily flows may significantly exceed the mean monthly flows, particularly during the snowmelt runoff period; thus, the bed may be mobile more often than is indicated by this analysis. The flows imported to the study reach through the Gunnison Tunnel and South Canal for irrigation purposes under existing (no-project) conditions and future imported flows associated

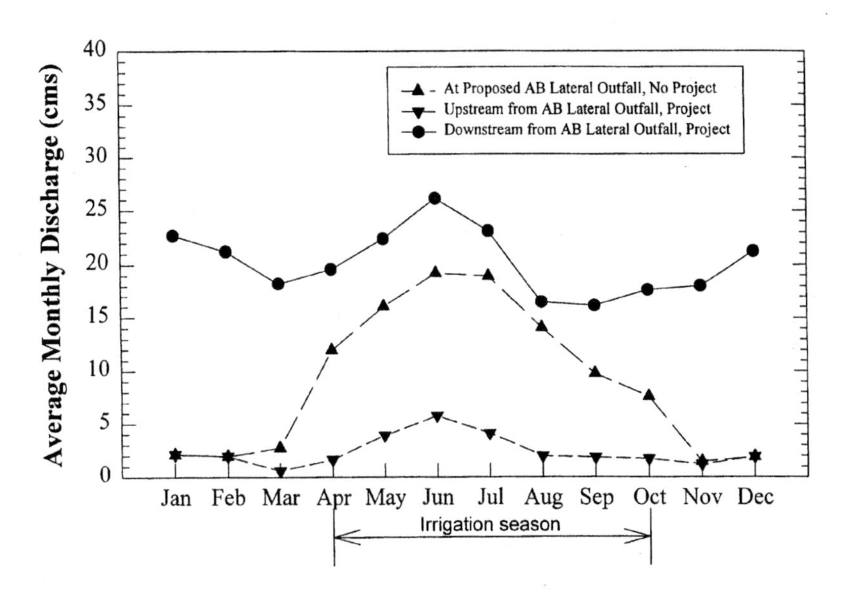

Fig. 2.2. Average monthly flows in the Uncompahgre River near the AB Lateral Outfall, no-project and project conditions.

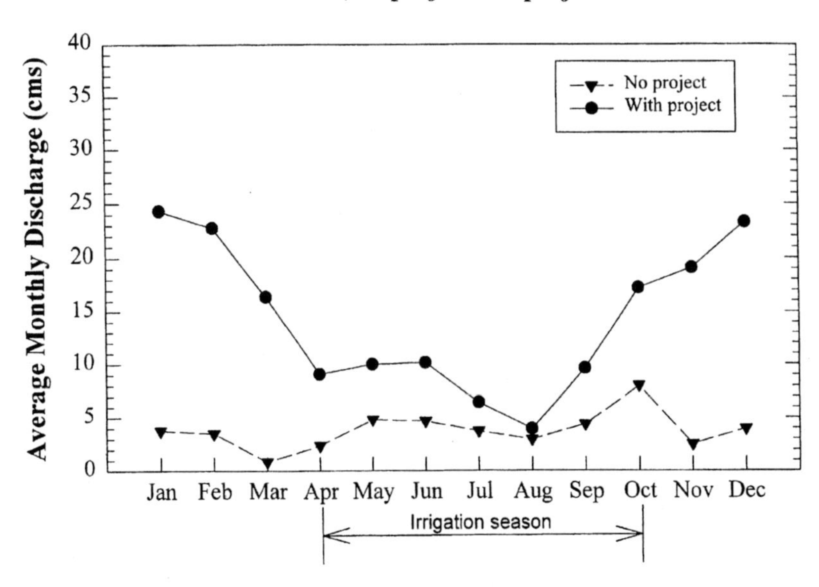

Fig. 2.3. Average monthly flow in the Uncompahgre River near Delta, Colorado, no-project and project conditions.

with project operations, however, are relatively constant. For this reason, results obtained by using mean monthly flows are believed to provide a reasonable evaluation of the effects of project operation on channel stability during the intervening times between larger flood flows.

3. GEOMORPHIC SETTING

The Uncompahgre Valley, which is about 48 km in length, lies along the western flank of the Rocky Mountains. The valley is bounded to the east by the predominantly Mancos Shale (Cretaceous age)-underlain Adobe Hills and to the west by the Uncompahgre Plateau that is underlain by Precambrian metamorphic rocks and Cretaceous-age Dakota Group. To the south, the valley heads in the San Juan Mountains that are primarily composed of Cretaceous-age sedimentary rocks and Precambrian metamorphosed sediments that are overlain by rhyolitic volcanic rocks. The river flows north to its confluence with the Gunnison River at Delta in a relatively wide and low-relief valley cut into the Mancos Shale (Chronic, 1980). From Montrose to the San Juan Mountains to the south, the valley contains extensive inset flights of Pleistocene age glacio-fluvial outwash terraces that are composed of sediments (sands to boulders) derived from glaciation of the San Juan Mountains. Present-day erosion of the upper watershed and reworking of the glacio-fluvial terrace sediments as a result of lateral stream erosion are the primary sources of the sediments that are transported as the bed-material load of the stream.

Downstream from Montrose, the valley widens somewhat and is bounded on both sides by soft rock pediments cut into Mancos Shale and capped by fluvial gravels. The river encounters the Mancos Shale bedrock in the valley walls upstream from Spring Creek and immediately downstream from the Garnet Diversion. Tributaries that drain to the river within the study reach deliver primarily fine-grained sediments (silts and clays) as a result of summer thunderstorm runoff. This material makes up the bulk of the wash load that is carried by the river.

Upstream from Montrose, the floodplain of the river, which is located about 1 m above the thalweg, is inset below two terraces. The lower and upper terraces are at elevations about 3 and 6 m above the thalweg, respectively. Lateral erosion of both terraces introduces coarse sediment to the river downstream from the South Canal outfall, which tends to compensate for the sediment deficit caused by the introduction of clear water from the South Canal.

Downstream from Montrose, the floodplain is located at an average elevation about 1.2 m above the thalweg, and is flanked by a terrace that averages about 3 m in height above the thalweg. In general, the floodplain is colonized by riparian vegetation and the terrace is farmed. As a result of considerable channel adjustment to the high-magnitude and long-duration flood discharges of 1984, the floodplain was severely dissected and eroded, which has led to the development of a wide expanse of low-relief sand and gravel bars. The tops of the gravel bars are located about 0.6 m above the thalweg. The bars have been colonized by even-aged stands of Russian olive as well as young cottonwoods and willows. In some locations, tamarisk is the dominant colonizing plant species.

Although the Uncompahgre River downstream from Montrose has a somewhat sinuous planform and appears to be a coarse-grained, meandering river, it is, in fact, better classified as a wandering, gravel-bed river (Desloges and Church, 1989). With the exception of the downstream-most reach at the Gunnison River confluence, bed slopes exceed 0.4 percent. In contrast to a truly meandering channel, where there is a systematic change in planform through time that results in a diminishing radius of curvature at individual bends that eventually leads to bend cutoff (Fisk, 1947; Harvey, 1988), the planform changes of a wandering channel are irregular and controlled by episodic flood flows. During periods between high-magnitude and long-duration, morphogenetically significant flows, the active channel develops a somewhat sinuous planform as a result of lateral erosion. The rate of lateral erosion and bank retreat is dependent on the resistance to erosion of the concave bank materials (Nanson and Hickin, 1986), the duration and magnitude of the flows (Odgaard, 1987: Harvey and Mussetter, 1994), and the capacity of the flows to transport bed material sediment (Neill, 1984). When relatively infrequent large floods occur, bends of almost any radius of curvature cut off and the channel sinuosity is significantly reduced. In contrast to meandering channels, this mode of channel adjustment ensures that there is a relatively frequent reworking of the floodplain sediments.

4. HYDRAULIC ANALYSIS

A detailed hydraulic analysis of the 43-km-long study reach (the confluence with the Gunnison River at Delta upstream to the proposed AB Lateral facility at Montrose), was carried out using the 1-D, HEC-2 water-surface profile program (USACOE, 1990). The hydraulic model contained 447 cross sections derived from a combination of field surveys, available topographic mapping, aerial photography, and field estimation (HDR, 1989).

The study reach contains numerous man-made structures that affect the river, including 12 bridge crossings, 4 major diversion weirs, areas of significant historical, and active in-channel gravel mining, and a considerable number of localized bank-protection measures. Due to limited topographic coverage of the overbank areas, the HEC-2 model is valid only for flows up to about 88 m^3/s (approximately 10-year return period flood peak), which encompasses the range of flows that would be affected by project operation. For this range of flows, all of the diversion weirs are hydraulic controls. The model was calibrated using two sets of measured water-surface elevations and the rating curve at the U.S. Geological Survey (USGS) stream gage at Delta by adjusting Manning's n values and, in some cases, the shape of estimated cross sections until the difference between the measured and modeled water-surface elevations matched within a reasonable tolerance (Mussetter Engineering, Inc., 1995).

5. CHANNEL STABILITY ANALYSIS

The expected response of the channel to the modified flow regime was evaluated by considering (1) the discharge necessary to mobilize the channel bed material (incipient motion analysis), (2) the quantity of sediment that can be carried through each reach of the river by the range of flows greater than those required to mobilize the bed material (bed material transport capacity analysis), (3) the approximate change in bed elevation associated with transport of the bed material, assuming that the channel does not widen (sediment continuity analysis), and (4) the approximate change in channel width that can be expected assuming no degradation and associated flattening of the channel gradient. The latter two analyses define two extremes for the expected channel changes; the actual change will probably be a combination of widening and bed lowering, with the relative amount of each being dependent on the local characteristics of the channel. Where the banks are erodible, the channel is more likely to widen than to degrade, and where the banks are not erodible, some degradation may occur, at least until the bed becomes armored.

To facilitate the channel stability analysis, the study reach between the AB Lateral outfall and the Gunnison River was divided into 14 subreaches based on similarity of geomorphic and hydraulic characteristics, and the location of hydraulic controls (Table 5.1). Reach-averaged values of main channel velocity, hydraulic depth, and width were computed for each subreach and were subsequently used in the channel stability analysis.

Table 5.1. Description of subreaches used in the channel stability analysis.

Subreach	Distance upstream from mouth (m)	Subreach Length (m)	Description
Supply			
	143600		Upstream end of study reach
1		8208	
	135392		Selig Canal
2		6777	
	128615		
3		8540	
	120075		
4		5915	
	114160		Spring Creek
5		13800	
	100360		Ironstone Canal
6		10764	
	89597		East Canal
7		9032	
	80565		Downstream edge of Olathe
8		14323	
	66242		Blossom Road
9		13038	
	53204		Garnet Diversion
10		6849	
	46355		B Road
11		13520	
	32835		
12		10355	
	22480		1575 Road
13		15364	
	7116		Upstream end of channelized reach
14		7116	
	0		Gunnison River

5.1. Bed Material Characteristics

Sediment gradation data were developed from 35 total bed surface, bed subsurface, and bank samples that were collected in 1989, 1993, and 1994. Representative gradations for each of the subreaches were developed from appropriate samples from this data set (Fig. 5.1). The median size (D_{50}) of the surface bed material along the reach averages about 50 mm, and varies from about 38 to 76 mm (Fig. 5.2). The bed material is coarser than the average in areas downstream from the Selig and Garnet Diversions where the sediment supply has been interrupted, and the river cuts into coarse-grained material on the valley sideslopes downstream from the Garnet Diversion.

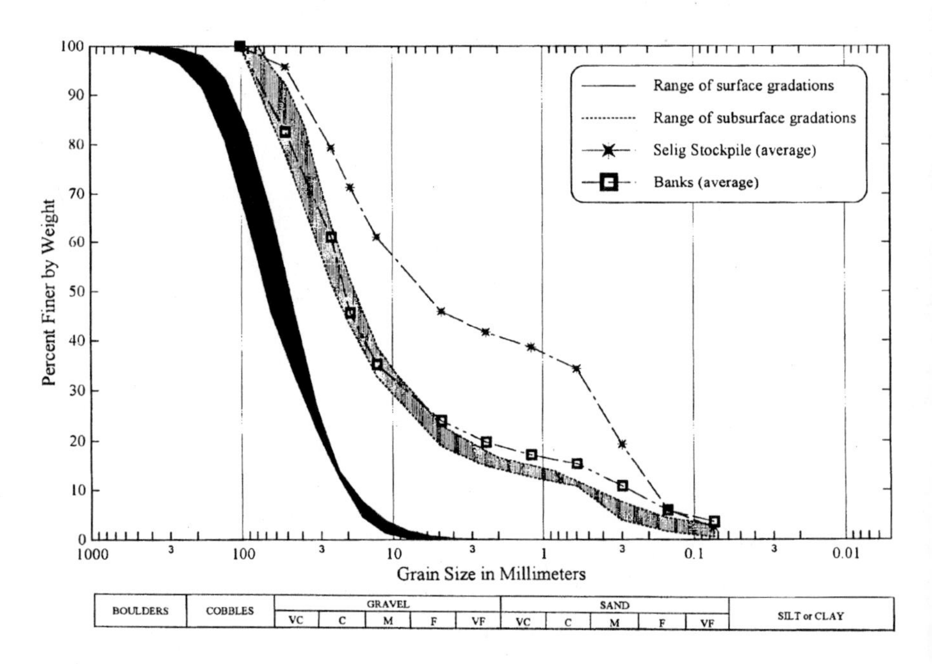

Fig. 5.1. Representative pavement gradation curves used in the channel stability analysis.

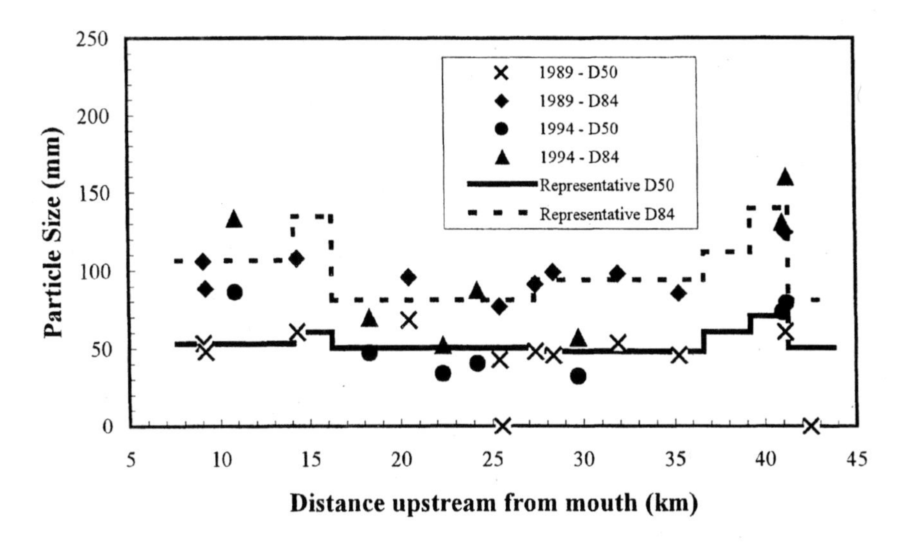

Fig. 5.2. **Variation in D_{50} and D_{84} sediment sizes in the Uncompahgre River with distance upstream of the Gunnison River for samples used to establish the representative sediment gradations.**

5.2. Incipient Motion Analysis

The incipient motion analysis was performed to estimate the range of discharges required to mobilize the existing bed material (i.e., the critical discharge), and thus, potentially cause degradation of the channel bed. At discharges less than the critical discharge, the bed is effectively armored. Sediment transported through the system under these conditions is fine in comparison to the bed material, and little or no vertical adjustment to the bed occurs. As the discharge rises above critical, the coarse-bed material is mobilized, and vertical adjustment is possible. The incipient motion analysis was performed by comparing the critical shear stress (τ_c) for the median particle size with the bed shear stress due to grain resistance (τ') for discharges up to 88 m^3/s using the Shields (1936) relation:

$$\tau_c = \tau_{*_c}(\gamma_s - \gamma)D_{50} \tag{5.1}$$

where τ_c is the critical shear stress, τ_{*c} is the dimensionless critical shear stress (often referred to as the Shields parameter), γ_s is the unit weight of sediment

(□2.65 g/cm^3), γ is the unit weight of water, and D_{50} is the median particle size. When the critical shear stress for the median particle size is exceeded, the bed is mobilized and all sizes up to about five times the median size can be transported by the flow (Parker et al., 1982; Andrews, 1984). A dimensionless critical shear stress (τ_{*c}) of 0.03 was used in this study based on the observed characteristics of similar gravel-bed streams (Andrews, 1984; Parker et al., 1982; Neill, 1968).

The bed shear stress due to grain resistance (τ') was used in the analysis because it is a better descriptor of near-bed hydraulic energy in gravel- and cobble-bed streams than the total shear stress, which includes the effects of flow resistance due to irregularities in the channel boundary, nonlinearity of the channel, variations in channel width, and other factors that contribute to the total flow depth, but not the energy available to move individual particles on the channel bed (Einstein, 1950; Mussetter, 1989; Nelson and Smith, 1989).

The grain shear stress (τ') is computed from $\gamma Y'S$, where γ is the unit weight of water, Y' is the portion of the total hydraulic depth associated with grain resistance (Einstein, 1950), and S is the energy slope at the cross section. The value of Y' is computed by iteratively solving the semilogarithmic velocity profile equation:

$$\frac{V}{V_*'} = 5.75 + 6.25 \log\left(\frac{Y'}{K_s}\right) \tag{5.2}$$

where V is the mean velocity at the cross section, K_s is the characteristic roughness height of the bed (taken here as $3.5D_{84}$ after Hey, 1979), and V_*' is the shear velocity ($\sqrt{gY'S}$).

The dimensionless grain shear stress (τ_*'), which is the ratio of the grain shear stress (τ') to the critical shear stress (τ_c) or

$$\tau_*' = \frac{\tau'}{\tau_c} = \frac{\gamma Y'S}{\tau_{*_c}(\gamma_s - \gamma)D_{50}} \tag{5.3}$$

provides a convenient means of evaluating the incipient conditions. When $\tau_*' < 1$, the shear stress is insufficient to mobilize the bed material (i.e., the bed is effectively armored), and when $\tau_*' > 1$, bed mobilization is indicated. When τ_*' is between 1.0 and approximately 1.5, bed material transport rates are low, but measurable (Neill, 1968) and bed adjustment will occur relatively slowly. At higher shear stresses, transport can be large and bed adjustment rapid.

The predicted discharges that are required for each reach to produce incipient conditions ($\tau_*'=1$) and measurable bed material transport ($\tau_*'=1.5$) are summarized in Table 5.2. The table also shows the approximate percentage of time the mean monthly discharge would equal or exceed the indicated discharge for no-project and project conditions.

Table 5.2. Summary of estimated critical discharges and their exceedance percentages.

Reach	Critical Discharge (m^3/s) ($\tau_{*c} = 0.03$)	Percentage of time average monthly flow is greater than or equal to indicated discharge		Discharge (m^3/s) at $\tau'_* = 1.5$	Percentage of time average monthly flow is greater than or equal to indicated discharge	
		No Project (%)	With Project (%)		No Project (%)	With Project (%)
Supply	19.1	9.6	1.7	53.2	0	0
1	19.1	9.6	62.8	53.2	0	0.6
2	15.7	34.5	84	50.4	0.4	3.3
3	8.3	54.2	90.9	50.1	0.2	1.7
4	12.7	9.6	65.3	43.1	0	0.9
5	14.4	2.8	54.8	59.9	0	0.1
6	9.7	1.7	52.7	58.0	0	0
7	15.9	0.9	41.4	44.3	0	0.7
8	36.3	0.2	1.1	>99.0	0	0
9	34.3	0.2	1.1	>99.0	0	0
10	10.0	6.4	55.9	37.5	0.2	1.4
11	16.5	1.1	42	54.7	0	0.4
12	17.8	1.1	40.4	54.9	0	0.4
13	15.7	1.1	43.4	58.3	0	0.2
14	17.3	1.1	40.8	41.8	0.2	1.1
average	17.4	8.9	48.3	57.5	0.1	0.8

The critical discharge to entrain the median particle size varies from about 8.3 m^3/s in Reach 3 to approximately 36.3 m^3/s in Reach 8 and averages about 17.4 m^3/s through the project reach. Reach 3, which has the lowest critical discharge and apparently highest percentage of time in which the bed material is mobile, is not representative of the long-term adjusted condition of the channel because it has been heavily disturbed by recent instream sand and gravel mining.

For the no-project condition, the mean monthly flow exceeds the critical discharge about 9 percent of the time (approximately 1 month per year) on average, and in most of the reaches downstream from Spring Creek (Reaches 5 through 14), less than 2 percent of the time (approximately 2.4 months in 10 years), increasing to over 56 percent of the time (excluding Reaches 8 and 9) for project conditions. The fan and valley constriction created by Loutzenhizer Wash at the downstream end of Reach 9 causes the upstream channel to be flatter than the typical gradient of the river which accounts for the low velocities and shear stress in this area. In the reach upstream from the AB Lateral outfall, the critical discharge is exceeded about 10 percent of the time (approximately 1.2 months per year) for no-project conditions, reducing to less than 2 percent of the time under project conditions. The proposed project will, therefore, essentially eliminate the supply of bed material to the study reach for the range of flows that will be affected by project operations. This, of course, will be mitigated to some extent by the fact that flood flows ($Q>54$ m^3/s) will not be affected by project operation.

The discharge required to produce measurable bed material transport (i.e., $\tau_*' = 1.5$) varies from about 36.8 m^3/s in Reach 10 to over 99 m^3/s in Reaches 8 and 9 and averages about 57 m^3/s through the project reach. The mean monthly flow rarely exceeds this discharge under no-project conditions and exceeds it less than 1 percent of the time (<1 month in 10 years) for project conditions.

5.3. Sediment Continuity Analysis

The incipient motion analysis provides an evaluation of the magnitude and duration of flows necessary to mobilize the bed material, but it does not provide a means of quantifying the potential response of the channel to the mobilizing flows. A sediment continuity analysis was, therefore, performed to estimate the average depth of degradation or aggradation associated with differences in bed material transport capacity along the study reach. It is important to note that, since the sediment continuity analysis is based on the existing channel geometry, it implicitly assumes that no channel widening occurs. Since widening is likely at higher flows in areas where the banks are not armored, the sediment continuity results represent an approximate upper limit of vertical adjustment that can be expected in response to the discharges that were analyzed.

The predicted average annual change in bed elevation was obtained by comparing the transport capacity of each reach, which was estimated using the

Parker bedload equation (Parker and Sutherland, 1990), with that of the next upstream reach. As expected, based on the results of the incipient motion analysis, the predicted changes in bed elevation are insignificant for no-project conditions. For project conditions, most of the reach is also relatively stable. Degradation of about 8 cm per year is indicated in Reaches 2 and 10 (downstream of the Selig and Garnet Diversions, respectively). Reaches 3 and 4 (Selig Diversion to Spring Creek) are slightly aggradational (less than 3 cm per year in Reach 3 and approximately 6 cm per year Reach 4). Reach 11 is also slightly aggradational (less than 3 cm per year).

5.4. Channel Widening

The existing bed material in Reaches 2, 3, and 10 (downstream from the Selig and Garnet Diversions, respectively) is coarser than the remainder of the study reach, which is the expected response to trapping of bed material upstream of the diversions. Since the parent bed material is similar throughout the reach (except in the vicinity of the Garnet Diversion where very coarse sediment is delivered to the channel from undercutting of a strath terrace along the left bank), it is probable that similar coarsening will occur in other reaches where there is a degradational tendency. Coarsening of the bed material will inhibit (or prevent) further bed degradation, and the river will respond by widening.

In evaluating the hydraulic geometry of gravel-bed rivers, Parker (1979) found that, for those reaches that appeared to be in an approximately graded (or equilibrium) condition with a normal upstream sediment supply, the hydraulic geometry had adjusted so that the dimensionless grain shear stress for the bed material, (τ_*') based on the average main channel hydraulics was about 1.33 assuming $\tau_{*c} = 0.03$. For channels with minimal upstream sediment supply, threshold channel concepts (Parker, 1979; Li et al., 1976) apply and the dimensionless shear stress should adjust to near incipient conditions (τ_{*c}□0.03).

The width to which the channel would need to adjust in response to the proposed project flows, in the absence of slope adjustments, to develop these shear stresses were computed for each of the reaches to bracket the expected changes in channel geometry (Table 5.3). The computations were performed using the existing bed-material gradations and the reach-averaged hydraulics for the effective discharge (Andrews, 1984). The average adjusted channel width predicted by this procedure varies from 21 m using a Shields value of 0.04 to 35 m using a value of 0.03. The existing average width is about 20 m.

An alternative method was also used to estimate the amount of width adjustment by considering the width/depth ratios of the channel. Based on the existing bankfull channel geometry in the project reach and the South Canal verification reach that will be discussed in the next section, as well as

Table 5.3. Summary of estimated channel widths under project conditions.

			$t_*'=0.03$		$t_*'=0.04$		W/D=35		
Reach	Existing Average Width (m)	Average Discharge (m^3/s)	Adjusted Average Width (m)	Change in Width (m)	Adjusted Average Width (m)	Change in Width (m)	Adjusted Average Width (m)	Change in Width (m)	t_*'
1	22.0	26.2	28.2	6.3	22.0	---	27.7	5.8	0.030
2	25.0	26.2	34.4	9.4	25.0	---	26.5	1.5	0.034
3	23.8	26.2	40.5	16.7	23.8	---	26.6	2.8	0.036
4	16.5	26.2	35.3	18.9	19.3	2.9	27.9	11.4	0.034
5	30.2	26.2	41.1	10.9	30.2	---	30.2	0.0	0.036
6	18.0	26.2	31.9	13.9	18.0	---	28.1	10.1	0.032
7	16.8	26.2	25.9	9.1	16.8	---	28.3	11.5	0.029
8	18.3	26.2	23.0	4.7	18.3	---	29.0	10.7	0.027
9	17.4	26.2	18.4	1.0	17.4	---	29.8	12.4	0.024
10	17.7	27.6	41.7	24.0	22.3	4.6	27.2	9.5	0.037
11	16.8	27.6	33.6	16.9	18.4	1.7	28.2	11.5	0.033
12	18.9	27.6	30.9	11.9	18.9	---	27.9	9.0	0.031
13	18.9	27.6	32.2	13.3	18.9	---	27.7	8.8	0.032
14	22.3	27.6	29.5	7.3	22.3	---	26.0	3.8	0.032
Length Weighted Average	20.1	26.7	31.3	11.2	20.6	0.5	28.2	8.0	0.032

published width/depth ratios for similar gravel-bed streams where the banks are erodible but moderately vegetated (Parker, 1979; Andrews, 1984), the width/depth ratio of the channel after adjustment is expected to be about 35. Under these conditions, the average adjusted width will be about 28 m, which is between the upper and lower widening limits predicted by the incipient motion criteria. The average Shields value for the reach at a width/depth ratio of 35 is about 0.032.

The actual width change that will occur in response to the project flows will obviously depend on the conditions in each specific segment of the river, including the supply of bed material from upstream and lateral sources, the amount of bank protection within the reach, the vegetation and sediment size characteristics of the banks, and the local hydraulic conditions. Where the banks are composed of relatively fine, noncohesive material (sands and gravels), are unprotected, and where there is little bed material supply from upstream, the adjusted width will probably be near the upper end of the predicted range. Where the banks are less erodible (i.e., contain cohesive sediments or are well vegetated), or a significant supply of bed material is available, the adjusted channel width will probably be near the lower end of the range.

6. SOUTH CANAL REACH VERIFICATION STUDY

The processes that cause channel change are numerous and complex; thus, the adequacy with which all of the processes are accounted for in the preceding analyses is uncertain. The approximately 3.2-km-long reach of the Uncompahgre River between the South Canal outfall and the M&D Diversion (the South Canal reach), which is located approximately 8 km upstream of Montrose, provided an opportunity to check the reasonableness of the predicted channel changes. This reach of the river has geomorphic characteristics that are similar to the project reach and it has been subjected to a historic range of flows that are similar to the proposed project flows.

Upstream from the South Canal, the flows are essentially natural, with little regulation or augmentation. Additionally, a significant supply of bed material-sized sediment is available for transport from the upstream channel. Between the South Canal outfall and M&D Diversion, the flows are a combination of natural flows from upstream and imported flows from the South Canal, with the essentially sediment-free, imported flows, comprising about 55 percent of the total volume passing through the reach (Fig. 2.1).

Evaluation of the basic data, and application of the previously described analytical procedures to the South Canal reach, confirm the magnitude of the changes that were predicted for the project reach. The bed material downstream from the South Canal outfall is considerably coarser, and the channel is wider than in the project reach or upstream from the outfall. Median bed material sizes in the riffles of the South Canal reach varies from 85 mm to over 117 mm as compared to about 86 mm in the project reach, and 72 mm upstream from the South Canal outfall. Incipient motion analysis showed that, in the upstream portion of the South Canal reach between the outfall and Uncompahgre Road (Fig. 1.1), average monthly discharges exceeded the critical discharge about once every two years and never exceeded that necessary for significant bed material transport. Upstream of the South Canal outfall, the critical discharge was exceeded about one month per year, and the discharge required for significant bed material transport was exceeded about once in 20 years.

The channel in the South Canal reach is considerably larger than in the project reach, with average bankfull topwidths varying from 26 to 33 m as compared to the average bankfull widths in the project reach which range from 17 to 30 m and averaging 20 m (Fig. 6.1). Bankfull width/depth ratios are also greater in the South Canal reach, ranging from 36 to 48 as compared to 18 to 44 with an average of 34 in the project reach. Based on the hydraulic modeling results, the average bankfull discharge is about 15.5 m^3/s upstream of the South Canal outfall, 18.1 m^3/s between the South Canal and Uncompahgre Road, and about 24.7 m^3/s between Uncompahgre Road and the M&D Diversion.

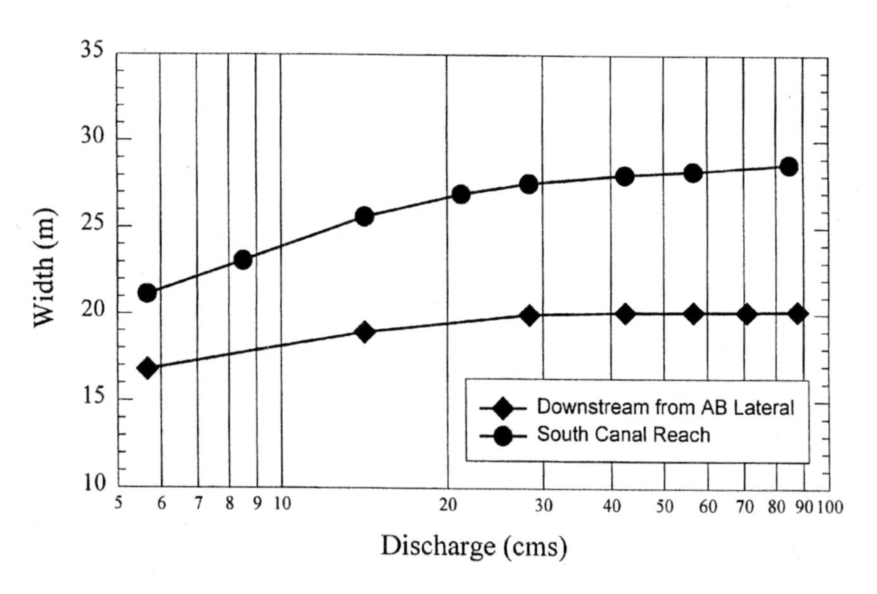

Fig. 6.1. Average main channel topwidth in the South Canal reach, compared with widths in the project reach.

Several sections of the upstream portion of the South Canal reach exhibit the characteristics of a threshold channel (Parker, 1979; Li et al., 1976)

These reaches have an approximately parabolic shape and very coarse bed material that is no longer mobile under the higher discharges that occur on a frequent basis.

7. EXPECTED CHANNEL ADJUSTMENT RATES

Considering the changes to the river that are expected to occur in response to the proposed project, the most effective means of preserving the environmental qualities of the river, while still protecting infrastructure and adjacent property, is to allow the river to adjust freely within the corridor defined by the terraces, wherever possible. Protection of floodplain surfaces from erosion should be limited to those areas where public safety or infrastructure would be endangered.

After the channel has adjusted to the project flows, long-term sediment transport rates should not be significantly different from the existing condition. During the adjustment period, however, the amount of erosion and, thus, the sediment transport rates will almost certainly be higher. The volume of material eroded during the predicted widening, using an adjusted width/depth ratio of 35, is approximately 171,000 m^3. The impact of the increased sediment loads on the Uncompahgre River, and ultimately the Gunnison River, depends primarily on the rate at which this material is eroded and transported downstream.

A simple sediment-routing model was developed to evaluate the approximate erosion rate, the potential for aggradation in the Uncompahgre River and the increased sediment loading to the Gunnison River during the adjustment period. The routing algorithm is based on three assumptions: (1) the shear stress acting on the channel boundary, and thus, the transport capacity will decrease as the channel widens, (2) the capacity of the river to transport particles smaller than 2 mm is significantly greater than the availability of that material, thus, when the coarser material is entrained, the sands and finer materials will also be entrained in proportion to their representation in the parent gradation, and (3) based on the average height of the banks in comparison to the bottom width of the channel, the percentage of the transport capacity that can be entrained from the banks at any given time (R_{bl}) will vary from a minimum of about 10 percent to a maximum of about 25 percent, with the actual amount at any specific location being dependant on the local bank characteristics and hydraulic conditions. The lower percentage represents the worst-case scenario in terms of the length of time over which the sediment loads to the Gunnison River will be elevated and the higher percentage represents the worst-case in terms of the initial erosion rates and sediment delivery to the Gunnison River.

The routing model predicts that between 30,800 m^3 (R_{bl}=10%) and 49,100 m^3 (R_{bl}=25%) of material will be eroded from the banks along the study reach during the first year of project operation (Fig. 7.1). For both assumed conditions, the rate of bank erosion and widening decreases significantly with time. For the higher percentage (R_{bl}=25%), the annual erosion rate reduces to approximately 9,600 m^3 (approximately 19 percent of the initial value) after 5 years, to about 2,900 m^3 (approximately 6 percent of the initial rate) after 10 years and is less than 1.5 percent of the initial value (800 m^3) after 30 years. For the lower rate (R_{bl}=10%), the annual bank erosion rate reduces to approximately 7,300 m^3 (24 percent of the initial rate) after 5 years, 3,400 m^3 (11 percent of the initial rate) after 10 years, and is less than 5 percent of the initial rate (approximately 1,400 m^3) after 30 years. Between 96,400 m^3 (R_{bl}=10%) and 118,600 m^3 (R_{bl}=25%) of material (of the estimated 171,400 m^3 total) will be eroded from the banks during the first 10 years of project operations.

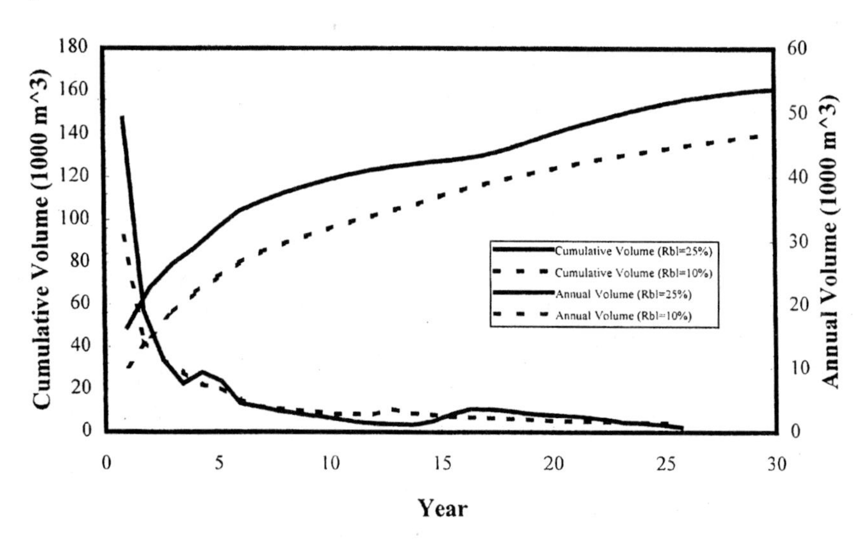

Fig. 7.1. Predicted average annual and cumulative bank erosion volume along the Uncompahgre River.

The predicted bank erosion will increase the sediment load in the Uncompahgre River, and the sediment yield to the Gunnison River, during the adjustment period. The maximum total increase in sediment concentration in the Uncompahgre River resulting from the increased sediment load was estimated to be between 127 ppm (R_{bl}=10%) and 190 ppm (R_{bl}=25%), based on the sediment routing results, and will occur during the first year of operation. Based on the average bank material gradation (Fig. 5.1), approximately 19 percent of the eroded bank material will be sand and smaller and about 4 percent silt and smaller, which indicates that the concentration of silt/clay will increase by about 8 ppm (R_{bl}=10%) to 13 ppm (R_{bl}=25%).

Some local aggradation is likely in the approximately 4-km-long reach upstream from Blossom Road (Reach 8) and in the channelized reach just upstream from the Gunnison River confluence (Reach 14) during the first few years of operation. The estimated maximum aggradation in Reach 8 occurs during the fourth and fifth year after the start of project operation and varies from 15 cm (R_{bl}=10%) to 21 cm (R_{bl}=25%). The aggraded material is completely removed after approximately 15 years. In the channelized reach upstream from the Gunnison River confluence (Reach 14), the maximum aggradation varies from a negligible amount for the lower assumed erosion percentage (R_{bl}=10%) to about 15 cm for the higher erosion percentage, and occurs during the first few years of operation. For the worst-case scenario (R_{bl}=25%), this material is completely removed by the fifth year of operation.

The increased sediment yield to the Gunnison River associated with the channel widening will range from about 16,800 m^3 (R_{bl}=10%) to 20,400 m^3 (R_{bl}=25%) during the initial year of operation. The annual suspended sediment load in the Gunnison River at Grand Junction was approximately 1.9 million metric tons during the period 1914 to 1957, prior to completion of the upstream dams (Aspenall project) (Iorns et al., 1965). Using the published USGS mean daily flow record and a rating curve developed from approximately 140 suspended sediment measurements taken between 1959 and 1993, the corresponding annual post-Aspenall suspended sediment yield is about 0.9 million metric tons (563,000 m^3 assuming bulked unit weight of 1.6 g/cm^3). Considering the generally accepted estimate that the bed material load in rivers similar to the Gunnison River is about 10 percent of the total sediment load (Gregory and Walling, 1973), the annual total load is about 1 million metric tons (624,000 m^3). The increased sediment loading from the Uncompahgre River, therefore, represents about 3 to 3.5 percent of the existing total load in the Gunnison River during the initial period when the adjustment rate is highest and diminishes to less than one percent after 10 years of operation.

8. CONCLUSIONS

The Uncompahgre River study illustrates that, with a proper understanding of the controlling hydraulic and geomorphic characteristics, relatively simple computational techniques are available to predict the response of a river to a proposed change in flow regime. While the quantitative estimates must be viewed as approximations, input parameters and assumptions can be selected that will provide results that should bracket the actual changes that are expected to occur. These results provide permitting agencies with information necessary to evaluate the possible adverse impacts of the project and provide the project proponent with a rational basis for selecting and designing measures to mitigate those impacts.

9. ACKNOWLEDGMENTS

This investigation was conducted for Montrose Partners, Inc. The authors thank Bill Fowler and Jim Hokit for permission to publish the results of the study.

10. REFERENCES

Andrews, E.D. 1984. Bed Material entrainment and hydraulic geometry of gravel-bed rivers in Colorado. *Geol. Soc. America Bulletin* 95, March, 371-378.

Chronic, H. 1980. *Roadside Geology of Colorado*, Mountain Press Publishing Co. Missoula, MT.

Desloges, J. R. and M. Church. 1989. Wandering Gravel Bed Rivers: Canadian Landform Examples, *Canadian Geographer*. 33:360-364.

Einstein, H.A. 1950. The bed load function for sediment transportation in open channel flows. Washington, D.C.: U.S. Soil Conservation Service. Technical Bulletin No. 1026.

Fisk, N.H. 1947. Fine-grained alluvial deposits and their effects on Mississippi River activity. U.S. Army Corps of Engineers, Waterways Experiment Station. Report 2:121.

Gregory, K.J. and D.E. Walling. 1973. *Drainage Basin Form and Process, a Geomorphological Approach*, John Wiley & Sons. 257-259.

Harvey, M.D. 1988. Meanderbelt dynamics of the Sacramento River, USDA Gen. Tech. Rpt. PSW-110:54-60.

Harvey, M.D. and R.A. Mussetter. 1994. Geologic, geomorphic and hydraulic controls at spawning locations for endangered Colorado squawfish, *EOS*. Trans. Amer. Geophys. Union. 75:269.

HDR. 1989. Preliminary Design Report, Uncompahgre River Bank Stabilization, AB Lateral Hydropower Facility, prepared for Uncompahgre Valley Water Users Association and Mitex, Inc. 1 and 2.

HDR. 1995. Technical Memorandum, Hydrology Study, Uncompahgre River Bank Stabilization, AB Lateral Hydropower Facility, prepared for Montrose Partners, Ltd.

Hey, R.D. 1979. Flow resistance in gravel-bed rivers. *Journal of the Hydraulics Division,* American Society of Civil Engineers. April. 105:Hy4:365-379.

Iorns, W.V., C.H. Hembree, and G.L. Oakland. 1965. Water Resources of the Upper Colorado River Basin--Technical Report. U.S. Geological Survey, Geological Survey Professional Paper 441, U.S. Government Printing Office, Washington, D.C.

Kellerhals, R., M. Church, and L.B. Davies. 1979. Morphological effects of interbasin river diversions. Canadian Journal of Civil Engineering. , 6:1:18-31.

Lane, E.W. 1955. The importance of fluvial morphology in hydraulic engineering, Proc., ASCE. 21:745:17.

Li, R.M., D.B. Simons, and M.A. Stevens. 1976. Morphology of cobble streams in small watersheds, *Journal of the Hydraulics Division*, ASCE. 102: HY8: 1101-1117.

Mussetter, R.A. 1989. *Dynamics of Mountain Streams*. Doctoral Dissertation. Colorado State University, Department of Civil Engineering.

Mussetter Engineering, Inc. 1995. Uncompahgre River Channel Stability Study, prepared for Montrose Partners, Ltd.

Nanson, G.C. and E.J. Hickin. 1986 . A statistical analysis of bank erosion and channel migration in Western Canada, *Geological Society of America Bulletin*. 97:497-504.

Neill, C.R. 1984. Bank erosion versus bedload transport in a gravel bed river. *Proceedings of ASCE, Hydraulics Division, Conference on River Meandering*, New Orleans. 204-211.

Neill, C.R. 1968. Note on initial movement of coarse uniform bed material. *Journal of Hydraulic Research*. 6:2:173-176.

Nelson, J.M. and J.D. Smith. 1989. Flow in Meandering Channels with Natural Topography, in *River Meandering*, American Geophysical Union, Water Resources Monograph 12, S. Ikeda and G. Parker, editors. 82-83.

Odgaard, A.J. 1987. Streambank erosion along two rivers in Iowa, *Water Resources Research*. 23:7,1225-1236.

Parker, G. 1979. Hydraulic geometry of active gravel rivers, ASCE, *Journal of the Hydraulic Division*, 105:HY9.

Parker, G. and A.J. Sutherland. 1990. Fluvial armor. *Journal of Hydraulic Research*, 28:25, 529-544, January.

Parker, G., P.C. Klingeman, and D.G. McLean. 1982. Bedload and size distribution in paved gravel-bed streams. *Journal of the Hydraulics Division*, American Society of Civil Engineers. 108:HY4, Proc. Paper 17009:544-571.

Schumm, S.A. 1977. *The Fluvial System*, New York, NY, Wiley.

Shields, A. 1936. Application of similarity principles and turbulence research to bed load movement. California Institute of Technology, Pasadena; Translation from German Original. Report 167.

Thorne, C. and N.K. Tovey. 1981. Stability of Composite River Banks. *Earth Surface Processes and Landforms*. 6:469-484.

U.S. Army Corps of Engineers (USACOE). 1990. HEC-2, Water Surface Profiles, User's Manual, U.S. Army Corps of Engineers, Hydrologic Engineering Center. Davis, California.

THE FLUVIAL GEOMORPHOLOGY OF SYDNEY AND ITS ROLE IN POLLUTION MANAGEMENT

Robin F. Warner[1]

ABSTRACT

The landforms of Sydney and adjacent catchments are composed of Triassic basin sediments, ranging from thick, coarse fluvial sands to a cap of finer fluvial and near-shore deposits. Tertiary uplift and warping further emphasize the basin structure. Higher peripheral lands were deeply dissected to produce the characteristic sandstone valleys. Near the present coast, dissection occurred well below present sea level, such that the post-Pleistocene transgression flooded valleys to produce deep, long ria-like harbors. Now some 3.7 million people have urbanized 2,400 km^2, or about 10 percent of the total catchments. Spatial attributes of these landforms and temporal spacing of flow events are now such that there is a generally unrecognized potential to accommodate part of the pollution loads threatening these environments. The intermittent tributaries can "store" pollutants until only periodic re-establishment of flow continuity. This allows "recovery time" for larger receiving waters. Of these, perennial rivers sometimes have large parts of their catchments above urban areas, allowing flushing by large flow events. Tidal waters are "flushed" twice daily but effectiveness diminishes upstream. It is suggested that there are means of working with these attributes rather than against them.

KEYWORDS

Landforms, urbanization, regime shifts, intermittent streams, perennial rivers, ria-like harbors

1. INTRODUCTION

Sydney is a beautiful city and has been compared favorably with Rio de Janeiro and San Francisco. It is also a lucky city in what was called "the lucky country" by ex-Prime Minister Keating. The luck pertains to the nature of the physical platform on which it stands and the processes which operate there.

---

[1] Department of Geography, The University of Sydney, NSW, 2006, Australia.

This paper suggests that Sydney has a unique set of geomorphic characteristics and process regimes, which are conducive to the maintenance of a sustainable aquatic environment. An attempt is made to show how these characteristics and regimes accommodate contaminants, so that receiving waters are not continuously loaded to the point where there are few opportunities for recovery. Indeed the major aim of this contribution is to show how Sydney's landforms and processes help assimilate considerable proportions of the wastewater loads introduced from 2,400 km^2 of urban surface. This is achieved by an examination of the sustaining roles of natural channels, catchment runoff and tidal water dynamics.

Environmental managers have not totally understood these characteristics. They have failed to work with them and often inadvertently perhaps they have worked against them. Many aspects of environmental management have been in the hands of water engineers and politicians. The former have sought effective storage, distribution and disposal of water; the latter have provided the funds and legislative structures. They both have failed to realize that the urban landscape and its hinterlands have been a geomorphic platform of both opportunity and inconvenience, depending on whether it accommodated or hindered engineering. This unique landscape, much modified by urbanization, provides spatial characteristics, which can accommodate some of the water contamination problems. There are also temporal opportunities in the process regimes for contaminant storage and flushing which need to be understood. These characteristics are quite different from other Australian cities (Bridgman et al., 1995) where characteristics are often less conducive to effective management of contaminants.

This paper analyses Sydney's landforms and processes, examines the impacts of urbanization on river systems, considers the nature of pollution in Sydney as well as processes of assimilation and reviews the potential for geomorphology to assist in pollution management.

2. SYDNEY'S LANDFORMS AND PROCESSES

This section considers the geological origins of and subsequent changes to this system, the role of the Holocene transgression and the impacts of human activity.

2.1. Triassic Sedimentation

Sydney occupies a sedimentary basin, which has elevated margins to the north, west and south; the eastern side being truncated by the ocean. It has been infilled from marginal erosion mainly by sand-sized facies (Branagan et al.,

1976); however, central parts of the basin have shales and local sandstones. The former were conducive to farming; whereas, the girdle of uplifted sandstones are infertile.

Subsequent earth movements further emphasized the basin with slopes elevated from the center of the basin to low sandstone cliffs at the coast, the Hornsby Plateau to the north, the Illawarra Ramp (Woronora Plateau) to the south and the more elevated Blue Mountains west of the Lapstone Monocline and Kurrajong faults (Fig. 2.1).

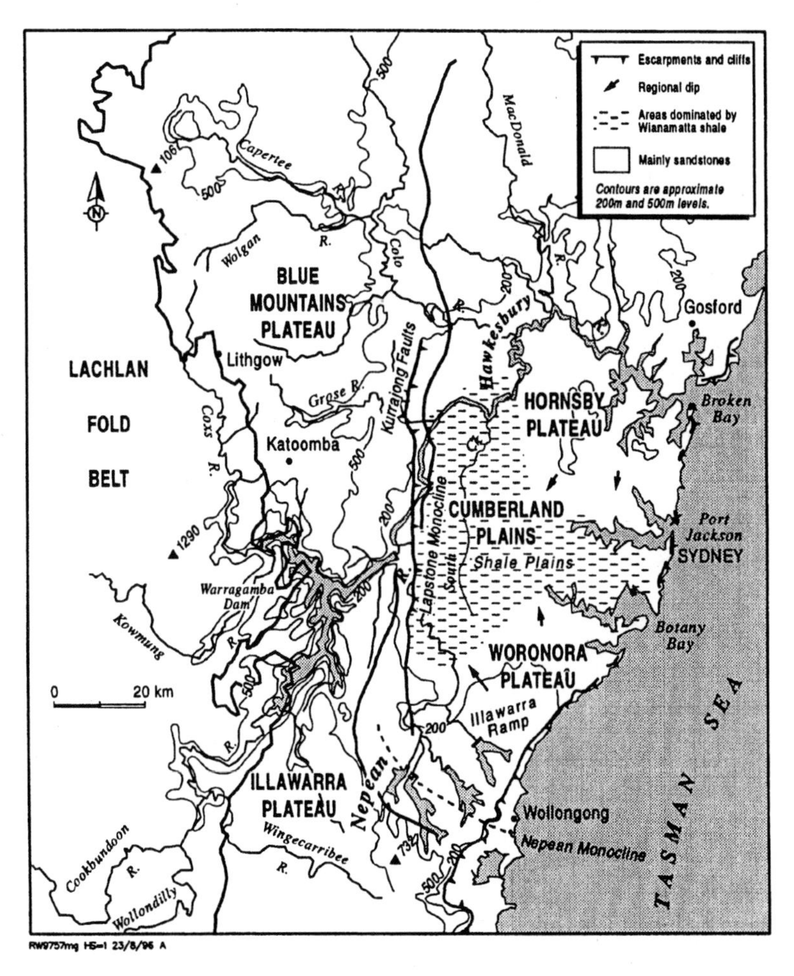

Fig. 2.1. Main landform units of the Sydney Basin (structures based on Branagan et al., 1976).

The basin was further emphasized by Tertiary movements and there has been extensive dissection of the marginal sandstones.

2.2. Pleistocene and Holocene

In low sea-level stages of the Pleistocene, valleys were dissected near the coastline to depths well below current sea level. Some 18,000-yr BP sea level stood at -200 m, but by 6,000-yr BP (Thom and Roy, 1981), it had returned to its present level, flooding the wide, deeply dissected valleys. This produced long, tidal reaches to the lower valleys (i.e., over 120 km in the Hawkesbury, 40 km in the Georges) with huge tidal volumes for flushing, due to the great widths of the "ria harbors". Unlike valleys outside the sandstone basin, tidal-reach infill has been slow because of low rates of denudation. Thus, although there has been deltaic sand deposition in upstream reaches and finer infill downstream, tidal prisms have remained large. The fact that sub-tidal capacities have been reduced has further enhanced flushing capabilities. Infills of 30 m or more are recorded at the Brookland Bridge over the Hawkesbury and 20 m at Tom Ugly's Bridge over the Georges (Warner and Pickup, 1978).

Most efficient flushing is in the lower parts of the tidal rivers because ebb tides (two per day) only run out for over 6 hours. Lower riverine inputs further reduce flushing (Warner and Smith, 1979a, 1979b), as do sand dredging operations, which increase sub-tidal capacities (Warner and Pickup, 1978).

2.3. Human Impacts on Landforms and Rivers

In the last 200 years of European occupation human activities must be added in any consideration of geomorphological characteristics and process regimes. Land clearance and urbanization in particular have increased both runoff and sediment yield (Wolman, 1967). Vast amounts of sediment have been moved to provide route ways, aggregate, concrete and building materials. Accelerated natural processes have changed natural waterways. Materials have been reworked so that deposition in urban streams in construction phases has now been moved into tidal inlets to infill margins and sub-tidal capacity (Warner and Pickup, 1978).

Large dams to the south of Sydney on the Woronora Plateau and in the Blue Mountains were built to maintain water supplies as the population grew. These effectively cut off sediment supplies to the Upper Nepean and Lower

Warragamba Rivers (Warner, 1983; Scholer, 1974), thereby promoting further accelerated erosion in the Lower Nepean and Hawkesbury.

The resulting changed hydrologic and geomorphic processes have been further complicated by the alternating flood- and drought-dominated regimes (Erskine and Warner, 1988; Warner, 1987a, 1991, 1994). These have, in flood-dominated periods, doubled mean annual floods (or channel-forming discharges), which added to the pressures on channels to adjust by enlargement. Much of the reworked bank material finds its way into the channel bed (Erskine and Melville, 1983) or into tidal zones (Warner, 1991). In drought-dominated regimes, channels decreased in area through the deposition of in-channel incipient floodplains.

Thus the uplifted sandstone elements of the basin have been deeply dissected, marginal valleys have been drowned and only partially infilled, and the Holocene landscape has been greatly modified by the impact of human activity and natural regime shifts.

3. URBANIZATION AND IMPACTS ON RIVER SYSTEMS

This section addresses the processes and magnitude of urbanization, the related impacts on rivers or waterway development and sand and gravel extraction from active channels and floodplains.

3.1. Processes and Magnitude of Urbanization

The First Fleet settled at Sydney Cove in 1788 in a small valley drained by the perennial Tank Stream which was used for drinking water. Its name came from pool enlargement to store more water. Disposal of waste in the same catchment soon required a new water source from Centennial Park (Lachlan Swamps). Sydney developed westward along the shores of Port Jackson and the Parramatta River (Fig.3.1). Easiest development was on the shale plains of the basin center, where land could be farmed and developed. In the uplifted marginal dissected sandstone areas around the periphery, soils were infertile and relief too steep.

Expansion in the 19th century was largely east and west on the south side of the Harbor. Following World War I, development on the north side created the need for the Harbor Bridge (1932). Rail and ferry networks enabled widespread settlement adjacent to as well as inland from numerous waterways.

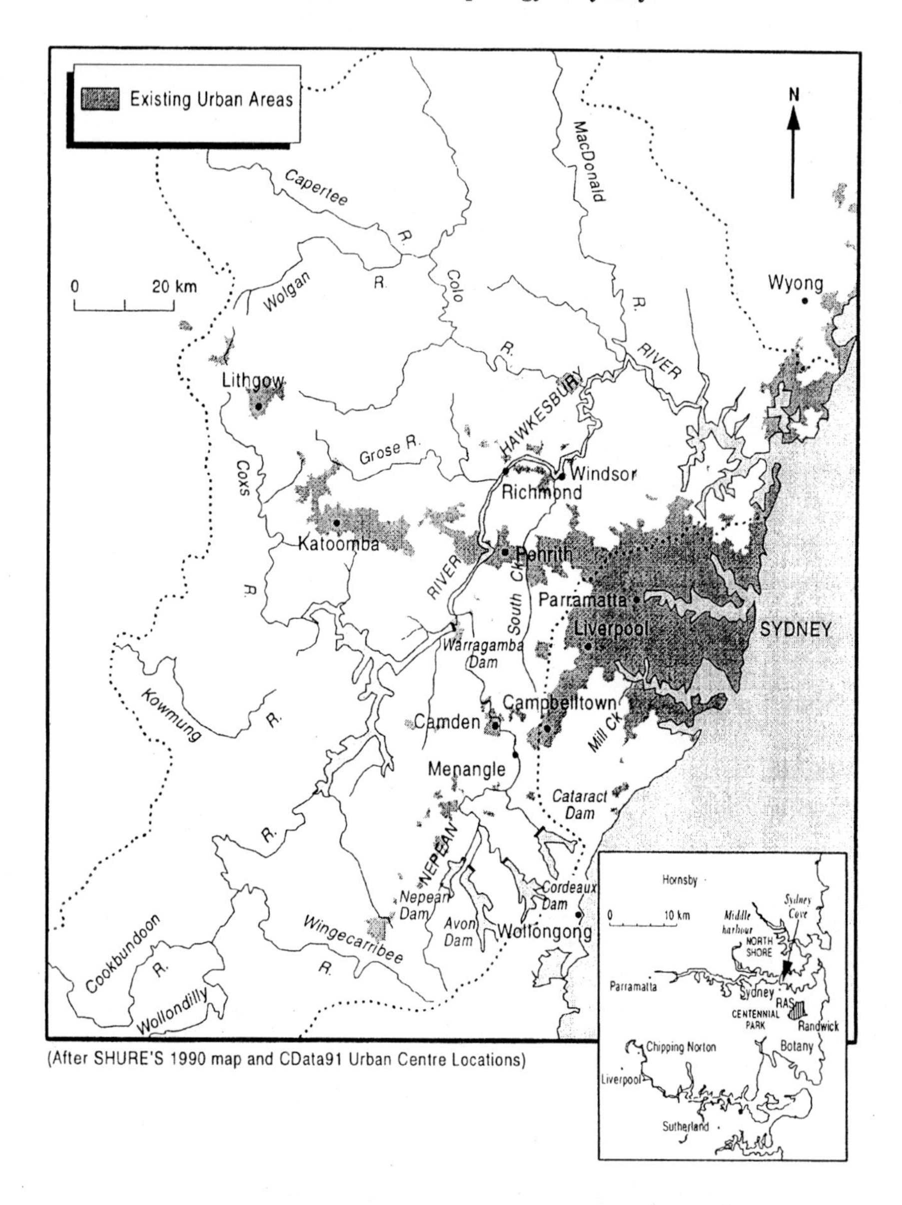

(After SHURE'S 1990 map and CData91 Urban Centre Locations)

Fig. 3.1. The urban spread of Sydney.

By 1994 (Table 3.1) 2,400 km^2 could be classified as urban, with the smaller coastal catchments being nearly fully urbanized (Cooks - 100 percent; Port Jackson - 87 percent) (Dowsett et al., 1995). The urban extent comprised a continuous built-up area across the shale plains to Penrith on the Nepean and to Windsor on its tidal extension, the Hawkesbury. Urban penetration onto adjacent plateaus was by way of route ways: Katoomba in the Blue Mountains, Sutherland on the Illawarra Ramp and the North Shore on the Hornsby Plateau.

Table 3.1. Catchment, urban and natural areas in the Sydney Basin.

Catchment	Area	Urban	Natural	Percent Urban
Hawkesbury-Nepean	22,000	1,540	14,930	7*
Georges	912	290	384	32**
Port Jackson	480	418	62	87
Cooks	102	102	0	100
Port Hacking	103	22	53	21
Narrabeen	42	10	31	24
Dee Why	4.8	4.8		100
Curl Curl	4.6	4.4	0.2	96
Manly	16.6	13.3	3.3	80
TOTAL	23,665	2,404.5	15,463.5[1]	10

Based on B. Dowsett et al. (1995).
*6 dams and 24 wastewater treatment plants (41 percent of catchment dammed).
**1 dam and 3 wastewater treatment plants.
[1]Area of cropping and grazing lands not given.

Urbanization has reduced infiltration and increased runoff to a variable extent. Shopping centers and commercial business development areas are virtually 100 percent impervious. High density housing on the shale plains has impervious areas of 70 to 90 percent of the total. Housing densities are generally lower in sandstone valleys on the plateaus, with impervious areas up to 25 percent.

Urbanization and the attendant reduction of infiltration have not only increased runoff; yields of sediments and contaminants are also high. The impacts of these on natural drainage have been at least threefold: (1) lined channels and stormwater drains have been required to convey surface runoff to natural channels; (2) natural channels have either been lined to reduce flooding potential or have been modified as part of a complex response to urban yields of additional runoff, sediments and solutes; and (3) tidal channels have accommodated the "construction" load (formerly accommodated in aggrading natural channels), which has now moved from tributaries into these low energy systems. In the tidal channels, sedimentation on the bed has reduced subtidal capacity and lateral accretion has lowered the tidal prism volume, thus changing the hydrodynamics of tidal channels (Warner et al., 1977). In the former case, this can increase flushing potential, but in the latter, lower prism volumes mean lower flushing. This latter set of processes began with urbanization when high yields in the construction phases were first stored in natural channels and then, when the supply reduced, it was reworked to move into tidal channels. In areas still being urbanized, the process is ongoing.

3.2. Impacts on Waterways

Growing urban requirements of water in the late 19th century required the damming of rivers for urban water and weirs for riparian supplies. The main location of the dams was on the Illawarra Ramp of the Upper Nepean and on the Georges River (Woronora) (Fig. 3.2). Because these were largely set in sandstone gorges, no great downstream changes have been noted. Sedimentation in these dams has been insignificant because of low denudation rates in the resistant and forested upper catchments.

In 1960 the Warragamba Dam (20,000,000 m^3) was opened (Fig. 3.2). This has sediments from over 8,000 km^2 of upper catchment and, in the post-1949 flood-dominated regime, sediment deficient overspills have had profound erosional effects on channels downstream (Warner, 1987a, 1987b).

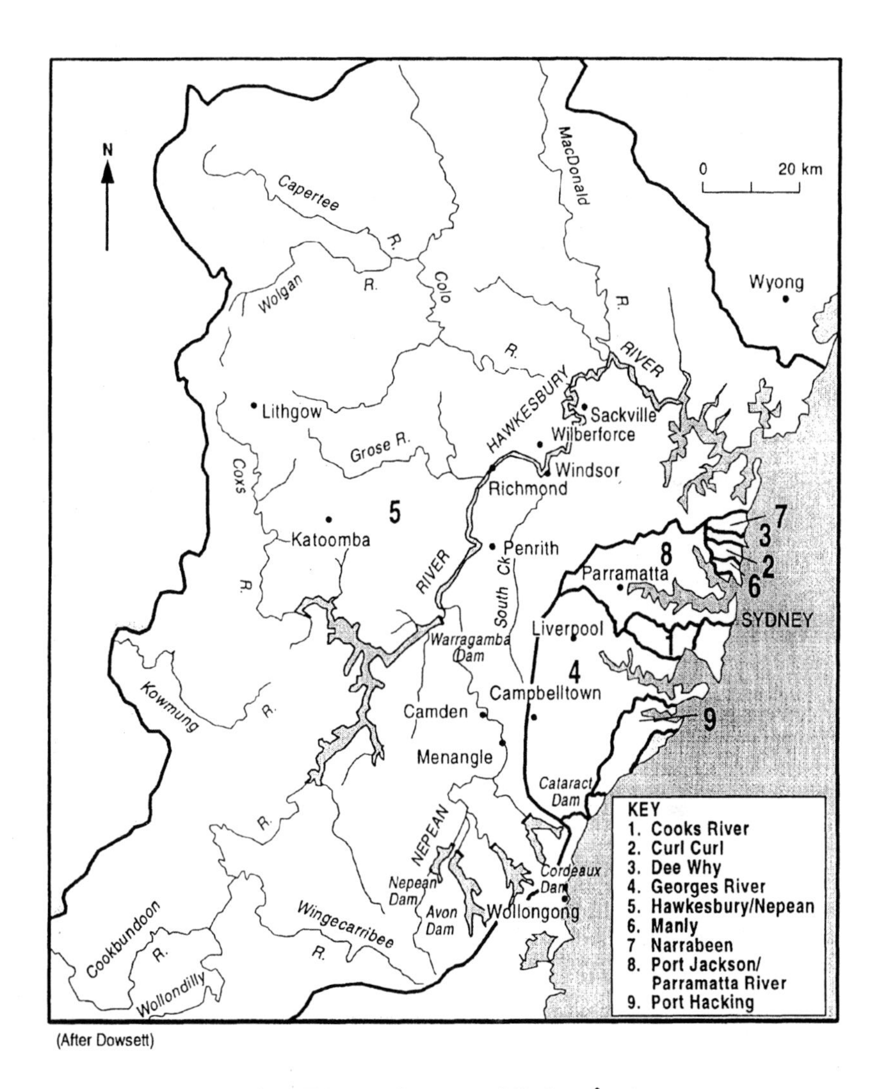

Fig. 3.2. The catchments of Sydney's rivers.

Extensive channelization was required in some rivers, especially where urbanization has increased flood peak frequency and magnitude. Much of the Cooks River has been lined in response to increased flooding impacts from upstream and upslope urbanization which added to the runoff, sediment load and pollution burdens.

3.3. Sand and Gravel Extraction

The need for aggregate in a booming city with no great fossil (terrace) spreads of sand and gravel, meant that in-channel dredging and meander-core floodplain exploitation were the main sources, especially to meet post-war needs for concrete and landfill.

In the Georges River, both bed and meander core exploitation at the head of the tidal river were used. The former reduced flushing, by increasing subtidal capacities. The latter greatly increased both the upper tidal prism as well as capacity in the top few kilometers. This completely changed tidal dynamics, and in dry times, has greatly reduced flushing potential, as a dye experiment revealed (Warner and Smith, 1979a, 1979b).

In the Hawkesbury-Nepean, much of the gravel and sand initially came from the bed. Although this did not affect tidal dynamics, bed lowering has threatened the weir foundations at Penrith (Warner, 1983). Subsequently the trend has been to quarry and dredge floodplains and Pleistocene terraces (Nanson, 1982; GHD, 1979).

Bank and bed erosion, particularly in the present flood-dominated regime, has been experienced both up- and downstream of these operations (Warner, 1987a, 1991, 1994). Lower energy in dredged reaches enhances sedimentation.

4. INDIRECT URBAN AND LAND-USE IMPACTS ON RIVERS

The indirect urban impacts on natural channels, both riverine and tidal, need to be considered more closely, together with examples of research substantiating these changes. Land-use changes are less severe, but not absent. All of these, including direct river modification or floodplain variation, need some consideration.

4.1. Urban Impacts on Rivers

In Sydney there are two kinds of natural channels: steep, short tributaries cut into sandstone and longer, flatter tributaries cut into shales. The former are characterized by pools and riffles, which can be composed of boulders or bedrock ledges (Petrozzi, 1994). The pools, in dry weather, can hold water and influent movement and evaporation can concentrate pollutants. Flora and fauna may also take up some of the nutrients. In the latter case, the original channel often had no linear continuity and discrete ponds along valley floors have been described as chains of ponds (Eyles, 1977; Gallagher, 1984; Bannerman, 1986).

Urban impacts on such channels have been variable. In the case of the sandstone valleys, no great morphological change has been noted. The greater runoff with high peak flows has merely reworked finer bed-load more rapidly. The high nutrient status of urban stormwater runoff (SWRO) has killed off native plants, which thrive in local phosphorus-deficient soils. These plants have been replaced by exotic species and vines and creepers, which can obliterate native vegetation (Dobson, 1986). Habitats can be destroyed in banks and on floodplains where they exist (Warner, 1995).

In the shale channels, metamorphosis (Schumm, 1977) has been more dramatic. Chains of ponds have been replaced by arroyo-like channels with the additional runoff resulting from removal of catchment vegetation, trampling by animals and other changes to upslope surfaces. High-energy flows, especially in confined valley-floor troughs, were able to incise the inter-pond depressions. The resulting channel has a gully-like appearance, with banks often in old colluvial and alluvial materials. Channel capacities were increased well above suggested $Q_{1.58}$ or $Q_{2.33}$ levels (Dury, 1969). In fact, a survey on part of the Cumberland Plain revealed channel capacities of $Q_{6\text{-}10}$, with incised channels up to Q_{20} (Pickup and Warner, 1976). Clearly, channel water storage in both types of intermittent system is high and part of this storage needs to be exceeded before flow continuity can be re-established temporarily. Only then can flushing of pools and other in-channel storage occur.

In older parts of the city, where flooding was a hazard with high runoff and higher flow peaks surcharging floodplains more frequently, there was often a perceived need to engineer the channel. Thus, lined channels reduced sinuosity, increased hydraulic radius, increased slope, decreased roughness and greatly increased the capacity. Any water added to such a channel could move at velocities of about 1 m/s or 3.6 km/hr towards receiving waters and there is little natural storage. Residence times are very short.

So now there are three types of channels above backwater tidal limits: the lined, engineered channel with no storage and rapid evacuation of stormwater and pollutants; the steep, discontinuous, cascading sandstone channel in which dry weather residence times of above 80 days have been estimated (Petrozzi, 1994); and long, flatter, shale channels, meandering below a floodplain which is seldom inundated because of increases in bankfull storage. Long residence times are conditioned by lower velocities, flatter gradients, bed-load storage and often backwater curve impacts induced on lower channels as they approach major perennial rivers.

4.2. Urban Impacts on Tidal Channels

Sydney's urban spread has been interrupted and influenced by long, west-east tidal channels in the major ria-like harbor systems. Many intermittent tributaries join these receiving waters from both shale and sandstone subcatchments.

This means that the streams of the last section have a backwater lower reach where energy is much lower. Indeed, because of the effects of the Holocene transgression and differential sedimentation in the tributary, such tributaries may be lower than the main channel. The wider, estuarine channels are subject to less water rise in floods and the lower energies allow delta forms in upstream parts and finer bed accretion downstream. So infill is two-phased: slow as well as progressive. Natural rates of infill in drought-dominated regimes were accelerated by urbanization. They have been further increased in the present flood-dominated regime. Nutrient-rich sediments from urban sources may be beneficial to bottom feeders, but the sediments may often contain contaminants less beneficial to the food chain. This has been enhanced in tidal areas by leachates from landfill sites which have decimated mangrove stands and infilled marginal reentrant valleys.

The impacts on urban tidal channels are such that bed accretion reduces capacities below the tidal prism and marginal sedimentation in the inter-tidal zone can reduce the volume of the flushing prism (Warner and Pickup, 1978; Warner et al., 1977).

4.3. Other Land-Use Impacts on Channels

These are similar to those documented by Wolman (1967), except that flood-dominated regimes (FDRs) (1799-1820, 1857-1900 and 1949-present) and drought-dominated regimes (DDRs) (1821-1856 and 1901-1948) have complicated the agro-urban colonizing sequence (Warner, 1994). Large, frequent floods in FDRs tend to promote width increases and depth decreases ($w^+ d^-$) and yet increase the channel capacity. Floods at Windsor above 9-10 m tend to promote post-settlement alluviation (PSA) which has partially buried fence posts. Depending on the stage in agricultural development, these impacts may or may not be present. In DDRs, for instance, the trend is for $w^- d^+$, but width reduction is dominated by the formation of an incipient floodplain inside the main channel to the height of the annual flood (about 6 m at Windsor). The last DDR (1900-1948) coincided with land abuse (overdevelopment, poor conservation, non-use of fertilizers, noxious weed spread, rabbits, etc.). Consequently, gullying added to channel load to help promote the incipient floodplains (Pickup, 1976).

Now in the current FDR, where conservation is pronounced, there is less introduced sediment. Subsequent sediment starvation induced by dams and extractions also means less sediment reaches downstream channels. These are promoting both bank erosion and bed degradation (not normally expected in a FDR) (Warner, 1994).

4.4. Links between Tributaries and Receiving Waters

Dry weather concentration of pollutants locked up in tributaries promotes local pollution problems. For instance, channels immediately below source areas have been found to be highly contaminated at such times. The types of junctions with receiving waters can sometimes further promote ponding of water and delays in moving pollutants and sediments into receiving waters.

Entrances may be normal, that is an open channel entering the mainstream. This may be subject to ponding through backwater flows. A big flood in the main channel will back up in the tributary causing ponding, additional flooding and, perhaps, sedimentation in the lower parts of shale channels and on their adjacent floodplains. This happens in South Creek, near Windsor.

Where the tributary is small and a levee can form across its entrance adjacent to the main channel, there may be ponding in the lower part of the tributary which may not enter the channel without drainage ditches and flood gates. Pollutants may accrete on the floor of such flood basins and may explain the high distribution of heavy metals found in some sedimentation environments (Thiel, 1993; Parker, 1992; Thoms and Thiel, 1995).

Tidal tributaries have broad entrances into the main channel, where both backwater effects and gradients are minimal. Floods would seldom rise much more than a meter, because of inlet and main channel widths.

The reverse-hung tributary, where local infill has not been able to keep up with sedimentation in the main stream, is often levee bound. Then again, ponding in the lower tributary can cause infilling from accelerated urban erosion. Sediments may contain nutrients, bacteria and heavy metals.

Flood mitigation structures (GHD, 1979) have similar impacts. The strengthening and raising of the levee to a design-flood height (Q_2 or Q_3) keeps all floods to these levels in the channel. There is no overbank dissipation of energy and confining it in-channel can promote erosion of bed and bank.

More importantly, the polluted, ponded tributary water will only be released when floodgates on main drainage channels open after the main flood has passed. Pollutants will then be introduced to receiving waters during the recession link when the hydrograph is low.

5. POLLUTION IN SYDNEY

There are many kinds of pollution of land, air and water. For Sydney, the two considered here are stormwater and sewage pollution of waterways.

5.1. Stormwater Runoff

Pollution in metropolitan areas can come from several sources: atmospheric (gases and nucleii), land-based (concentrated in landfills, diffuse elsewhere), water-based (from earlier runoff, pipes, sewer leaks), and from point sources (stormwater runoff from drains, sewage overflow valves, leaking septic systems).

Land-based and atmospheric-based pollutants cover the surface to be affected by rapid surface runoff from impervious urban surfaces. This, in general, will form stormwater runoff, which can find its way into pipes, drains, lined channels, modified normal and tidal channels. Its quality is often worse than sewage (Ellis, 1979; Ellis et al., 1995).

By and large stormwater consists of: excess rainfall, outside water use runoff, clean leaks above user points and contaminated leaks below these. Clean source areas can reduce its impact, but usually management is attempted in channels or pathways above receiving waters. In lined, engineered channels, there is only rapid evacuation of water. However, in natural channels, assimilation may be induced by channel form and internal morphology, as well as by fauna, by vegetation, by sedimentation, etc. This is true of both shale and sandstone channels.

5.2. Sewage

In a reticulated sewage system, domestic and industrial wastes should move safely from water-use points to treatment plants where discharge quality would depend on the level of treatment.

In Sydney, 1.3 million of the 1.4 million dwellings are on sewer lines, which total 21,000 km in length. In 1994, 27,000 properties were still unsewered, with about 16,000 of these in the Hawkesbury-Nepean Valley (Dowsett et al., 1995). Many of these were still septic although new aerated wastewater treatment systems (AWTS) are making inroads in this area. Ninety percent of the reticulated load is handled by ocean outfalls in 10 plants with 3 major offshore (new) discharge points. These handle 1.03 million m^3 per day but 95 percent is subjected to primary treatment or less and only 5 percent to secondary treatment. The other 10 percent or 110,0000 m^3 is handled by 24 inland plants many of which discharge into the Hawkesbury-Nepean River. All have tertiary treatment but nutrient loads are high and variable (Dowsett et al., 1995).

Thus, the total sewage volume is 1,140,000 m^3 per day or about 308 L per person. Total water use per day averages 1,700,000 m^3 with 20 percent of that unaccounted for in leaks and unknown connections, etc. (Dowsett et al., 1995). These volumes are large, with an annual sewer load of 367 million m^3 = 150 mm depth equivalent over the 2,400 km^2 of metropolitan area. Rainfall on Sydney averages 890 mm or 2,136 million $m^{3,}$ which is higher than total demands for storage (620 million m^3) by a factor of more than 3. Most of this water becomes stormwater, other than that evaporated or lost to groundwater.

The problem with Sydney's sewage is that at times leakages are considerable (Table 5.1). Smalls (1995) suggested that as much as 40 percent of the raw sewage reached Sydney Harbor and did not reach the treatment plants. In considering overflows, Dowsett et al. (1995) suggested that there are more than 3,000 relief points. These are overflow points in wet weather, with the main problems at Middle Harbor, the Upper Parramatta River, Cooks River and the Georges River at Chipping Norton. After a particularly heavy storm in February 1992, it was estimated that about 6.3 million m^3 of sewage overflowed into Port Jackson; this is about 6 times the load going to the ocean per day (Dowsett et al., 1995).

Often, flow interchange occurs between groundwater and sewage. In dry weather, up to 30 percent of the average sewage load may be groundwater leaking into the system. On the other hand there can be sewage leakage to groundwater environments, notably in the aeolian Botany Sands of the Botany-Randwick area when water tables fall in dry weather.

Table 5.1. Stormwater and sewage overflow percentage contributions.

	Total Coliform Overflow		Total Nitrogen (TN)		Total Phosphorous (TP)	
	Overflow	Stormwater	Overflow	Stormwater	Overflow	Stormwater
Port Jackson	84	16	20	53	26	68[1]
Upper Georges	32	68	7	93	19	81
Lower Georges	79	21	34	65	53	47[2]
Botany Foreshore	85	15	13	40	24	69[3]
Cooks River	92	8	54	46	55	45
Narrabeen	11	89	8	92	3	97
Dee Why	4	96	5	95	3	97
Curl Curl	56	44	4	95	5	97[4]
Manly	80	20	19	80	19	81[5]

Based on B. Dowsett et al. (1995).
[1]TN includes 9 percent direct rainfall and 18 percent industries etc.;
TP includes 4 percent rain and 2 percent industries.
[2]TN includes 1 percent direct rainfall.
[3]TN includes 47 percent rainfall; TP includes 7 percent rain.
[4]TN includes 1 percent rainfall.
[5]TN includes 1 percent rainfall.

Lack of infrastructure improvement as housing evolves from low to medium or high density can promote leaks by overcharging of the system, as can the high frequency of low intensity earthquakes which can crack pipes. In sandstone areas, sewer lines are either along slopes, where they are subject to slip and creep, or in creek beds, where bed-load abrasion can crack main pipes.

Non-sewered areas may suffer groundwater and channel contamination by leakage from septic systems, especially after a wet spell. Pumpout systems where there was no soil for excess septic tank absorption depended on weekly tanker services to remove effluent. However, in rain, some owners released effluent into gutters and drains. Nutrient-rich effluent is irrigated on unused parts of the garden or passive parts of home blocks (Martens and Warner,

1991, 1992, 1995) from the new aerated wastewater treatment systems (AWTS). All these may add to stormwater pollution.

6. ACCOMMODATION OF POLLUTION BY WATERWAYS

A neglected aspect of fluvial hydrology is the role of the channel storage in temporarily storing or accommodating part of the contaminant load. Its behavior in dry and wet weather is so different that it is worthwhile to consider each case in turn.

6.1. Dry Weather Pollution

In dry periods, when precipitation is insufficient to cause effective runoff, or where the runoff is insufficient to re-establish continuity throughout the drainage network, contaminants are retained locally in pools, ponds and other minor storages, like farm dams. Here, evaporation and influent conditions can reduce water levels and concentrate contaminants. Some may be stripped by nutrient-using vegetation or other life forms.

This form of pollution is essentially local, where concentrations of contamination may increase steadily through time and where they decrease with channel distance away from source areas. Where water bodies such as these are not part of the public domain, they may present few problems. However, ornamental ponds in one of Sydney's most used parks have become highly polluted by continuous injection of polluted stormwater wastes, as well as those from horses at the Royal Agricultural Society's Showground. They then become a health threat to 3.5 million people, 220,000 dogs and 60,000 horses who use the park annually. The major threat is blue-green algal blooms where contact or ingestion can cause health problems.

Small falls of rain may cause runoff from streets and impervious surfaces, but the volume of water may be insufficient to establish flow continuity in more than a short distance of the channel. For instance, at Lucas Heights solid waste disposal site, runoff from the State's largest landfill site may be enough to cause a significant flow in the upper parts of Mill Creek. However, two km downstream, no flow increase may be experienced because the flow has been taken up by available pool storage upstream (M. Petrozzi, pers. comm.).

6.2. Wet Weather Pollution

When rain events are longer and more intense, infiltration capacities throughout subcatchments may be overcome. Then runoff is more general and channel storage is satisfied so that there is flow continuity throughout the system. Pollution then comes from three sources: rain, the ground surface, and that already stored in the channel. However, the volume of runoff in wet periods may be enough to dilute the high concentrations of contaminants reached in shrinking pools. Thus, in wet weather, contamination ceases to be local and can affect the whole system, but this widespread impact should be at lower concentration levels. However, in Sydney this is not the case, especially in areas where sewage leaks are pronounced and where obsolete sewers are surcharged. In the 1970s, some of the highest coliform readings were obtained during floods in the Georges River. Comments by Smalls (1995) already referred to confirm this, as do sewer overflow values given in Table 5.1.

Thus the role of dry-weather pollution in restricting the frequency of introducing pollutants to receiving waters to 6 to 12 occasions a year is offset to an extent by the level of contamination in wet-weather flows. However, since continuity may not exist for 300 or more days a year in "natural" streams, there are opportunities for large, dynamic receiving waters to recover. In the case of lined channels, there is no storage or accommodation of contaminants. Consequently, runoff of polluted waters into receiving waters, especially if they are small and stagnant like Centennial Ponds, is more continuous (Warner, 1996).

7. GEOMORPHOLOGY AND POLLUTION MANAGEMENT

Landforms and geomorphic processes in the Sydney region do offer some natural advantages in the mitigation of pollution, but there are also some limitations. Both need to be considered.

7.1. Positive Attributes

These may be considered in three ways:

1. **Tidal flushing**
   The large ria-like harbors of Port Hacking, Port Jackson and Broken Bay, together with the once well-flushed Botany Bay, have huge tidal

prisms which twice daily can flush the lower parts of tidal systems with ocean water. This flushing extends well inland with the Georges River extending to Liverpool, the Parramatta River to Parramatta and the Hawkesbury to Richmond. However, the total distances are too great in river estuaries to allow the flushing experienced in the lower, more open harbors, unless there is upstream catchment discharge into the river. Even the narrower Georges River has a spring tide prism of more than 30 million m^3 (Warner and Pickup, 1978). Sydney Harbor and Broken Bay would have much larger volumes, probably 100 million m^3 in the case of the former and several 100 million m^3 in the latter.

2. **Catchment flushing**
Some rivers have large parts of their catchments above urban areas (Fig. 3.2; Table 3.1). Thus, periodic high discharges can dilute contaminated urban runoff. However, large floods drive saltwater out of the estuaries with their large volumes of "fresh" water, thereby increasing coliform counts and their ability to survive. It then takes time to re-establish salinity profiles back in the estuaries. Those catchments which are totally urbanized or nearly so do not have these advantages, i.e., Cooks River (100 percent), Port Jackson (87 percent) (Dowsett et al., 1995).

3. **Low, infrequent tributary inputs**
"Natural" tributaries, whose catchments occupy a large part of the 2,400 km^2 of urbanized surfaces, have been shown to accommodate small flows within their storage potential, only releasing contaminated water when flow continuity is periodically re-established throughout the system. Thus, for 300 days or so a year, no local pollution is added to large receiving waters, such as perennial rivers, riverine estuaries and ria-like harbors. This is not the case with near-shore environments because the highly organized stormwater drainage in beachside suburbs can deliver polluted waters rapidly to the beaches.

These three factors: twice daily flushing by large volumes of sea water, periodic flushing by runoff from forested catchments and only infrequent injection of polluted stormwater from intermittent tributaries, can be related to the present landscape and the patterns of its recent evolution, even to the modifications effected by urbanization and short-term natural regime shifts. In combination, and utilized positively, they could further achieve some environmental sustainability.

7.2. Limitations and Problems

The effectiveness of these positive attributes has been limited in several ways, often inadvertently. For instance, tidal prism volumes have been reduced in lower systems by marginal, inter-tidal sedimentation associated with urbanization (Warner et al., 1977; Warner and Pickup, 1978). Sand and gravel removals at the top of the Georges River tidal channel have created lakes attached to the tidal river which require many times the former volume to satisfy tidal dynamics. So, reduced incoming volumes from Botany Bay have to travel faster in middle reaches to fill dredged areas, thereby increasing bank erosion risks, with much greater velocities. A dye experiment (Warner and Smith, 1979a, 1979b) revealed that ebb tide flushing only reached Milperra Bridge and these conditions of poor flushing were likely to prevail for up to 40 percent of the time (i.e., virtually no river runoff). So tidal flushing is impaired by sand and gravel removals with increases in both capacity and prism volumes in the upper parts of a tidal river. In the 140 km of tidal channel, Hawkesbury flushing is obviously less effective throughout the upper reaches. This may explain the persistence of blue-green algae in the vicinity of Sackville.

The water quality of the Hawkesbury should be good with only 7 percent of its 22,000 km^2 catchment urbanized. However, 41 percent of the catchment is below Windsor and other urban areas. The downstream tributaries of the Colo and MacDonald Rivers cause significant backwater effects. These can pond floodwaters and promote upstream sedimentation, and the backwater effects of a 6-km wide floodplain reducing to 200 m in width 30 km downstream at Sackville create hysteresis loops where recession velocities at Windsor are exceedingly low, promoting sedimentation and fallout of heavy metals (Warner, 1994). A further 43 percent of the catchment is above the urban area but is upstream of Sydney's water-supply dams. Consequently, runoff from only 9 percent of the catchment is available for flushing runoff from the urban 7 percent. This has been further compromised by huge numbers of farm dams that are required in the rain shadow area of the basin. Consequently, runoff from the shale plains may be from only half or less of that limited catchment area. Added to this (Table 3.1) there are 24 sewerage treatment plants in the catchment releasing large volumes of nutrients. The water requirements upstream and backwater impacts from downstream, coupled with limited runoff and polluted runoff from non-urban surfaces, do little to help flush and dilute urban stormwater runoff.

In the Cooks River, urbanization is old-well over 100 years in places- and so is the late 19th Century sewage infrastructure. Much has been overbuilt by medium to high density housing. So the 100 percent urban catchment, which has mainly lined channels, suffers 92 percent sewage overflows of total coliforms, 54 percent TN (total nitrogen) and 55 percent TP (total phosphorus) (Table 5.1) in a narrow, engineered channel which has low tidal flushing from Botany Bay. There, flushing circulation has been compromised by impacts of parts of two airport runways and Port Botany developments, which project into the bay and have affected tidal circulation.

8. CONCLUSIONS

Natural landforms and process combinations do play some part in the flushing of pollutants introduced to receiving waters storage along all kinds of transmission paths or channels. The strong engineering and political bases of former decision-making have not always taken these spatial and temporal aspects into consideration. Thus, many of the advantages may have been partially lost. However, it may be possible, with better planning and replacement of dated infrastructure, to work with these potential advantages for environmental, if not ecological, sustainability.

9. REFERENCES

Bannerman, S.M. 1987. An investigation of chains of ponds as a distinctive hydrogeomorphic form. Unpublished B.A. Honors Thesis, Department of Geography, The University of Sydney.

Branagan, D. (ed.), C. Herbert, and T. Langford-Smith. 1976. An outline of the geology and geomorphology of the Sydney Basin. Local Excursion Guide for 25th International Geological Congress, Sydney.

Bridgman, H., R.F. Warner, and J. Dobson. 1995. *Urban Biophysical Environments*. Oxford University Press, Melbourne.

Dobson, J.R. 1986. Urbanization and vegetation changes. In Dragovich, D. (ed.) *The Changing Face of Sydney*. Geographical Society of NSW, Sydney, 42-60.

Dowsett, B., G. Mathew, C. Mercer, B. Paterson, and D. Vincent. 1995. *A New Course for Sydney Water*. Final Report of the Sydney Water Project.

Dury, G.H. 1969. Hydraulic geometry. In Chorley, R. J. (ed.) *Water, Earth and Man.* Methuen, London, 319-329.

Ellis, J.B. 1979. The nature and sources of urban sediments and their relation to water quality: a case study from north-west London. In Hollis, G. E. (ed.) *Man's Impact on the Hydrological Cycle in the United Kingdom.* Geo Abstracts, Norwich, 199-216.

Ellis, J.B., R.B. Shutes, and D.M. Revitt. 1995. Ecotoxicological approaches and criteria for the assessment of urban runoff impacts on receiving waters. In Herricks, E. E. (ed.) *Stormwater Runoff and Receiving Systems.* CRC Press, Boca Raton, 113-126.

Erskine, W.D. and M.D. Melville. 1983. Impact of the 1978 floods on the channel and floodplain of the Lower Macdonald River, NSW. *Australian Geographer,* 15:204-292.

Erskine, W. D. and R.F. Warner. 1988. Geomorphic effects of alternating flood-and drought-dominated regimes on New South Wales coastal rivers. In Warner, R. F. (ed.) *Fluvial Geomorphology of Australia.* Academic Press, Sydney, 303-322.

Eyles, R. J. 1977. Changes in drainage networks since 1820, Southern Tablelands, New South Wales. *Australian Geographer,* 13:377-386.

Gallagher, W. 1984. Changes in drainage forms on the Cumberland Plain since 1788. Unpublished B.Sc. Honors Thesis, Department of Geography, The University of Sydney.

Gutteridge, Haskins and Davey (GHD). 1979. Hawkesbury-Nepean floodplains study - strategy report. Report for NSW Department of Public Works.

Martens, D.M. and R.F. Warner. 1991. Evaluation of environmental impacts of aerated wastewater treatment systems. Teaching Company Scheme Mid-Project Report, 92 pp.

Martens, D.M. and R.F. Warner. 1992. Sewage treatment and urban runoff quality. *Proceedings Joint Meeting of New Zealand Geographic Society and Institute of Australian Geographers*, Auckland, 473-484.

Martens, D.M. and R.F. Warner. 1995. Impacts of on-site domestic wastewater disposal in Sydney's unsewered urban areas. Department of Geography Monograph, The University of Sydney, 102 pp.

Nanson, G.C. 1982. An investigation of possible river erosion of the high bank of the Nepean River from Victoria Bridge to McCann's Island. Report for Penrith Lakes Development Corporation, 22 pp.

Parker, C. 1992. Heavy metals in the sediments of the Hawkesbury Valley: an assessment of the spatial and temporal distribution. Unpublished B.Sc. Thesis, Department of Geography, The University of Sydney.

Petrozzi, M. 1994. Fluvial controls on water quality: potential leachate migration in a steep, discontinuous, bedrock stream. Unpublished B.A. Honors Thesis, Department of Geography, The University of Sydney.

Pickup, G. 1976. Geomorphic effects of changes in river runoff, Cumberland Basin, New South Wales. *Australian Geographer,* 13:188-194.

Pickup, G. and R.F. Warner. 1976. Effects of hydrologic regime on magnitude and frequency of dominant discharge. *Journal of Hydrology,* 29:51-75.

Scholer, H.A. 1974. Geomorphology of New South Wales rivers. *University of NSW, Water Research Laboratory Report No. 139.*

Schumm, S.A. 1977. *The Fluvial System.* Wiley, New York.

Smalls, I. 1995. Cited in Anon. *Green Cities.* Australian Urban and Regional Development Review, Strategy Paper No. 3, 119.

Thiel, P. 1993. The impacts of urbanization on the bed sediments of South Creek. Unpublished B.A. Honors Thesis, Department of Geography, The University of Sydney.

Thom, B.G. and P.S. Roy. 1981. Relative sea levels and coastal sedimentation in Southeastern Australia in the Holocene. *Journal of Sedimentary Petrology,* 55:257-264.

Thoms, M.C. and P. Thiel. 1995. The impact of urbanization on the bed sediments of South Creek, New South Wales. *Australian Geographical Studies,* 33:31-43.

Warner, R. F. 1983. Channel change in the sandstone and shale reaches of the Nepean River, New South Wales. In Young, R. W. and Nanson, G. C. *Aspects of Australian Sandstone Landscapes.* Australian and New Zealand Geomorphology Group, Special Publication No. 1, 106-119.

Warner, R.F. 1987a. The impacts of alternating flood- and drought-dominated regimes on channel morphology at Penrith, New South Wales, Australia. *International Association for Scientific Hydrology*, Publication No. 168, 327-338.

Warner, R.F. 1987b. Spatial adjustments to temporal variations in flood regime in some Australian rivers. In Richard, K. S. (ed.) *Rivers: Environmental Process and Morphology*, Blackwell, Oxford, 14-40.

Warner, R.F. 1991. Impacts of environmental degradation of rivers, with some examples from the Hawkesbury-Nepean system. *Australian Geographer,* 22:1-13.

Warner, R. F. 1994. A theory of channel and floodplain responses to alternating regimes and its application to actual adjustments in the Hawkesbury River, Australia. In Kirkby, M. J. (ed.) *Process Models and Theoretical Geomorphology*, Wiley, Chichester, 173-200.

Warner, R. F. 1995. The role of stormwater management in urban rivers. Paper presented at International Association of Geomorphologists Regional Congress, Singapore.

Warner, R.F. 1996. Do we really understand our rivers? Or rivers in the pooh - semper in excreta. *Australian Geographical Studies,* 34:3-17.

Warner, R.F. and G. Pickup. 1978. Channel changes in the Georges River between 1959 and 1973/76 and their implications. Report for Bankstown Municipal Council, 84 pp.

Warner, R.F. and D.I. Smith. I. 1979a. The use of fluorometric tracers to study effluent movement in the Georges River, New South Wales, Australia. *Proceedings 10th New Zealand Geographical Society and 49th ANZAAS Congress,* 48-53.

Warner, R.F. and D.I. Smith. 1979b. The simulation of effluent movement in the tidal Georges River, New South Wales, Australia, using the dye tracer Rhodamine W.T. Report for New South Wales State Pollution Control Commission, 75 pp.

Warner, R. F., E.J. McLean, and G. Pickup. 1977. Changes in an urban water resource, an example from Sydney, Australia. *Earth Surfaces Processes,* 2:29-38.

Wolman, M.G. 1967. A cycle of sedimentation and erosion in urban river channels. *Geografiska Annaler,* 49A:385-395.

ARROYO CHANGES IN SELECTED WATERSHEDS OF NEW MEXICO, UNITED STATES

Allen C. Gellis and John G. Elliott[1]

ABSTRACT

Whether climate, land use, or intrinsic variables is the main control of arroyo incision in the American west has been debated for more than a century. The climatic position for example, has maintained that periods of low rainfall followed by intense storm activity promotes arroyo incision. Average annual rainfall intensity for Santa Fe, New Mexico, shows more frequent high-intensity rainfall events in the period 1868 to 1880 and an increase in rainfall intensity beginning about 1967. This precipitation data could indicate that conditions are similar to the late 1800s when a period of incision occurred; although precipitation records at Santa Fe are not representative of the entire American west. By contrast, recent studies in the American Southwest indicate that many arroyos currently are aggrading. The Rio Puerco, New Mexico, incised around 1885 and, based on aerial photographic comparisons and replicate surveys, the arroyo is continuing to evolve. Inner-channel width-to-depth ratios have decreased since 1977, and the inner floodplain has aggraded 0.3 to at least 1.0 meter by overbank deposition. In one cross section the channel-bed elevation increased 2.55 meters (m), the inner-channel width decreased from 40.6 m to 23.4 m, and the cross-sectional area decreased by 125 square meters (m^2) from 1936 to 1995. Results of resurveys between 1992 and 1994 of 85 channel cross sections in three arroyos near Zuni, New Mexico, indicated that 72 percent of the cross sections are aggrading. This apparent discrepancy may be related to the intrinsic thresholds of arroyo systems, since the modern arroyos examined here are at a different stage of their development than they were in the late 1800s.

KEYWORDS

Arroyo changes, climate, watersheds, high-intensity rainfall events

---

[1] Hydrologist, U.S. Geological Survey, 4501 Indian School Road, Suite 200, Albuquerque, New Mexico 87110

1. INTRODUCTION

Arroyos are channels incised into valley alluvium that can be tens to hundreds of meters wide and several meters to tens of meters deep. An episode of arroyo incision that occurred in the late 1800's in the American west has intrigued geomorphologists for more than a century. A widespread debate about the cause of this incision centered on climate change, land use, and more recently, intrinsic geomorphic conditions. This debate was litigated when the Zuni Tribe of New Mexico sued the United States government in the 1970s, claiming that mismanagement of their land, overgrazing, and timber harvesting led to erosion, of which arroyo incision was part (Landon, 1990).

The position that climatic change causes arroyo incision postulates a relation between changes in precipitation frequency and storm intensity and changes in erosion. For example, periods of low rainfall that are followed by intense summer storm activity may provide the impetus for arroyo incision (Leopold, 1951; Cooke, 1974; Balling and Wells, 1990). Analysis of climatic records in Santa Fe, New Mexico, by Leopold (1951) indicated that the period 1850 to 1880 was characterized by a decrease in low-intensity rainfall and by an increase in high-intensity rainfall. Frequent low-intensity rainfall sustains vegetative growth, but low-intensity rainfall was deficient between 1850 and 1880. High-intensity rainfall which was above average in the late 19th century resulted in rapid runoff and promoted erosion. Schoenwetter and Dittert (1968) noted that periods of erosion are characterized by a summer moisture regime where most annual precipitation falls during a long summer of intense thunderstorms, whereas periods of alluviation are characterized by a winter moisture regime where precipitation accumulates as snow, runs off slowly, and infiltrates more readily. Leopold and others (1966) suggested that arroyo incision is associated with increasing aridity and aggradation with increasing humidity.

The position for climate as a control on arroyo incision is strongly supported by several Quaternary incision/aggradation cycles in the Western United States that have been documented in the literature (Leopold and Snyder, 1951; Cooley, 1962; Cooke and Reeves, 1976; Hall, 1977; Waters, 1985). Because changes in land use could not have caused incision episodes prior to human habitation, climate seems to be a logical explanation.

The position that land use causes arroyo incision has centered on human activities such as overgrazing, farming, and timber harvesting, which reduce vegetative cover and infiltration rates (Thornwaite and others, 1942; Antevs, 1952; Cooke and Reeves, 1976). These activities promote increases in surface runoff or channelization of streamflow, which can initiate arroyo incision. Land-use practices may have been a factor in at least one episode of pre-19th century arroyo incision. Judd (1964) theorized that residents of Pueblo Bonito in New Mexico may have inadvertently concentrated Chaco River flows in the center of the valley where these flows converged to form a 12th century arroyo.

A third position is that arroyo incision is intrinsic or has a natural cyclicity (Schumm and Hadley, 1957; Patton and Schumm, 1975). Aggradation over time causes the valley gradient to steepen until a threshold slope is exceeded and the valley incises. This argument contends that arroyo incision would occur regardless of climatic or land-use changes, although climate and human activities might affect the timing of incision and aggradation. In other words, alluvial valleys in the late 1800s were ready to incise.

This paper discusses arroyo formation and evolution, and describes recent arroyo changes in the southwestern United States based on field observations and information presented by other investigators. Analyses of geomorphic and climatic data are presented to determine possible links between arroyo changes and climate.

2. ARROYO CHANGES IN THE HISTORICAL RECORD

2.1. Arroyo Evolution

Sufficient time has passed since arroyo incision began in the late 19th and early 20th century to observe a pattern of change in arroyo morphology. Arroyos initially are formed by vertical incision of a streambed, by upstream migration of headcuts, or by coalescing of discontinuous gullies. Following initiation, the maximum depth of an arroyo is quickly attained (Leopold and Miller, 1954). Removal of valley fill continues by lateral erosion of valley alluvium rather than by continued downcutting.

Through time, the morphology of arroyos in larger watersheds evolves as it responds to changes in upstream erosion and sediment delivery. Parker (1977) observed the evolution of arroyo-like features in a rejuvenated watershed in an experimental flume. As headcuts and the locus of sediment production advanced up tributaries, sediment delivered to the lower main-stem reach increased geometrically and exceeded the transport capacity of the channel. Deposition, braiding, and channel widening resulted as the channel aggraded within the incised reach.

Elliott (1979) and Gellis and others (1991) documented channel changes in arroyos and termed these changes arroyo evolution. Arroyo evolution involves sequential channel deepening, widening, inner floodplain formation, and, in larger arroyos, channel pattern adjustment. These arroyo changes generally proceed over time from the mouth of arroyos to upstream reaches. Leighly (1936) and Cooke and Reeves (1976) noted a general tendency for creation of a new channel and floodplain within an arroyo. Concurrent with these arroyo changes is a decrease in the amount of sediment delivered from the watershed as sediment is deposited in the arroyo (Gellis et al., 1991; Gellis, 1992). Annual suspended-sediment concentrations at the U.S. Geological Survey streamflow-gaging station Rio Puerco near Bernardo, New Mexico, have decreased from greater than 0.13 to less than 0.06 metric ton per cubic meter of runoff from 1948 to 1994 (Fig. 2.1).

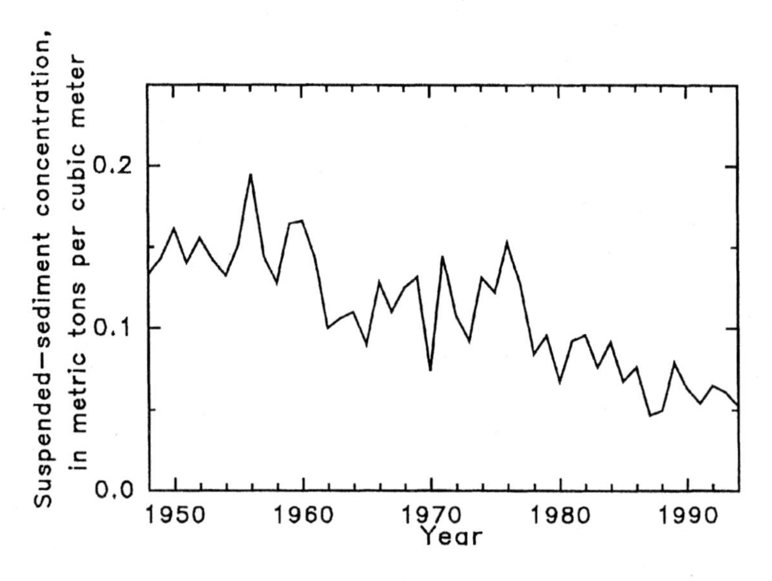

Fig. 2.1. Suspended-sediment concentration, 1948-1994, Rio Puerco near Bernardo, New Mexico.

Elliott (1979) described a five-stage sequence of arroyo cross-section and channel pattern evolution for the Rio Puerco. Presently (1996), reaches of the lower Rio Puerco are exhibiting relatively stable channel patterns and inner floodplain vegetation growth. However, a general decrease in channel width-to-depth ratios and overbank floodplain aggradation observed in 1994 and 1996 in the lower reaches suggests that Rio Puerco evolution is continuing.

Morphologic evolution similar to that of arroyos was observed in incised channels in the southeastern United States (Schumm and others, 1984; Simon, 1989). Schumm and others described a five-stage model of channel evolution from disequilibrium to stability for Oaklimeter Creek, Mississippi. In this model, attainment of channel stability involved a reduction in slope and an increase in width after incision. For the wide, gently-sloping stream, unit stream power decreased and the channel aggraded. Sinuosity changes and vegetation establishment in the channel caused further vertical accretion. Although Schumm and others (1984) were describing incised channels in Mississippi, similar channel morphologic trends are observed in arroyos of the southwestern United States.

Gellis (1988) developed a seven-stage arroyo evolution model based on studies in northeastern Arizona (Fig. 2.2). The arroyo is formed

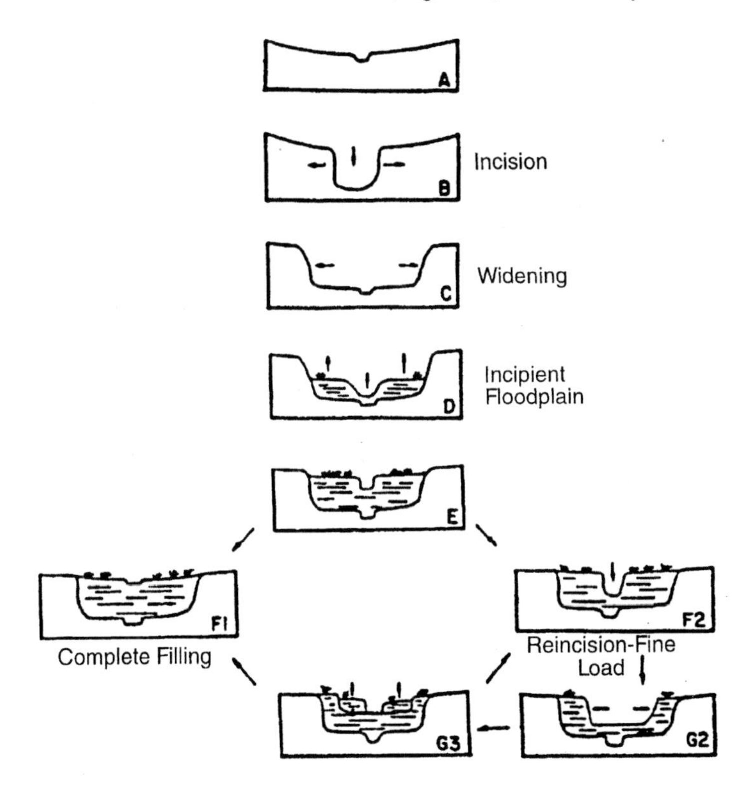

Fig. 2.2. Sequence of cross sections illustrating arroyo evolution.

by the upstream migration of a headcut. Width-to-depth ratios at the top of the channel cross section are typically lower in reaches immediately downstream from the headcut than in reaches upstream from the headcut where the channel has not yet deepened.

2.2. Recent Aggradation

Several researchers have presented evidence suggesting widespread arroyo aggradation in the Western United States in the 20th century. Leopold and others (1966) concluded, in their study of sediment sources in an ephemeral drainage network, that channels near Santa Fe, New Mexico, were aggrading from 1958 to 1964. Average net aggradation for channels in their study ranged from 0.0009 to 0.030 meters per year. The main source of sediment for channel aggradation was sheetwash erosion. Continued monitoring showed aggradation continuing to 1974 (Leopold, 1976), although the rate of aggradation in these channels was slower from 1968 to 1974 than from 1961 to 1968. Reports on channel changes from eight other sites in the Western United States showed either aggradation or equilibrium; none showed degradation (Emmett, 1974).

Graf (1987) identified a young stratigraphic unit deposited between 1943 and 1980 in 10 drainage basins of the Colorado Plateau. Rates of sedimentation for this unit were almost twice the rate of sedimentation for the next oldest unit, which was deposited between 1250 and 1880. The younger unit identified by Graf documents the aggradation cycle reported in the southwestern United States. Modern aggradation in the Little Colorado River began about 1952 (Hereford, 1984). Basinwide aggradation in the Paria River, Utah and Arizona, began in the early 1940s (Hereford, 1986).

2.3. Rio Puerco, New Mexico

The Rio Puerco in central New Mexico is a tributary of the Rio Grande with a drainage area of 15,180 km^2 (Fig. 2.3), and is an arroyo that formed in the 1880s (Bryan and Post, 1927). Evidence of several earlier cut-and-fill cycles is preserved in the Quaternary valley fill of the Rio Puerco.

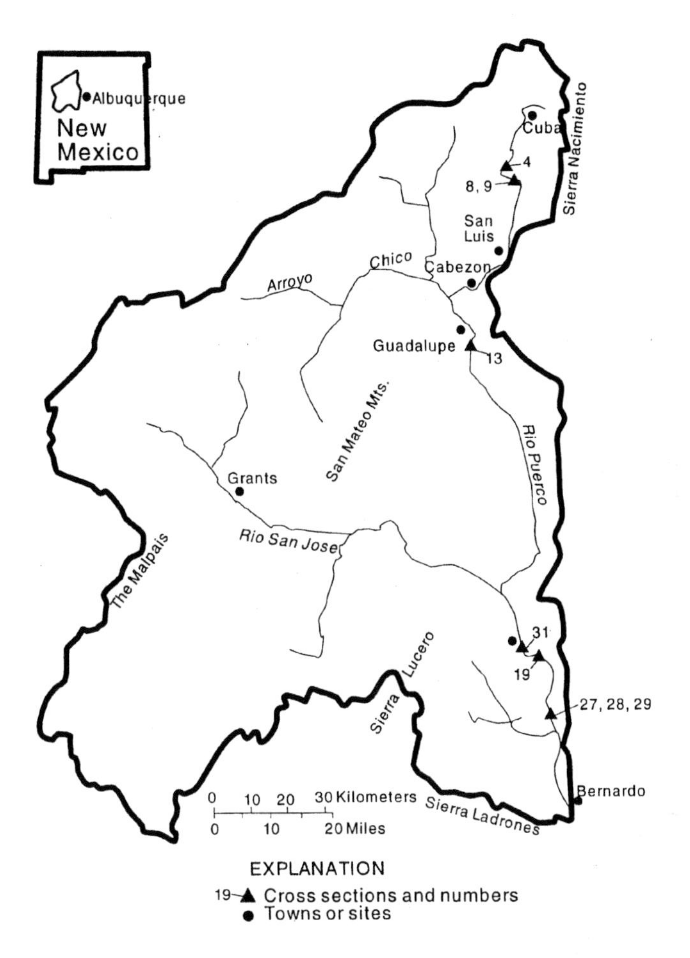

Fig. 2.3. Rio Puerco watershed, New Mexico, and location of resurveyed cross sections.

The most recent cutting cycle in the Rio Puerco occurred after European descendants settled the valley. Prior to 1885, the Rio Puerco flowed through a discontinuous arroyo, as much as 6 meters deep in places, but elsewhere with inconspicuous banks. The valley in non-incised reaches was inundated by flood flows and supported a fairly continuous plant cover and several agriculturally based communities. A combination of climatic, hydrologic, and anthropogenic factors caused the Rio Puerco to incise in the late 1880s (Elliott, 1979). Bryan (1928) dated the incision at various locations in the Rio Puerco Valley by the abandonment of several communities and concluded that arroyo incision progressed headward from the confluence with the Rio Grande. The entire length of the Rio Puerco had incised by 1928 (Bryan, 1928), and approximately $4.87 \times 10^8\ m^3$ of valley alluvium had eroded as the arroyo formed (Bryan and Post, 1927).

Change in the Rio Puerco has continued since the Bryan and Post (1927) study, documented by several replicated cross-section surveys and aerial photographs. Since 1885, the Rio Puerco has widened and now can be described at two scales: (1) at the scale of the entire arroyo width, and (2) at the scale of the inner channel (Fig. 2.4). The arroyo is tens to hundreds of meters wide between the two highest terrace scarps, which are remnants of the pre-19th century valley floor. Flood flows no longer occupy the entire arroyo width. An inner channel, meters to tens of meters wide, and an inner floodplain have developed on the arroyo floor (Fig. 2.4).

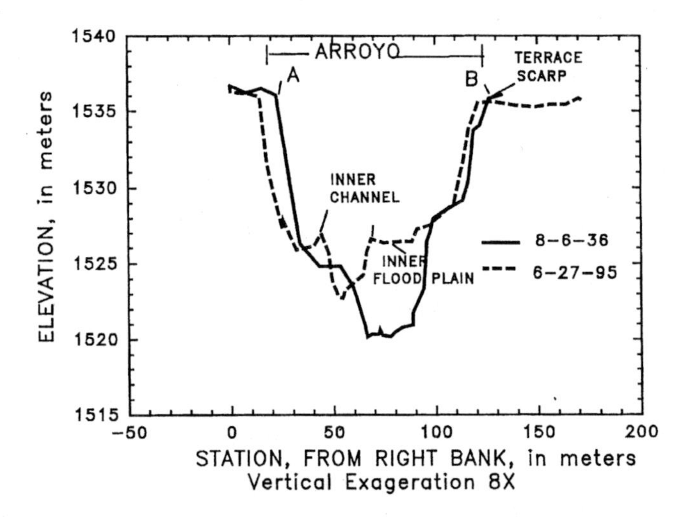

Fig. 2.4. **Resurvey of Kirk Bryan's 1936 Rio Puerco cross section downstream from Highway 6, near Los Lunas, New Mexico, showing aggradation and channel morphologic terms.**

Elliott (1979) analyzed arroyo and inner-channel characteristics measured from aerial photographs taken in 1935, 1953, and 1954 and from 29 inner-channel cross sections in the upper, middle, and lower reaches of the Rio Puerco surveyed in 1977 and 1978. In the late 1970s, specific reaches of the Rio Puerco could be identified as displaying one of two general morphologies on the basis of geomorphic and sedimentologic variables. The upper reaches of the channel were characterized by Type I morphology: the inner channel typically had a sandy bed, a high width-to-depth ratio, unstable and poorly defined banks, and a laterally shifting channel pattern that was actively eroding valley fill. The middle and lower reaches of the channel were characterized by Type II morphology: the inner channel typically had a silt- and clay-rich bed and banks, a relatively low width-to-depth ratio, steep banks, and a relatively stable channel pattern. In the 1935 aerial photographs, all reaches of the Rio Puerco were characterized by Type I morphology (Elliott, 1979).

The change from Type I to Type II morphology in the middle and lower reaches of the Rio Puerco between the 1930s and the 1970s reflected the development and stabilization of a floodplain within the 19th century terrace scarps. Evidence of this floodplain development in the middle and lower arroyo reaches between the middle 1930's and the late 1970s included channel-width reduction and pattern stabilization, dense vegetation encroachment, and overbank sedimentation. Floodplain development and channel stabilization appeared to be progressing from downstream to upstream (Elliott, 1979).

Several of Elliott's arroyo reaches were resurveyed by the authors in November 1994 and June 1996 to assess arroyo, inner-floodplain, and inner-channel changes since the late 1970s (Fig. 2.3). The Guadalupe reach (cross section 13), below the confluence of Arroyo Chico and a reach south of Cuba (cross sections 8 and 9) were characterized by Type I morphology in 1977 and continued to display Type I morphology in 1994. These reaches were characterized by wide and shallow inner-channel cross sections, lateral inner-channel instability within the arroyo, and an increase in arroyo width by lateral erosion into the Quaternary valley fill. No vertical aggradation was indicated within the arroyo, nor was there evidence that Type II characteristics were beginning to succeed Type I characteristics in these reaches. However, a reach south of Cuba (cross section 4) had changed significantly since 1977. The inner-channel width had decreased by 36 percent, and a vegetated inner floodplain had developed on the right side of the previously braided streambed. This reach appeared to be in transition from a Type I to a Type II morphology.

Two reaches in the lower Rio Puerco (cross section 19; cross sections 27, 28, and 29) that were characterized by Type II morphology in 1977 continue to display Type II morphology (Fig. 2.3). These reaches had narrow and deep, well-defined inner channels and vegetated inner floodplains; however, two significant changes were observed. Channel width-to-depth ratios had decreased since 1977 and the inner floodplain had aggraded 0.3 to 1.0 meter by overbank deposition.

Cross section 31 (Fig. 2.4) in the middle Rio Puerco downstream from Highway 6 west of Los Lunas, was originally surveyed on August 6, 1936, by Kirk Bryan and was resurveyed on June 27, 1995. The cross section is located at river km 67.6 (river distance measured upstream from km 0 at the mouth) and 830 m downstream from the railroad crossing near Highway 6. Two steel bars from the original survey were found on the left bank, substantiating good elevation control at this cross section. Comparison of the two surveys indicates that the Rio Puerco streambed has aggraded 2.55 m, and the inner channel has narrowed from a width of 40.6 m in 1936 to 23.4 m in 1995 (Fig. 2.4). The area of the cross section defined from the top edge of the left terrace scarp to the top edge of the right terrace was 1,064 m^2 in 1936 and 939 m^2 in 1995. The rate of cross-sectional filling between these two surveys is 2.12 m^2/yr.

Although there has not been adequate date control on past cut-and-fill cycles in the Rio Puerco, Love and Young (1983) listed two filling cycles from 900 to 1250 A.D and from 1325 to 1450 A.D. An incisional episode was assumed to have occurred between the two fill cycles, from 1250 to 1325 A.D. This incisional episode would correlate with other regional cutting dates, such as that in Chaco Canyon, New Mexico (850 to 600 B.P.) (Hall, 1977). After this incisional episode filling began about 650 B.P. and reached the same elevation in the valley floor that the channel occupied prior to the incision. If the contemporary Rio Puerco fills to the elevation of its pre-incision valley floor, the channel bed could occupy the elevation labeled A and B in Fig. 2.4. At a filling rate of 2.12 m^2/yr, the area of the 1995 channel could be completely filled in approximately 440 years. This value plus the 110 years that has passed since incision started around 1885 equals 550 years. Emmett (1974) estimated the rate at which arroyos in New Mexico would fill to the former valley floor at 200 to 700 years.

2.4. Zuni River, New Mexico

The Zuni River at the Arizona-New Mexico border drains an area of 3,403 km^2 (Fig. 2.5a) and is a tributary of the Little Colorado River. Arroyos on the Zuni Reservation that drain into the Zuni River incised between 1905 and 1920 (Balling and Wells, 1990). Changes in 85 monumented channel cross sections were examined in three subbasins of the Rio Nutria, a tributary of the Zuni River, between January 1992 and December 1994 (Fig. 2.5b). Drainage areas of the three subbasins are Y-Unit Draw, 24.55 km^2; Benny Draw, 3.0 km^2; and Conservation Draw, 6.40 km^2.

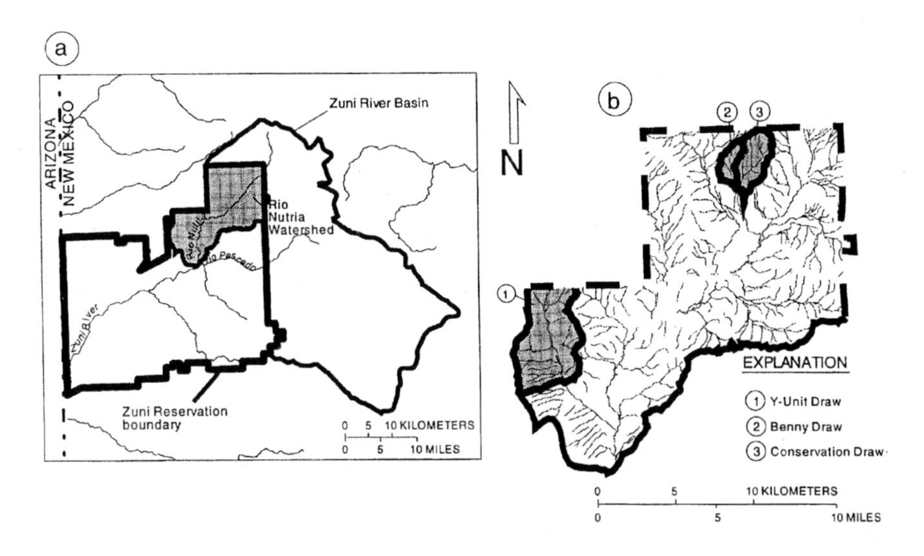

Fig. 2.5. (a) Location of major watersheds of the Zuni River Basin on the Zuni Reservation; (b) Location of subbasins in the Rio Nutria watershed.

Results of Zuni Reservation channel resurveys of 85 cross sections indicated that 72 percent of cross sections are aggrading and that 52 percent show a decrease in cross-sectional area. More cross sections show an increase in cross-sectional area (n=41) than show scour (n=23) because some cross sections that are aggrading are also increasing in cross-sectional area. For all cross sections, 43 percent are aggrading but are also increasing in cross-sectional area because of an accompanying increase in width.

In the Zuni study area, most arroyos are aggrading and widening. This may indicate that most resurveyed cross sections are in Stage C to E (Fig. 2.2).

2.5. Discussion

Case studies and previously published work support the conclusion that many channels that were incised starting in the late 19th century and early 20th century are currently aggrading. Since incision, many arroyos continue to evolve. The sequential stages of arroyo evolution--channel deepening, channel widening, floodplain formation, and the establishment of vegetation--generally proceed upstream through the watershed and ultimately lead to channel aggradation and reduced sediment yields. Therefore, channel aggradation observed in channels in the Western United States may be due in part to arroyo evolution.

3. RECENT PRECIPITATION TRENDS

Hack's (1942) climatic interpretations of arroyo cut-and-fill cycles suggested that aggradation occurs during wet periods and incision during dry periods. This generalization of cut-and-fill cycles has continued to the present (Leopold, 1994). Based on climatic variations during the most recent arroyo cutting episode (1880-1920), a further assumption for channel cutting is that following dry periods, a high frequency of high-intensity rainfall events occurred that ultimately lead to channel incision (Leopold, 1951; Balling and Wells, 1990).

Leopold (1951) analyzed precipitation records at Santa Fe that extended back to 1849. Only records extending back to 1868 could be obtained and were reanalyzed by the authors in this paper. No trend was detected in average annual rainfall (Leopold, 1951) (Table 3.1). The average annual

Table 3.1. Average frequency, in days per year, of rains of various sizes during six time periods.

	Average number of days having rainfall of indicated amount			
Time Period	Daily Rainfall Less than 2.7 mm (days)	Daily Rainfall 12.7 - 25.4 mm (days)	Daily Rainfall Greater than 25.4 mm (days)	Mean Annual Rainfall (mm)
1868-1880	68.1	4.6	2.1	345
1881-1910	82.2	5.5	1.1	360
1911-1930	83.0	5.4	1.2	364
1931-1950	76.6	4.8	1.1	357
1951-1970	63.5	5.3	1.4	350
1971-1993	58.6	5.2	1.6	355

frequency of daily rainfall in three classes, less than 12.7 mm, 12.7 to 25.4 mm, and greater than 25.4 mm, was calculated for selected time periods from 1868 to 1993. The data showed an increase in frequency in the less than 12.7 mm class from the period 1868-1880 to the time periods spanning 1881-1930 (Table 3.1). The frequency of daily rainfall greater than 25.4 mm decreased after 1880. After 1930 the frequency of daily rainfall in the less than 12.7 mm class decreased, and after 1950 daily rainfall in the greater than 25.4 mm class increased slightly. According to Leopold (1951), vegetation utilizes low-intensity rainfall. Periods characterized by low-intensity rainfall tends to be associated with an increase in vegetation density and a decrease in erosion (Leopold, 1951), assuming no change in land use. Conversely, high-intensity rainfall tends to be associated with increased runoff and erosion. From 1868 to 1880 low-intensity rainfall occurred relatively infrequently and high-intensity rainfall occurred slightly more frequently (Leopold, 1951); therefore climatic conditions in the southwestern United States may have been conducive to the widespread arroyo incision that occurred in the late 19th and early 20th century. Analysis of the climatic record by Leopold in 1951 led him to conclude that arroyo incision occurred during periods of aridity.

Leopold and others (1966) calculated average rainfall intensity by dividing annual rainfall by the number of days having rainfall in that year. For a constant annual precipitation, higher values of average rainfall intensity would indicate that annual rainfall fell in fewer days, presumably with a greater intensity that could have resulted in accelerated erosion. A plot of average annual rainfall intensity (Fig. 3.1) shows more intense rainfall in the period 1853-1880 that may have led to late 19th century arroyo incision. Based on the manner in which Leopold (1951) and Leopold and others (1966) interpreted average annual rainfall intensity, rainfall since 1967 may indicate that climatic factors are favorable for arroyo incision in the southwestern United States.

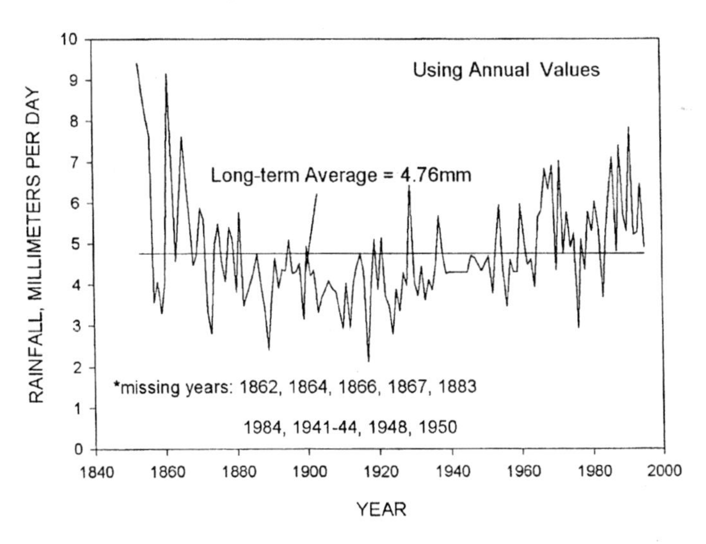

Fig. 3.1. Average annual rainfall intensity, Santa Fe, New Mexico, 1853-1993.

4. SUMMARY AND CONCLUSIONS

Recent studies in the southwestern United States indicate that many incised channels are aggrading. Graf (1987) indicated aggradation beginning about 1940. Studies of the Rio Puerco and Zuni River tributaries suggest inner floodplain aggradation within arroyos that developed in the late 19th and early 20th centuries. Beginning about 1930, climatic records at Santa Fe, New Mexico, show a decrease in the frequency of low-intensity rainfall, the type of rainfall associated with increased vegetation growth and channel aggradation. Average annual rainfall intensity increased and has generally remained above normal since 1967. The decrease in frequency of low-intensity rainfall at Santa Fe indicates that climatic conditions may be suitable for a period of degradation, contrary to onsite evidence from the Rio Puerco and Zuni River tributaries. Because precipitation records for Santa Fe may not be representative of the entire southwestern United States, the conclusion remains tentative pending climatic analysis for other areas.

The arroyo incision in the late 19th and early 20th century may have involved climate change as a triggering mechanism. Valleys in the southwestern United States had filled since an earlier cutting episode 400-800 B.P. and probably were at the threshold of incision in the late 19th century. Evidence that arroyos currently are aggrading may suggest that the initiation of later-stage aggradation in arroyos is partially controlled by intrinsic conditions. These intrinsic conditions are determined by hydraulic properties related to channel geometry and the stage of arroyo evolution. Although climate may have been the triggering mechanism for arroyo incision that began in the late 19th century, climate may not be the main factor in channel aggradation. Intrinsic conditions determined by hydraulic properties related to channel geometry and the stage of arroyo evolution may be a major influence on channel aggradation, whereas climate may exert a major influence on the rate at which channels aggrade. The preliminary conclusions in this paper are based on initial data and further data sets and interpretations are forthcoming.

5. REFERENCES

Antevs, E. 1952. Arroyo-cutting and filling. Journal of Geology, 60:375-378.

Balling, R.C. and S.G. Wells. 1990. Historical rainfall patterns and arroyo activity within the Zuni River drainage basin, New Mexico. Annals of the American Geographers 80:603-617.

Bryan, K. 1928. Historic evidence of changes in the channel of the Rio Puerco, a tributary of the Rio Grande, New Mexico. Journal of Geology 36:265-282.

Bryan, K. and G. Post. 1927. Erosion and control of silt on the Rio Puerco, New Mexico. Unpublished Report to the Chief Engineer, Middle Rio Grande Conservancy District, Albuquerque, New Mexico.

Cooke, R.U. 1974. The rainfall context of arroyo initiation in southern Arizona. Zeitschrift fur Geomorphologie Supplementband 21:63-75.

Cooke, R.U. and R.W. Reeves. 1976. Arroyos and environmental change in the American southwest. Oxford Research Studies in Geography. Clarendon Press.

Cooley, M.E. 1962. Late Pleistocene and recent erosion and alluviation in parts of the Colorado River system, Arizona and Utah. U.S. Geological Survey Professional Paper 450-B:48-50.

Elliott, J.G. 1979. Evolution of large arroyos; The Rio Puerco of New Mexico. Fort Collins, Colo., Colorado State University, unpublished M.S. thesis.

Emmett, W.W. 1974. Channel aggradation in western United States as indicated by observations at Vigil Network sites: Zeitschrift fur Geomorphologie Supplementband 21:52-62.

Gellis, A.C. 1988, Decreasing sediment and salt loads in the Colorado River basin--A response to arroyo evolution: Fort Collins, Colo., Colorado State University, unpublished M.S. thesis.

Gellis, A.C., R. Hereford, S.A. Schumm, and B.R. Hayes. 1991, Channel evolution and hydrologic variations in the Colorado River basin--Factors influencing sediment and salt loads. Journal of Hydrology 124:317-344.

Gellis, A.C. 1992, Decreasing trends of suspended-sediment loads in selected streamflow stations in New Mexico: New Mexico Water Resources Research Institute Report 265:77-93.

Graf, W.L. 1987, Late Holocene sediment storage in canyons of the Colorado Plateau. Geological Society of America Bulletin 99:261-271.

Hack, J.T. 1942. The changing physical environment of the Hopi Indians of Arizona. Peabody Museum Papers 35:1.

Hall, S.A. 1977. Late Quaternary sedimentation and paleoecologic history of Chaco Canyon, New Mexico. Geological Society of America Bulletin 88:1593-1618.

Heede, B.H. 1977, Case study of a watershed rehabilitation project: Alkali Creek, Colorado: U.S.D.A. Forest Service Research Paper RM-189.

Hereford, R. 1984, Climate and ephemeral-stream processes--Twentieth-century geomorphology and alluvial stratigraphy of the Little Colorado River, Arizona. Geological Society of American Bulletin 95:654-668.

Hereford, R. 1986, Modern alluvial history of the Paria River drainage basin, Southern Utah. Quaternary Research 25:293-311.

Judd, N.M. 1964. The architecture of Pueblo Bonito, Smithsonian Miscellaneous Collections 147:1.

Landon, S. 1990. Eroding a way of life. Albuquerque Journal, July 1, 1990, p. 1, SectionF.

Leighly, J. 1936. Meandering arroyos of the dry southwest. Geographical Review 36:270-282.

Leopold, L.B. 1951. Rainfall frequency--An aspect of climatic variation. American Geophysical Union Transactions 32:3:347-357.

Leopold, L.B. 1994. A view of the river. Harvard University Press.

Leopold, L.B. and C.I. Snyder. 1951. Alluvial fills near Gallup, New Mexico. U.S. Geological Survey Water-Supply Paper :1110-A.

Leopold, L.B. and J.P. Miller. 1954. A postglacial chronology for some alluvial valleys in Wyoming.U.S. Geological Survey Water-Supply Paper :1261

Leopold, L.B., W.W. Emmett, and R.M. Myrick. 1966. Channel and hillslope processes in a semiarid area, New Mexico. U.S. Geological Survey Professional Paper :352-G.

Leopold, L.B. 1976, Reversal of erosion cycle and climatic change. Quaternary Research 6:557-562.

Love, D.W. and J.D. Young. 1983, Progress report on the late Cenozoic geologic evolution of the lower Rio Puerco. New Mexico Geological Society Guidebook, 34th field conference, Socorro Region II.

Parker, R.S. 1977. Experimental study of drainage basin evolution and its hydrologic implications. Hydrology Papers, Colorado State University, Fort Collins, Colorado 90:58.

Patton, P.C. and S.A. Schumm. 1975. Gully erosion, northwestern Colorado--A threshold phenomenon. Geology 3:88-90.

Schoenwetter, J. and A.E. Dittert. 1968. An ecological interpretation of Anasazi settlement patterns. The Anthropological Society of Washington, Anthropological Archaeology in the Americas:41-66.

Schumm, S.A. and R.F. Hadley. 1957. Arroyos and the semiarid cycle of erosion. American Journal of Science 25:161-174.

Schumm, S.A., M.D. Harvey, and C.C. Watson. 1984, Incised channels: Littleton Press, Littleton, Colorado.

Simon, A. 1989, A model of channel response in disturbed alluvial channels. Earth Surface Processes 14:11-26.

Thornwaite, C.W., C.F.S. Sharpe, and E.F. Dosch. 1942, Climate and accelerated erosion in the arid and semiarid southwest, with special reference to the Polacca Wash drainage basin, Arizona. U.S. Department of Agriculture Technical Bulletin 808.

Waters, M.R. 1985, Late Quaternary alluvial stratigraphy of Whitewater Draw, Arizona--Implications for regional correlation of fluvial deposits in the American Southwest. Geology 13:705-708.

CHANNEL INSTABILITY IN THE LOESS AREA OF THE MIDWESTERN UNITED STATES: A COMBINATION OF ERODIBLE SOILS AND HUMAN DISTURBANCE

Andrew Simon[1] and Massimo Rinaldi[2]

ABSTRACT

The loess area of the midwestern United States contains thousands of miles of unstable stream channels that are undergoing systemwide channel-adjustment processes as a result of (1) modifications to drainage basins dating back to the turn of the 20th century: land clearing and poor soil-conservation practices, which caused the filling of stream channels, and consequently (2) direct, human modifications to stream channels such as dredging and straightening to improve drainage conditions and reduce the frequency of out-of-bank flows. Today, at the turn of the 21st century, many of these channels are still highly unstable and threaten bridges, other structures, and land adjacent to the channels. The most severe, widespread instabilities are in western Iowa where a thick cap of loess and a lack of sand- and gravel-sized bed sediments in many channels hinders downstream aggradation, bed-level recovery and the consequent reduction of bank heights, and renewed bank stability. In contrast, streams draining west-central Illinois, east-central Iowa, and other areas where the loess cap is relatively thin and there are ample supplies of sand- and gravel-sized material, are closer to recovery. Throughout the region, however, channel widening by mass-wasting processes is the dominant adjustment process.

KEYWORDS

Channel instability, erodible soils, channel degradation

[1] Agricultural Research Service, National Sedimentation Laboratory, Oxford, Mississippi, USA.

[2] Dipartimento di Scienzo della Terra. Universita degli Studi di Firenze, Italia.

1. INTRODUCTION

The dynamic nature of alluvial streams signifies the ability to adjust to changes imposed on a fluvial system, be they natural or due to human activities. Channel adjustments migrate up- and downstream to offset the disturbance by altering aspects of their morphology, sediment load, and hydraulic characteristics. Under "natural" conditions, in geologically stable areas such as the midwestern United States, the processes of erosion and deposition might occur at such low rates and over such extended periods of time, that they can be virtually imperceptible. Human and natural factors or disturbances, however, may accelerate and exacerbate these processes such that rapid and observable morphologic changes occur as the channel responds to the disturbance and returns to an equilibrium condition. Adjustments to human disturbances can involve short time scales (days) and limited spatial extent (a stream reach), or longer periods of time (scores to hundreds of years) and entire fluvial systems, depending on the magnitude, extent, and type of disturbance (Williams and Wolman, 1984; Simon, 1994).

In the highly erodible loess area of the midwestern United States (Fig. 1.1), human disturbances to floodplains and upland areas culminating near the turn of the 20th century resulted in channels being choked with sediment and debris. Beginning about 1910, channels were enlarged and straightened throughout the region to alleviate frequent and prolonged flooding of bottomlands (Speer et al., 1965). Over the next 80 years, accelerated channel erosion and the formation of canyon-like stream channels have resulted in severe damage to highway structures, utilities, and land adjacent to the stream channels. Accelerated stream-channel degradation has resulted in an estimated $1.1 billion in damages to infrastructure and the loss of agricultural lands since the turn of the century in western Iowa (Baumel, 1994). A survey of 15 counties in northwestern Missouri identified 957 highway structures as damaged by channel degradation. Degradation and channel widening in the loess area led to the collapse of several bridges in West Tennessee (Robbins and Simon, 1983), southwest Mississippi (Wilson, 1979), Missouri (Emerson, 1971), and southeast Nebraska. The 1993 floods in the midwestern United States focused additional attention on channel-stability problems in the loess area because scores of bridges were either closed or failed during and immediately after the floods.

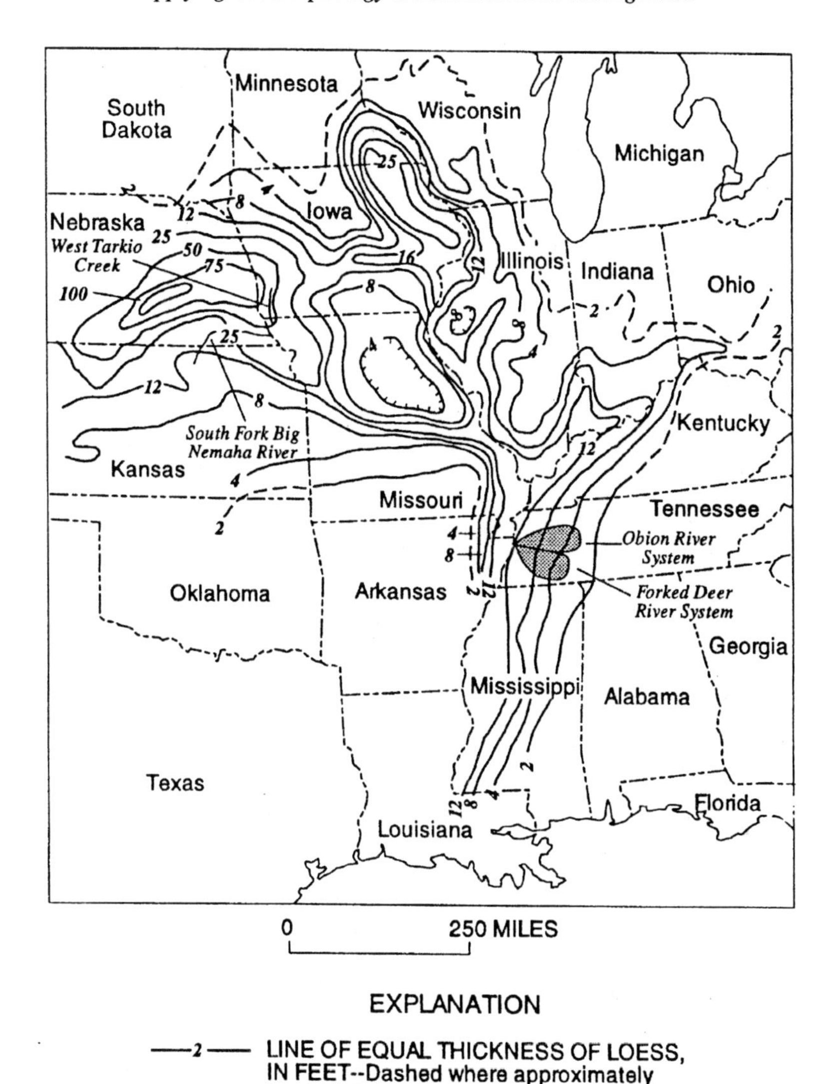

Fig. 1.1. Location map of loess area of midwestern United States and thickness of loess, in feet.

Channel degradation in some streams has resulted in a fourfold increase in channel depth (20 feet) and an almost fivefold increase in channel width (100 feet) since the middle of the 19th century (Piest et al., 1976; 1977).

Erosion rates from two severely degraded silt-bed streams in western Iowa were calculated at 131,000 and 145,000 tons/year (Ruhe and Daniels, 1965; Piest et al., 1976). About 190,000 tons/year of channel materials were discharged from 7.7 miles of Hotophia Creek, northern Mississippi between 1958 and 1976 (Little and Murphey, 1981). In the 20 years of channel adjustment following modification of about 90 miles of channels in the Obion River System, West Tennessee, an estimated 221 million ft^3 of channel sediments were eroded and transported out of the system (11 million tons/year; Simon, 1989a). About 10 million tons/year were discharged from the Forked Deer River System, West Tennessee between early 1970 and 1987. On average, about 18 percent of this material was eroded from the channel bed, the remainder coming from the channel banks (Simon, 1989a).

The purpose of this study, sponsored by the Federal Highway Administration, was to assess the general magnitude and extent of channel-stability problems in the loess area of the midwestern United States (Fig. 1.1). This region represents a "worst-case" scenario where easily erodible soil has combined with extensive human disturbance to produce highly unstable stream-channel systems. Reconnaissance-level observations were made from a low-flying aircraft and/or on the ground noting stability problems along unstable streams in west, east-central, and southeastern Iowa, southeastern Nebraska, west-central Illinois, northeastern Kansas, and northwestern Missouri. Silt-bedded West Tarkio Creek, southwestern Iowa and northwestern Missouri, and to a lesser extent, 2 sites on the South Fork Big Nemaha River, Nebraska were investigated in more detail to show typical channel changes. The sand-bedded Obion River System, West Tennessee, was studied previously (Simon, 1989a; 1989b; 1989c; Hupp, 1992; Simon and Hupp 1992; Simon, 1994).

2. SYSTEMWIDE CHANNEL INSTABILITY

Channel instability in the loess area of the midwestern United States involves entire drainage systems. Common adjustment processes affecting these fluvial systems include: upstream progressing degradation; downstream aggradation; channel widening and to a limited extent, channel narrowing in the downstream-most reaches; and changes in the quantity and character of the sediment load. These processes differ from localized processes such as scour and fill, which can be limited in magnitude as well as in spatial and temporal scale. Scour and fill can occur over periods of hours to days and affect localized areas and bridges in response to storm flow. In contrast, the processes of aggradation and degradation, which represent systematic changes in bed elevation over a period of years (Mackin, 1948), can affect long stream reaches, entire stream lengths, or whole stream systems. In fact, longer-termed adjustment processes can, by themselves, instigate or exacerbate local scour problems (Robbins and Simon, 1983). Whether bed erosion occurs as scour, as degradation, or in combination, sufficient bed-level lowering can lead to bank instabilities and to changes in channel pattern. Thus, mitigation of channel instability should begin with consideration of how the site or reach in question fits into the broader spatial and temporal scheme of systemwide adjustment and channel evolution. Addressing only local symptoms at a site will often be insufficient to mitigate an instability problem where systemwide adjustment processes are active.

3. FACTORS CONTROLLING CHANNEL STABILITY IN THE LOESS AREA

Factors that affect channel stability can be conceptualized in terms of (1) the resistance of the channel boundary to erosion and (2) the forces acting on the channel to erode the boundary. If these opposing forces are balanced, the channel is considered to be in equilibrium and no net erosion or deposition will occur with time.

Vertical channel instabilities may result from a disruption of the equilibrium between available stream power (the discharge-gradient product) and the discharge of bed-material sediment (Lane, 1955; Bull, 1979):

$$Q\,S \propto Q_s\,d_{50} \tag{3.1}$$

where

Q	=	bankfull discharge,
S	=	channel gradient,
Q_s	=	bed-material discharge, and
d_{50}	=	median grain size of bed material.

Equation 3.1 indicates that if available stream power is increased by an increase in the bankfull discharge or in the gradient of the stream, there would be an excess amount of stream power relative to the discharge of bed-material sediment. Additional sediment would be eroded from the channel bed resulting

in (1) an increase in bed-material discharge to an amount commensurate with the heightened stream power, and (2) a decrease in channel gradient and, consequently, in stream power as the elevation of the channel bed is lowered. A similar response would be expected from a decrease in the erosional resistance of the channel boundary or a decrease in the size of bed-material sediment (assuming it is not cohesive). Because channel banks in the loess area of the midwestern United States are composed predominantly of silt, non-cohesive sand or gravel banks contributing bed-material sediment are not considered here. In contrast, a decrease in available stream power or an increase in the size or discharge of bed-material sediment would lead to aggradation on the channel bed. Factors that increase hydraulic roughness will decrease flow velocity and, therefore, discharge and stream power. For example, proliferating vegetation on streambanks causes slower velocities and reduced stream power in the near-bank zone.

Lateral (bank) instabilities are also considered in terms of force and resistance. Because loess-channel banks are fine-grained (silty), low to moderately cohesive (Lohnes and Handy, 1968; Lutenegger, 1987; Simon, 1989b), and generally erode by mass failure, the shear strength of the bank material represents the resistance of the boundary to erosion. Shear strength comprises two components — cohesive strength and frictional strength. For the simple case of a planar failure of unit length the Coulomb equation is applicable:

$$S_r = c + (N - \mu) \tan \phi \tag{3.2}$$

where

S_r	=	shear stress at failure,
c	=	cohesion,
N	=	normal stress on the failure plane,
μ	=	pore pressure, and
φ	=	friction angle.

Also,

$$N = W (\cos \theta) \tag{3.2a}$$

where	W	=	weight of the failure block, and
	θ	=	angle of the failure plane.

The gravitational force acting on the bank is $W \sin \theta$. Factors that decrease the erosional resistance (S_r) such as excess pore pressure from saturation and the development of vertical tension cracks favor bank instabilities. Similarly, increases in bank height by bed degradation and bank angle by undercutting favor bank failure by causing the gravitational component to increase.

3.1. Natural Factors

The predominant natural factors regarding channel-stability problems in the loess area of the midwestern United States are related to the relative resistance to erosion of the loess, and the distribution of loess around and beneath the channel boundary. Although loess can stand in tall vertical cliffs when dry, it tends to be highly erodible when wetted by streamflow or by raindrop impact. Once disturbed, channels with beds of loess-derived alluvium generally incise rapidly and, without an ample supply of sand or gravel, tend to be some of the deepest in the region. Because the fluvial transport of fine-grained sediments such as silt is not controlled by the hydraulic properties of the streamflow, but by the supply of silt delivered to the flow from channel or upland sources, little silt is deposited in downstream reaches to aid in bed-level recovery. Channels that cut through the loess cap and entrain coarser underlying sand and gravel deposits below will, on average, recover quicker from disturbances because there is a supply of hydraulically-controlled material that is likely to be deposited in downstream reaches, causing aggradation and the consequent decrease in bank heights.

The relatively low resistance (shear strength) of the loess bank material to erosion by mass failure is an important factor in channel evolution in the region. The mean values of cohesion and friction angle for 22 sites throughout the region are 203 lbs/ft^2 and 35.4°, respectively (Lutenegger, 1987). The mean values for 14 shear-strength tests in Iowa bluff-line streams are 187 lbs/ft^2 and 24.9° (Lohnes and Handy, 1968). Data indicate that, on average, loess in Kansas and Nebraska is considerably more cohesive, ranging from 720 to 1,440 lbs/ft^2 (Turnbull, 1948; Lohnes and Handy, 1968). The mean value of cohesion is also relatively high for loess-derived sediments (1,120 lbs/ft^2) for 23 borehole shear-strength tests conducted by Thorne et al. (1981) in northern Mississippi. The mean friction angle in these Mississippi tests is 21°. Cohesion and friction-angle values obtained in Mississippi for loess-derived sediments range from 192 to 719 lbs/ft^2, and from 16 to 35°, respectively (Turnipseed and Wilson, 1992; Wilson and Turnipseed, 1993; 1994).

A study of streambank stability in West Tennessee indicated a mean cohesive strength of 182 lbs/ft^2, and a mean friction angle of 30.1° (168 tests; Simon, 1989c). The West Tennessee tests were taken during summer low-flow conditions when the channel banks were driest and, therefore, the most resistant to mass failure. Nevertheless, the mean ambient degree of saturation was 86 percent, leaving the channel banks vulnerable to complete saturation during wet periods. Upon saturation and the generation of excess pore pressures (values of μ in Eq. 3.2 greater than the steady-state values as determined by the "normal" position of the water table), shear strength values can decrease. By Eq. 3.2, a continued increase in μ can result in the frictional component of shear strength becoming zero, leaving only the cohesion component to resist mass failure. In West Tennessee, the cohesion component comprises, on average, only about 10 percent of the strength of the loess-derived channel banks (Simon, 1989b) and, following saturation, the loss of suction caused by negative pore pressures, and the generation of excess pore pressures, is often insufficient to resist mass failure. Along channels deepened by degradation, mass failure occurs on the recession of river stage as the banks lose the support afforded by the water in the channel.

Established woody vegetation on streambanks can enhance bank stability by affecting hydraulic characteristics of the near-bank zone as well as the stability of the bank mass. Hydraulic roughness related to stem density decreases flow velocity in the near-bank zone and ameliorates bank erosional processes (Kouwen, 1970; Hupp and Simon, 1986). When compared to unvegetated banks, banks with vegetation are drier and better drained (Thorne, 1989), often providing for greater values of suction (negative pore pressures) within the capillary fringe. Plant roots increase the tensile strength of the soil resulting in reinforced earth (Vidal, 1969). Increases in bank stability due to woody vegetation are applicable only if (1) degradation does not create bank heights and angles in excess of the critical shear-strength conditions of the bank material, and (2) rooting depths extend past the depth of the failure plane. The added weight of woody vegetation on a bank acts as a surcharge and can also have negative effects on bank stability by increasing the downslope component of weight, particularly on steep banks.

3.2. Human Factors

Human factors related to channel-stability problems can often be considered as disruptions or disturbances to the "natural" balance between the available erosional force acting on the channel boundaries and the erosional resistance provided by those boundaries. Such disturbances can be imposed directly on the channel as in the case of dredging or channel straightening, or can be indirect as in the case of land clearing. In the loess area of the midwestern United States, direct and indirect disturbances have contributed to the channel-stability problems of the region.

Large tracts of land were cleared for cultivation during settlement of the region prior to and after the Civil War (Ashley, 1910; Brice, 1966; Piest et al., 1977). Stream courses were tortuous with sinuosities ranging from about 3 to 4, with valley slopes in the order of 10^{-4} to 10^{-3} (Moore, 1917; Speer et al., 1965). Channel gradients from trunk streams in southeastern Nebraska and West Tennessee were about 1.2 and 1.1 x 10^{-4}, respectively (U.S. Army Corps of Engineers, 1907; Moore, 1917). The removal of grasses and woody vegetation resulted in reduced water interception and storage, increased rates of surface runoff, erosion of uplands, and gullying of terraces. Surface runoff and peak-flow rates in western Iowa are estimated to have increased 2 to 3 times and 10 to 50 times, respectively, when compared to estimates of presettlement amounts (Piest et al., 1976; 1977). The removal of woody vegetation from streambanks resulted in decreased hydraulic roughness, increased flow velocities and stream power, and contributed to increased peak discharges. In combination, these factors caused extensive downcutting (Piest et al., 1976). Much eroded material was deposited in channels and on floodplains resulting in loss of channel capacity and frequent and prolonged flooding of agricultural lands (Morgan and McCrory, 1910; Moore, 1917; Piest et al., 1976). Moore (1917) reports that aggradation was almost continuous along the trunk streams of southeastern Nebraska.

As a result of ubiquitous channel filling, local drainage districts implemented programs to dredge, straighten, and shorten stream channels (channelize) to reduce flooding and thereby increase agricultural productivity (Hidinger and Morgan, 1912; Moore, 1917). Work was undertaken in southeastern Nebraska, West Tennessee and west-central Illinois around 1910 (Moore, 1917; Speer et al., 1965); and around 1920 in western Iowa (Lohnes, et al., 1980) and in parts of northern Mississippi (Schumm et al., 1984). In many areas, this work increased channel gradients by about an order of magnitude (Moore, 1917; Simon, 1994). Entire lengths of trunk streams were channelized in southeastern Nebraska and in West Tennessee. By the 1930s, most streams tributary to the Missouri and Mississippi Rivers in the loess area of the midwestern United States had been dredged and straightened (Speer et al., 1965; Piest et al., 1976; Lohnes et al., 1980; Simon, 1994). In many parts of the region these activities were conducted periodically throughout the 1960s and 1970s as additional tributaries were channelized and because some previously dredged channels filled with eroded sediment from upstream reaches (Lohnes et al., 1980; Simon, 1994).

Dredging and straightening significantly increased bankfull discharge and channel gradient, resulting in a proportionate increase in bed-material discharge (Eq. 3.1) and rapid morphologic changes. These changes included upstream degradation, downstream aggradation (in the sand-bedded streams), and bank instabilities along altered streams and adjacent tributaries. In combination with the low resistance to erosion exhibited by the loess-derived channel materials, it is the increase in the erosional force (stream power) by channel dredging and straightening near the turn of the 20th century that caused the deep entrenchment, general states of instability, and present day problems in the channel systems in the loess area of the midwestern United States.

4. STAGES OF CHANNEL EVOLUTION

Geomorphologists have noted that alluvial channels in various environments, destabilized by different natural and human-induced disturbances, pass through a sequence of channel forms over time (Davis, 1902; Ireland et al., 1939; Schumm and Hadley, 1957; Daniels, 1960; Emerson, 1971; Keller, 1972; Elliott, 1979; Schumm et al., 1984; Simon and Hupp, 1986; Simon, 1989a; Fig. 4.1; Table 4.1). These systematic temporal adjustments are collectively termed "channel evolution" and permit interpretation of past and present channel processes, and prediction of future channel processes. One of these schemes is the 5-stage channel evolution model of Schumm et al., (1984), which was developed from morphometric data acquired on Oaklimiter Creek, northern Mississippi (Fig. 4.2). Another channel evolution model was developed independently by the U.S. Geological Survey at about the same time from data collected north of the Mississippi-Tennessee state line from a 10,600 mi^2 area of West Tennessee (Simon and Hupp, 1986; Simon, 1989b, 1994; Fig. 4.1a). The West Tennessee model has 6 stages, is based on shifts in dominant adjustment processes, and is associated with a model of bank-slope development (Fig. 4.1b). Differences in the Schumm et al. (1984) model and the Simon and Hupp (1986) model include (Figs. 4.1a and 4.2):

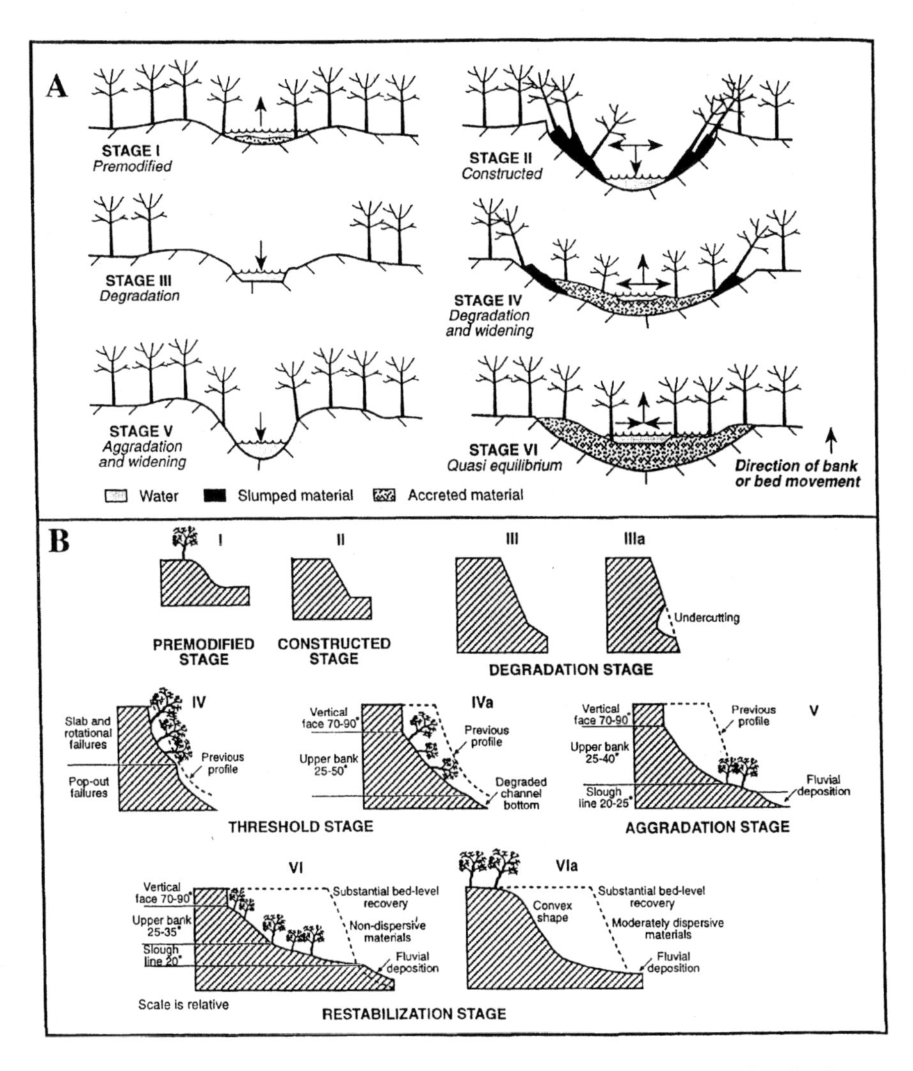

Fig. 4.1. **Six-stage models of (A) channel evolution and (B) bank-slope development for disturbed alluvial channels developed in West Tennessee (Modified from Simon and Hupp, 1986; Simon, 1989b).**

Table 4.1. Stages of channel evolution (from Simons and Hupp, 1986; Simon, 1989b).

Stage		Dominant Processes		Characteristic Forms	Geobotanical Evidence
Number	Name	Fluvial	Hillslope		
I	Premodified	Sediment transport-mild aggradation; basal erosion on outside bends; deposition on inside bends	—	Stable, alternate channel bars; convex top-bank shape; flow line high relative to top bank; channel straight or meandering.	Vegetated banks to flow line
II	Constructed	—	—	Trapezoidal cross section; linear bank surfaces; flow line lower relative to top bank	Removal of vegetation(?)
III	Degradation	Degradation; basal erosion on banks	Pop-out failures	Heightening and steepening of banks; alternate bars eroded; flow line lower relative to top bank	Riparian vegetation high relative to flow line and may lean towards channel
IV	Threshold	Degradation; basal erosion on banks	Slab, rotational and pop-out failure	Large scallops and bank retreat; vertical-face and upper-bank surfaces; failure blocks on upper bank; some reduction in bank angles; flow line very low relative to top bank	Tilted and fallen riparian vegetation
IV	Aggradation	Aggradation; development of meandering thalweg; initial deposition of alternate bars; reworking of failed material on lower banks	Slab, rotational and pop-out failures; low-angle slides of previously failed material	Large scallops and bank retreat; vertical face, upper bank, and slough line; flattening of bank angles; flow line low relative to top bank; development of new floodplain (?)	Tilted and fallen riparian vegetation; reestablishing vegetation on slough line; deposition of material above root collars of slough-line vegetation
VI	Restabilization	Aggradation; further development of meandering thalweg; further deposition of failed material; some basal erosion on outside bends; deposition of floodplain and bank surfaces	Low-angle slides; some pop-out failures near flow line	Stable, alternate channel bars; convex-short vertical face on top bank; flattening of bank angles; development of new floodplain (?); flow line high relative to top bank	Reestablishing vegetation extends up slough line and upper bank; deposition of material above root collars of slough-line and upper-bank vegetation; some vegetation establishing on bars

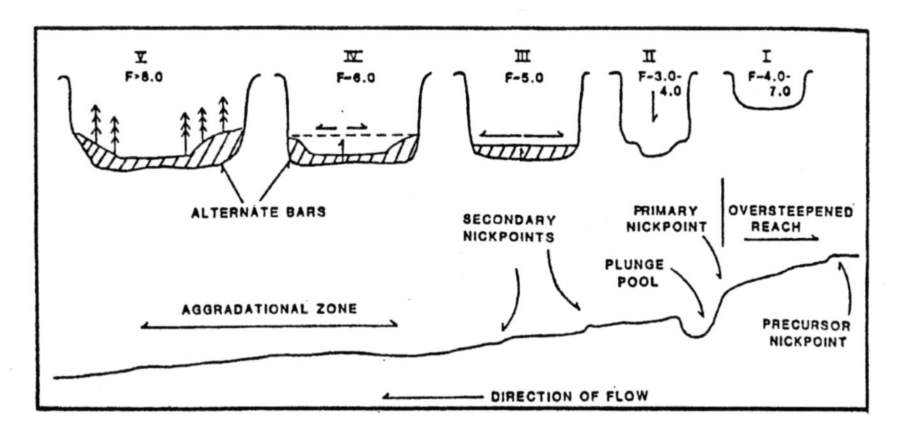

Fig. 4.2. Five-stage model of channel evolution developed on Oaklimiter Creek, northern Mississippi (from Schumm et al., 1984).

1. Stage II of the Simon and Hupp (1986) model represents the constructed/disturbed state and can be considered as an almost instantaneous condition prior to adjustment; more importantly,

2. The onset of channel widening by mass-wasting processes is associated with aggradation on the channel bed in the Schumm et al. (1984) model (stage III; Fig. 6-7, p128). In the Simon and Hupp (1986) model, mass failures of bank material are identified earlier in the adjustment sequence (stage IV), prior to the onset of aggradation when the channel is still degrading its bed.

Additionally, the Simon and Hupp model provides for integration of woody-vegetation recovery patterns and processes (Simon and Hupp 1986; Hupp, 1992; Simon and Hupp, 1992).

In alluvial channels, disruption of the dynamic equilibrium often results in some amount of upstream channel degradation and downstream aggradation. Using the Simon and Hupp (1986), model we can consider the equilibrium channel as the initial, predisturbed stage (I) of channel evolution, and the initially disturbed channel as an instantaneous condition (stage II). Rapid channel degradation of the channel bed ensues as the channel begins to adjust (stage III, Fig. 4.1a). Degradation flattens channel gradients and consequently reduces the available stream power for given discharges with time. Concurrently, bank heights are increased and bank angles are often steepened by fluvial undercutting and by pore-pressure induced bank failures near the base of the bank. Thus, the degradation stage (III) is directly related to destabilization of the channel banks and typically leads to channel widening by mass-wasting processes (stage IV) once bank heights and angles exceed the critical shear-strength of the bank material. The aggradation stage (V) becomes the dominant trend in previously degraded downstream sites as degradation migrates further upstream because the flatter gradient at the degraded site cannot transport the increased sediment loads emanating from degrading reaches upstream. This secondary aggradation occurs at rates roughly 60 percent less than the previous degradation rate associated with the site (Simon, 1992). These milder aggradation rates indicate that bed-level recovery will not be complete and that attainment of a new dynamic equilibrium (stage VI) will take place through further (1) bank widening and the consequent flattening of bank slopes, (2) the establishment and proliferation of riparian vegetation that adds roughness elements, enhances bank accretion, and reduces the stream power for given discharges, and (3) further gradient reduction by meander extension and elongation.

The lack of complete bed-level recovery results in a two-tiered channel configuration with the original floodplain surface becoming a terrace. Most storm flows are, therefore, constrained within this enlarged channel below the terrace level, resulting in higher flows having greater erosive power than when flood flows could dissipate energy by spreading across the floodplain. This scenario is common in the loess area of the midwestern United States and poses serious problems for the stability and maintenance of bridges and other river-crossing structures.

4.1. Channel Evolution in the Loess Area of the Midwestern United States

In western Iowa, stages of channel evolution were identified by Hadish (1994) from aerial reconnaissance of 551 stream miles in 1993 and 988 miles in 1994 using the Simon and Hupp (1986) model. Results show that for both years, 56 percent of the stream lengths were classified as stage IV (widening and mild degradation). Bed-level recovery (stage V; aggradation and widening) is occurring along about 24 percent of the stream lengths, predominantly along the downstream-most reaches. This indicates that channel widening by mass-wasting processes is currently the dominant adjustment process in the degraded streams of western Iowa, occurring along about 80 percent of the observed stream reaches. Only 6 percent of the stream reaches were classified as being stable, either premodified (stage I) or restabilized (stage VI) (Hadish, 1994), indicating that about 94 percent of the stream lengths in western Iowa can be considered unstable.

Stage V conditions (aggradation and mass failures) appear to dominate much of the mainstems of the easterly-flowing sand-bedded streams in southeastern Nebraska. However, because stage IV conditions (mass failures with bed degradation) are common along tributary streams, channel widening by mass failures is also the overall dominant adjustment process in southeastern Nebraska. A reconnaissance-level study conducted by the U.S. Army Corps of Engineers (1995) using the Simon and Hupp (1986) model confirmed and mapped these observations. Although the percentage of stream reaches in each stage of evolution was not calculated, a conservative estimate of the percentage of stream reaches experiencing mass failures is 75 percent. The north-facing banks on the easterly-flowing streams are generally more unstable as indicated by fresh failure surfaces and a lack of established woody-riparian vegetation than on south-facing banks in the same reach. This is probably related to (1) higher moisture contents and consequently lower values of suction (negative pore pressure) in the unsaturated zone of the north facing banks, and (2) a greater incidence of freezing and thawing of the silty bank materials on the north-facing banks. In contrast to western Iowa where stage III degrading conditions can be found in the upstream reaches of most tributary streams, in southeast Nebraska only small upstream tributaries near basin divides can be classified as stage III.

In West Tennessee, about 65 percent of the 1,645 studied sites are unstable, with channel widening occurring at about 60 percent of them (Bryan et al., 1995). Similar comparisons can be made in this part of the loess area between channel evolution in streams with silt beds (that dominate western Iowa) and those with sand beds (that dominate southeastern Nebraska). Tributaries of the largest, modified West Tennessee streams rise in unconsolidated sand-bearing formations that supply sand to the channels as bed material. Aggradation in downstream reaches occurs after 10 to 15 years of incision and channel widening occurs at moderate rates (Simon, 1989b). In contrast, smaller tributary streams near the Mississippi River bluff have cut only into loess materials, have silty beds, and no source of coarse-grained material. To reduce erosional forces and stream power for a given discharge without a coarse-grained sediment supply for downstream aggradation, channel widening (stage IV) may be the only mechanism for the silt-bed streams to recover (Simon, 1994). The silt-bed channels are the deepest and most rapidly-widening channels in West Tennessee and may take hundreds of years to recover. This is similar to the western Iowa silt-bed streams, such as West

Tarkio Creek, where downcutting (stages III and IV) has lasted for as much as 70 years and channel widening is widespread. These findings indicate that:

1. There is more coarse-grained sediment (sand and gravel) available for bed-material transport in southeastern Nebraska and West Tennessee than in western Iowa;
2. Channel evolution and recovery in southeastern Nebraska and West Tennessee is further advanced than in western Iowa because of plentiful supplies of sand;
3. Channel widening by mass-wasting processes dominates all areas; and
4. Degradation can be expected to continue to migrate upstream in western Iowa tributaries.

This latter point is supported by data from western Iowa reported by Antosch and Joens (1979) that show depths below channel beds to coarse-grained material (sand and gravel) to exceed 16 feet along the major drainageways. This is further supported by bed-material data collected during this study along two unstable creeks in western Iowa and along West Tarkio Creek in western Iowa and northwestern Missouri (Fig. 1.1). They show a preponderance of silty bed material, except in the downstream-most reaches where some sand can be found.

Reconnaissance in west-central Illinois and east-central and southeastern Iowa revealed many sinuous streams with bank failures and meander extension occurring on outside bends from the growth of point bars on the opposite, inside bank. This is typical of late stage V and stage VI channels, particularly where aggradation rates have been very high. Bank heights did not appear to be nearly as high as in southeast Nebraska, or western Iowa where bank heights between 27 and 35 feet are common. Assuming that channel modifications occurred in west-central Illinois and east-central and southeastern Iowa at about the same time as in western Iowa, it seems that many of the eastern loess-area streams have recovered more quickly. This is probably because (1) the initial direct disturbance to the channel (dredging and/or straightening) may not have been as extensive as in other parts of the loess area, and (2) the thinner loess cap has been penetrated by the streams, exposing transportable sand and gravel deposits.

4.1.1. Regional Summary

Sequences of channel evolution in the loess area of the midwestern United States are regionally consistent in that similar disturbances near the turn of the 20th century initiated channel adjustment of entire fluvial systems. Although sequences of channel evolution are similar throughout the region, distinct differences in the amount of time required for streams to recover, and to attain a new equilibrium condition vary with the class of sediments comprising the channel bed and banks. For example, the degradation stage (III) accounts for 10 to 15 years in the sand-bedded streams of West Tennessee but about 70 years in the silt-bedded streams of western Iowa. Time-based differences in sequences of channel evolution are conceptualized on the basis of boundary sediments. A schematic representation of changes in bed elevation and channel width, with the associated stage of channel evolution for loess-area streams composed of different bed-material sediments, is shown in Fig. 4.3 for the period 1850 to 2000. Note that sand-bed streams filled with sediments during the 1930s through the 1950s, instigating redredging and the clearing and snagging of downstream reaches. This additional channel work rejuvenated trunk and tributary streams causing the sequence of channel evolution to begin again (Fig. 4.3).

The sequence of channel evolution is less complicated for the silt-bedded streams of western Iowa and those smaller tributary streams of the region that cut only into loess sediments. In these cases, deep incision along trunk and tributary streams resulting from the initial channel work near the turn of the 20th century rapidly created bank heights in excess of the critical conditions of the material, causing a long period of bank instability and channel widening (Fig. 4.3). Without significant reductions in bank heights by aggradation, channel widening by mass-wasting processes will continue into the next century until such time as (1) bank angles are reduced by successive failures in the same location, and (2) the channel becomes so wide that the frequency of bank-toe removal by fluvial action is reduced considerably.

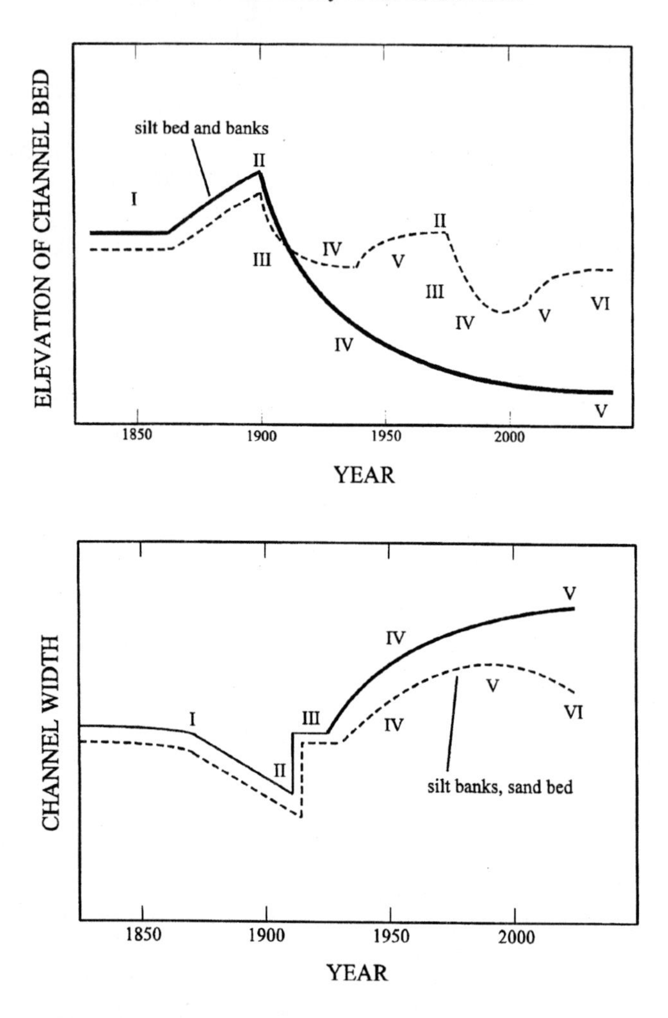

Fig. 4.3. **Schematic representation of typical changes in channel-bed elevation and channel width for silt- and sand-bedded streams in the loess area of the midwestern United States for the period 1850-2000.**

4.1.2. 1993 Midwest Floods

The 1993 floods in the midwestern United States inundated a large part of the upper Mississippi River Basin. The magnitude of the damages related to the flood exceeded $10 billion. Recurrence intervals of peak discharges were commonly greater than 100 years in western Iowa and eastern Nebraska (U.S. Geological Survey, 1993). The 1993 Midwest floods did not result in a rejuvenation of previously unstable channel systems or accelerated bed-level adjustments. Gully extension into agricultural fields occurred. The predominant effect of the 1993 floods on channel instability of tributary streams was channel widening. This occurred because of removal of toe material and prolonged saturation of channel banks, resulting in reduced shear strength and failures upon recession of storm flows. In straight reaches, low-cohesion channel banks up to 45 feet high failed readily at average rates of almost 7 ft/yr. In reaches characterized by alternate bars, moderate flows eroded low-bank surfaces on the opposite side of the channel resulting in steep bank angles and accelerated rates of channel widening. The combination of alternate bar growth and accelerated widening on the opposite bank represent incipient meandering of the channel within the straightened alignment. Outside meander bends generally erode more rapidly than in straight reaches because the buttressing effect of failed bank material is removed from the bank toe by fluvial action. This process was exacerbated by the 1993 floods.

5. WEST TARKIO CREEK, IOWA AND MISSOURI

Trends of channel change on West Tarkio Creek, Iowa and Missouri were determined to provide a quantitative measure of the rates and magnitudes of channel-adjustment processes in the region. West Tarkio Creek was selected because (1) it represents a typically unstable silt-bedded stream in the loess area of the midwestern United States that has undergone human disturbance, and (2) of the availability of historical information. Trends of channel adjustment were determined using historical bed-elevations, bank profiles, shear-strength tests, bed-material particle sizes, stages of channel evolution and dendrochronologic evidence. Historical bed-level data were compiled from cross-section surveys and longitudinal profiles (Fig. 5.1; Piest et al., 1977; Lohnes, R.A., Iowa State University, written comm., 1994). Bed-elevation data for 1994 and the additional field data listed above were collected during the summer of 1994.

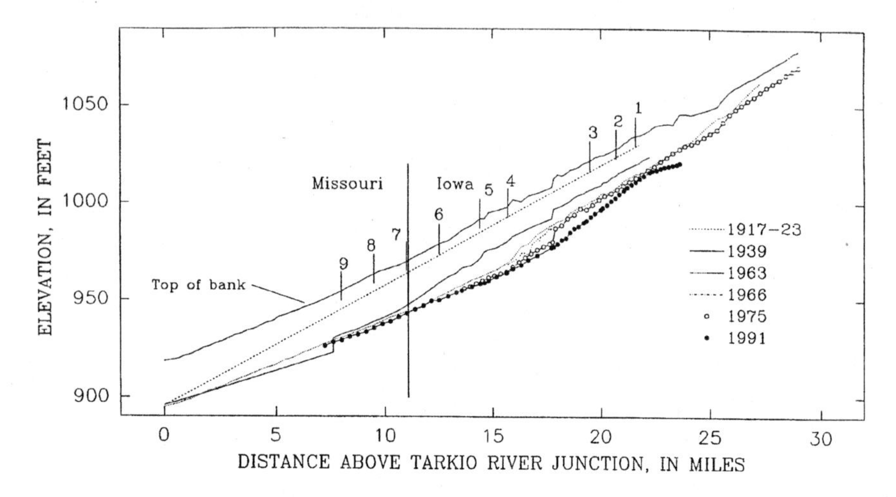

Fig. 5.1. Profiles of West Tarkio Creek, southwestern Iowa and northwestern Missouri.

West Tarkio Creek incised in the middle 1800s as a result of increased runoff rates emanating from cleared lands (Piest et al., 1977) and has been undergoing renewed channel adjustment since being straightened in about 1920. The channel displays the systematic variation of stage of channel evolution with distance upstream; aggradation (stage V) in its downstream-most reaches merging to widening with mild degradation (stage IV), to rapid degradation (stage III), with increasing distance upstream (Fig. 5.2). The dominant bed-material size class generally varies systematically with the stage of channel evolution and distance upstream; sand beds in aggrading reaches and silt beds in degrading reaches. Piest et al. (1977) describe stable reaches of West Tarkio Creek in 1975 near its junction with the Tarkio River at a channel gradient of 0.0006. It is assumed that by stability the authors were referring to a lack of further downcutting. This implies that degradation lasted for about 55 years. By 1994, aggrading conditions on West Tarkio Creek extended at least 5 miles into Iowa to the J52 bridge (Fig. 5.2). Assuming that stable conditions in 1975 were at the confluence with the Tarkio River and were about 15.5 miles farther upstream in 1994 than in 1975 (19 years), a rate of migration of 0.81 miles/yr is estimated.

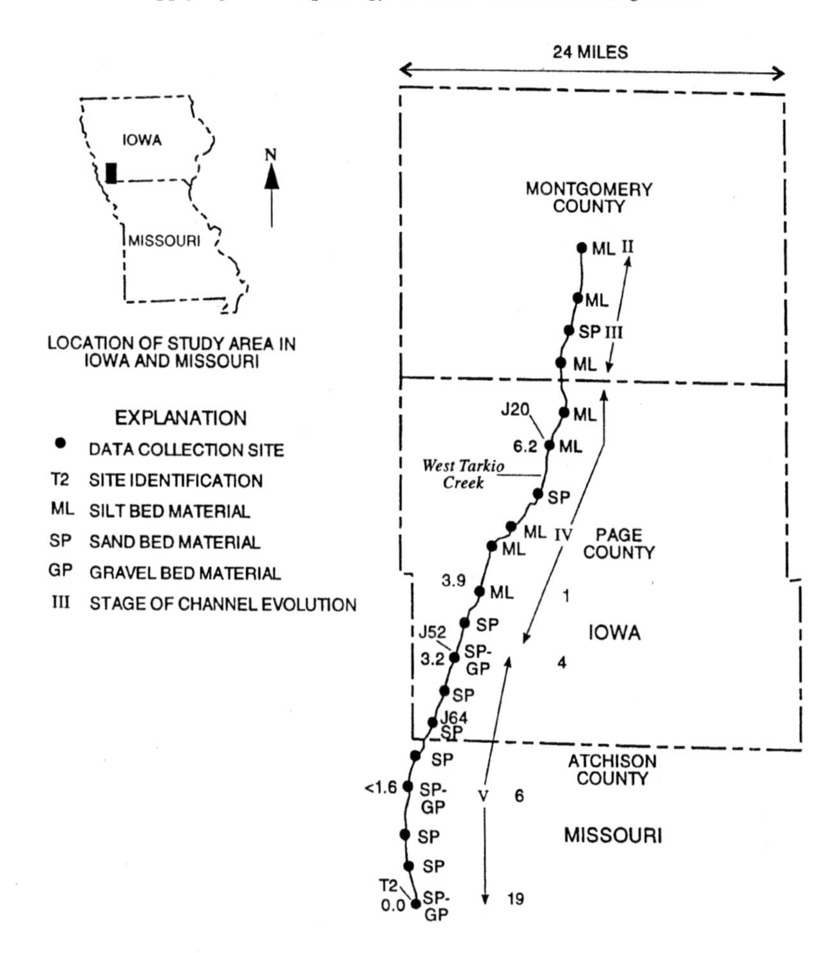

Fig. 5.2. Map of West Tarkio Creek, southwestern Iowa and northwestern Missouri, showing stages of channel evolution, dominant bed-material size class, age of oldest riparian tree, (years; numbers on right), and widening rate (feet per year; numbers on left).

The age of woody vegetation growing on streambanks is used as an indicator of bank conditions and the timing of renewed bank stability (Hupp and Simon, 1986; 1991; Hupp, 1992). The oldest riparian tree found at site T2 (Fig. 5.2) is 19 years old indicating that sections of streambanks have been stable for some time (since 1975) at this site. Ages and germination dates of the oldest streambank trees at other sites are shown in Fig. 5.2, indicating the timing of renewed bank stability along at least some low-bank surfaces. Note that average, recent widening rates, as estimated from dendrochronologic evidence generally increase moving from stage V conditions in downstream reaches, to stage IV conditions farther upstream (Fig. 5.2).

5.1. *Bed-Level Adjustments*

Trends of bed-level changes from 1920 to present were obtained by fitting historical data to an exponential decay equation (Simon, 1992):

$$z / z_o = a + b\, e^{(-k\, t)} \tag{5.1}$$

where

z = elevation of the channel bed (at time t);

z_o = elevation of the channel bed at (t_o=0);

a = dimensionless coefficient, determined by regression and equal to the dimensionless elevation (z/z_o) when Eq. 5.1 becomes asymptotic. a is greater than 1.0 for aggradation, a is less than 1.0 for degradation;

b = dimensionless coefficient, determined by regression and equal to the total change in the dimensionless elevation (z/z_o) when Eq. 5.1 becomes asymptotic, b is greater than 0.0 for degradation, b is less than 0.0 for aggradation;

k = coefficient determined by regression, indicative of the rate of change on the channel bed per unit time; and

t = time since the year prior to the onset of the adjustment process, in years (t_o=0).

Although aggradation rates along the downstream-most 8 miles of channel cannot be estimated using this method because of a lack of recent survey data, accretion measured around trees on bank surfaces, particularly at the sites in Missouri show that deposition at about 0.13 ft/yr is an ongoing process. Degradation trends at 9 sites are shown in Fig. 5.3. The 3 plots in the bottom row of the figure show that degradation along the downstream-most 11 miles of channel is virtually complete. Considerable degradation, continuing beyond the year 2000 can be expected along upstream reaches or until the curves become asymptotic (Fig. 5.3). An empirical model of bed-level adjustment for West Tarkio Creek is obtained by plotting the a-value coefficient (from Eq. 5.1) against distance from the junction with Tarkio River (Fig. 5.4). The future elevation of the channel bed, when degradation will have reached a minimum, can be determined by multiplying the a-value at a site (Fig. 5.4) by the initial bed elevation.

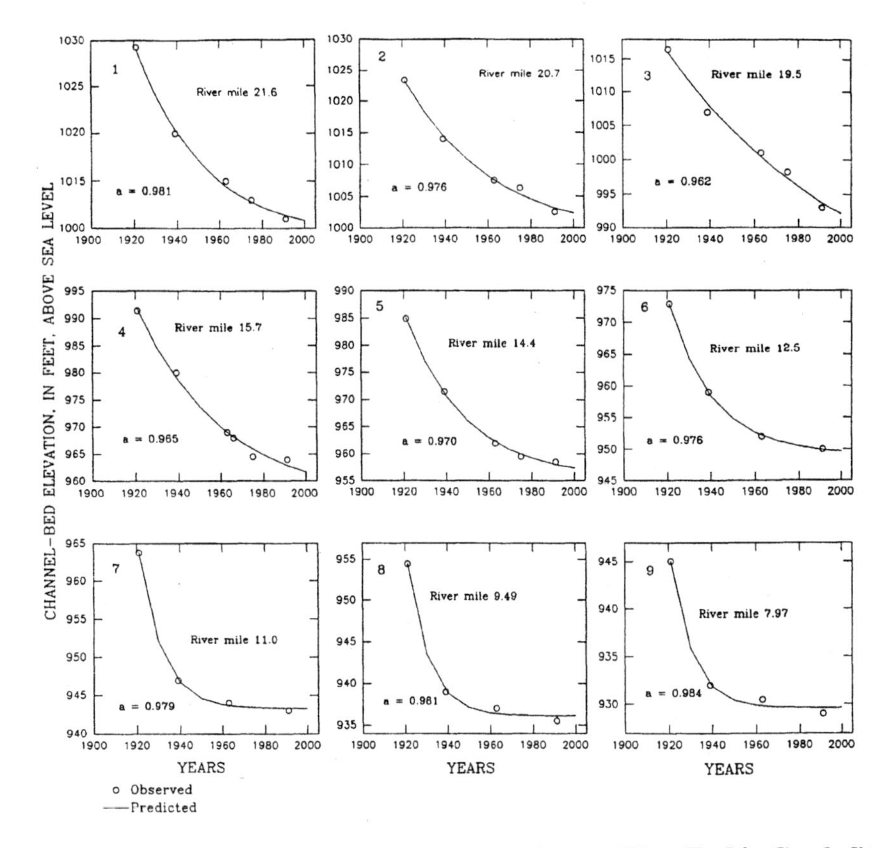

Fig. 5.3. **Trends of degradation at nine sites on West Tarkio Creek fit with Eq. 5.1.**

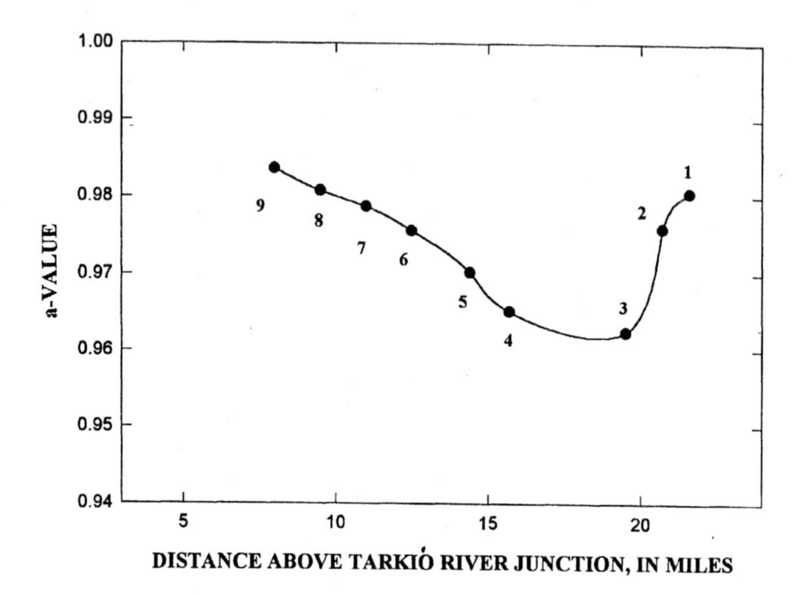

Fig. 5.4. Longitudinal distribution of *a*-values (from Eq. 5.1) representing an empirical model of bed-level response for West Tarkio Creek.

Maximum degradation (minimum *a*-values) along West Tarkio Creek is estimated to occur near river mile 20 (Fig. 5.4), in comparison to Piest et al.'s (1977) observation that it would occur near the Iowa-Missouri state line (river mile 11). This is not a particularly important distinction. However, that maximum degradation occurs in the middle reaches of West Tarkio Creek as opposed to downstream reaches as is typical of silt-bedded streams disturbed throughout their length (Simon, 1994) is important to note. Piest et al. (1977) attributed this to several factors including inferred differences in soil erodibility, steepness of the longitudinal profile and the "*natural evolutionary sequence*" (p.487). Degrading reaches of the channel do not have a statistically significant difference in the amount of clay or sand in the bed material to indicate different potentials for entrainment and erosion. Differences in (1) the gradient of the longitudinal profile and (2) channel morphology (stages of channel evolution) are merely a function of the migration of adjustment processes with time and reflect conditions at the time of observation. It seems probable, however, that maximum degradation should occur in the middle reaches of West Tarkio Creek because of the longitudinal distribution of available stream power in the basin. Increases in runoff rates and consequently, stream discharges and total stream power would be most effective in the middle reaches of drainage basins where available stream power generally is greatest (Graf, 1982; Lewin, 1982; Schumm et al., 1984). The middle reaches would then represent the "*area of maximum disturbance*," where maximum amounts of degradation can be expected (Simon, 1989b; 1992; 1994) for both disturbances; the indirect changes to rainfall-runoff relations in the 1850s, and the channel straightening in about 1920.

5.2. Channel-Width Adjustments

Changes in channel width have been and continue to be dramatic on West Tarkio Creek. Figure 5.5 shows increases in channel width for the period 1845-1994 at a site approximately 2 miles south of the Iowa-Missouri state line where bed-level recovery has started. This does represent a present-day worst-case condition because channel widening has slowed as aggradation became the dominant process on the channel bed.

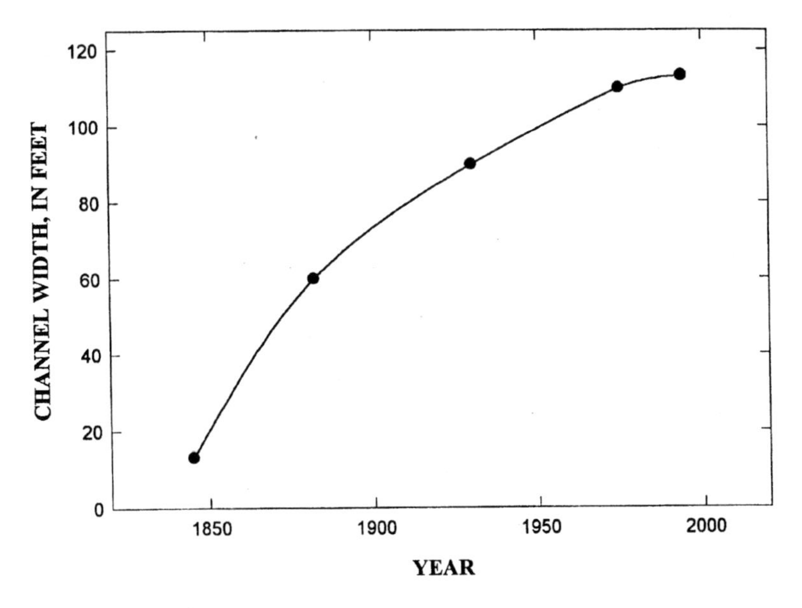

Fig. 5.5. Changes in channel width during the period 1845-1994 for a site on West Tarkio Creek about 2 miles south of the Iowa-Missouri state line.

More severe bank instabilities are located further upstream. The J52 bridge over West Tarkio Creek, located 5 miles north of the Iowa-Missouri state line (Fig. 5.6a), is used as an example for applying analytic methods regarding bank instability and estimation of stable bank geometries. In-situ shear-strength tests at 2.8- and 4.9-foot depths were taken about 160 ft downstream of the bridge. A mean cohesive strength (c) of 187 lbs/ft^2 and a mean friction angle (φ) of 29.8° were obtained. The mean, bulk unit weight (γ) of two undisturbed samples of the bank material was 138 lbs/ft^3. A stability number (N_s) of 6.90 was calculated using $\varphi = 29.8°$, assuming a vertical (90°) bank slope (i), and (Lohnes and Handy (1968):

$$N_s = (4 \sin i \cos \varphi) \; / \; [1 \; \pm \; \cos (i \; \pm \; \varphi)] \tag{5.2}$$

Critical bank height (H_c; above which there would be mass failure) is obtained by substituting N_s into the following equation from Carson and Kirkby (1972):

$$H_c = N_s \, (c \, / \, \varphi) \tag{5.3}$$

For $i = 90°$, $H_c = 9.35$ feet. Iterating for bank angles of 80, 70, 60, 50 and 40°, results in a bank-stability chart for ambient field conditions at the J52 site (upper line; Fig. 5.6a). This entire procedure is then repeated assuming that the banks are saturated and that $\varphi = 0.0$ (Lutton, 1974) to obtain H_c values under saturated conditions, resulting in the lower line of Fig. 5.6a. The H_c values obtained range from 5.4 feet at a bank angle of 90° to 14.9 feet at a bank angle of 40° (Fig. 5.6a). The effect of saturated conditions on decreasing H_c can be seen by drawing a vertical line anywhere on Figure 5.6a and comparing the difference in values at the intersection of the ambient- and saturated-condition lines.

The frequency of bank failure for the three stability classes is subjective, but is based on empirical field data from southeastern Nebraska, northern Mississippi, western Iowa, and particularly, from West Tennessee. An "unstable" channel bank can be expected to fail at least annually, and possibly after each major flow event (assuming that there is at least one in a given year). "At-risk" conditions indicate that bank failure can be expected every 2 to 5 years, again assuming that there is a runoff event that is sufficient to saturate the channel banks. "Stable" banks, by definition, do not fail by mass-wasting processes. Although channel banks on the outside of meander bends may widen by particle-by-particle erosion and may ultimately lead to collapse of the upper part of the bank, for the purposes of this discussion, stable-bank conditions refer to the absence of mass wasting.

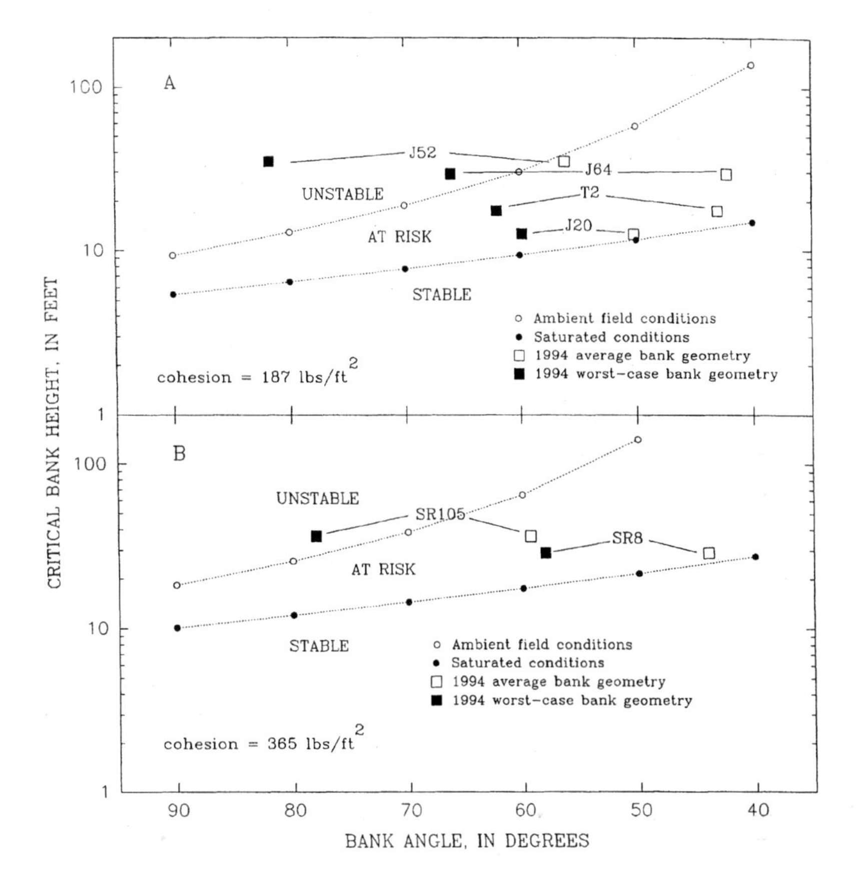

Fig. 5.6. Bank-stability chart for shear-strength conditions in the (a) vicinity of the J52 bridge over West Tarkio Creek, Page County, Iowa; also used as representative of other West Tarkio Creek sites and (b) bank-stability chart in the vicinity of the State Route 105 bridge over South Fork Big Nemaha River, southeastern Nebraska; also used as representative of site at State Route 8.

Using the slope angles of the vertical face and upper bank to obtain an average bank angle of 62°, and using a surveyed bank height of 35 feet (vertical distance from bank top to thalweg), channel banks at J52 plot between the at-risk and unstable zones (Fig. 5.6a). If the angle of the "vertical face" (81°) is used to represent worst-case conditions with bank toes eroded, the banks are clearly unstable, even when relatively dry. This is probably an unrealistic scenario, but provides the end-member stability case for unstable banks. Because the shear-strength characteristics at the J52 site are similar to average values obtained in studies of western Iowa (Lohnes and Handy, 1968) and West Tennessee (Simon, 1989c; Simon and Hupp, 1992), it seems reasonable to use the bank-stability chart (Fig. 5.6a) to approximate stability conditions at other sites on West Tarkio Creek. Average and worst-case bank geometries are plotted for several other sites on West Tarkio Creek to show current (1994) bank-stability relations.

Shear strength and bank-geometry data for two sites on the South Fork Big Nemaha River (SFBNR), southeastern Nebraska, are provided in Fig. 5.6b to show the effects of greater cohesion on critical bank height. Note that at a bank angle of 90°, H_c at saturated conditions for the SFBNR is greater than for a comparable bank slope under ambient field conditions for West Tarkio Creek. The cohesive strength of 365 lbs/ft^2 for the SFBNR banks (almost two times greater than for West Tarkio Creek) is largely responsible for the maintenance of stable banks at greater bank heights. The range of cohesive strengths tested in situ in southeastern Nebraska is from 328 to 864 lbs/ft^2, considerably higher than values obtained in West Tennessee (Simon, 1989c; Simon and Hupp, 1992) or in western Iowa (Lohnes and Handy, 1968), and supports the observation made earlier that the southeastern Nebraska channels maintain some of the highest banks in the region.

Generalizations about critical-bank heights (H_c) and angles can be made with knowledge of the variability in cohesive strengths. Five categories of mean cohesive strength of the channel banks (in pounds per square foot) are created: 0 to 72, 73 to 144, 145 to 216, 217 to 288, and 289 to 533. H_c above the mean low-water level and saturated conditions are used to construct Fig. 5.7 because failures typically occur during or after the recession of peak flows. The result is a nomograph (Fig. 5.7) giving critical bank-heights for a range of bank angles and cohesive strengths that can be used to estimate stable-bank configurations for worst-case conditions (saturation during rapid decline in river stage) at a given cohesive strength (Simon and Hupp, 1992). For example, a saturated bank at an angle of 55° and a cohesive strength of 252 lbs/ft^2 could support a bank of no more than about 10 feet (Fig. 5.7). Because changes in bed-elevation and bank heights (floodplain elevation minus thalweg elevation) can be obtained with Eq. 5.1, this type of analytic solution is useful in determining the (1) timing of the initiation of general bank instabilities (in the case of degradation and increasing bank heights), (2) timing of renewed bank stability (in the case of aggradation and decreasing bank heights), and (3) bank height and angle that need to be engineered to attain a stable-bank configuration under a range of moisture conditions.

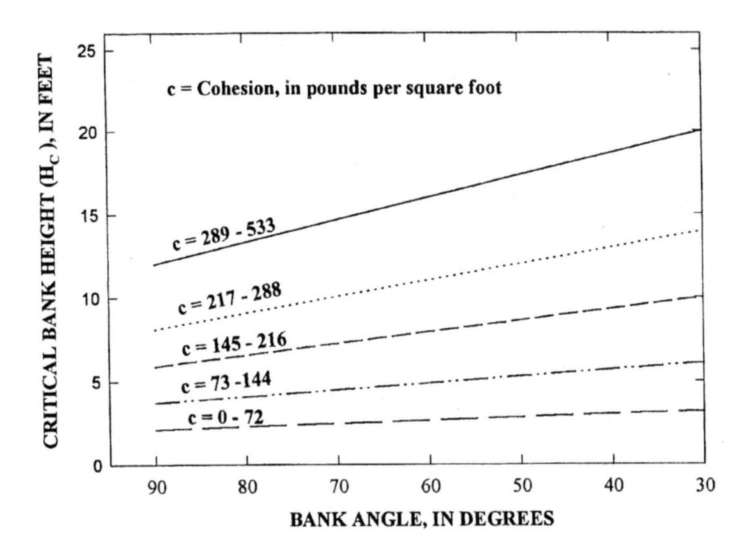

Fig. 5.7. Critical bank-slope configurations for various ranges of cohesive strengths under saturated conditions (modified from Simon and Hupp, 1992).

In most stage IV or V channels of the loess area of the midwestern United States the re-shaping of bank slopes to regain bank stability is probably an unacceptable option because flattening of bank angles results in additional loss of floodplain (or terrace) lands. Similarly, it is probable that bank heights cannot be reduced sufficiently by channel infilling to values less than H_c along the thousands of miles streams that are still experiencing bank failures. The proliferation of woody vegetation on the channel banks in combination with aggradation rates that are about 60 percent less than the initial degradation rate may slowly alleviate some of this problem naturally. In places where floodplain or terrace resources need to be protected, further human intervention may be necessary.

6. SUMMARY AND CONCLUSIONS

The loess area of the midwestern United States contains thousands of miles of unstable streams, predominantly responding to human modifications imposed near the turn of the 20th century. The dominant process appears to be channel widening and meander extension by mass failure. The most severe, widespread instabilities are in western Iowa where a thick loess cap and a limited supply of coarse-grained (sand and gravel) material restricts bed-level recovery by aggradation. Here relatively high streambanks combine with low cohesive strengths to sustain channel widening. Channel adjustments in West Tennessee and southeastern Nebraska are not as severe as in western Iowa. In general, channels in west-central Illinois and east-central Iowa appear to be closer to recovery than in other parts of the region.

Degradation rates for streams draining the loess hills of western Iowa and eastern Nebraska have decreased nonlinearly since 1920, approaching minimal values. In some downstream reaches, sandy alternate bars and fluvially deposited sand on low-bank surfaces indicate the beginning of bed-level recovery and the "aggradation" stage of channel evolution. This "natural" recovery process, which often occurs in other unstable stream systems following 10 to 15 years of degradation, has apparently been delayed for an additional 55 years along some reaches in the loess area because of the lack of hydraulically-controlled (sand- or gravel-sized) source material. Finer-grained silts from eroding beds and banks are efficiently transported through tributary systems to the Missouri River. With loess thickness generally decreasing with distance upstream and with the degradation process migrating upstream with time, incision through the loess ultimately results in the exposure of coarser-grained glacial till. These deposits, thus, provide the coarse material necessary for downstream aggradation and for the initial phase of channel recovery.

7. ACKNOWLEDGMENTS

The authors wish to express their sincere gratitude to Roy Trent, Federal Highway Administration, for sponsoring this work and for taking the time to join us in the field to see these channels first hand; to Gregg Hadish, Golden Hills Resource Conservation and Development, for assistance in organizing field operations, sharing data, and for assistance in the field; to Robert Lohnes, Iowa State University, for insightful discussion and for sharing his wealth of experience in the loess area, to Stephen Darby, Agricultural Research Service, for field assistance; and to Cliff Hupp, U.S. Geological Survey, for assistance with collecting dendrochronologic information on West Tarkio Creek.

The second author thanks Paolo Canuti, Universita degli Studi di Firenze, for making funds available for him to participate in this research. Finally, the authors are indebted to the pilots of the single-engine aircraft who safely flew reconnaissance for us from the Council Bluffs, Iowa, and the Moline, Illinois, airports.

8. REFERENCES

Antosch, L. and Joens, J. 1979. Iowa's lowhead dams: their past, present and future roles. *Iowa State Water Resources Research Institute*, No. 96, Special Report.

Ashley, G.H. 1910. Drainage problems in Tennessee. *Tennessee State Geological Survey Bulletin* 3-A, pp. 1-15.

Baumel, C.P. 1994. Impact of degrading western Iowa streams on private and public infrastructure costs. In. Hadish, G. A., (Ed.) *Stream Stabilization in Western Iowa*, Iowa DOT HR-352, Golden Hills Resource Conservation and Development, Oakland, Iowa, pp. 4-1 to 4-39.

Brice, J.C. 1966. Erosion and deposition in the loess mantled Great Plains, Medicine Creek Drainage Basin, Nebraska. *U.S. Geological Survey Professional Paper* 352-H, pp. 255-339.

Bryan, B.A., A. Simon, G.S. Outlaw, and R. Thomas. 1995. Methodology for assessing channel conditions at bridges in Tennessee related to scour critical conditions. *U.S. Geological Survey Water-Resources Investigations Report*, 94-4229, 54 p.

Bull, W.B. 1979. Threshold of critical power in streams. *Geological Society of America Bulletin*, v. 90, p. 453-464.

Carson, M.A. and M.J. Kirkby. 1972. *Hillslope Form and Process*. Cambridge University Press, Oxford, 475 p.

Daniels, R.B. 1960. Entrenchment of the Willow Drainage Ditch, Harrison County, Iowa. *American Journal of Science*. v. 258, pp. 161-176.

Davis, W.M. 1902. Baselevel, grade, and peneplain. *Journal of Geology*, v. 10, p. 77-111.

Elliott, J.G. 1979. Evolution of large arroyos, the Rio Puerco of New Mexico. Colorado State University, Fort Collins, Colorado, Master of Science thesis, 106 p.

Emerson, J.W. 1971. Channelization: A case study. *Science*, v. 172, pp. 325-326.

Graf, W.L. 1982. Spatial variations in of fluvial processes in semi-arid lands. In *Space and Time in Geomorphology*, C.E. Thorn (Ed.), Publications in Geomorphology, 12. State University of New York at Binghamton, pp. 193-217.

Hadish, G.A. 1994. *Stream Stabilization in Western Iowa*. Iowa DOT HR-352, Golden Hills Resource Conservation and Development, Oakland, Iowa. 198 p.

Hidinger, L.L. and A.E. Morgan. 1912. Drainage problems of the Wolf, Hatchie, and South Fork Forked Deer Rivers, in West Tennessee. In. *The Resources of Tennessee*, Tennessee Geological Survey, v. 2, no. 6, pp. 231-249.

Hupp, C.R. and A. Simon. 1986. Vegetation and bank-slope development. *Proceedings, Fourth Federal Interagency Sedimentation Conference*, Las Vegas, v. 2, pp. 5-83 to 5-92.

Hupp, C.R. and A. Simon. 1991. Bank accretion and the development of vegetated depositional surfaces along modified alluvial channels. *Geomorphology*, v. 4, pp. 111-124.

Hupp, C.R. 1992, Riparian vegetation recovery patterns following stream channelization: A geomorphic perspective. *Ecology*, v. 73, pp. 1209-1226.

Ireland, H.A., C.F.S. Sharpe, and D.H. Eargle. 1939. Principles of gully erosion in the Piedmont of South Carolina. *U.S. Department of Agriculture Technical Bulletin* 633, 142 p.

Keller, ____, 1972. Development of alluvial stream channels: A five-stage model. Geological Society of America Bulletin, v. 83, p. 1531-1536.

Kouwen, N. 1970. Flow retardance in vegetated open channels. PhD thesis, University of Waterloo, Ontario, Canada.

Lane, E.W. 1955. The importance of fluvial morphology in hydraulic engineering. *Proceedings, American Society of Civil Engineers*, v. 81, no. 795, p. 1-17.

Lewin, J. 1982. British floodplains. In *Papers in Earth Studies,* B.H. Aldman, C.R. Fenn, and L. Morris (Eds.), GeoBooks, Norwich, pp. 21-37.

Little, W.C. and J.B. Murphey. 1981. Evaluation of streambank erosion control demonstration projects in the bluffline streams of northwest Mississippi. Appendix A, *Stream Stability Report* to Vicksburg District, U.S. Army Corps of Engineers, AD-A101-386, National Technical Information Service, 342 p.

Lohnes, R.A. 1994. Stream stabilization research. In *Stream Stabilization in Western Iowa*, Hadish, G. A., (Ed.), Iowa DOT HR-352, Golden Hills Resource Conservation and Development, Oakland, Iowa., pp. 3-1 to 3-54.

Lohnes, R.A. and R.L. Handy. 1968. Slope angles in friable loess. *Journal of Geology*, v. 76, no. 3, pp. 247-258.

Lohnes, R.A., F.W. Klaiber, and M.D. Dougal. 1980. *Alternate Methods of Stabilizing Degrading Stream Channels in Western Iowa*. Iowa DOT HR-208, Department of Civil Engineering, Engineering Research Institute, Iowa State University, Ames, Iowa, 132 p.

Lutenegger, A.J. 1987. In situ shear strength of friable loess. In. *Loess and Environment*, M. Pesci (Ed.), Catena Supplement 9, pp. 27-34.

Lutton, R.J. 1974. Use of loess soil for modeling rock mechanics. *U.S. Army Corps of Engineers Waterways Experiment Station Reports* S-74-28, Vicksburg.

Mackin, J.H. 1948. Concept of a graded river. *Geological Society of America Bulletin*, v. 59, p. 463-511.

Moore, C.T. 1917. Drainage districts in southeastern Nebraska. In *Nebraska Geological Survey,* Miscellaneous Papers, Barbour, E.H., (Ed)., v. 7, part 17, p. 125-164.

Morgan, A.E. and S.H. McCrory. 1910. Drainage of the lands overflowed by the North and Middle Forks of the Forked Deer River, and the Rutherford Fork Obion River in Gibson County, Tennessee. *Tennessee State Geological Survey Bulletin* 3-B, pp. 17-43.

Piest, R.F., L.S. Elliott, and R.G. Spomer. 1977. Erosion of the Tarkio drainage system,1845-1976. *Transactions, American Society of Agricultural Engineers*, v.20, pp.485-488.

Piest, R.F., C.E. Beer, and R.G. Spomer. 1976. Entrenchment of drainage systems in western Iowa and northwestern Missouri. *Proceedings, Third Federal Interagency Sedimentation Conference*, pp. 5-48 to 5-60.

Robbins, C.H. and A. Simon. 1983. Man-induced channel adjustment of Tennessee streams. *U.S. Geological Survey Water-Resources Investigations Report* 82-4098, 129 p.

Ruhe, R.V. and R.B. Daniels. 1965. Landscape erosion-geologic and historic. *Journal of Soil and Water Conservation*, v. 20, pp. 52-57.

Schumm, S.A. and R.F. Hadley. 1957. Arroyos and the semiarid cycle of erosion. *American Journal of Science*, v. 225, pp. 161-174.

Schumm, S.A., M.D. Harvey, and C.C. Watson. 1984. *Incised Channels, Morphology, Dynamics and Control.* Water Resources Publications, Littleton, Colorado, 200 p.

Simon, A. 1989a. The discharge of sediment in channelized alluvial streams. *Water Resources Bulletin*, v. 25, no. 6, pp. 1177-1188.

Simon, A. 1989b. A model of channel response in disturbed alluvial channels. *Earth Surface Processes and Landforms*, v. 14, pp. 11-26.

Simon, A. 1989c. Shear-strength determination and stream-bank instability in loess-derived alluvium, West Tennessee, USA. In: *Applied Quaternary Research*, E.J. DeMulder and B.P. Hageman (Eds.), A.A. Balkema Publishers, Rotterdam, pp. 129-146.

Simon, A. 1992. Energy, time, and channel evolution in catastrophically disturbed fluvial systems. In *Geomorphic Systems, Geomorphology*, J.D. Phillips and W.H. Renwick, (Eds.), v. 5, pp. 345-372.

Simon, A. 1994. Gradation processes and channel evolution in modified West Tennessee streams: process, response, and form. *U.S. Geological Survey Professional Paper* 1470, 84 p.

Simon, A. and C.R. Hupp. 1986. Channel evolution in modified Tennessee channels. *Proceedings, Fourth Federal Interagency Sedimentation Conference*, Las Vegas, March 24-27, 1986, v. 2, p. 5-71 to 5-82.

Simon, A. and C.R. Hupp. 1992. Geomorphic and vegetative recovery processes along modified stream channels of West Tennessee. *U.S. Geological Survey Open-File Report* 91-502, 142 p.

Speer, P.R., W.J. Perry, J.A. McCabe, O.G. Lara, and others. 1965. Low-flow characteristics of streams in the Mississippi Embayment in Tennessee, Kentucky, and Illinois. With a section on Quality of water, by H.G. Jeffery: *U.S. Geological Survey Professional Paper* 448-H, 36 p.

Thorne, C.R. 1989. Effects of vegetation on riverbank erosion and stability. In *Vegetation and Erosion*, Thornes, J.B., (Ed.), Chichester, England, John Wiley & Sons, p. 125-144.

Thorne, C.R., J.B. Murphey, and W.C. Little. 1981. *Stream Channel Stability*. Appendix D, Bank stability and bank material properties in the bluffline streams of northwest Mississippi: Oxford, Mississippi, U.S. Department of Agriculture Sedimentation Laboratory, 257 p.

Turnbull, W.J. 1948. Utility of loess as a construction material. *International Conference on Soil Mechanics and Foundation Engineering*, 2nd, Rotterdam, Proceedings, v.5, sec. 4d, pp. 97-103.

Turnipseed, D.P. and K.W. Wilson, Jr. 1992. Channel and bank stability of Standing Pine Creek at State Highway 488 near Freeny, Leake County, Mississippi. *U.S. Geological Survey Open-File Report* 92-112, 18 p.

U.S. Army Corps of Engineers. 1907. Index map of the South Fork of Forked Deer River, Tennessee from Jackson to the mouth. Nashville, Tennessee State Archives, scale 1:48,000, 1 sheet.

U.S. Army Corps of Engineers. 1995. *Southeast Nebraska Streambank Erosion and Streambed Degradation Study*, Omaha District.

U.S. Geological Survey. 1993. Flood discharges in the upper Mississippi River Basin. *U.S. Geological Survey Circular*, 1120-A.

Vidal, H. 1969. The principle of reinforced earth. *Highway Research Record*, v. 282, p. 1-16.

Williams, G.P. and M.G. Wolman. 1984. Downstream effects of dams on alluvial rivers. *U.S. Geological Survey Professional Paper* 1286, 83 p.

Wilson, K.V. 1979. Changes in channel characteristics, 1938-1974, of the Homochitto River and tributaries, Mississippi. *U.S. Geological Survey Open-File Report* 79-554, 18 p.

Wilson, K.V., Jr. and D.P. Turnipseed. 1994. Geomorphic response to channel modifications of Skuna River at the State Highway 9 crossing at Bruce, Calhoun County, Mississippi. *U.S. Geological Survey Water-Resources Investigations Report*, 94-4000, 43 p.

Wilson, K.V., Jr. and D. P. Turnipseed. 1993. Channel-bed and channel-bank stability of Standing Pine Creek tributary at State Highway 488 at Free Trade, Leake County, Mississippi. *U.S. Geological Survey Open-File Report* 93-37, 20 p.

Section 4

APPLIED FLUVIAL GEOMORPHOLOGY

CHANGES IN BED SEDIMENT MOBILITY DUE TO HYPOTHETICAL CLIMATE CHANGE IN THE EAST RIVER BASIN, COLORADO

R.S. Parker, J.M. Nelson, and G. Kuhn[1]

ABSTRACT

A number of studies have suggested that changes in climate may affect the timing and supply of water, but little work has been done to assess the effect of these hydrograph alterations on processes dependent on the timing of streamflow. For example, bed sediment flux is dependent on flow characteristics within a stream reach and changes in the streamflow regime may significantly alter the sediment flux. This paper examines the sensitivity of bed sediment discharge of the East River, a tributary of the Gunnison River in west-central Colorado, to changes in climate (i.e., air temperature). Linking a watershed model, a hydraulic model, and a sediment-transport relation simulates these effects. Using data for water year 1984 the effects of three climate scenarios were investigated. These scenarios included an increase in daily mean air temperature of 2°C, 4°C, and 6°C. These increases altered the timing of the hydrograph but reduced the annual volume of water by only 2, 4, and 7 percent, respectively. This modest reduction in streamflow has been observed by other water resource investigations. Changes in the timing of the annual hydrograph have a substantial effect on the bed sediment discharge, with the sediment flux reduced by 86 percent for the climate scenario with an increase of 6°C in daily mean air temperature. The linking of a series of models provides a way to quantitatively describe changes within a specific channel reach that result from diffuse alterations in the upstream drainage basin.

KEYWORDS

bed sediment discharge, climate change, watershed model, hydraulic model

---

[1] Hydrologists, U. S. Geological Survey, MS 415, Denver Federal Center, Denver, Colorado 80225

1. INTRODUCTION

There has been a number of studies indicating that changes in climate may affect the timing and supply of water, and identifying components of the water balance that may be altered, such as precipitation, snowpack accumulation and melt, evapotranspiration, and streamflow (Gleick, 1987, 1989; Rango and van Katwijk, 1990; Kuhn and Parker, 1992). Many of these studies focus on examining the sensitivity of the water resource by altering climatic parameters and examining the resultant water resource. Because potential changes in the water supply have been an important issue, these studies have concentrated on streamflow volume. Although previous studies indicate alterations in the timing of the streamflow hydrograph, little work has been done to assess the effect of these hydrograph alterations on processes dependent on the timing of streamflow. For example, bed sediment flux is dependent on flow characteristics within a reach of stream. Changes in the peak streamflow regime may significantly alter the sediment flux even for cases where little or no change in streamflow volume occurs.

Concern about the sensitivity of the water resource in the Colorado River basin to changes in climate has produced a series of studies. Several of these studies have focused on the East River basin, a tributary of the Gunnison River, which is a substantial contributor to the water resource of the upper Colorado River. These studies have examined changes in annual and seasonal streamflow volumes resulting from changes in air temperature and precipitation. Nash and Gleick (1991, 1993) showed that annual runoff in the East River basin decreased 16.5 percent with a hypothetical temperature increase of 4°C. McCabe and Hay (1995) showed a 4 percent decrease in annual runoff with a 4°C increase in temperature and suggested that the difference between the two studies may be due, in part, to different algorithms used to compute the water balance. These previous studies also evaluated the effects of temperature increase on seasonal runoff in the East River basin and both concluded that there was a shift in seasonal runoff to earlier in the year, generally producing decreases in summer runoff and increases in winter/spring runoff. These studies concluded that temperature had a minor effect on the magnitude of annual runoff, but that temperature strongly affected the timing of snowmelt during the spring and summer seasons. These results were similar to those of other studies that examined effects of potential climate change on water resources in the western United States (Gleick, 1987; Lettenmaier and Gan, 1990; Rango and van Katwijk, 1990).

Although previous studies have indicated sensitivity of runoff volume and timing to changes in air temperature and precipitation, there have been few attempts to quantify the sensitivity of riverine processes that are dependent on streamflow. It would seem that if altered climatic conditions influence streamflow, a process such as bed sediment discharge that is dependent on streamflow magnitude and timing, might also be greatly altered.

This paper examines changes in bed sediment discharge or flux to alterations in the timing of streamflow resulting from an increase in air temperature. In order to examine these changes, there must be a linkage from the watershed (and changes in streamflow from alterations in air temperature within the basin), to the hydraulics and sediment transport within a given reach of the channel. Such an analysis must be performed in a broad spatial scale to define and incorporate changes in climate and the water balance within a watershed. Analysis must provide information on a sufficiently fine resolution to define the hydraulic characteristics within a river reach to evaluate the changes in climate and streamflow on sediment flux. In this study, several models are linked together in order to perform the analyses at these various spatial scales.

2. THE STUDY AREA

The East River basin is an important contributor of water to the Gunnison River in western Colorado (Fig. 2.1). In this 748 sq. km basin, elevations range from 2,440 to 4,359 m. Mean monthly temperatures in the basin range from below -10°C during January to almost 13°C during July, and remain below 0°C from November through March. Precipitation ranges from 25 mm to almost 60 mm per month, and is highest in November through March, primarily in the form of snow. Streamflow increases in April as a result of snowmelt, and the peak discharge occurs in late May or early June. Streamflow during the remainder of the summer and fall seasons are primarily a recession from the snowmelt peak.

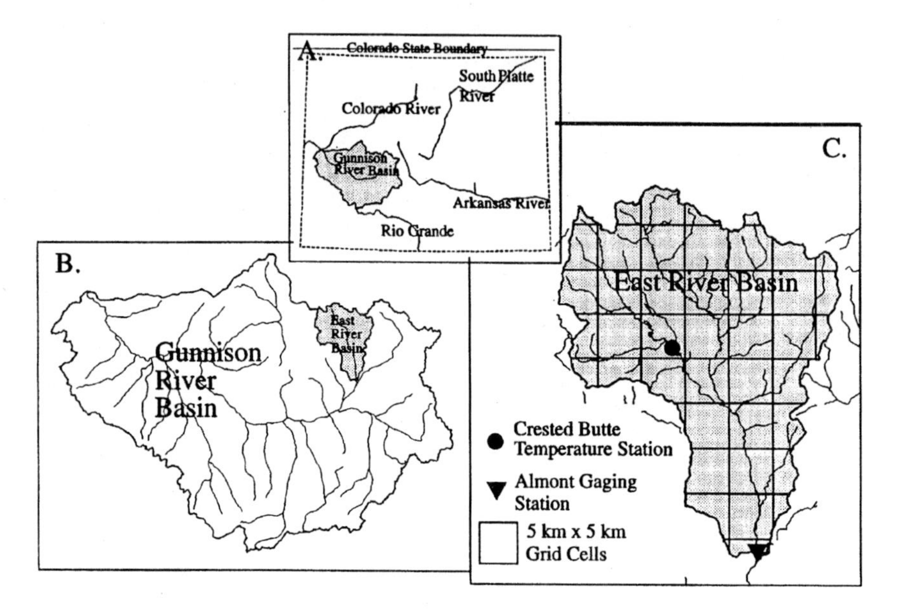

Fig. 2.1. Locations of the Gunnison River basin and the East River basin, Colorado (adapted from McCabe and Hay, 1995).

A stream gage at Almont is near the outlet of the basin, and the reach in which it is located is used to investigate the sensitivity of hydraulic properties and sediment flux to increases in air temperature. The cobble bed has a d_{50} of 112 mm for the surficial material and the slope of the channel is 0.0048. Williams (1978, Table 2) lists the bankfull discharge at this gaging station as 89.2 m/s; at this discharge the channel width is 25.9 m and the mean depth is 1.58 m.

3. MODEL SEQUENCE

Three models are used in sequence to examine the sensitivity of changes in bed sediment discharge to increases in air temperature. A watershed model incorporates the increases in temperature, distributes these changes over the drainage basin, and tracks changes in the water and energy balance. Daily streamflow from the watershed model is used as input to a 1-dimensional hydraulics model to provide hydraulic characteristics of the reach near the basin outlet, and a bed sediment-transport equation yields sediment discharge for given hydraulic conditions. Each model is discussed separately in the following sections.

3.1 Watershed Model

The U.S. Geological Survey's Precipitation Runoff Modeling System (PRMS) (Leavesley et al., 1983) is a modular, distributed-parameter, watershed-modeling system useful in evaluating hydrologic response to changes in climate or land use. Parameters are spatially distributed by partitioning a watershed into hydrologic response units (HRUs) that are assumed to be homogeneous in their hydrologic response and are characterized using information on topography, soils, vegetation, and precipitation distribution. A daily water balance is computed for each HRU on the basis of model inputs (daily values of air temperature, precipitation, and solar radiation). A snowpack is maintained for each HRU and modified on the basis of a calculation of the energy balance of the snowpack twice daily. The responses of all HRUs are summed on a unit-area basis and provide daily streamflow at the outlet of the drainage basin. In this study HRUs were defined and parameter values assigned using a geographical information system (GIS) with a 5 x 5 km grid that covered the East River basin (Battaglin et al., 1993).

Temperature and precipitation data were obtained from the National Weather Service meteorological station at Crested Butte (Fig. 2.1) and extrapolated to each HRU based primarily on the average elevation of each unit. Solar radiation data were not available at this station so solar radiation was estimated as a function of maximum daily air temperature through an algorithm provided in PRMS. Measured runoff was available from the gaging station at Almont, Colorado (Fig. 2.1).

To provide an indication of the goodness-of-fit of the modeled to measured daily streamflow of the East River at Almont for water year 1984, the coefficient of efficiency (E) is used (Nash and Sutcliffe, 1970). This coefficient, E, provides a measure of the variance of the measured streamflow that is accounted for by the model. The coefficient includes systematic errors between the simulated and measured daily streamflows and, therefore, is always smaller than the coefficient of determination (R). Values for E range from +1 to negative infinity where a negative value indicates that the mean of the measured streamflow is a better predictor than the simulated streamflow. The computed model efficiency of the daily flows for 1984 is 0.93. Thus, the fit between simulated and measured streamflow is good. The watershed model was not calibrated and the coefficient of efficiency is used here as an independent verification of the quality of prediction for water year 1984.

The watershed model was used to simulate runoff for water year 1984 from the East River basin for the current climate condition and for hypothetical conditions that may represent a range of possible climate changes. The 1984 water year had an instantaneous peak discharge that was the fifth highest in the 64-year history of the gage (through 1994) and provides a good example of changes in bed sediment discharge. Only a limited range of possible climate scenarios are given here and include current conditions and changes in daily mean air temperatures of +2°C, +4°C, and +6°C. These scenarios were developed by altering the current time series of daily maximum and minimum air temperatures to provide the annual changes specified. This method preserves the daily variability in temperature. The present lapse rates were used in each climate scenario and no attempt was made to modify the lapse rates. No climate scenarios that include changes in precipitation are reported here, as the focus in this report is to examine the effects of streamflow timing changes on sediment transport. Scenarios that include changes in precipitation in the East River resulted primarily in altering the volume of the hydrograph (McCabe and Hay, 1995) and are outside the scope of this paper.

The resulting hydrograph for current conditions and the three temperature scenarios are shown in Fig. 3.1 and changes in annual streamflow volume are listed in Table 3.1. The general changes in the 1984 hydrograph show a shift to earlier in the year accompanied by a decrease in the peak of the hydrograph due primarily to an earlier melting of the snow pack from increased air temperatures. These results are generally consistent with analyses made by others for this basin (Nash and Gleick, 1991, 1993; McCabe and Hay, 1995). The volume of water does not change substantially (Table 3.1) and shows a decrease of only 2 percent in the +2°C temperature scenario and 5 percent in the +4°C temperature scenario. An examination of the water balance suggests that most of the water volume losses come from increases in evapotranspiration due to the higher temperatures. This change in annual water volume is similar to that reported by McCabe and Hay (1995) for the East River in which they modeled the period 1973-89 and the +4°C temperature scenario yielded an average decrease in annual streamflow volume of 4 percent for the 17 years that were modeled. However, the daily peak discharge for water year 1984 decreases from 105.2 m^3/s for the current conditions, to 100.7 m^3/s for the +2°C temperature scenario, to 91.4 m^3/s for the +4°C temperature scenario, and to 87.6 m^3/s for the +6°C temperature scenario (Fig. 3.1).

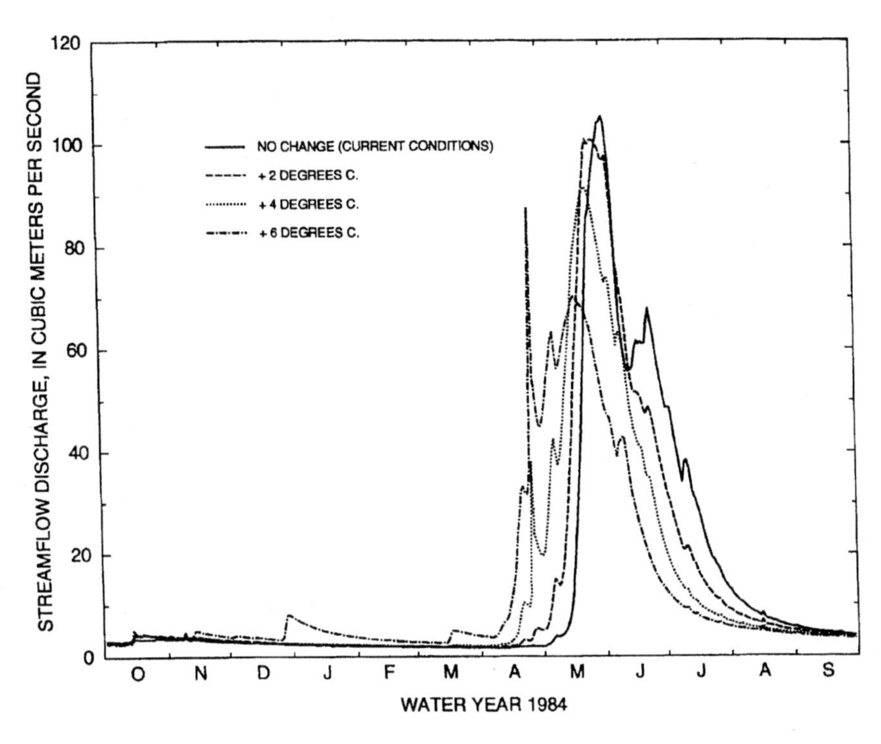

Fig. 3.1. **Streamflow discharge for East River at Almont, Colorado, for water year 1984 and model results for selected increases in air temperature.**

Table 3.1. **Changes in streamflow and bed sediment flux for East River at Almont, Colorado, for water year 1984 for selected increases in air temperatures.**

Scenario (Changes in Air Temperature, °C)	Streamflow			Bed Sediment Flux		
	Annual Daily Peak Discharge (m^3/s)	Annual Volume (million m^3)	Percent of Current Volume	Annual Daily Peak (tons/day)	Annual Load (tons/yr)	Percent of Current Annual Load
Current modeled	105.2	434.3	---	42.4	253.6	---
+2°C	100.7	425.4	98	30.0	317.6	125
+4°C	91.5	411.2	95	13.9	148.2	58
+6°C	87.6	405.5	93	9.5	36.1	14

3.2. Hydraulic Model

The calculations of the hydraulics are done for the reach at the gage at Almont where the d_{50} of the bed material is 112 mm. In such streams the size of bed material particles may be comparable to flow depth and the relation between flow velocity and bed stress varies strongly and in a complex manner with flow depth (Limerinos, 1970; Jarrett, 1989). To address these problems a one-dimensional hydraulics model is used that computes drag from bed roughness (Nelson et al., 1991). This model is similar to that described by Wiberg and Smith (1991) and differs primarily in assumptions with respect to turbulent structure near the bed. The model predicts the vertical distributions of velocity and stress that are spatially averaged over the reach using detailed grain geometry information. By treating the drag on grains making up the bed, the approach yields depth-discharge relations. These are used to develop stage-discharge and stress-discharge relations that provide the necessary hydraulic input to a bedload equation. The stage-discharge relation can be compared against the measurements at the gaging station to assess the goodness of fit of the model.

Input to this hydraulic model is streamflow discharge, the grain geometry of the bed, the slope of the channel, and the channel cross section. Daily streamflow discharge is obtained from the watershed model. The grain geometry of the bed was obtained by using a modified Wolman (1954) sampling technique. Particles were selected every meter through a 100-m distance along the channel. Particles were measured in place to measure downstream, cross-stream, and vertical protrusion of each particle in the flow. This allowed the computation by size class of the cross-sectional area of particles perpendicular to the flow and the area of the bed occupied by the particles. Channel cross-section geometry and slope were determined by doing a level survey.

As a measure of goodness-of-fit of the hydraulic model to the reach, a comparison between the measured and modeled stage-discharge relation for the gage at Almont is shown in Fig. 3.2. In general, the relation is predicted well by the model; however, there is a discrepancy at the highest flows with the measurements showing higher stage at a given discharge than the model predicts. Inspection of field notes and the site indicates that there are some backwater effects during high flows from the confluence with the Taylor River immediately downstream of the measuring site on the East River. The model assumes that the flow is both steady and uniform on average, meaning that channel slopes are in equilibrium with the local roughness and are unaffected by backwater effects.

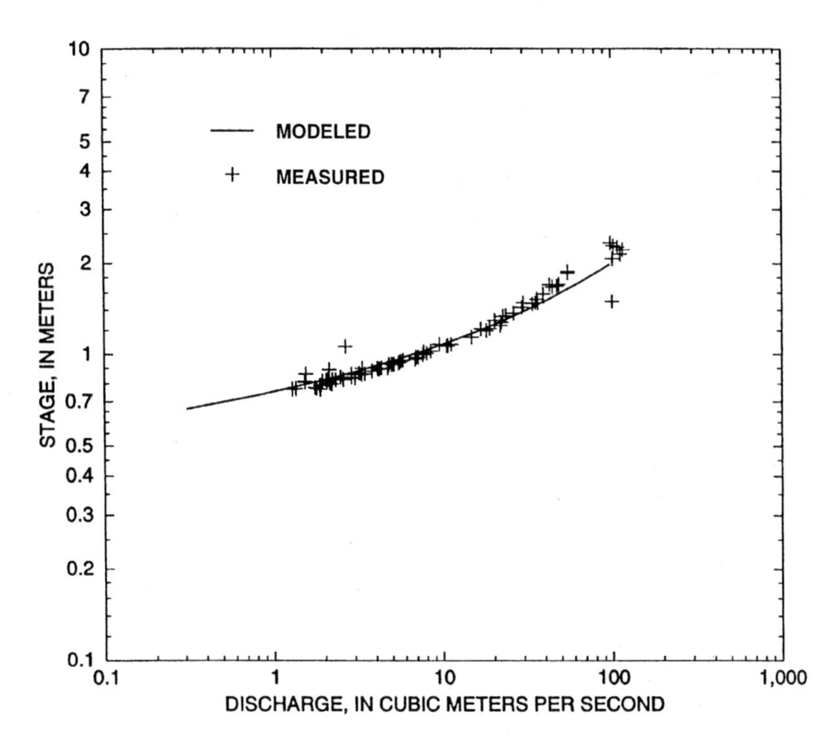

Fig. 3.2. Stage-discharge relation for the East River at Almont, Colorado.

3.3. Sediment Transport

In steep mountain channels such as this reach of the East River at Almont, the boundary shear stress usually reaches only slightly above the critical value for the initiation of sediment motion. The result is that these channels have marginal transport conditions where significant sediment transport tends to occur only for relatively large flow events, often at or near bankfull flow. Because of these conditions, the bedload-transport relation developed by Parker and others (1982) is used. This technique allows computation of the sediment flux in a gravel-bed channel using mean grain size by applying a similarity argument for the fluxes in various grain-size classes. Size of the bed material was determined using measurements of the particle b-axis from the grain geometry sample described above. Because the surficial bed material is measured rather than the sub-surface material, the critical Shields stress of 0.0299 is used as suggested by Parker et al. (1982, p. 561).

The volumetric bedload fluxes per unit width calculated from the Parker equation are multiplied by sediment density and integrated across the channel width for each discharge using the surveyed channel cross section. This yields the total mass of bedload discharge as a function of flow discharge.

Figure 3.3 shows the bed sediment discharge for current climatic conditions and for the air temperature scenarios previously discussed. Table 3.1 shows the annual bed sediment flux for each of the climate scenarios. Overall, there is a substantial decrease in the sediment flux with increases in daily air temperature. In the scenario with a 6°C change in air temperature, the annual sediment flux is reduced to 36.1 metric tons per year. That is only 14 percent of the transport computed for current conditions. The transport conditions in this reach of stream are significantly affected by the reductions in peak discharges that result from the climate scenarios.

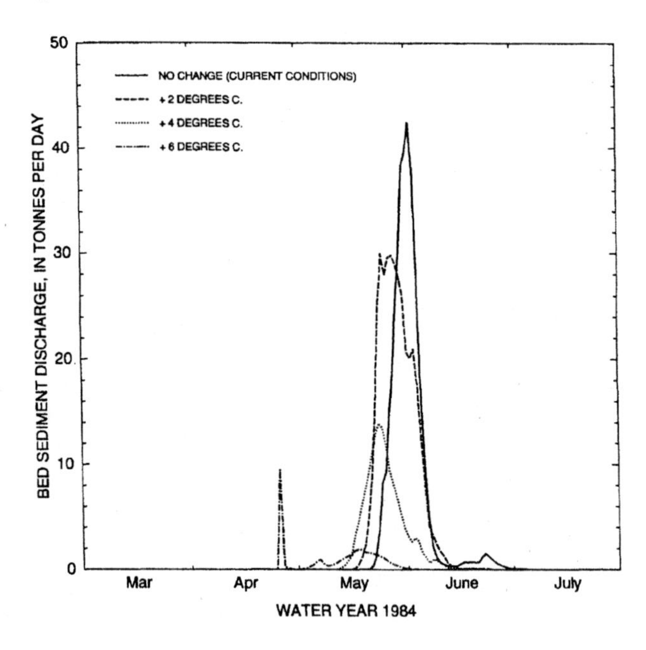

Fig. 3.3. Bed-sediment discharge for East River at Almont, Colorado, for water year 1984 and for selected increases in air temperature.

4. DISCUSSION

Bed sediment transport occurs on only a limited number of days, during high flow. For the current climatic conditions there is some bed sediment movement during 45 days of the year. The number of days with sediment movement does not change much for three temperature scenarios. For the 2°C, 4°C, and 6°C scenarios the number of days with sediment movement is 40, 34, and 40 days respectively. Thus, between 9 and 12 percent of the days of the year have some bed sediment movement, although there are many days of minimal transport in the 6°C scenario.

Changes in the timing of the annual hydrograph have a substantial effect on the bed sediment flux, with decreases to 14 percent of the current annual flux for an increase in air temperature of 6°C. However, an increase in bed sediment discharge is seen in the change from current conditions to the scenario of +2°C (Table 3.1). The reason for this increase is seen by examination of the streamflow hydrograph (Fig. 3.1) where the peak discharge decreases from 105.2 to 100.7 but the streamflow remains near the peak for an additional three days in the +2°C scenario. In both the current and +2°C scenario the first 16 days of the rising limb of the hydrograph provides an average bed sediment discharge of 21.1 metric tons per day. This average daily bedload transport rate continues for an additional three days in the +2°C scenario which contributes an additional 63 metric tons of sediment flux and accounts for the 25 percent increased flux for the +2°C scenario. Thus, changes in the daily streamflow can be subtle and result in substantial changes to the sediment flux.

Some simplifying assumptions have been used in linking the models. Output from the watershed model (PRMS) is daily mean discharge and input to the hydraulics model is an instantaneous discharge, which is assumed to be the instantaneous peak discharge for the day. An assumption is made that the daily mean discharge and the instantaneous peak discharge for the day are the same for the study reach and the values used are assumed to be daily values to compute daily sediment flux. A comparison of the annual peak of daily mean discharge and the instantaneous peak discharge for the same day for the 60 years of record shows that the daily mean peak is 9.2 percent lower than the instantaneous peak on average but the relation between the two discharges are similar. A simple linear regression between the instantaneous peak and daily mean discharge for the period of record yields an R^2 of 0.99 and a standard error of regression of 2.1 m^3/s. The intercept of the equation is not significantly different from zero but the slope of the regression equation is 1.09 and is significantly different from 1.0 ($p < 0.05$). As this study focuses on a comparison between climate scenarios, the difference between these two discharges was not rectified.

The method used in this paper of linking a series of models may provide a powerful method to assess the impact of diffuse changes in drainage basins (e.g., air temperature changes or land use changes) on specific hydraulic and sediment-transport conditions within a channel reach. These linkages are not only important in examining the sensitivity of climate but in a variety of land use evaluations. However, model error that may be easily defined for one model cannot be transferred and summed to the next model in the series. In this case, the goodness-of-fit of the watershed model is defined by the coefficient of efficiency using measured and modeled daily streamflow, but there is no mechanism to pass this error assessment to the hydraulics model. The goodness-of-fit of the hydraulics model is assessed separately by examination of the measured and modeled stage-discharge relation. Such error assessment may not be so important in a comparative study as was done in this paper but the need for cumulative error assessment when dealing with land use changes will be important.

5. SUMMARY

Evaluating the effects of climate change on water resources often focuses on streamflow volumes, although some seasonal shifts in the hydrograph may be noted. Unfortunately, those issues that are dependent on streamflow such as channel maintenance, habitat, and sediment transport often are overlooked and may be highly sensitive to small fluctuations in streamflow. A promising method for investigating these issues dependent on streamflow is to use a series or cascade of models.

For the reach investigated here, sediment transport (and therefore channel processes dependent on sediment transport) is sensitive to changes in daily air temperature. For water year 1984 three climate scenarios were investigated. These scenarios included an increase in daily air temperature of 2°C, 4°C, and 6°C. These increases altered the timing of the hydrograph but reduced the annual volume of water by only 2, 4, and 7 percent, respectively. This modest reduction in streamflow has been observed in other water resource investigations. Changes in the timing of the annual hydrograph have a substantial effect on the bed sediment discharge with the sediment flux reduced by 86 percent for the climate scenario of increased daily air temperature of 6°C.

This method of linking a series of models provides a way to quantitatively describe the changes within a specific reach of the channel from diffuse alterations in the drainage basin. These linkages represent some of the most difficult and important tasks carried out by geomorphologists in providing information to the land manager. Unfortunately, the most common methods for understanding these linkages have relied on developing empirical relations requiring the acquisition of large amounts of data over a long period of time in order to characterize existing conditions and extrapolate changes. This represents a serious shortcoming in our ability to manage riverine environments in a rational and scientific manner. The method described here of linking process models may yield an understanding of channel response in situations where basic data are not available.

6. REFERENCES

Battaglin, W.A., L.E. Hay, R.S. Parker, and G.H. Leavesley. 1993. Applications of a GIS for modeling the sensitivity of water resources to alterations in climate in the Gunnison River basin. Colorado: Water Resources Bulletin, 29: 6: 1021-1028.

Gleick, P.H. 1987. Regional hydrological consequences of increases in atmospheric CO2 and other trace gages. Climatic Change, 10: 137-161.

Gleick, P.H. 1989. Climate change, hydrology, and water resources. Review of Geophysics, 27: 329-344.

Jarrett, R.D. 1989. Hydraulics Research in mountain rivers. in: B. C. Yen (ed.), Proceedings of the International Conference on Channel Flow and Catchment Runoff, p. 599-608.

Kuhn, G. and R.S. Parker. 1992. Transfer of watershed model parameter values to noncalibrated basins in the Gunnison River basin, Colorado. in Hermann, R., ed., Managing Water Resources During Global Change, Proceedings from American Water Resources Association 28th Annual Conference and Symposia, Reno, Nevada, November 1-5, 1992, p. 741-751.

Leavesley, G.H., R.W. Lichty, B.M. Troutman, and L.G. Saindon. 1983. Precipitation-runoff modeling system. Users manual: U.S. Geological Survey Water Resource Investigations Report 83-4238, 207 p.

Leavesley, G.H., P.J. Restrepo, S.L. Markstrom, M. Dixon, and L.G. Stannard. 1996. The modular modeling system (MMS). User's manual: U.S. Geological Survey Open-file Report 96-151, 142 p.

Lettenmaier, D.P. and T.Y. Gan. 1990. Hydrological sensitivities of the Sacramento-San Joaquin River basin, California, to global warming. Water Resources Research, 26: 69-86.

Limerinos, J.T. 1970. Relation of the Manning coefficient to measured bed roughness in stable natural channels. U.S. Geological Survey Professional Paper 650D, p. D215-D221.

McCabe, G.J., Jr. and L.E. Hay. 1995. Hydrological effects of hypothetical climate change in the East River basin, Colorado, USA. Hydrological Sciences, 40: 3: 303-318.

Nash, J.E. and J.V. Sutcliffe. 1970. River flow forecasting through conceptual models. 1, a discussion of principles: Journal of Hydrology, 10: 282-290.

Nash, L.L. and P.H. Gleick. 1991. Sensitivity of streamflow in the Colorado basin to climate changes. Journal of Hydrology, 125: 221-241.

Nash, L.L. and P.H. Gleick. 1993. The Colorado River basin and climate change: the sensitivity of streamflow and water supply to variations in temperature and precipitation. US Environmental Protection Agency Report EPA 230-R-93-009, Washington, D.C.

Nelson, J.M., W.W. Emmett, and J.D. Smith. 1991. Flow and sediment transport in rough channels. in Shou-Shen Fan and Yung-Huang Kuo, eds., Proceedings of the Fifth Federal Interagency Sedimentation Conference, vol. 1, p. 4-55 to 4-62.

Parker, G., P.C. Klingeman, and D.G. McLean. 1982. Bedload and size distribution in paved gravel bed streams. Journal of Hydraulics, Div. ASCE, 108(HY4), p. 544-571.

Rango, A. and V. van Katwijk. 1990. Water supply implications of climate change in western North American basins. Proceedings from American Water Resources Association Symposium on International and Transboundary Water Resource Issues, Toronto, Canada, April, p. 577-586.

Wiberg, P.L. and J.D. Smith. 1991. Velocity distribution and bed roughness in high-gradient streams. Water Resources Research, 27: 825-38.

Williams, G.P. 1978. Bank-full discharge of rivers. Water Resources Research, 14: 6: 1141-1154.

Wolman, M.G. 1954. A method of sampling coarse riverbed material. Transactions of the American Geophysical Union, 35:6:951-56.

NAPIAS CREEK FALLS, IDAHO: A NATURAL OR MAN-MADE BARRIER FOR ENDANGERED CHINOOK SALMON

Michael D. Harvey[1]

ABSTRACT

Napias Creek Falls is a barrier to upstream migration of federally-listed endangered Chinook salmon. Potential spawning and rearing habitat is located upstream of the Falls in Napias Creek. Critical habitat designation under the Endangered Species Act depends on whether the barrier is natural or man-made. This study of the Falls, which have a vertical drop of about 37 m over a distance of about 200 m, investigated their genesis. The results indicate that the Falls are a natural barrier because (1) a post-Pleistocene-age fault with about 40 m of vertical offset crosses the valley at the head of the Falls, (2) the boulder-step morphology is the result of deposition of boulders (1.4 to 3.7 m) transported by a landslide-induced dam-break flood ($Q_{peak} = 630\ m^3/s$) that occurred about 200 years ago, and (3) road construction on the north side of the creek has not affected the morphology of the Falls.

KEYWORDS

Fish habitat, endangered species, morphology

1. INTRODUCTION

In 1993 the National Marine Fisheries Service (NMFS), under the Endangered Species Act (ESA), identified critical habitat for federally-listed endangered Snake River salmon. Salmon races affected included Snake River Spring-, Summer-, and Fall-run Chinook salmon. Critical habitat was defined as all river reaches (including the bottom and water of the waterways and adjacent riparian zone) presently or historically accessible within the Salmon River in

[1] Principal Geomorphologist, Mussetter Engineering, Inc., P.O. Box 270785, Fort Collins, Colorado 80527.

Idaho. Approximately 1.7 km upstream of the confluence of Napias and Panther Creeks in the Salmon River Basin (Fig. 1.1) is a reach of Napias Creek that is referred to as Napias Creek Falls. The Falls, which have a horizontal dimension of about 200 m, and a vertical drop of about 37 m, are located about mid-point in a steeper reach of Napias Creek that extends for a distance of about 1,700 m and has an average slope of about 10.4 percent (Fig.1.2). The Falls are believed to be a barrier to the upstream passage of salmon to potential

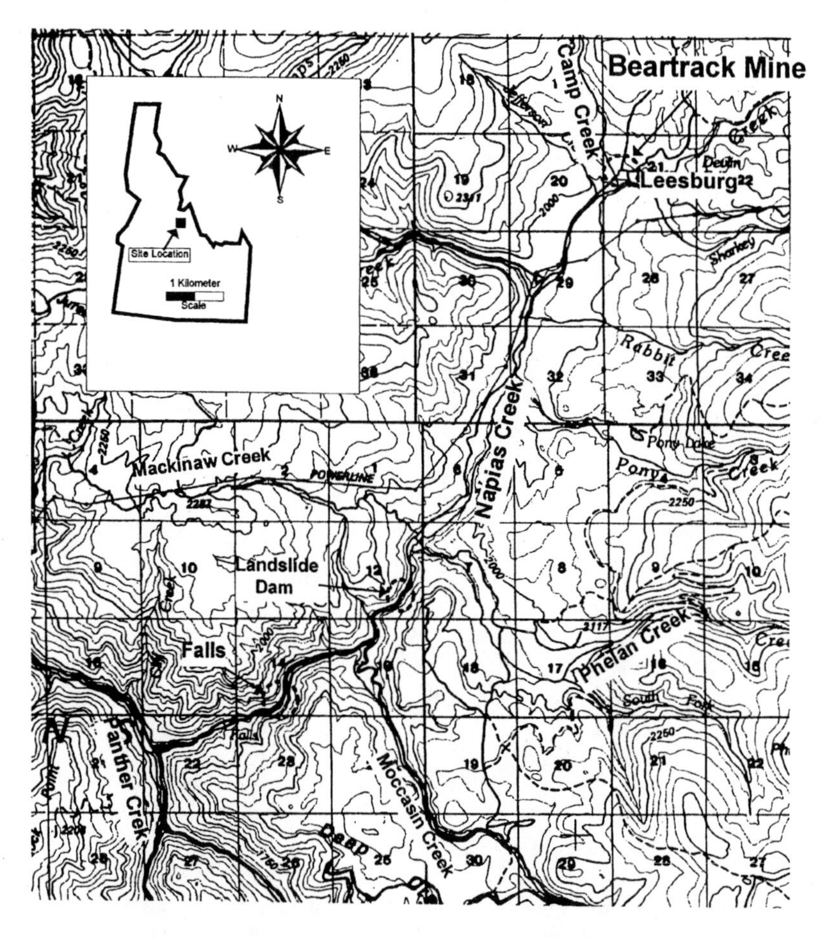

Fig.1.1. Map showing the location of Napias Creek Falls and Napias Creek and its tributaries.

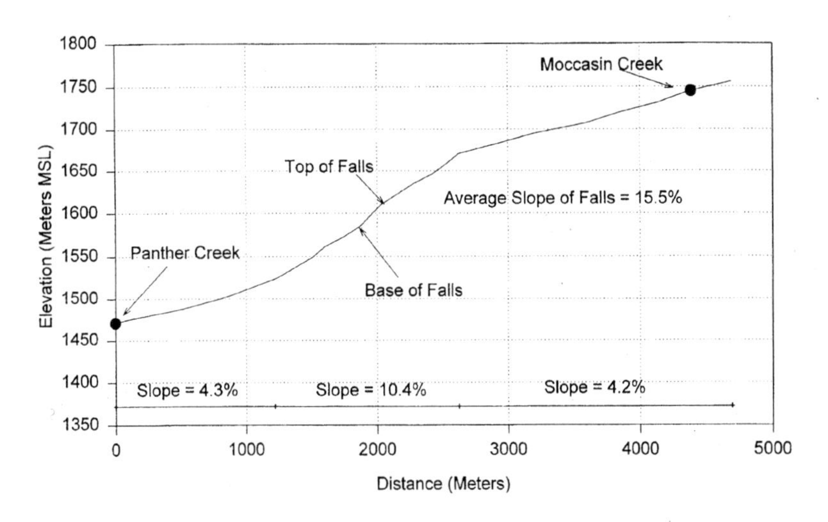

Fig.1.2. Longitudinal profile of lower Napias Creek from Panther Creek to the Moccasin Creek confluence.

spawning and rearing habitat located upstream (Bonneville Power, Administration, 1986; U.S. Forest Service, 1993). A biological opinion prepared by the NMFS under Section 7 of the ESA questioned whether the barrier was natural or was created by the construction of the Napias Creek roads in the late 1880s and 1930s. On the basis of their uncertainty with regard to the origin of the Falls NMFS assumed the upstream habitat to be accessible until conclusive data were provided to confirm that the Falls were historically impassable.

A study of fish passage of the Falls at a discharge of about 0.5 m^3/s by EA Engineering (1995) identified four steps within the Falls where the drop exceeded 2.4 m and the slope exceeded 50 percent. Maximum salmon leaping heights and horizontal distances for trajectories of 40, 60, and 80 degrees for a range of condition factors were computed using the methods developed by Powers and Orsborn (1985) (Table 1.1). Using a very conservative condition factor of 0.75 because the Falls are located about 1,100 km from the ocean (Powers and Orsborn, 1985), the maximum leaping height is 1.3 m and the maximum horizontal distance is 0.5 m. The results of the analysis indicated, therefore, that the Falls are a definite barrier to salmon passage. However, salmon are more likely to be attempting passage of the Falls at a discharge of about 4 m^3/s in the Spring. In order to assess whether passage would be possible at the higher discharge because the steps might be drowned out, a weir equation (Chow, 1959) was used to evaluate the change in hydraulic head (Harvey, 1996). Using very conservative assumptions, the head differential for flows between 0.5 and 4 m^3/s at the four steps was computed to be 0.4 m. If it is assumed that the velocity head component of the total head is zero, then the maximum increase in tailwater elevation that could occur in the pools at the bases of the steps is only about 0.4 m. Therefore, the steps would still be barriers at the higher discharges.

Table 1.1. Computed leaping heights and horizontal distances for three condition factors for Chinook salmon. Condition factors are related to the length of time a fish has been migrating upstream from seawater; Condition Factor = 1.0 at seawater; Condition Factor = 0.5 at 5-7 months in freshwater (Data from E.A. Engineering, 1995).

	Condition Factor = 0.50		Condition Factor = 0.75		Condition Factor = 1.0	
Leaping Angle (deg)	HL Maximum Height (m)	XL Horizontal Distance (m)	HL Maximum Height (m)	XL Horizontal Distance (m)	HL Maximum Height (m)	XL Horizontal Distance (m)
40	0.3	0.6	0.6	1.5	1	2.3
60	0.5	0.5	1	1.2	1.8	2
80	0.6	0.2	1.3	0.5	2.3	0.8

The objective of this investigation of Napias Creek Falls was to determine whether the Falls are a natural feature and were, therefore, historically impassable to salmon, and to identify the factors associated with their genesis.

2. WATERSHED GEOLOGY AND LAND USE HISTORY

The valley of Napias Creek downstream of the Phelan Creek confluence is underlain primarily by Precambrian-age augen gneiss. Pleistocene-age glacial deposits are also located within the upper portion of the valley (Mitchell and Bennett, 1979). The north-northeast trending Panther Creek Fault determines the course of Napias Creek. Shear zones in the augen gneiss located in roadcuts along the lower part of Napias Creek are probably related to this fault. Pleistocene-age glacial till deposits are offset by the fault. In the Falls reach, there is rapid change in elevation over a very short reach of channel. Augen gneiss bedrock crops out at water's edge at the top of the Falls and there is outcrop exposed at the base of the Falls on both the left and right valley walls. Further, the width of the valley increases locally downstream of the head of the Falls. A very limited exposure of a high angle fault was located in bedrock outcrop on the north side of the valley at the elevation of the head of the Falls. Based on the bedrock geometry, it is apparent that there has been about 37 m of vertical offset across the fault which indicates that this is the primary reason for the existence of the Falls. Similar post-Pleistocene-age, fault-induced waterfalls have been mapped in nearby locations (Bennett, 1977).

Napias Creek watershed has been severely impacted by historic gold mining, primarily upstream of the Phelan Creek confluence. Mining has occurred in the watershed for over 100 years. Considerable portions of the relatively low gradient alluvial valley floor from Leesburg downstream to about the Phelan Creek confluence were placer and dredge mined. However, there is no evidence of direct mining effects on the stream channel in the vicinity of the Falls. Road construction along Napias Creek and within the watershed has had some impacts on the stream. Roads built within the watershed during the early mining days probably had more effect on the stream than the modern roads because sidecasting of slope cut materials was a common practice.

Within the vicinity of the Falls, two roads have been constructed. The original Napias Creek stagecoach road from Leesburg to Cobalt was constructed in the late 1800s with a steeper grade than the modern road. The present road constructed by the Civilian Conservation Corps (CCC) in the 1930s has a flatter grade and is located above the old road for most of the length of the Falls. The two road grades merge at the head of the Falls where a greater than 200-year old Douglas fir is located between the road and the right bank of the channel. For most of the length of the Falls, the old road traversed talus deposits downslope of the rock outcrops. It is apparent that for ease of construction, the early road builders attempted to place the road between the upslope bedrock outcrops on the north side of the creek and the largest talus boulders (gray color, Munsell 7.5 YR 6/0 with lichen diameters exceeding 20 cm) that are located at the base of the slope. The sizes of the boulders sidecast and stacked on the downslope side of the old road are consistent with a mid to lower slope talus gradation. The old road deposits tend to have developed a slightly weathered surface, are light brown in color (Munsell 7.5 YR 6/4) and small lichens (< 2 cm diameter) have colonized some of the boulders. In contrast, the newer road was located at a higher elevation above the streambed at a flatter grade and construction involved drilling and blasting of rock outcrop near the downstream end of the Falls reach. Freshly broken and unweathered surfaces on sidecast boulders below the blasted outcrop identify these deposits. No lichen colonization was observed on the newer deposits.

Timber harvesting and associated road construction have occurred historically within the Napias Creek watershed, and the rate of harvesting has increased since 1989. There is no evidence that timber harvesting has occurred in the vicinity of the Falls, but there is evidence in the form of burnt tree stumps that the area may have been affected by fire.

3. FIELD INVESTIGATION AND DATA COLLECTION

A field investigation of Napias Creek and the Falls was conducted in April 1995. The entire length of Napias Creek was observed from its confluence with Phelan Creek to the confluence with Panther Creek (Fig. 1.1). At the Falls, a detailed data collection program was developed to identify the processes responsible for the formation of the Falls. Data collected included information on the local bedrock outcrop pattern, lithology and faulting, the sizes of the boulders that form the Falls, imbrication orientations of the boulders, the sizes of sidecast material for both the original road (old road) and current (new road) road alignments on the north (right) side of the creek, sizes of lichens on different boulder deposits, and cores from trees on either side of the Falls.

The thalweg through the Falls was surveyed with a level and rod and a baseline for stationing purposes was established along the right side of the channel. The valley bottom was subdivided into five linear subzones (looking downstream) that were oriented parallel to the channel: (1) left valley wall (LVW) outside of the line of trees that marked the left margin of the channel, (2) between the tree line and water's edge (inundated at higher flows) on the left side of the channel (TLCH), (3) the channel itself between the left and right water's edges (CH), (4) between the right edge of the water (inundated at higher flows) and the tree line on the right side of the channel (TRCH), and (5) the right valley wall outside of the line of trees that marked the right margin of the channel (RVW).

4. MORPHOLOGY AND SEDIMENTOLOGY OF THE FALLS

The Falls are composed of a number of steps and pools (Fig. 4.1) that are formed in large boulders, the majority of which exhibit an imbricated fabric. The upstream imbrication of many of the boulders suggests a fluvial origin for the deposits. The presence of bedrock outcrops marginal to the channel and coarse-grained, upslope-fining talus deposits also suggests that colluvial processes were involved in the genesis of the Falls.

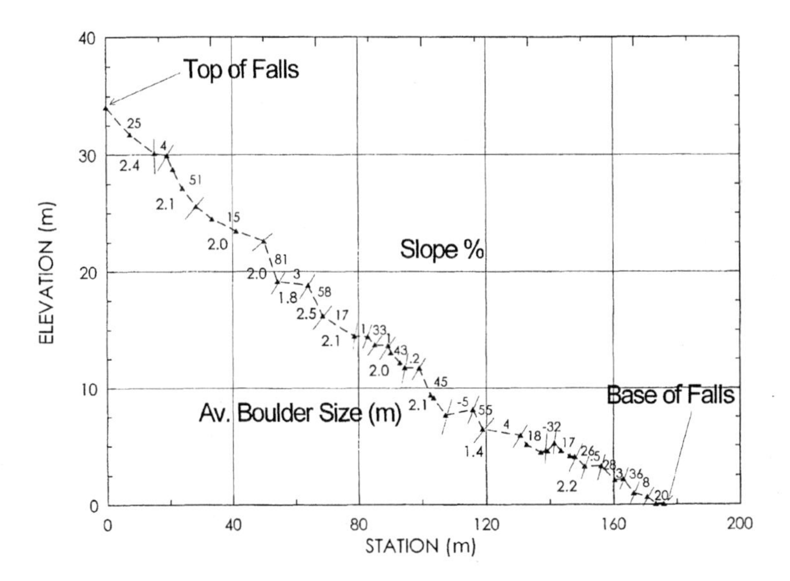

Fig. 4.1. Surveyed longitudinal profile of the Falls showing slopes and average boulder sizes for individual segments.

During the field reconnaissance of the watershed, a breached landslide dam was identified about 680 m downstream of the Phelan Creek confluence (Fig. 1.1). Landslide dams occur as a result of slope failures in mountainous terrain in many locations of the world and they are generally triggered by excessive rainfall, snowmelt and earthquakes (Schuster and Costa, 1986). Most landslide dams are fairly short-lived (91 percent fail within 1 year of formation) and fail as a result of overtopping. Failure of the dams results in severe downstream flooding (Costa, 1985). In narrow mountain valleys the dam-break flood peaks are not subject to much attenuation (Costa, 1985). Landslide dam-break floods produce characteristic flood deposits in wider valley reaches, boulder berms and step-pool morphology (Scott and Gravlee, 1968; Jarrett and Costa, 1984). Boulder berms are generally deposited in flow expansion areas (Church and Desloges, 1984; Eisbacher, 1982). Deposition of the large boulders transported during a dam-break flood is generally restricted to lower gradient (<7 percent) and wider reaches (Jarrett and Costa, 1984). The step-pool morphology formed during the extreme flood is very stable and, therefore, persists during the subsequent normal range of flows (Whittaker and Jaegi, 1982; Chinn, 1989; Grant et al., 1990; Grant and Mizuyuma, 1991).

The observed landslide originated in augen gneiss bedrock on the south side of Napias Creek, and was probably triggered by an earthquake (Schuster and Costa, 1986). The height of the landslide dam crest above the bed of the present stream was 14.5 m. The bottom and top width dimensions of the breach were 13.6 and 20.4 m, respectively. The upstream extension of the landslide-dammed lake and its pre-failure volume (1.2×10^6 m^3) were estimated from the 7.5-minute topographic quadrangle. The above were required to develop estimates of the peak discharge for the dam-break flood. Dam-break flood deposits (boulder berms) were identified immediately downstream of the breached section of the landslide dam. Further dam-break flood deposits were identified in wider valley reaches located just upstream of the Mackinaw Creek confluence, upstream of the Moccasin Creek Road bridge, at a shear zone-influenced wider reach located about 1,200 m downstream of the Moccasin Creek confluence, and at the Falls (Fig 1.1).

Talus deposits are composed, in general, of coarse angular rock fragments (Bates and Jackson, 1984). They accumulate below the rock outcrops from which they are derived and they generally have slopes between 28 and 46 degrees (Selby, 1982). Commonly, the talus deposit has an open-work fabric and exhibits an imbricate structure in which elongate particles pack against each other and dip back into the slope (Selby, 1982). Most commonly the longitudinal profile of a talus deposit is straight in the upper segments and concave at the base. Upper talus slopes are usually composed of the smaller rock fragments and particle size increases downslope which accounts for the basal concavity. At any given position on the slope profile, the particle sizes tend to be similar because smaller particles cannot roll beyond the gaps between the larger particles. The sieving action of the sorted talus slope is also supplemented by the greater kinetic energy of the larger particles (Kirkby and Statham, 1975). Talus deposits are located downslope of a number of rock outcrops that exist on both sides of the channel of Napias Creek within the length of the Falls. Rock outcrops are located on the left valley wall at approximate Stations 0, 50, 100, and 153 m. On the right valley wall outcrops are located at Stations 0, 68, and 170 m.

The intermediate axis of boulders within each of the five valley floor subzones were measured. One hundred boulders were measured in each of the subzones (TLV, TLCH, TRCH, TRV) except for the channel (CH) where 150 boulders were measured. The means and standard deviations for each subzone for 34 m long segments of the Falls are presented in Table 4.1.

Table 4.1. Summary of Intermediate Axis Boulder Measurements at Napias Creek Falls for 30 m Reaches.

Station (m)	Boulder Size (mean and standard deviation) (m)				
	TLV	TLCH	CH	TRCH	TRV
0 - 30	1.7 ± 0.9	1.9 ± 0.8	2.2 ± 0.4	1.6 ± 0.4	1.0 ± 0.3
30 - 60	0.9 ± 0.4	1.3 ± 0.4	2.2 ± 0.7	1.9 ± 0.2	1.7 ± 0.6
60 - 90	1.3 ± 0.3	2.0 ± 0.6	2.1 ± 0.8	1.6 ± 0.2	2.0 ± 0.4
90 - 122	1.4 ± 0.5	2.0 ± 0.6	1.9 ± 0.4	1.5 ± 0.1	1.6 ± 0.4
122 - 152	1.8 ± 0.6	1.5 ± 0.3	1.6 ± 0.8	1.8 ± 0.6	1.9 ± 0.6
152 - 177	0.9 ± 0.4	1.6 ± 0.3	2.2 ± 0.4	1.9 ± 0.4	2.6 ± 0.1

In the left valley wall subzone (TLV) boulder sizes range from 2.6 to 0.5 m with the largest boulders being located in the vicinity of the bedrock outcrops on the valley wall. Along the right valley wall (TRV) boulder sizes range from 2.9 to 0.5 m. The smaller sizes within the first 30 m of the profile are due to the presence of boulders that were very obviously sidecast (on the basis of boulder color, orientation, and absence of lichens) during road construction. For the remainder of the profile the distance between the channel margin (tree line) and the lower road progressively increases and, therefore, the possibility of road effects on the size distribution of the boulders diminishes.

Along the left margin of the channel (TLCH) boulder sizes range from 3.4 to 0.8 m. There are no particular trends in the sizes of the boulders along the length of the Falls. Along the right margin of the channel (TRCH) boulder sizes range from 2.6 to 1 m. There appears to be a slight coarsening trend in the downstream direction with the smaller boulder sizes being located near the head of the Falls and the largest towards the base of the Falls. The smaller boulders within the first 30 m are related to road construction, and the largest boulders located in the lower 55 m are very coarse talus deposits immediately below rock outcrops on the right valley wall.

Within the channel (CH) boulder sizes range from about 1.4 to 3.7 m. The boulder sizes that comprise the channel do show a slight fining trend in the downstream direction. The downstream fining trend is not associated with road construction because the potential road effects are limited to the upper 60 m of the Falls since the road diverges away from the channel downstream of this location. Further, the 4 steps that exceed 2.4 m in height, and are the barriers to salmon migration, are all located upstream of Station 102 m.

The boulder measurements from the margins of the channel and on the valley walls outside of the channel margin indicate that the size of the boulders is primarily correlated with the presence of rock outcrops upslope of the measured deposits. With the exception of the top 30 m of the profile where sidecast road material influences the sizes of the boulders on the right valley wall and right channel margin, the boulder deposits in the five subzones across the valley floor were emplaced by either fluvial (flood) or colluvial (gravity) processes.

Some sidecasting of talus deposits occurred during construction of both roads. In order to determine the range of boulder sizes that were sidecast as a result of road construction a modified Wolman pebble count (Wolman, 1954) was performed on both the old and new road sidecast materials. The intermediate axes of 100 boulders identified as sidecast material were measured on the downslope side of each road and the boulders were assigned to 0.17 m size intervals from <0.17 to >1.2 m. The new road materials were located within the first 17 m from the top of the Falls and the old road materials were located between 20 and 41 m from the top of the Falls. Fig. 4.2 presents the gradation curves developed from both sets of measurements. The gradations are almost identical with a median (D_{50}) boulder size of about 0.6 m. The similarity in the curves is to be expected since the roads traverse the talus deposits at the same approximate elevation. Farther downstream from the top of the Falls the gradations would diverge with the old road gradation becoming coarser since the boulders are located farther downslope. However, downstream of about Station 41 m both roads are so far away from the channel that they would be very unlikely to affect the distribution of the materials that comprise the Falls. The gradation curves show that about 97 percent of the sidecast boulders from both roads are less than 1.2 m in diameter. Also plotted on Fig. 4.2 are the means and standard deviations of the boulder measurements for the five subzones over the length of the Falls. It is apparent from the gradation curves and plotted statistical parameters that the sidecast materials generally fall within the fine tails of the distributions in each of the five subzones except the channel. Ninety-seven percent of the sidecast materials

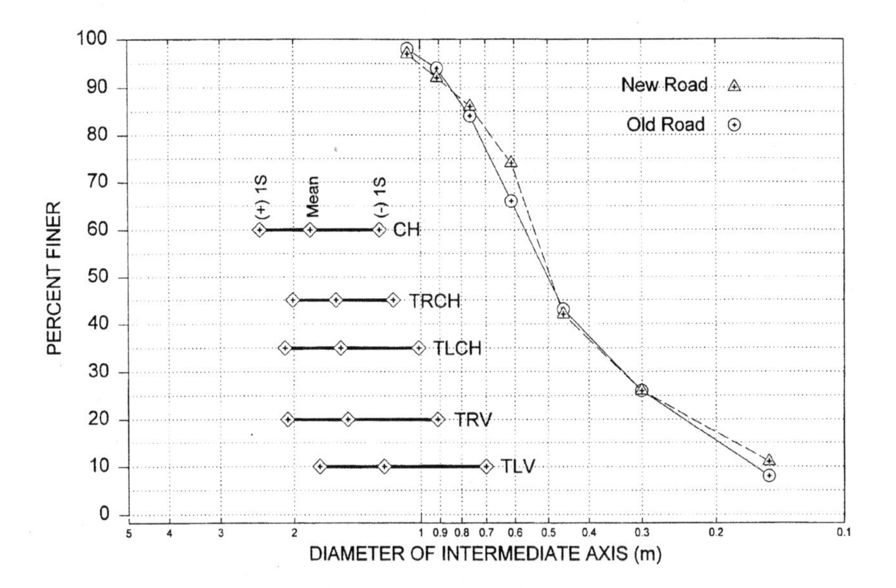

Fig. 4.2. Gradation curves for sidecast materials below both road grades on the north side of the creek. Also shown are the means and standard deviations for the boulders that comprise the five valley floor subzones.

(1.2 m) are smaller than -1 standard deviation (1.4 m) from the mean (2.0 m) of the channel deposits (CH) which indicates that sidecasting cannot have had much effect on the general size of the materials that form the Falls.

5. GENESIS OF THE FALLS

Within the upper 135 m of the Falls, the boulders that comprise the channel tend to be larger than those on the channel margins and lower valley walls. The boulders in the channel tend to have upstream imbrication which suggests a fluvial origin. Therefore, it was necessary to develop a dam-break flood discharge and determine the magnitude of the flood that would have been capable of transporting boulders in the size range of about 2 to 3.4 m.

5.1. Dam-Break Flood Analysis

Costa (1985) and Schuster and Costa (1986) developed a series of relationships to predict the magnitude of the peak discharge for landslide dam failures. Variables used to predict the peak discharges include the height of the dam:

$$Q_{max} = 6.3H^{1.59} \tag{5.1}$$

the reservoir volume at time of failure:

$$Q_{max} = 672V^{0.56} \tag{5.2}$$

and a dam factor which is the product of the dam height and reservoir volume.

$$Q_{max} = 181(HV)^{0.43} \tag{5.3}$$

Price et al. (1977) developed a relationship for predicting the dam-break flood peak based on the reservoir depth upstream of the dam before failure (Y), the base width of the breach (b) and the top width of the breach (T):

$$Q_{\max} = \frac{8}{27} g^{1/2} Y^{3/2} (0.4b + 0.6T) \tag{5.4}$$

Using Eq. 5.1, the peak discharge (Q_{max}) was predicted to be about 387 m^3/s. Eq. 5.2 yielded a peak discharge of 744 m^3/s and Eq. 5.3 yielded a peak discharge of 590 m^3/s. Eq. 5.4 yielded a peak discharge of 630 m^3/s. The standard errors for all of the relationships are relatively high (129 to 147 percent), but the lowest standard error is that associated with Eq. 5.4. Therefore, for the purposes of subsequent analysis the predicted peak discharge of 630 m^3/s was used.

5.2. Incipient Motion Analysis

Within the channel are boulders with intermediate axes of up to 3.4 m. Therefore, to address the question of the origins of the boulders that comprise the Falls it was necessary to evaluate the magnitude of the discharge that would be capable of transporting the boulders. The estimated 100-year peak discharge for Napias Creek at the Falls based on the U.S. Geological Survey regional bivariate regression relationship (Harenberg, 1980) is about 36 m^3/s. The current hydrology of the Napias Creek watershed that is entirely located at elevations above 1870 m is snowmelt-dominated. At elevations above 1870 m in Idaho extreme thunderstorm-generated flood discharges do not occur (Jarrett, 1993) and, therefore, peak discharges greatly in excess of the 100-year flood peak (36 m^3/s) are also unlikely to occur. Using a log Pearson III distribution for the Panther Creek at Shoup gage and correcting for watershed size the estimated 500-year peak flow is approximately 44 m^3/s which is only 23 percent larger than the estimated 100-year peak flow.

The critical particle size (D_c), defined as the particle size that is on the verge of motion for a given discharge is estimated with the Shields (1936) relation:

$$\tau_c = \tau_{*_c}(\gamma_s - \gamma) D \qquad (5.5)$$

where τ_c is the critical shear stress, τ_{*c} is the dimensionless critical shear stress, γ_s is the unit weight of sediment (26,000 N/m^3), γ is the unit weight of water (9,800 N/m^3), and D is the particle size. D_c is obtained by substituting D_c for D in Eq. 5.5 and rearranging into the following form:

$$D_c = \tau_c \,/\, \tau_{*_c}(\gamma_s - \gamma) \qquad (5.6)$$

Reported values for τ_{*c}, for the median particle size range from 0.03 (Meyer-Peter and Muller, 1948; Neill, 1968) to 0.06 (Shields, 1936). A value of 0.047 is commonly used and represents a low, but measurable transport rate; whereas, 0.03 probably represents true incipient motion (Neill, 1968).

For the purposes of this analysis of the dam-break flood's competence to transport boulders the bed slope upstream of the Falls was assumed to be 4 percent and the flow width and Manning's *n* value were assumed to be 27 m and 0.08, respectively. For the estimated peak discharge of 630 m^3/s the critical boulder sizes (D_c) are 2.2 and 3.6 m for τ_{*c} values of 0.047 and 0.03, respectively. For the 100-year flood peak (36 m^3/s), using a flow width of 13.6 m, the D_c values are 0.6 and 1.0 m for τ_{*c} values of 0.047 and 0.03, respectively. For the 500-year flood peak (44 m^3/s) using the same flow width, the D_c values are 0.6 and 1.2 m for τ_{*c} values of 0.047 and 0.03, respectively. Comparison of the critical boulder sizes for the three flood events indicates that the very coarse-grained boulder deposits that contain many boulders in the >1.7 m range were most likely entrained and deposited by the dam-break flood. The inability of the 500- and 100-year flood events to mobilize the boulders that comprise the Falls indicates that the step-pool morphology is a relic feature from the dam-break flood.

5.3. Chronology

The age of Napias Creek Falls in its present configuration cannot be determined absolutely. However, dendrochronologic data (tree cores) can provide a minimum age constraint. It is likely that the faulting that set up the basic configuration of the Falls is post-Pleistocene in age (Bennett, 1977). The margins of the channel along the profile of Napias Creek Falls are well defined on both sides by Douglas fir and cottonwood trees. If a flood of the magnitude of the dam-break flood had occurred within the lifetime of these trees the upstream sides of the tree trunks would be very heavily scarred by boulder impacts (Sigafoos, 1964; Jarrett and Costa, 1984). Examination of the trunks of the trees lining both sides of the channel did not reveal any flood scars, which suggests that the trees post-date the dam-break flood. In fact, a dam-break flood of similar magnitude (510 m^3/s) removed a swath of mature trees along the entire length of Roaring River following the Lawn Lake Dam failure in Rocky Mountain National Park, Colorado, in 1982 (Jarrett and Costa, 1984).

Cores were extracted from four Douglas fir trees on each side of the channel (Table 5.1). On the left side of the channel, all but one (L1) of the cores reached the center of the tree. Excluding L1, the average age of the trees on the left side of the channel is about 173 years. On the right side of the channel only two of the cores reached the center of the tree (R2, R4). The diameters of trees R1 and R3 were too great for the corer to reach the center of the tree and, therefore, it is not possible to determine the true age of the trees. Based on trees R2 and R4, the average age is about 141 years, but tree R3 has a minimum age of 208 years. The tree core data in Table 5.1 indicate that the dam-break flood occurred at least 132 years ago and most probably occurred

Table 5.1. Tree Core Data from Napias Creek Falls.

Tree Identification	Station (m)	Diameter Breast Height (cm)	Circumference (cm)	Core Length (cm)	Age (years)
L1	184	58.4	183	22.9	155*
L2	150	45.2	142	21.8	177
L3	88	39.6	125	21.6	176
L4	55	39.6	125	21.6	165
R1	197	58.4	183	22.6	97*
R2	150	39.6	124	22.6	132
R3	102	54.1	170	22.6	208*
R4	54	41.1	130	22.9	149

*The corer did not reach center of tree.

over 200 years ago. Therefore, the boulder-step morphology of the channel at the Napias Creek Falls was most likely established at least 200 years ago.

Since the old stagecoach road was constructed in the 1880s, and the present road was constructed in the 1930s, the basic structure of the Falls was in place prior to any road construction.

6. DISCUSSION

The primary objective of the investigation of Napias Creek was to determine whether the Falls, which currently are a barrier to upstream migration of federally-listed endangered Snake River salmon (BPA, 1986; NMFS, 1994; EA Engineering, 1995), are natural features that historically prevented the upstream migration of the salmon to potential spawning and rearing habitat located upstream of the Falls.

Field investigation at the Falls determined the presence of a post-Pleistocene-age, approximately northwest-trending, high-angle fault within the augen gneiss bedrock at the top of the Falls. Vertical offset across the fault is at least 37 m, and is the primary reason for the vertical elevation difference between the top and base of the Falls. The rapid change in elevation also permits narrow valley-confined flows from above the Falls to expand and drop their sediment load.

During the field reconnaissance, a breached landslide dam was identified downstream of the confluence of Napias and Phelan Creeks (Fig. 1.1). Characteristic flood deposits were recognized downstream of the landslide at a number of locations, including the Falls. Based on the geometry of the breached section of the landslide dam an estimate of the dam-break peak discharge is 630 m^3/s (Eq. 5.4). Incipient motion analysis indicated that this flood would be capable of transporting boulders in a size range of 2.0 to 3.7 m. The dam-break flood peak exceeded the estimated 100-year flood peak (36 m^3/s) by a factor of about 17 and the estimated 500-year flood peak (44 m^3/s) by a factor of about 15. The 500- and 100-year flood peak have the ability to entrain a range of boulder sizes from about 0.7 to 1.2 m. The step-pool channel morphology of the Falls (Fig. 4.1) is characteristic of a channel pattern formed by extreme flood events. The sizes of the boulders that form the channel within the Falls correlate with the sizes that could be transported by the dam-break flood (Table 4.1).

Channel margin and valley-wall boulder sizes are influenced by colluvial processes. Inverse grading of talus deposits located below bedrock outcrops on the valley walls results in the largest boulders being located at the base of the slope where the channel and channel margins are located (Table 4.1). Comparison of boulder sizes included in sidecast materials from construction of the old (1880s) and new (1930s) roads with the sizes of boulders found in the five valley-floor subzones indicates that the road construction activities have had very little or no effect on the channel.

Dam-break floods of the magnitude of the Napias Creek flood tend to destroy all trees located along the channel margins. At the least, any trees that existed during the flood and survived should bear severe trunk scars as a result of boulder impacts. No scarring was observed on the trunks of any of the trees and, therefore, it is reasonable to conclude that the trees post-date the dam-break flood. Dendrochronologic data from tree cores obtained from four trees on each side of the channel margin indicate that the present morphology of the Falls, which is a barrier to fish passage, is at least 132 years old and most probably greater than 200 years old.

7. CONCLUSIONS

A wide range of environmental problems are faced by society today at global, regional, and local scales and the solutions to many of the problems involve an understanding of surficial processes and their interactions with geological and biological processes (Morris and Harbor, 1995). This investigation of Napias Creek provides an example of how an integrated study of geology, geomorphology, hydrology, sedimentology, and botany was used to solve an environmental problem. The following conclusions about the genesis of Napias Creek Falls can be drawn from the integrated study: (1) the vertical elevation difference (about 37 m) between the top and base of the Falls is due to the presence of a high angle fault that crosses the valley at the top of the Falls, (2) the boulders that form the step-pool morphology that characterizes the longitudinal profile of the Falls were primarily emplaced by a dam-break flood that most likely occurred over 200 years ago, (3) talus deposits that exhibit inverse grading that results in the largest boulders being located at the base of the slope below bedrock outcrops have contributed to the overall coarseness of both the channel and channel margin areas, (4) road construction in both the 1880s and 1930s did not materially affect the form of the channel within the Falls, and (5) Napias Creek Falls in their present configuration are a natural feature, and since they are presently impassable to salmon, it follows that they were also historically impassable.

8. ACKNOWLEDGMENTS

This investigation was conducted for Meridian Gold Company. The author thanks John Lawson, Environmental Superintendent for the Beartrack Mine, for permission to publish this paper.

9. REFERENCES

Bates, R. L. and J. A. Jackson. 1984. *Dictionary of Geological Terms*. American Geological Institute, Anchor Press, Garden City, New York.

Bennett, E. H. 1977. Reconnaissance Geology and Geochemistry of the Blackbird Mountain Panther Creek Region, Lemhi Co., Idaho. *Idaho Bureau of Mines and Geology Pamphlet No. 167*.

Bonneville Power Administration. 1986. Habitat rehabilitation, Panther Creek, Idaho. Report, January.

Chinn, A. 1989. Step-pools in stream channels. *Progress in Physical Geography*, 13:391-407.

Chow. V T. 1959. *Open-Channel Hydraulics*. McGraw-Hill Book Co. New York.

Church, M. and J. R. Desloges. 1984. Debris torrents and natural hazards of steep mountain channels-- east shore of Howe Sound: Canadian Association of Geographers, *Field Trip Guidebook No. 7*, Department of Geography, Univ. British Columbia, Vancouver.

Costa, J. E. 1985. Floods from dam failures. *U.S. Geological Survey Open-File Report 85-560*.

E.A. Engineering, Inc. 1995, Analysis of Napias Falls as a barrier to chinook salmon. Report to FMC Gold, Beartrack Mine, November.

Eisbacher, G. H. 1982. Mountain torrents and debris flows. Episodes 1982:4:12-17.

Grant, R. E., and T. Mizuyuma. 1991. Origin of step-pool sequences in high gradient streams: A flume experiment in Proc. Japan-US Workshop on Snow Avalanche , Landslide, Debris Flow Protection, 523-532.

Grant, G. E., F. J. Swanson and M. G. Wolman. 1990. Pattern and origin of stepped-bed morphology in high gradient streams, Western Cascades, Oregon. *Geological Society of America Bull*. 102:340-352.

Harenberg, W. A. 1980. Using channel geometry to estimate flood flows at ungaged sites in Idaho. *U.S. Geological Survey Water Resources Investigation 80-32*.

Harvey , M. D. 1996. Evaluation of the genesis of Napias Creek Falls, Napias Creek drainage basin, Idaho, Report to FMC Gold, Beartrack Mine, May.

Jarrett, R.D. 1993. Flood elevation limits in the Rocky Mountains. in Kuo, C.Y. ed. Engineering Hydrology, Symposium Proceedings sponsored by the Hydraulics Div. ASCE, 180-185.

Jarrett, R. D. and J. E. Costa. 1984. Hydrology, geomorphology, and dam-break modeling of the July 15, 1982, Lawn Lake Dam and Cascade Lake Dam failures, Larimer County, Colorado. *U.S. Geological Survey Open-File report 84-612*.

Kirkby, M .J. and I. Statham. 1975. Surface stone movement and scree formation. *Journal of Geology*. 83:349-362.

Meyer-Peter, E. and R. Muller. 1948. Formulas for bedload transport. Proceedings of the 2nd Congress of the International Association for Hydraulic Research, Stockholm, Paper No. 2, 39-64.

Mitchell, V.E. and E. H. Bennett. 1979. Geologic Map of the Elk City Quadrangle, Idaho. Idaho Bureau of Mines and Geology.

Morris, S. E. and J. M. Harbor. 1995. Geomorphology applied to environmental problems: forward. Physical Geography. 16(5):357-358.

National Marine Fisheries Service. 1994. Endangered Species Act- Section 7 Consultation, Biological Opinion Beartrack Mine, March.

Neill, C. R. 1968. Note on initial movement of coarse uniform bed material. *Journal of Hydraulic Research*. 6:2:173-176.

Powers, P. D. and J. F. Orsborn. 1985. Analysis of barriers to upstream fish migration. Report to Bonneville Power Administration, Project No. 82-14.

Price, J. T., G. W. Lowe, and J. M. Garrison. 1977. Unsteady flow modeling of dam-break waves. in Proceedings of Dam-Break Flood Routing Model Workshop, Bethseda, MD., U.S. Water Resources Council. PB-275-437.

Schuster, R. L. and J. E. Costa. 1986. A perspective on landslide dams. Landslide Dams: Processes, Risk, and Mitigation. *Geotechnical Special Publication No.3*. American Society of Civil Engineers. New York.

Scott, K. M. and G. C. Gravlee. 1968. Flood surge on the Rubicon River, California -- hydrology, hydraulics and boulder transport. *U.S. Geological Survey Prof. Paper 422*.

Selby, M. J. 1982. Hillslope Materials and Processes. Oxford Univ. Press, Oxford.

Shields, A. 1936. Application of similarity principles and turbulence research to bed load movement. California Institute of Technology, Pasadena; Translation from original German. Report No. 167.

Sigafoos, R.S. 1964. Botanical evidence of floods and floodplain deposition. *U.S. Geological Survey Prof. Paper 485-A*.

U.S. Forest Service. 1993. Biological assessment for the Panther Creek Sub-basin . Section 1. General Description of Section 7 Watershed Panther Creek Sub-basin. February.

Whittaker, J. G. and M. N. Jaegi. 1982. Origin of step-pool systems in mountain streams. *Journal of the Hydraulics Div*. ASCE. 108:HY6:758-773.

Wolman, M. G. 1954. A method for sampling coarse river-bed materials. Trans. American Geophysical Union. 35:951-956.

STORAGE AND MOVEMENT OF SLUGS OF SAND IN A LARGE CATCHMENT: DEVELOPING A PLAN TO REHABILITATE THE GLENELG RIVER, SE AUSTRALIA

Ian Rutherfurd[1]

ABSTRACT

Like many streams in southeastern Australia, the Glenelg River (12,700 km^2) has been invaded by slugs of sand following catchment and gully erosion. About 6 million m^3 of sand has entered the stream network over the last century. The slugs of sand are the major stream management problem in the catchment because they have devastated the ecology of the streams, and threaten the high-value estuary of the river. To develop a rehabilitation plan for the stream system stream managers need to know the distribution and future movement of this sand.

The main source of sand to the Glenelg River is currently from large deposits in the downstream reaches of the major tributaries. These deposits feed sand into the Glenelg River in such quantities that the river has been dammed by the sand, producing several backwater lakes. Downstream of the major slugs the sand progrades by successively aggrading riffles, leaving the pools free of sand.

Widening of the incised tributaries means that up to half of the sand has been stored in 'abandoned thalwegs' and might not move into the trunk streams. At present rates of sand transport, the first major slugs of sand will reach the estuary in a few decades, but it will take centuries for the sand to move through the system.

The goal of the rehabilitation strategy is to keep sand out of the backwater lakes, out of the remaining stretches of stream with little sand, and out of the valuable estuarine reach. The preferred management option is to encourage strategic commercial sand extraction at the downstream end of specific sand slugs.

This study demonstrates the value of considering sources, sinks and fluxes of sand at the catchment scale in a sand slug rehabilitation project. It also demonstrates some of the complex interactions between sand slugs, incised tributary streams and trunk streams.

KEYWORDS

River, management, slugs, sand, aggradation, rehabilitation, restoration, sediment, Australia

[1] Cooperative Research Center for Catchment Hydrology, Department of Civil Engineering, Monash University, Clayton, Victoria, 3168.

1. INTRODUCTION

It is common for streams to receive large injections of sediment that aggrade the bed beyond the normal range of annual cut-and-fill. Since the mid 1800s, many Australian streams have received such large injections of sediment following European disturbance of channels and catchments (see review in Rutherfurd, 1996). These slugs tend to dramatically reduce the geomorphic and biological complexity of a stream system by filling stream pools, and transforming a complex bed into a featureless sheet of sand. Such sand slugs have low value for macroinvertebrates (O'Connor and Lake, 1994), and destroy fish habitat. They also reduce the recreational and aesthetic value of the stream.

Sediment slugs can originate from many anthropogenic and natural sources, such as mining, fires, gullying, large floods, or volcanoes (see review in Nicholas et al., 1995). The defining feature of these sediment slugs is that the supply of sediment declines, or ceases over time. The slug can then be considered as a transitory wave of sediment diffusing through the stream system, with the wave becoming longer and flatter as it progresses (Gilbert, 1917; Pickup et al., 1983; James, 1989). More complexity can be added to the wave by tributary contributions of sediment and water (Knighton, 1989, 1991).

As anthropogenic erosion rates decline toward the latter half of the twentieth century, stream managers and fluvial geomorphologists are increasingly turning their attention to the rehabilitation of disturbed streams. While a great deal of attention has been paid to the rehabilitation and natural recovery of incised streams (e.g., Schumm et al., 1984; Simon, 1992), less attention has been paid to anthropogenically aggraded streams. This paper describes the distribution and dynamics of sand slugs in the Glenelg River, southeastern Australia. The source of the sand in this catchment was general sheet and gully erosion from granite portions of the catchment. The paper also describes a proposed management plan for the sand. The study is of general interest because:

1. It considers sand movement through the entire stream network of a large catchment.
2. The movement and storage of sand through the catchment has been made more complex because the sand has aggraded tributary streams that were incising when the sand arrived.
3. The large sand loads from these tributaries have dammed the trunk streams, producing a series of lakes.
4. Targetted commercial sand extraction, combined with controls over grazing, are recommended as the major options for managing sand slugs in the Glenelg River.

2. THE STUDY AREA

The Glenelg River catchment (12,700 km^2) lies in SW Victoria, on the South Australian border (Fig. 2.1).

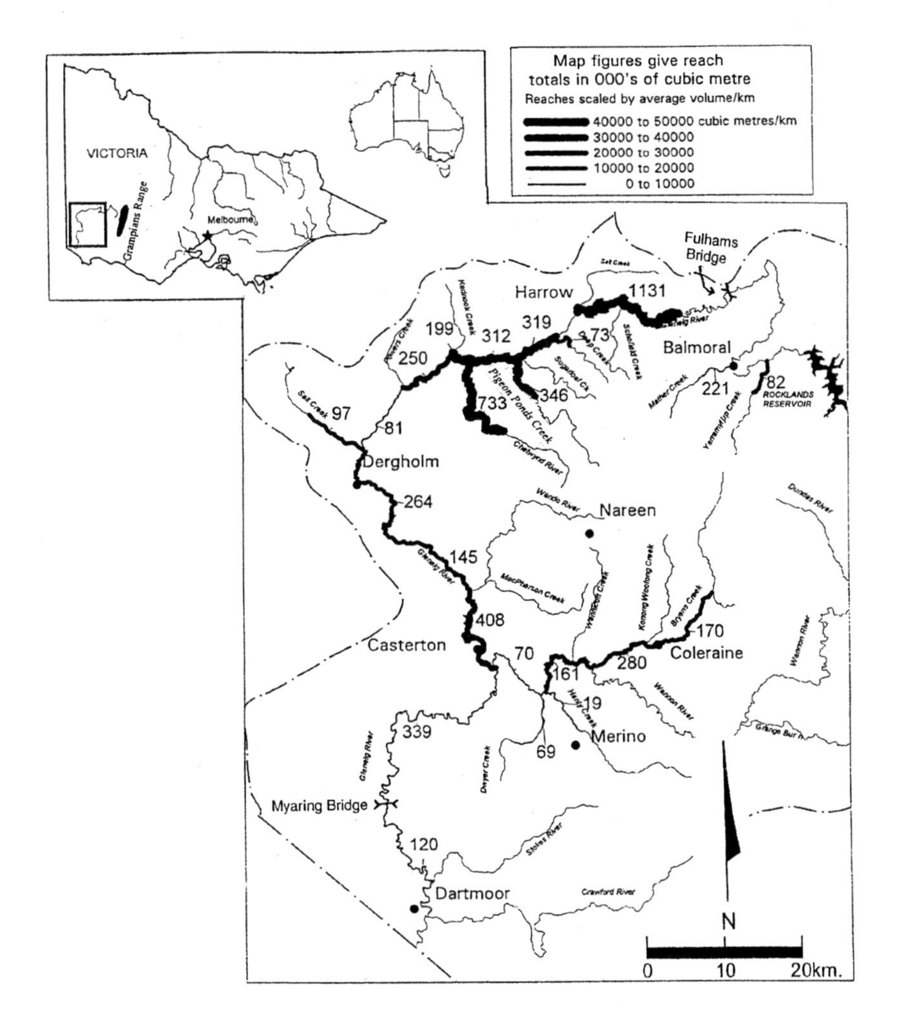

Fig. 2.1. Location of the Glenelg River Catchment, localities mentioned in the text, and volumes of sand in storage in the channels of the catchment.

The river rises in the Grampians Ranges at an elevation of 600 m. It flows to the west for about 50 river kilometres (km) before turning to the south, where it flows for 70 km to meet its largest tributary, the Wannon River (4,000 km^2). About 90 km downstream of the Wannon Junction, the river reaches the head of its estuary, which is a 65 km limestone gorge. The Glenelg skirts an uplifted palaeoplain, known as the Dundas Tablelands, that are dissected by a radial drainage network centred near Nareen (Fig 2.1). This drainage network comprises the major tributaries to the Glenelg River, including Pigeon Ponds Creek, and the Chetwynd and Wando Rivers. To the north and west of the stream are low plains with poorly developed drainage.

The Glenelg basin was one of the first areas of Victoria to be settled and cleared by Europeans in the 1830s and 1840s. At this time tributaries were characterised by a chain-of-ponds morphology. That is, the drainage lines consisted of deep pools separated by densely vegetated zones. This type of channel morphology was common across much of humid Australia (Eyles, 1977; Rutherfurd, In Press). The main trunk of the Glenelg also had deep pools. In the upper reaches, between the Grampians and Fulhams Bridge (Fig 2.1) the river had an anastomosing morphology, with pools over 6 m deep, and over a km long, separated by multiple shallow channels. The anastomosing channels coalesce into a single meandering channel below Fulhams Bridge, but the bed downstream of this point was still characterised by occasional deep pools, up to 9 m deep. The original channel was about 30 to 40 m wide, and 5 to 6 m deep, with a sandy bed and steep sandy-clay banks. Historical descriptions of the stream suggest that the bed of the pools were clay.

The mean annual flow of the basin is approximately 725 gigalitres (Department of Water Resources Victoria, 1989). The upper third of the Glenelg River catchment is regulated from the Rocklands Reservoir (closed 1953), from which two thirds of the mean annual flow of the river is diverted to the Wimmera-Avon Rivers Basin for the Wimmera-Mallee Stock and Domestic system. Average releases to the Glenelg River, out of the irrigation season, are now limited to 0.3 m^3/s. Flows are strongly seasonal with winter flood peaks falling to negligible summer flows. This study considers the river below Rocklands Reservoir.

3. HISTORY OF SAND MOVEMENT IN THE GLENELG CATCHMENT

The Glenelg River catchment has long had a reputation as one of the most severely eroded catchments in Victoria. Severe sheet and gully erosion was reported in the Glenelg catchment as early as 1853 (Mitchell, 1978), with the most severe erosion occurring in the granodiorite portions of the Dundas Tablelands (Gibbons and Downes, 1964). At the same time the larger tributaries draining the granodiorite areas (such as Pigeon Ponds Creek, Bryans Creek, and Wando Vale Ponds) were transformed from chains-of-ponds to incising streams. By the 1930s, sand choked the lower reaches of most of the major tributaries and much of the Glenelg River itself (McIlroy et al., 1938; Strom, 1947). Erosion in the Glenelg River itself is not severe, and its major source of sand is from the larger tributaries. Since the 1960s, tributary catchments have been treated by the Soil Conservation Authority, and more recently by the Department of Conservation and Natural Resources, and Landcare groups. As a result, the supply of sand to the upstream end of the tributaries has declined (Erskine, 1994a).

Management plans have concluded that sand slugs are the major stream management problem facing the Glenelg River catchment (Mitchell, 1990; Ian Drummond and Associates et al., 1992). In part this is because the 65 km estuary of the Glenelg River has been classified as a *Heritage River* under the Heritage Rivers Act of Victoria, 1984. This means that the estuary must be protected from all forms of damage, including invasion by sediment slugs. In order to develop a rehabilitation plan for the catchment, I investigated the volumes and distribution of sand stored in the streams of the catchment, and then investigated how the sand would move through the stream system if there were no intervention.

4. DISTRIBUTION OF SAND STORED IN THE GLENELG CATCHMENT

The distribution of sand in the Glenelg River was mapped from 1946 and 1992 aerial photographs (scale 1:15,840), and from field inspection. Fifty cross sections were surveyed in the Glenelg River and its tributaries. Sand depths were probed across each cross-section, and at numerous other points around the catchment, using a steel probe that could be extended to up to 7 m. It was usually easy to identify the contact between the sand and the underlying clay sediments with the probe. Historical descriptions of the streams suggest that there was limited sand in the bed, so the great majority of the sand volume in the present river is assumed to be of recent, anthropogenic origin. The volume of sand stored in the stream system was estimated by extrapolating from cross section and spot depths using a geographical information system. The range of possible volumes was estimated by using the coefficient of variation of measured depths. Only means are shown in Fig. 2.1.

There is between four and eight million cubic meters of sand stored in the bed of the Glenelg River and its tributaries (Figs. 2.1 and 4.1). The sand comes from left-bank tributaries draining the granodiorite portions of the catchment (Fig. 2.1). The major volumes of sand occur in three distinct types of deposit: tributary bed deposits, tributary junction plugs, and as small deposits on riffles (here called 'sluglettes').

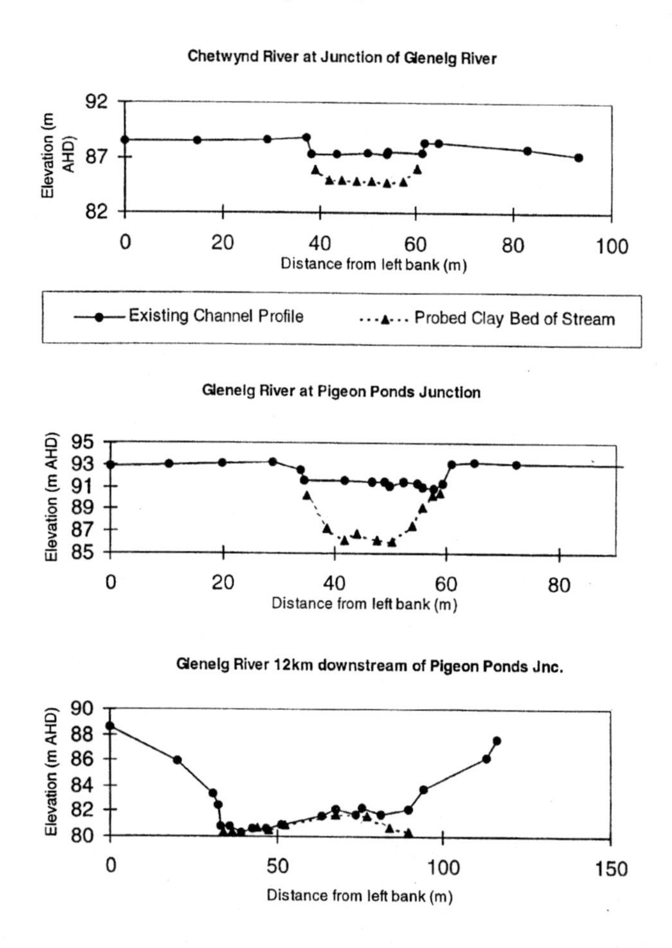

Fig. 4.1. **Surveyed cross-sections of the Glenelg River and tributaries showing the depth of sand in storage. Survey done in 1995.**

4.1. Tributary Bed Deposits

The lower 10 to 30 km of the major tributaries are aggraded with 1 to 5 m of sand. These tributaries drain the granodiorite portions of the catchment. The first such tributary is Yarramyljup Creek directly below the Rocklands Reservoir. There is no evidence of sand above the reservoir. The other tributary streams with major deposits of sand in the bed are: Mathers Creek, Deep Creek, Pigeon Ponds Creek, Bryans Creek and the Chetwynd River (Figures 2.1, 4.1 and 4.2). The major deposits of sand are concentrated in

Fig. 4.2. A sand slug in the lower end of a typical tributary (Bryans Creek oblique aerial photograph, taken in 1940s). Flow is towards the observer.

the lower 10 km or so of these streams, with the exception of Bryans Creek which has sand for the lower 30 km of its course. Sand typically fills about half of the original cross-sectional area of the channel, but it can be up to 70 percent (Fig. 4.1).

4.2. Tributary Junction Plugs

Sand from the tributaries has moved into the Glenelg River (and its major tributary the Wannon River) and blocked the channel, producing backwater pools up to 5 km in length (Figs. 4.3 and 4.4). These 'tributary junction

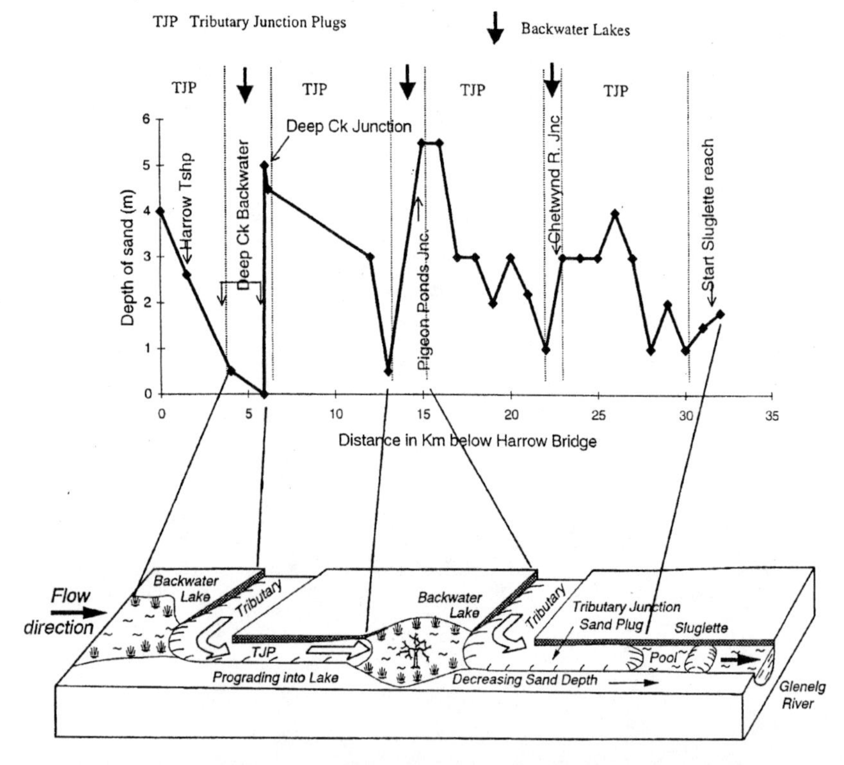

Fig. 4.3. Maximum depth of sand in cross-sections surveyed in the Glenelg River for 33 km below Harrow Bridge, showing the plugs of sand (TJPs) formed below the tributary junctions. Also a schematic figure showing the damming of the trunk stream by sand from the tributaries, backwater slug above the TJP, and the degeneration of a slug into a series of sluglettes.

Fig. 4.4. A lake formed in the Glenelg River upstream of a tributary junction plug of sand coming out of Deep Creek (flow is toward the lower left in photo).

plugs' (TJPs) have existed at the junction of the Glenelg River with the following streams since at least the 1940s: Yarramyljup Creek, Mathers Creek, Deep Creek, Pigeon Ponds Creek, and the Chetwynd and Wando Rivers (Fig. 4.2). Sand from Bryans Creek has also dammed the Wannon River. Figure 4.3 shows four TJPs that are less than 10 km long, with sand depth decreasing from 4-5 m at the tributary junction (up to 70 percent of the original channel area), to about 1 m depth downstream. In these examples, the sand from each slug progrades into the backwater formed by the next slug downstream. For example, the sand from Deep Creek progrades into the backwater lake formed by the TJP from Pigeon Ponds Creek (Fig. 4.3). Where there is a long distance between TJPs, then the TJP breaks down into what I call 'sluglettes' (described below).

The genesis of the TJPs has not been investigated here, but I hypothesise that the TJPs dam the trunk because of the sudden high load of sand from the tributaries, combined with a large gap between the tributary and trunk flood peaks. It is important to note that the TJPs dammed the trunk streams before the flooding regime of the stream was altered by Rocklands Reservoir.

4.3. Sluglettes

Sand from the Chetwynd River fills the channel of the Glenelg River to 4 m depth at the junction of the two streams, reducing to about 1 m depth 8 km below the junction (Fig. 4.3). At this point the continuous sheet of sand breaks down into a series of discrete 'sluglettes' about 200-500 m long, separated by about the same distance of open pool where there is little sand (Figs. 4.3 and 4.5). Sluglettes are about 1.5 m deep, and are usually superimposed on the

Fig. 4.5. The 'snout' of a sluglette in the Lower Glenelg River. Water depth is 0.5 m on the sluglette, and over 3 m off the end of the snout (flow is away from observer).

existing riffle areas, with their snouts prograding into pools (Fig. 4.5). It is interesting to note that when we began our investigations into the distribution of sand in the Glenelg River, many local people enthusiastically reported that the front of the sand wave in the river system was in their property. It was only after investigating several of these alleged sand fronts that it was discovered that there were numerous small sand fronts throughout the stream system, not just one – these are the sluglettes.

The sluglette reach below the Chetwynd River extends for some 50 km to the Wando River Junction. Sluglettes also occur in the 57 km of the Glenelg River below the Wannon River Junction (Fig. 2.1), with the last sluglette found some 28 km above the estuary at Dartmoor. For future reference I refer to this last sluglette as the 'end-slug'. The pools between the sluglettes are some of the only fish habitat remaining in the lower Glenelg River.

The TJPs are classical attenuating sediment waves, but the sluglettes appear to represent a different type of sand migration. The genesis of sluglettes has yet to be investigated in detail, but I have made the following observations:

1. The sluglettes are stable features. For example, the seven sluglettes shown in Fig. 4.6, directly below the Wannon Junction, have not changed their position for at least 44 years.
2. Scour chains placed in the sluglettes show that only the top 0.5m or less of sand in the deposit moved in a typical bankfull flow in 1995.
3. During the same 1995 flood, three new sluglettes were deposited downstream of the end-slug above Dartmoor (one of these is shown in Fig. 4.5).

From the three observations made above, I hypothesise that the sand moving in front of the major sand slug (the TJP) migrates downstream by aggrading existing riffles in the bed. The pools between the riffles may remain free of sand due to a velocity reversal mechanism (Lane and Borland, 1954; Keller, 1971; Keller and Florsheim, 1993), but this is yet to be investigated. Once the deposits on the riffles reach a threshold size the sand moves through them and progressively aggrades the next riffle downstream. In this way the sand wave on the Glenelg River migrates downstream. In the 1995 flood the sand front migrated about one kilometre by aggrading three riffles.

Thus, the major deposits of sand in the catchment are in the lower ends of major tributaries, in TJPs, and in sluglettes. The next requirement for the management plan is to understand how these deposits of sand will change over time. This is essentially the 'do nothing' management scenario. The next section describes sand movement and long-term storage in the streams of the catchment.

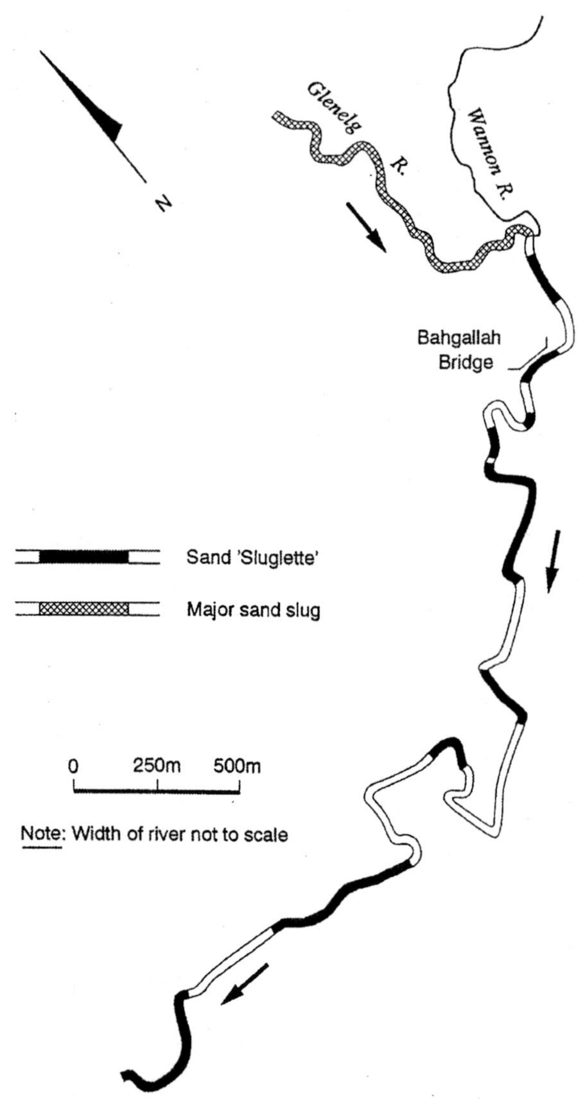

Fig. 4.6. Stable sluglettes in the Glenelg River below the Wannon Junction, 1992.

5. CHANGES IN SAND STORAGE WITH NO INTERVENTION

5.1 The Present Bedload Transport Rate

The Glenelg River and its tributaries have a low rate of sediment transport relative to the volume of sand in storage. Several lines of evidence (including scour chains, specific gauge records, bedload measurement, sediment transport equations, and sand extraction rates below weirs) suggest that the tributaries transport less than 10,000 m^3 of sand per year, on average, and that the Glenelg River has a low annual bedload transport rate (between 10,000 and 30,000 m^3 per year) relative to the volume of sand in storage (see full discussion in Rutherfurd and Budahazy, 1996). This transport rate would be higher if Rocklands Reservoir did not reduce flood frequency. The low bedload transport rate in the Glenelg River, combined with the reported rate at which the sand front is moving past Myaring Bridge (Fig. 2.1) suggests that the main sluglettes of sand will not reach the Heritage River reach of the river for 40 to 50 years.

It will take perhaps two decades to clear sand from the minor tributaries, 3 to 4 decades for the larger tributaries, and certainly more than a century for the Glenelg River itself. Note from Fig. 2.1 that there are between 10,000 and 40,000 m^3 of sand per kilometre of channel in the Glenelg River and its major tributaries. Thus, the total annual bedload transport rate will only remove sand from the equivalent of about one kilometre of stream per year.

5.2 Future Supply of Sand to the Glenelg River System

The major source of sand to the Glenelg River and the Wannon River is from the larger tributary streams. Sand supply to these tributary streams has declined as better land management practices have developed (Erskine, 1994a), but also as the yield from gullies naturally declines (e.g., Trimble, 1981). Thus the main source of sand is from the large stores in the lower end of the large tributary streams.

Already the supply of sand from some of these tributary streams has declined. Evidence for this are the large reed beds that have consolidated the sand deposits at the mouth of several tributaries (Fig. 5.1). Another factor contributing to the declining sand supply is the long-term storage of sand in the tributaries.

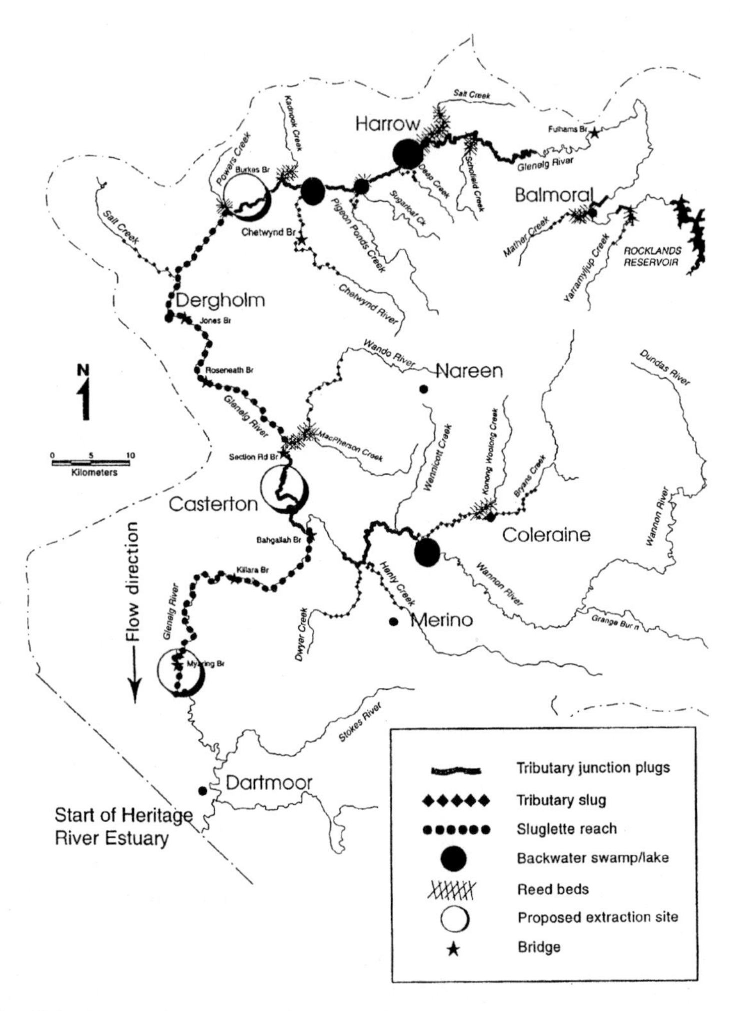

Fig. 5.1. Types of sand deposit in the Glenelg River and its tributaries.

5.2.1. Sand Storage in the Tributaries

Many studies describe the 'locking-up' of sediment in long-term storage sites which means that the long-term output from the system is less than the input (see Meade, 1982). For example, up to 90 percent of the sediment in the classical mining slugs of northern California have been deposited in floodplain areas from which it will be only very slowly remobilised (James, 1989).

Storage of sand in benches has been documented in several rivers (Erskine, 1994b). In the tributaries to the Glenelg River, sand is being stored in benches, point bars, overbank deposits and in former channel beds. These last deposits, that I call 'abandoned thalwegs', do not appear to have been described before.

As mentioned above, many of the tributaries to the Glenelg River had incised up to five metres before being filled with sand to up to 70 percent of their cross-sectional area. Widening is a predictable response to a sudden increase in bedload because "*At any given discharge Q and gradient S an alluvial river can transport a bed load of a given mean grain size at a greater rate the shallower the flow*" (Bagnold, 1977, p.322).

In a normal incised stream (i.e., one that has not been filled with sand), the entire enlarged trench would widen (Simon, 1992). When the trench has filled with sand the widening occurs near to the top of the trench. The result of this is that the new thalweg of the channel (which is in the deposited sand) moves progressively away from the deep thalweg of the original incised channel. In most streams that were probed the deep store of sand was located beneath the middle of the point bar. This is the 'abandoned thalweg'. In an example from Pigeon Ponds Creek (Fig. 5.2) the new thalweg has migrated about 35 m away from the original thalweg, and the abandoned thalweg of the

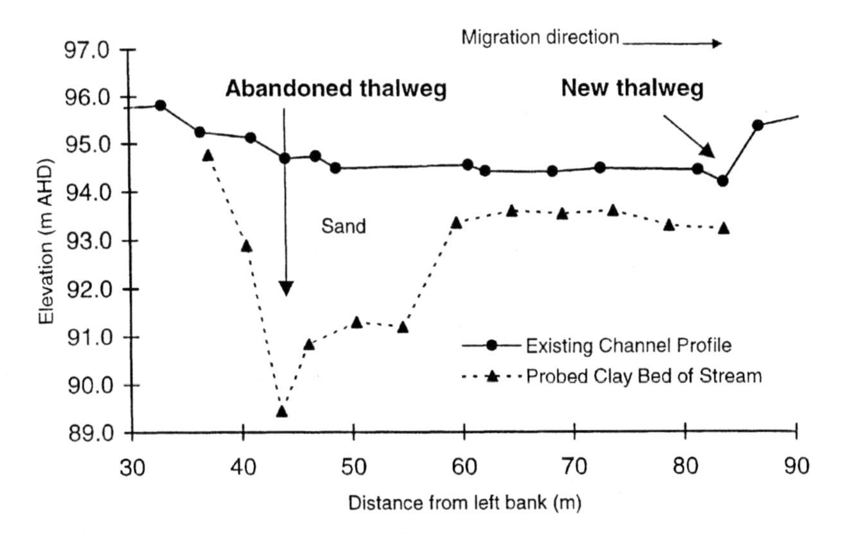

Fig. 5.2. **Cross section surveyed across Pigeon Ponds Creek 1 km upstream from the Glenelg Junction. Eroding outer bank is on the right side, sand store is on the left (AHD = Australian Height Datum).**

channel is over 5 m deep, compared with less than 1.5 m for the new thalweg. Over 90 percent of the sand in the cross section is now stored in the abandoned thalweg. Much of this will be stored indefinitely because the point bar can quickly be colonised with vegetation if cattle are kept out of the streambed. Even without vegetation, there is little scour of the point bar sediments. Scour chain measurements following a bankfull flow show that the stream can scour to the basal clays at the present thalweg, but that scour over the point bar is less than 0.5 m. Abandoned thalwegs were found at most bends probed for sand depth in the tributaries, but they were not present at inflection points of the channel.

The main implication of abandoned thalwegs is that they represent a large storage site for sediment. This type of interaction between incised streams and sand slugs might not have been described before. Sediment stored in benches (many benches can be seen in Fig. 4.2), overbank, and in abandoned thalwegs, is likely to remain in storage, especially if the deposit is stabilised by vegetation. This stored sediment is likely to be more than half of the sand stored in the tributary bed. This means that the high sediment load from the tributaries is likely to decline earlier than would be predicted from the large volume of sand stored in the beds of the tributaries. Alternatively, the sand that is stored in the benches and abandoned thalwegs could form a future source of sand as the channel migrates laterally back into these deposits. James (1989) has reported this type of erosion for sand slug streams in the Sacramento Valley.

5.3 Response of the TJPs to Reduced Sediment Load

As the sediment load from the tributaries declines what will happen to the tributary junction plugs of sand? Will they remain as permanent dams in the trunk streams? Pigeon Ponds Creek provides a test of this question. Following considerable commercial sand extraction from the creek, a bankfull flow in the Glenelg River in 1996 removed the plug of sand (along with several of our 2m scour chains) at the tributary mouth. This plug had been in position since at least the 1940s, and the sand had not returned by late 1998. The implication of this is that the progressive reduction in sand supply from the tributaries will lead to the removal of the TJPs by flow in the trunk stream. In other words, the TJP will behave like a classic sand slug, moving downstream as an attenuating wave (Gilbert, 1917; Pickup et al., 1983). Both the amplitude (i.e., depth) and velocity of the sediment waves decline as they move downstream. In the process the backwater lakes will be progressively drained, allowing the sand from the next TJP upstream to move downstream.

5.4 Changes in the Sluglettes Over Time

Declining sand supply from the tributaries is unlikely to have any effect on the rate at which sluglettes occupy new riffles, and so migrate downstream. This is because the sluglettes are supplied by sand from the TJPs. In fact, most of the sand for the sluglettes above the Wando River Junction (Fig. 5.1) comes entirely from the Chetwynd River TJP. If the supply of sand from the TJP began to decline, say because sand was stored in vegetated benches and point bars, I propose that the sluglettes would continue to migrate downstream, with the most upstream sluglettes becoming the source of sand. The result would be that the sluglettes would be degraded progressively downstream.

6. A CONCEPTUAL MODEL OF FUTURE SAND MOVEMENT

The mechanisms of sand movement, combined with information about the average rate of sand movement in the river provide a general conceptual model of the complex movement of sand through the system. The proposed temporal sequence of change is shown in Table 6.1. To summarise that model: the tributaries to the Glenelg River were incising when they were invaded by slugs of sand. These slugs have migrated into the Glenelg River where they have dammed the larger channel, producing backwater lakes that have limited the movement of sand through the trunk stream. Where sand can move for long distances, it does not migrate as a simple wave, but instead as discrete sluglettes that aggrade riffles.

Sand in the trunk channel will only move through the stream network when the plugs of sand blocking the channel are removed, and this will only happen when the supply of sand from the tributaries is exhausted. When tributary sediment supply is exhausted the TJPs will begin to erode from the upstream end, gradually draining the backwater lakes. As these drain, the sand slugs that are presently infilling the lakes, will begin to move downstream. Eventually there will be a single long sheet of sand from upstream of Harrow down to Burkes Bridge (Fig. 2.1), and abundant sand to supply the sluglettes migrating toward the Wando River junction.

Table 6.1. A conceptual model of response to sand slugs in the Glenelg River and its tributaries.

Dates	Tributaries	Trunk stream
Pre 1880s	Gullying, incision of larger tributaries.	Little change.
1880s – 1930s	Aggradation of the lower reaches in response to gully and sheet erosion.	Slugs of sand from tributaries dam the Glenelg and Wannon Rivers creating lakes upstream. Sand migrates below the TJPs as discrete sluglettes.
1940s – 1990s	Tributaries widen, storing large volumes of sand in abandoned thalwegs.	Tributary junction plugs are maintained. Sluglettes migrate tens of kilometres per decade.
1980s – 1990s	Supply of sand from the smaller tributaries declines.	TJPs begin to erode from the upstream end (as has recently occurred in Pigeon Ponds Creek).
2000 - 2050	Much reduced sand supply from the smaller tributaries (esp. Deep Creek, Pigeon Ponds Creek, and Henty Creek).	TJPs above the Chetwynd River begin to coallesce as backwater lakes are drained. Sluglettes reach the estuary of the Glenelg River.
2050 - 2100	Sand supply from the Chetwynd River will decline leading to removal of last TJP.	Sand between Harrow and Burkes Bridge will coalesce into a single slug as the backwater lakes drain. Declining habitat quality in the river. Sluglettes continue to migrate.

7. DEVELOPING A REHABILITATION PLAN FOR THE GLENELG RIVER

The aggradation of the bed of the Glenelg River and its tributaries by sand slugs has been identified as the major stream management issue facing the Glenelg River (Mitchell, 1990; Ian Drummond and Associates et al., 1992). The sand destroys the complex structure of the stream, and dramatically reduces habitat in the stream. The conceptual model of sand movement and storage described in Table 6.1 provides the framework for a rehabilitation strategy that will be described in this section

The first aim of rehabilitation is to protect the remnants of stream that remain in good condition (Rutherfurd et al., 1999). In the case of the Glenelg River, the key parts of the system to protect are:

1. The remaining pools between the end of the Chetwynd River and the Wando River Junction, and downstream of the Wannon River junction.
2. The Heritage River estuary.
3. The artificial backwater lakes formed above the tributary junction plugs (in the absence of natural stream habitat, these lakes now form valuable wetland habitat and are considered worthy of protection).

A management strategy has been devised that aims to reduce the supply of sand for downstream transport. The plan has two aims: to stabilise the sand that has been stored in benches and abandoned thalwegs, and to remove sand from the active channel floor. Clearly these strategies have to be carried-out together because removing the sand rapidly from the bed could destabilise the benches and other long-term sand stores.

Another option that was examined was the possibility of managing sand transport rates by altering flows from the Rocklands Reservoir. For example, higher competent flows, combined with sand extraction, could be used to more quickly clean out sluglettes of sand. Unfortunately there is little opportunity to regulate flows because the outlet structures on the dam can only release a maximum of about 12 m^3/s of water. Bedload transport measurements show that substantial bedload transport only begins at flows of about 80 m^3/s. Thus, there was limited opportunity to either increase or decrease bedload transport rates.

7.1. Stabilization

The large volumes of sand that have been stored in benches and abandoned thalwegs could be remobilised by later erosion (as described by James, 1989). The first strategy of the rehabilitation plan is to stabilise these deposits so that they cannot provide more sand to the active channel. The abandoned thalwegs were incised streams that have been filled with sand, and there is little ecological benefit in removing the large volume of sand from the former incised streams. The best way to stabilise the abandoned thalwegs and benches is by natural regeneration of vegetation. Where cattle have been excluded from the streambed, the benches and point bars of sand are rapidly colonised by grass and *Eucalyptus* species. Thus, fencing the stream and allowing natural regeneration is a priority task. Once the benches and point bars have been stabilised, the active channel might erode a new, stable, narrower channel.

7.2. Artificial Sand Extraction

The second strategy for protecting the remaining natural assets of the Glenelg River is artificial sand extraction. In Western Victoria there is a shortage of the good quality quartz sand such as that found in the Glenelg River, so there is a ready market for the product. In common with most streams in SE Australia, commercial sand extraction from streambeds was being phased out in the Glenelg catchment due to the perceived environmental damage associated with many extraction projects (DC&E, 1990; Erskine, 1990). Where, however, the problem is rapid anthropogenically-induced aggradation of streams by sand, then controlled extraction can be a useful management tool.

Licensed extraction from the Glenelg River itself has been less than 1,500 m^3 per year (from Department of Natural Resources and Environment records). This is far less than the 10–30,000 m^3 of bedload transported each year, and certainly trivial when compared with the 4-8 million m^3 of sand in storage. The major extraction from the catchment has been the 400,000 m^3 removed from Bryans Creek, mostly in the 1980s (Rutherfurd and Budahazy, 1996). This represents half of the sand stored in the lower reaches of that stream.

7.3. Where Should Sand be Extracted?

Sand can be extracted from the lower end of tributaries or from the channel of the Glenelg River itself. Rapid extraction from the downstream sections of tributaries is not recommended. Rapid extraction of sand from Bryans Creek has been followed by up to 3m of degradation into the clay bed underlying the sand. This incision has in turn, destabilised the channel banks, and mobilised the sand stored in abandoned thalwegs. There are several processes that could explain this bed erosion following extraction (see Rutherfurd and Budahazy, 1996), but the key point is that major extraction from the tributary streams is not recommended. Another reason why extraction from the tributary streams is not recommended is that it will lead to downstream degradation of the TJPs in the Glenelg and Wannon Rivers, and accelerate the loss of the backwater lakes. The best management strategy for the tributaries is to encourage the growth of vegetation in the channel.

Stopping the movement of sand out of the tributary streams will, in turn, trigger downstream migration of the TJPs in the Glenelg and Wannon Rivers. The major implication of this will be that the backwater lakes and swamps formed above the TJPs will eventually be drained as the surface of the TJP is eroded. This is undesirable because these swamps, even though they are a product of human disturbance, now represent some of the best biological habitat remaining along the river. For this reason, we have recommended that sand extraction be concentrated at points in the stream system where it will have least impact on the backwater swamps. As other reaches of the stream gradually improve in condition, the loss of the backwater swamps will be acceptable.

Therefore, we have recommended three sites for sand extraction. The first two sites are at the downstream end of the two long sand slugs, just before the slugs break into 'sluglettes': at Burkes Bridge (at the downstream end of the Chetwynd River slug), and near the town of Casterton (downstream end of the slug from the Wannon River). These sites are ideal for extraction for four reasons:

1. Extraction at these sites will stop the movement of sand into the sluglette reaches, and this will preserve the remaining deep pools in the middle and lower reaches of the river. The slugettes will gradually migrate out of the reach, from the top end down.
2. Extraction in these reaches will not have any impact upon the rate at which TJP backwaters will be drained.
3. The channel of the Glenelg River is large enough that extraction will not lead to accelerated erosion of either the bed or banks. That is, sand occupies about 20 percent of the channel here, in comparison to over 70 percent in some of the tributary streams.
4. Removing sand around Casterton (the largest town along the river) will improve the aesthetic and recreational values of the stream in this reach. It is also possible that extraction could be combined with a weir at Casterton that would provide a recreation pool for the town, as well as a convenient extraction site.

The third extraction site is downstream near the estuary reach at the Myaring Bridge. The aim of this extraction site is simply to intercept the sand moving as sluglettes into the Heritage River reach.

The artificial extraction rate will have to be substantial to have much impact on the 4 to 8 million m^3 of sand stored in the stream system. Extraction rates from the Glenelg River would need to exceed the annual bedload transport rate of 10,000 to 30,000 m^3 of sand per year if they are to improve the condition of downstream reaches.

8. SUMMARY AND CONCLUSIONS

The major stream management problem facing the Glenelg River catchment is the 4 to 8 million m^3 of sand that has been deposited in the stream system as a result of historical sheet and gully erosion. Schumm (1973) described the complex response of streams to a single disturbance, such as a baselevel fall. The Glenelg River and its tributaries provide another example of a complex response triggered by an initial perturbation. The complexity comes from the interaction between sand and incised streams, and between tributaries and the trunk streams.

Sand has moved through the stream network in the following way. The tributaries to the Glenelg River were incising when they were invaded by slugs of sand. These slugs have migrated into the Glenelg River where they have dammed the larger channel, producing backwater lakes that have limited the movement of sand through the trunk stream. Each of the tributary junction plugs (TJPs) that has dammed the Glenelg and Wannon Rivers can be considered as a discrete wave of sediment in the sense of Gilbert (1917). However, where the sand wave can move for long distances (i.e., it does not reach another backwater lake), I hypothesise that the sand does not migrate as a simple wave, but instead it migrates by progressively aggrading successive riffles. These deposits are labelled sluglettes here.

The sand slugs have been identified as the major stream management problem in the catchment. The conceptual model of sand movement, distribution and interactions in the Glenelg River provides a basis for planning a rehabilitation strategy for the streams. The goal of the strategy is to protect remaining pools in the stream, the valuable estuary reach, and the artificial backwater lakes formed by TJPs.

The supply of sand from the tributaries controls the movement of sand. The sand supply from the tributaries is declining rapidly because a large proportion of the sand in the tributaries is stored in abandoned thalwegs formed when the incised tributaries widened. As the TJPs in the trunk stream are starved of sand they will degrade and the backwater lakes will be drained. The several TJPs will then coalesce into a single long sand slug. At bedload transport rates of tens of thousands of cubic metres of sand per year, the rate of change in the streams will be slow. The first sluglettes of sand are likely to reach the estuary reach in a few decades, but the great bulk of the sand will continue to move slowly through the stream system for perhaps centuries.

The two major management strategies are, first, the stabilisation of benches and point bars with vegetation. This can be greatly accelerated by removing cattle from the stream. The second strategy is to artificially remove the sand that remains in the active channel bed. Commercial sand extraction from the Glenelg River is the preferred option, with only restricted extraction from the tributaries. Present commercial rates of sand extraction would have to exceed the annual bedload transport rates to protect the remaining assets. This represents a volume of 10,000 to 30,000 m^3 of sand per year. Three sites have been identified in the Glenelg River that will optimise the effectiveness of the extraction strategy. The management strategy is being implemented by the management agency, and the results are being monitored.

This study has demonstrated the value of developing a conceptual model of sand movement at a catchment scale as the basis for developing a rehabilitation plan. In the process, some interesting aspects of the complex movement of sand through the catchment have been identified. These include: the damming of the trunk streams by sand from tributaries, the long-term storage of sand in abandoned thalwegs in the tributaries, and the migration of sand as sluglettes rather than as large discrete bodies of sand. The processes involved in these patterns of sand movement deserve more study.

Although managing sand is a key to rehabilitating the Glenelg River system, it is not the only task required. Of the many other problems facing the stream, a key issue is the loss of water to inter-basin transfers, and the absence of an adequate environmental flow regime from Rocklands Reservoir. When such a flow regime is introduced it will have its own implications for sand transport in the river.

9. ACKNOWLEDGMENTS

Many thanks to John Oates (Victorian Dept of Natural Resources and Environment) who initiated the project; and to Mike Budahazy and the Glenelg Waterways Team (Nicole Davidson, Graeme Jeffrey, and Cathy Wagg) for field assistance. Rob Alexander and Mark Pearse assisted with the figures. Thanks also goes to Bruce Abernethy, Paul Mosley and an anonymous reviewer for many constructive comments on an earlier draft of the paper.

Finally, I would like to take this opportunity to thank Stan Schumm for the guidance, inspiration and friendship that he has provided for me when I spent an ITT International Fellowship in Fort Collins, Colorado and in Vicksburg, Mississippi in 1988-1989.

10. REFERENCES

Bagnold, R. A. 1977. Bed load transport by natural rivers. Water Resources Research 13(2): 303-312.

DC&E, 1990. Environmental Guidelines for River Management Works, Department of Conservation and Environment, Office of Water Resources, Victoria.

Department of Water Resources Victoria 1989. A Resource Handbook, Melbourne.

Erskine, W. 1994a. River response to accelerated soil erosion in the Glenelg River Catchment, Victoria. Australian Journal of Soil and Water Conservation 7(2): 39-47.

Erskine, W. D., 1994b. Sand slugs generated by catastrophic floods on the Goulburn River, New South Wales. , Olive, L.J., Loughran, R.J., & Kesby, J.A., Variability in stream erosion and sediment transport. Proceedings of an IAHS symposium, Canberra, pp. 143-151.

Erskine, W. D. 1990. Environmental impacts of sand and gravel extraction on river systems. In Davie, P. & D. Low Choy. The Brisbane River: A Source Book for the Future, Brisbane. Australian Littoral Society and Queensland Museum: 295-302.

Eyles, R. J. 1977. Changes in drainage networks since 1820, Southern Tablelands, N.S.W. Australian Geographer 13: 377-386.

Gibbons, F. R. and R. G. Downes, 1964. A study of the land in south-western Victoria, Soil Conservation Authority of Victoria.

Gilbert, G. K. 1917. Hydraulic mining debris in the Sierra Nevada. United States Geological Survey, Professional Paper 105: 1-154.

Ian Drummond and Associates, W. D. Erskine, et al., 1992. Glenelg catchment waterway management study, Shire of Dundas, April 1992, Final Report.

James, L. A. 1989. Sustained storage and transport of hydraulic gold mining sediment in the Bear River, California. Annals of the Association of American Geographers 79(4): 570-592.

James, L. A. 1991. Incision and morphologic evolution of an alluvial channel recovering from hydraulic mining sediment. Geological Society of America Bulletin 103 (June): 723-736.

Keller, E. A. 1971. Areal sorting of bedload material: the hypothesis of velocity reversal. Geological Society of America Bulletin 82: 753-756.

Keller, E. A. and J. L. Florsheim 1993. Velocity-reversal hypothesis: A model approach. Earth Surface Processes and Landforms 18: 733-740.

Knighton, A. D. 1989. River adjustment to changes in sediment load: the effects of tin mining on the Ringarooma River, Tasmania, 1875-1984. Earth Surface Processes and Landforms 14: 333-359.

Knighton, A. D. 1991. Channel bed adjustment along mine-affected rivers of northeast Tasmania. Geomorphology 4: 205-219.

Lane, E. W. and W. M. Borland 1954. River bed scour during floods. Trans, Am. Soc. of Civil Engineers 119: 1069-1079.

McIlroy, W., J. Brake, H. G. Strom, W. Lakeland and J. D. O'Carroll, 1938. Report of Committee appointed to investigate erosion in Victoria, Victorian Govt. Printer.

Meade, R. H. 1982. Sources, sinks and storage of river sediment in the Atlantic drainage of the United States. Journal of Geology 90: 235-252.

Mitchell, A. 1978. Development of soil conservation in Victoria. Journal of Soil Conservation, NSW 34: 117-123.

Mitchell, P., 1990. The environmental condition of Victorian streams, Department of Water Resources, Feb 1990.

Nicholas, A. P., P. J. Ashworth, M. J. Kirkby, M. G. Macklin and T. Murray 1995. Sediment Slugs: Large-Scale Fluctuations in Fluvial Sediment Transport Rates and Storage Volumes. Progress in Physical Geography 19(4): 500-519.

O'Connor, N. A. and P. S. Lake 1994. Long-term and seasonal large-scale disturbances of a small lowland stream. Australian Journal of Marine and Freshwater Research 45: 243-55.

Pickup, G., R. J. Higgins and I. Grant 1983. Modelling sediment transport as a moving wave - the transfer and deposition of mining waste. Journal of Hydrology 60: 281-301.

Rutherfurd, I. D., 1996. Sand slugs in SE Australian streams: origins, distribution and management. In: Rutherfurd, I.D. & Walker, M.R., First National Conference on River Management in Australia, Merrijig. Cooperative Research Centre for Catchment Hydrology, pp. 29-34.

Rutherfurd, I. D. In Press. Some human impacts of Australian stream channel morphology. Brizga, S.O. and Finlayson, B. Rive management: The Australasian Experience. Wiley and Sons, Chichester.

Rutherfurd, I. D. and M. Budahazy, 1996. A sand management strategy for the Glenelg River, Western Victoria, Melbourne, Cooperative Research Centre for Catchment Hydrology, Dec 1996, Report to Southern Rural Water and the Department of Natural Resources and Environment, Victoria.

Rutherfurd, I. D., K. Jerie, M. Walker and N. Marsh, 1999. Don't raise the Titanic: How to set priorities for stream rehabilitation. In: I. Rutherfurd, and R.Bartley, Second Australian Stream Management Conference, Adelaide, South Australia, Cooperative Research Centre for Catchment Hydrology. , .

Schumm, S. A. 1973. Geomorphic thresholds and complex response of drainage systems. Morisawa, M.E. Fluvial Geomorphology, New York. Publications in Geomorphology, State University of New York: pp. 299-310.

Schumm, S. A., M. D. Harvey and C. C. Watson 1984. Incised Channels: Morphology, Dynamics and Control. Colorado, Water Resources Publications.

Simon, A. 1992. Energy, time and channel evolution in catastrophically disturbed fluvial systems. Geomorphology 5: 345-372.

Strom, H. G., 1947. Siltation of the Glenelg River and its tributaries, State Rivers and Water Supply Commission, Victoria, Preliminary Memorandum.

Trimble, S. W. 1981. Changes in sediment storage in the Coon Creek Basin, Driftless Area, Wisconsin, 1853 to 1975 Science 214: 181-83.

GEOMORPHIC ASSESSMENT OF THE POTENTIAL FOR EXPANDING THE RANGE OF HABITAT USED BY NATIVE FISHES IN THE UPPER COLORADO RIVER

John Pitlick, Robert Cress, and Mark M. Van Steeter[1]

ABSTRACT

This paper examines long-term changes in the hydrology and geomorphology of the Colorado River between Rifle and Palisade, Colorado. Historically this reach provided habitat for the endangered Colorado squawfish and razorback sucker, but two diversion dams now prevent the fishes from migrating upstream. Analyses of discharge records from several gauging stations in the area indicate that in the last 30 years annual peak discharges of the upper Colorado River have decreased by 20 to 30 percent, and average annual sediment loads have decreased by 30 to 40 percent. The planform area of the main channel has not changed significantly in the last 60 years, but there has been a 30 percent decrease in the area of side channels and backwaters, two habitats that are important for the endangered fishes. Many side channels are still present, however, and preliminary sediment transport calculations suggest that the bed material in the reach may be mobile for periods of several weeks each year. The potential for expanding the range of the endangered fishes into this area seems good because spawning bars and side channel habitats are relatively abundant.

KEYWORDS

Geomorphology, hydrology, fish habitat, Colorado River

1. INTRODUCTION

Conflicts over the management and allocation of water resources in the Colorado River basin have been common throughout much of the 20th century.

---

[1] University of Colorado, Department of Geography, Boulder, Colorado 80309.

Early on there was debate over the appropriation of water between the upper basin states of Colorado, Wyoming, Utah, and New Mexico, and the lower basin states of Arizona, California and Nevada. In mid-century attention turned to the hydrologic effects of land-use practices such as grazing (Schumm and Lusby, 1963), timber harvesting, and other forms of vegetation management (Hibbert, 1979). In the 1970s there was widespread interest in the problem of diffuse- and point-source salinity (Laronne and Shen, 1982). Now the issue is "ecological integrity," and the key question is, how should the Colorado River be managed to protect endangered species? There are four federally listed endangered fish in the upper Colorado River basin: the Colorado squawfish (*Ptychocheilus lucius*), the razorback sucker (*Xyrauchen texanus*), the humpback chub (*Gila cypha*), and the bonytail chub (*Gila elegans*). These species were once plentiful in rivers of western Colorado and eastern Utah, but their populations are now very small. The decline of native fishes in the Colorado River basin is attributed to the combined effects of competition with non-native species, changes in the flow hydrograph due to reservoir operations, and reductions in the amount of riverine habitat (Stanford, 1994). It is not known whether any one of these factors is more important than another, but it seems reasonable that all have contributed in some way to reduce native fish populations.

The U.S. Fish and Wildlife Service (USFWS) oversees the Recovery Implementation Program for Endangered Fish Species in the upper Colorado River Basin (USFWS, 1987). This program has the ambitious goal of reestablishing self-sustaining fish populations while allowing the upper basin states of Colorado, Wyoming and Utah to further develop their water resources in accordance with interstate compacts. This goal obviously reflects a compromise between various interests (Federal and State agencies, water districts, and environmental organizations), but it was certainly not clear when this program started how species recovery could go hand in hand with continued water development. That situation has changed in the last decade as biologists and geomorphologists have begun to understand how native fish utilize various habitats. Since 1993 we have been studying long-term and on-going changes in the geomorphology of the Colorado River in western Colorado and eastern Utah. The area near Grand Junction is of particular interest because of its importance to the razorback sucker and the Colorado squawfish. Although the razorback sucker appears to be on the verge of extinction here, more adult Colorado squawfish are found in the reaches near Grand Junction than in any other portion of the Colorado-Green River system (Osmundson et al., 1995). Observations of radio-tagged adults suggest that they use a wide variety of habitats. Side channels and backwaters are considered important habitats for both adult and larval squawfish, presumably because these areas provide food, lower flow velocities, and refuge from predators (Stanford, 1994). Compared to other reaches of the Colorado River, side channels and backwaters are relatively common in the reaches near Grand Junction, and this may explain why squawfish congregate here. Gravel bars, which are important as spawning habitat, are also common in this area. Optimum conditions for spawning include clean, loose gravel particles, and water temperatures of 18°C to 22°C (Tyus and Karp, 1989; Harvey et al., 1994). Movement patterns of adult squawfish, and subsequent capture of larvae, suggest that spawning sites are widely distributed throughout the Grand Junction area (MaCada and Kaeding, 1991). However, it doesn't appear that the fish spawn at the same sites routinely, as is the case on the Yampa River (Harvey et al., 1994). Other habitats frequently used by Colorado squawfish include eddies, pools, and runs (Osmundson et al., 1995). It is not clear why the fish are found in some of these areas, particularly runs, because runs are the least diverse of all habitats; it may be that runs are just the most abundant habitat, rather than the most useful. Whatever the case, habitats used by native fish in the upper Colorado River differ widely in their abundance and function.

The present paper reports on work completed upstream of Grand Junction in the reach between Rifle and Palisade, Colorado. Historically, this reach also supported populations of Colorado squawfish and razorback sucker (Quartarone, 1993), but two diversion dams now prevent the fishes from migrating upstream. The idea of modifying these structures to allow fish passage has been discussed, but the characteristics of the reach have not been studied in detail. The channel is multi-thread in many areas, and consequently habitats formed by side channels, backwaters, and gravel bars are relatively abundant here. Peak flows are regulated by reservoirs upstream, but it is not known whether this has caused important geomorphic changes. There are two questions to answer with respect to this reach's potential for restoration: (1) how much habitat is present at the moment, and (2) what flows will be required to maintain or expand existing habitats? To answer these questions, we first look at how the reach has changed historically by analyzing discharge records and aerial photographs. From there we use field data to evaluate thresholds for coarse-sediment transport and channel change. Our approach is built on the premise that no single habitat serves a more important purpose than another, and that, at this stage, broad-based solutions to the problems of habitat restoration and management are most desirable.

2. STUDY AREA

The present study examines historic changes and existing conditions along an 88 km reach of the Colorado River between Rifle and Palisade, Colorado (Fig. 2.1). The upper segment of the reach is characterized as a broad alluvial valley

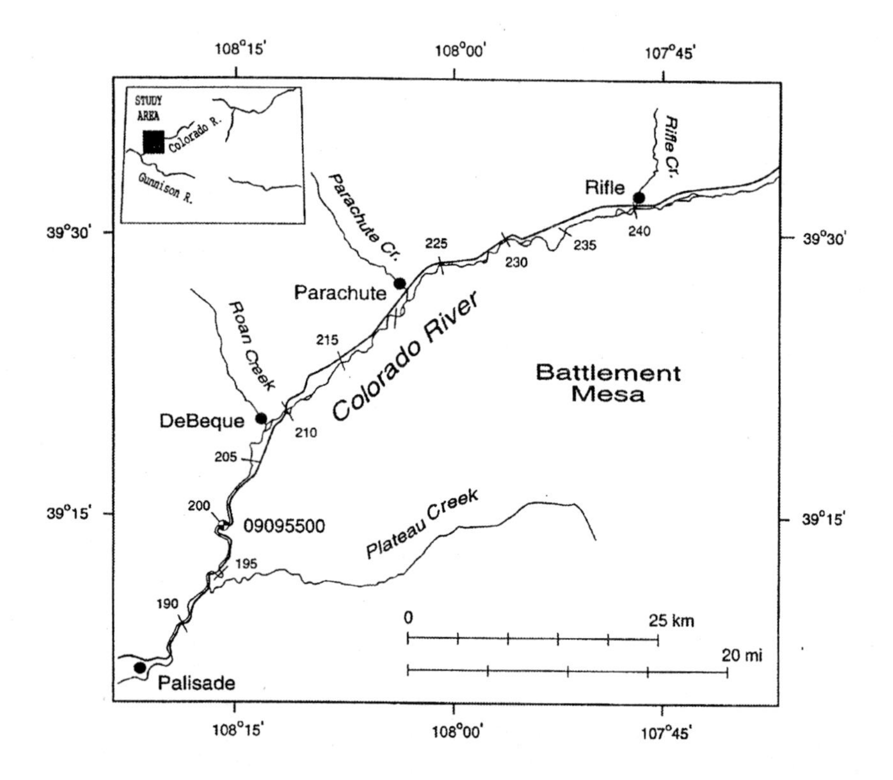

Fig. 2.1. General location map of study area. The number 9095500 indicates the location of the USGS gauging station near Cameo. The smaller numbers indicate river miles (RM) measured upstream from the Green River-Colorado River confluence.

bounded by Pleistocene terraces and pediments. The floodplain is up to one kilometer wide and supports numerous groves of mature cottonwood (*Populus deltoides*). The non-native shrub tamarisk (*Tamarisk chinensis*) is much less abundant in this reach than it is further downstream. Through most of this upper segment the river alternates between single-thread and multi-thread reaches, suggesting it is geomorphically active. Below the town of DeBeque (Fig. 2.1), the river enters DeBeque Canyon, a shallow canyon bounded by sandstone cliffs. A narrow floodplain is present along much of this reach, but in many areas rip-rap has been placed along the banks to protect the highway and railroad lines. The river is gravel- and cobble-bed throughout. The average slope between Rifle and Palisade varies from 0.0020 to 0.0016 and the median grain size of the bed material D_{50} varies from 30 to 60 mm.

Perhaps the most distinctive attribute of the reach between Rifle and Palisade is that here the Colorado River goes through a transition from a relatively clear, cold-water river to a turbid, warm-water river. This transition is due to the decrease in elevation and the greatly increased contribution of dissolved and suspended solids from tributaries such as Parachute Creek and Roan Creek (Fig. 2.1) that drain erodible shale and sandstone bedrock. Between Glenwood Springs and Cameo, Colorado, the annual discharge of the Colorado River increases by only about 70 percent, but the dissolved load increases by over 150 percent (Liebermann et al., 1989), and the suspended sediment load increases by about 500 percent (see below). Thus, most of the runoff in the upper Colorado River is derived from snowmelt in high elevation basins underlain by resistant crystalline rocks, whereas most of the sediment is derived from erodible sedimentary rocks in this area. There are about a dozen small- to moderate-sized dams in the upper basin, and although these dams collectively have significant effects on the annual hydrograph (see below), they are far upstream and have little influence on the amount of sediment supplied to the these reaches.

3. METHODS

In studying the geomorphology of this reach we have taken a general rather than site-specific approach, the rationale being that the endangered fishes have not inhabited the Rifle-Palisade reach for some time, and thus there aren't particular sites that are known to be good for spawning or some other ecological purpose. At this point, our goals are to determine whether there have been significant changes in the geomorphology of this reach over time, and whether the existing channel characteristics can be used to infer something about the reach's suitability as potential habitat for the endangered fishes.

3.1. Flow and Sediment Records

Data from the U.S. Geological Survey (USGS) gaging stations on the Colorado River at Glenwood Springs (station 9072500) and the Colorado River near Cameo (station 9095500) were used to evaluate long-term changes in streamflow and suspended sediment loads. The period of record for these gauges varies as does the amount of sediment data available. The flow record from the Glenwood Springs gauge is relatively long but complicated by the fact that the gauge was located above the Roaring Fork River until 1966, then it was moved downstream. Thus, the more recent record (1966-pres.) includes flow from the Roaring Fork River, whereas the earlier record (1899-1965) does not. However, the two records can be combined by subtracting daily discharges of the Roaring Fork River, which is also gauged, from the same-day discharges of the Colorado River. The discharge record at Cameo begins in 1934 and is continuous through the present. There is another gauge at DeBeque (station 9093700) but this record is short in comparison to the other two and not much different from the record at Cameo, which is less than 20 km downstream. Suspended sediment samples were taken at Glenwood Springs from 1951 through 1965 (Iorns et al., 1964; USGS Water Supply Papers). Suspended sediment samples were also taken at Cameo in the 1950s and then again from 1983 through the present. There were several attempts to measure bedload at DeBeque in 1984 (Butler, 1986), but these samples are very few in number.

3.2. Aerial Photograph Analysis

Changes in the geomorphology of the reach were analyzed using black and white stereo photographs taken in 1937 and 1995. These photographs are 1:32,000 and 1:20,000 in scale, respectively. They were also taken at different times of the year, so the river discharge varied between the two sets of photographs. In 1937 the discharge at the Cameo gauging station was 55 m^3/s, and in 1995 the discharge was 79 m^3/s. Because the higher flow occurs in the more recent set of photographs, the channel might actually appear larger in 1995 than in 1937, even if nothing had changed. It turns out that this is not a particularly important problem in comparison to errors associated with differences in scale and photograph distortion, which we discuss below.

The procedure for analyzing the photographs was to first adjust them to a common scale. This was done by finding four or five common points on each photograph and on a 1:24,000 topographic map, and then establishing the coordinates of these common points. The outlines of the water's edge, islands, emergent bars, side channels and backwaters were then digitized as polygons referenced to specific coordinates. Fig. 3.1 shows an example of how these features were differentiated. Side channels that form where the flow splits around an island or bar were distinguished from the main channel on the basis of their smaller size, although in a few cases this distinction was somewhat arbitrary. Backwaters and other areas of stagnant or slow moving water were often associated with side channels, thus we grouped them as one feature. The digitized images were exported to ARC INFO, a vector-based Geographic Information System (GIS). The GIS software allowed us to further analyze and compare measurements of instream water area, island area, and side channel-backwater area.

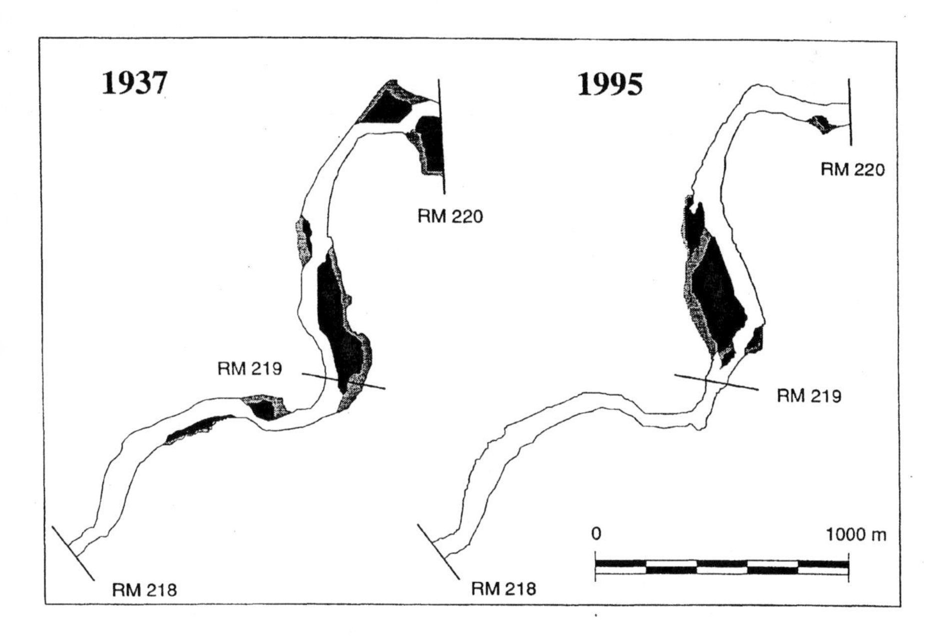

Fig. 3.1. Examples of digitized channel maps showing the outline of the water's edge, side channels and backwaters (shaded areas), and islands (black areas). The discharge at the time of the 1937 photographs was about 55 m^3/s, and in 1995 the discharge was 79 m^3/s.

The accuracy of these measurements is affected to varying degrees by differences in scale, photograph quality, flow level, and amount of distortion. Differences in photograph scale and quality can lead to problems in the interpretation of features and the accuracy with which they can be digitized. Differences in flow level, if large enough, can affect the planform area of the channel. Distortion can make objects appear larger or smaller than they really are, particularly if they are at the edge of a photograph. Using similar photographs and the same techniques, Van Steeter (1996) determined that differences in scale, flow level, and distortion, if additive, could introduce as much as 8% error in the area of large features, such as the main channel, and 16 percent error in the area of small features, such as side channels. If the error was always in the same direction, meaning the size of features was always overestimated in one set of photographs and always underestimated in another, then the errors could be twice as large. Although the latter scenario is highly unlikely, we nonetheless use this as a basis for evaluating whether the measured changes are indeed real.

3.3. Field Studies

During the summer of 1996 we surveyed 31 cross sections at one mile (1.6 km) intervals from river mile (RM) 227 to RM 196 (Fig. 2.1). Cross sections were surveyed with an electronic total station and a motorized rubber raft outfitted with a depth sounder. To survey a cross section, the total station was set up over one of the endpoints, distance readings were taken along the line of the section by targeting a prism on the rubber raft, and individual depth readings were taken off the chart recorder, and relayed by radio to the person on shore.

Samples of the surface and sub-surface bed material were taken every 8 km along the reach. The surface samples were taken with a point count of 200 particles. The sub-surface samples were taken by first removing the surface layer of particles, then collecting approximately 100 kg of the subsurface sediment. The coarse fraction was then sieved in the field, and the fine fraction was brought back and sieved in the laboratory.

3.4. Bedload Transport

Flows that will produce two different phases of bed-load transport (initial motion and significant motion) were estimated from the Shields' parameter or dimensionless shear stress:

$$\tau^* = \frac{\tau}{(\rho_s - \rho)\, g\, D} \tag{3.1}$$

where τ is the boundary shear stress, ρ_s and ρ are the densities of sediment and water, respectively, g is the acceleration due to gravity, and D is the particle size. The onset of bedload transport, and the beginning of the "initial motion" transport phase, occurs at $\tau > \tau_c$, where the subscript c refers to the critical condition. In the initial motion phase very few particles on the streambed are moving and transport rates are very low (Wilcock and McArdell, 1992; Andrews, 1994). This phase is nonetheless very important with respect to maintaining spawning bar habitats because it marks the point at which particles start moving and fine sediment begins to be flushed from the bed (Kondolf and Wilcock, 1996). The lower limit for this phase is $\tau^* \sim 0.030$, where (Eqn. 3.1) is formulated in terms of the local D_{50} and local τ (Wilcock et al., 1996). The "significant motion" transport phase is characterized by continuous movement of particles and much higher transport rates. A lower limit for this phase is not well established, however, Andrews (1994) suggests an approximate value of $\tau^* \sim 0.060$. At values higher than this, transport is so vigorous that the armor layer is completely broken and bed forms start to develop (Wilcock and Southard, 1988; Pitlick, 1992). Whether D_{50} is representative of the entire range of particle sizes is also debatable, but this is perhaps not as important as accurately specifying D and τ. For this analysis we use the average of several measurements of the surface particle size to determine D_{50}. The shear stress was determined from

$$\tau = \rho g H S \tag{3.2}$$

where H is the mean depth at the cross section and S is the reach-average slope determined from topographic maps. Although we would prefer to use local field-measured values of S that was not done in this study. In previous work we used a global positioning system (GPS) to measure slopes in the field, but found that when considering reaches greater than about 1 km in length, slopes measured with the GPS were virtually the same as slopes measured off of topographic maps (Cress, 1997).

It was also of interest to us to examine whether the existing channel geometry (width and depth) bears a relation to the gravel transport thresholds described above. Parker (1979) pointed out the paradox inherent in threshold channel theory that gravel-bed rivers could not both transport bedload and maintain stable widths. This is a paradox because if bedload transport occurs in the region near the banks, particles on the bank will move downward to replace particles on the bed, resulting in bank erosion and channel widening. However, as Parker pointed out, there are many gravel-bed rivers that transport bedload, yet do not appear to be widening. He resolved this paradox by showing that even in a channel of uniform depth there is a lateral diffusion of downstream momentum such that τ in the near-bank region is less than that given by (2). As a result it is possible for τ to exceed τ_c in the center of the channel, and be below τ_c in the region near the banks. This would allow the river to transport bedload without widening. It follows that the bankfull width and depth of the channel would be set by a shear stress somewhat greater than the critical value. Parker showed theoretically that this point would be reached at $\tau \sim 1.2\tau_c$. Subsequently, Andrews (1984) found that the bankfull shear stress τ_b was approximately $2\tau_c$. Whether Andrews' value is generally applicable has yet to be determined, but by knowing the range in τ_b , it is possible to make some estimate of the flows required to widen the existing channel, form new bars and create additional side-channel habitat.

4. RESULTS

4.1. Long-Term Trends in Streamflow

We have shown in previous studies (Pitlick and Van Steeter, 1994; Van Steeter, 1996) that in spite of natural fluctuations in precipitation and runoff, peak and average annual discharges of unregulated rivers in the upper Colorado River basin have not changed significantly this century. However, streamflows of the mainstem Colorado River and a number of other rivers in the basin have clearly been affected by the construction and operation of reservoirs, especially since about 1950. Figure 4.1 presents composite annual hydrographs of the Colorado River near Cameo for pre- and post-development periods. These data show the typical shift in seasonal flow patterns caused by reservoirs--base flows are higher now, and peak flows lower, than in the period prior to water development (Fig. 4.1). Some water is diverted out of the Colorado River basin, but in most years, the amount is small, and consequently the total volume of runoff at this site is only slightly different now than it was before (Van Steeter, 1996).

The hydrologic record of the Colorado River at Glenwood Springs is among the longest in the region, and the time series of annual peak discharges at this site provides a good illustration of how streamflows have changed this century (Fig.4.2a). The relatively high peak discharges in the early part of the century (1900-1929) reflect a regional climate anomaly in which runoff was well above average throughout the upper Colorado River basin (Stockton and

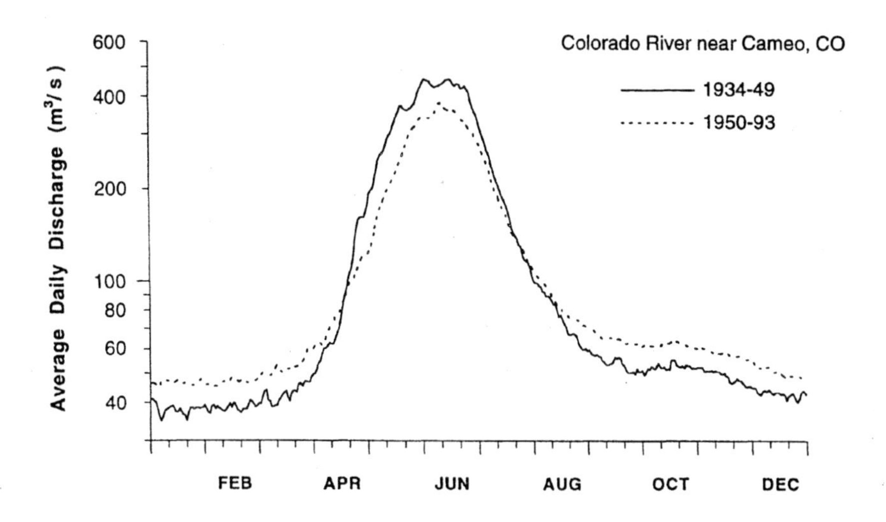

Fig. 4.1. Average annual hydrographs of the Colorado River for periods prior to and after the time (1950) when most of the reservoirs in the upper basin were built.

Jacoby, 1976; Graf, 1992). Pitlick and Van Steeter (1994) noted that the same trends are evident in the peak flow records of the Yampa River at Steamboat Springs, and the East River at Almont, an unregulated tributary of the Gunnison River. Lower peak flows at Glenwood Springs in the middle part of the century (1930-1961) indicate a return to drier conditions and the start of the period when reservoirs begin to affect peak discharges. Three of the main reservoirs in the upper basin were completed between 1943 and 1950. The construction and filling of several more reservoirs in the 1950s and 1960s, and their continued operation through the present, results in further decreases in the average annual peak discharge of the Colorado River (Fig. 4.2a). Peak discharges at Glenwood Springs averaged 381 m^3/s for the period 1930-1961, and 271 m^3/s for the period 1962-1993. To evaluate the significance of this difference, we used a Mann-Whitney test (a nonparametric test employed when the data do not satisfy the assumption of equal variance as required in a t-test), and found that the difference in peak discharge is significant at the 0.025 level. Similar trends are evident in the peak discharge record from the Cameo gauging station (Fig. 4.2b). From 1934 to 1961 peak discharges at Cameo averaged 629 m^3/s, and from 1962 to 1994, peak discharges averaged 494 m^3/s. The change in peak discharge is statistically significant here as well. This is noteworthy considering that the Cameo gauge is 150 to 250 km downstream from most of the upper basin reservoirs. On the basis of these and other results (Van Steeter, 1996), we conclude that reservoirs constructed in the upper basin in the last 50 years have had significant system-wide impacts on the natural flow regime of the Colorado River.

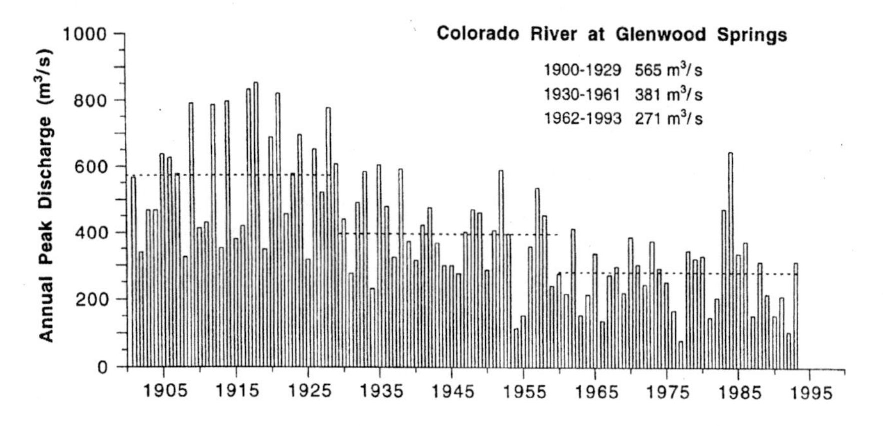

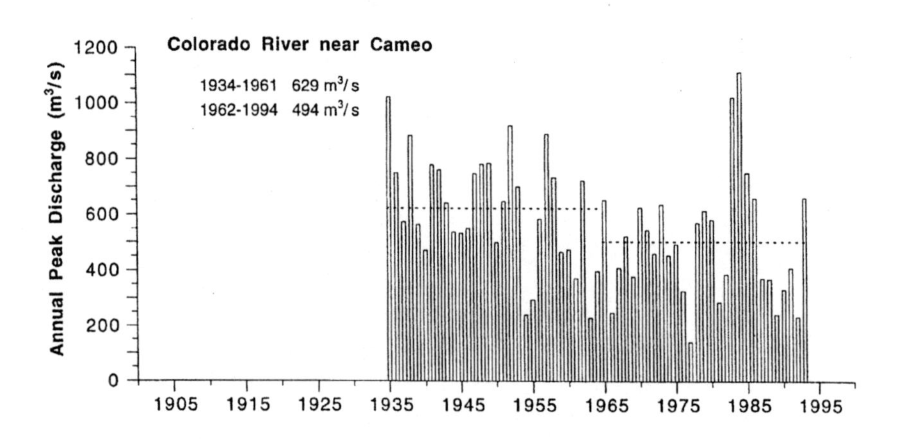

Fig. 4.2. **Annual peak discharges of the Colorado River at Glenwood Springs and the Colorado River near Cameo. Dashed lines indicate averages for individual periods. At both stations the differences in average annual peak discharge are statistically significant at the 0.025 level.**

4.2. Trends in Suspended Sediment Loads

The suspended sediment rating curve for the Colorado River at Glenwood Springs is based on 79 samples taken from 1951 through 1965 (Fig. 4.3a). The relation between suspended sediment concentration and discharge at this station appears to be log-linear with no obvious seasonal trends or hysteresis effects (Fig. 4.3a). These data were fit by eye with the following power function:

$$C_s = 0.10\,Q^{1.5} \tag{4.1}$$

where C_s is the suspended sediment concentration in milligrams per liter and Q is the water discharge in cubic meters per second.

The suspended sediment rating curve for the Colorado River near Cameo is based on 350 measurements made from 1951 to 1953, and from 1983 to 1993 (Fig. 4.3b). We have examined these data separately and found no

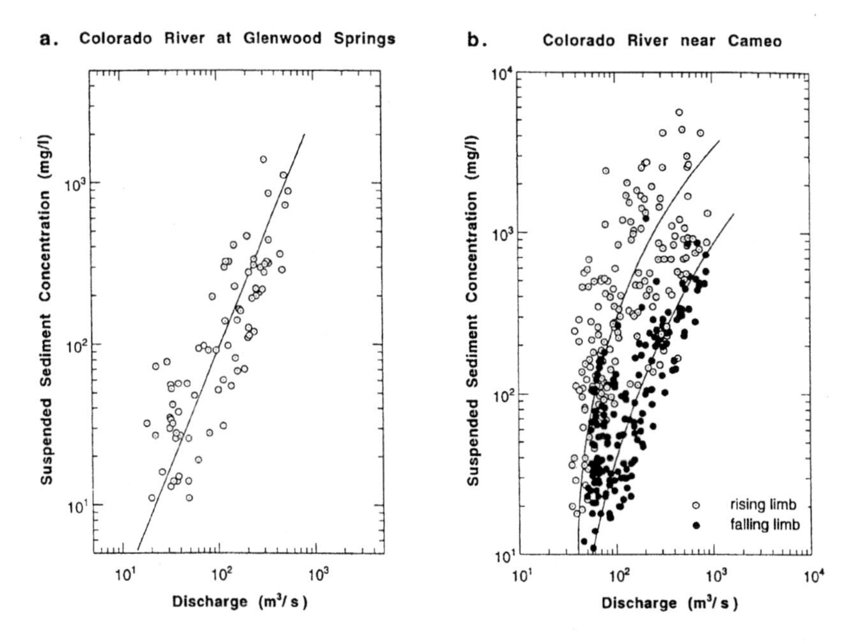

Fig. 4.3. Suspended sediment rating curves for (a) the Colorado River at Glenwood Springs and (b) the Colorado River near Cameo.

obvious difference between the two time periods, which leads us to believe that sediment concentrations have remained essentially stationary over time. Unlike the rating curve at Glenwood Springs, the rating curve at Cameo exhibits a very clear hysteresis, with sediment concentrations being much higher on the rising limb of the hydrograph than on the falling limb (Fig. 4.3b). This trend is presumably related to differences in the sources of water and sediment that enter this reach. Accordingly, these data were fit with the following relations:

rising limb:

$$C_S = \frac{5(Q-12)^4}{Q^3} \tag{4.2a}$$

falling limb:

$$C_S = \frac{(Q-12)^4}{Q^3} \tag{4.2b}$$

Equations 4.1 and 4.2 were used along with measured discharges at the respective stations to calculate daily sediment concentrations and loads for each day in the period from 1934 to 1994. The daily sediment loads were then summed to give the annual load.

Figure 4.4 shows long-term trends in suspended sediment load and average annual discharge of the Colorado River at Glenwood Springs and the Colorado River near Cameo. These trends indicate that, except for the record flow years of 1983 and 1984, suspended sediment loads of the Colorado River have been consistently lower in the last three decades than they were in the three decades prior to about 1960. At Glenwood Springs suspended sediment loads averaged 0.57 x 10^6 Mt/yr for the period 1930-1961 (Fig. 4.4a), and 0.32 x 10^6 Mt/yr for the period 1962-1994. Based on the Mann-Whitney test we conclude that the difference in sediment loads is significant at the 0.025 level. At the Cameo gauging station, suspended sediment loads averaged 2.55 x 10^6 Mt/yr from 1934 to 1961, and 1.81 x 10^6 Mt/yr from 1962 to 1994. The Mann-Whitney test again indicates that the difference in sediment loads is significant at the 0.025 level. For both stations, the difference in average annual discharge between the two time periods is comparatively small (Fig. 4.4). This indicates that most of the difference in sediment loads before and after 1961 is due to reductions in the frequency of high flows caused by reservoir operations.

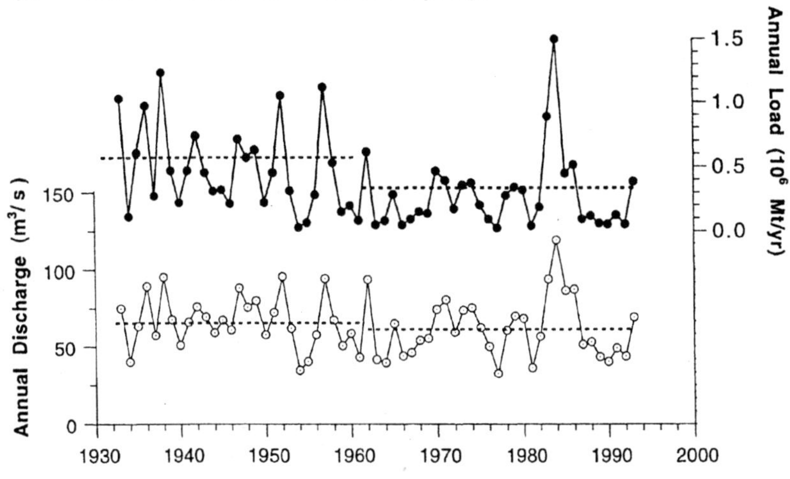

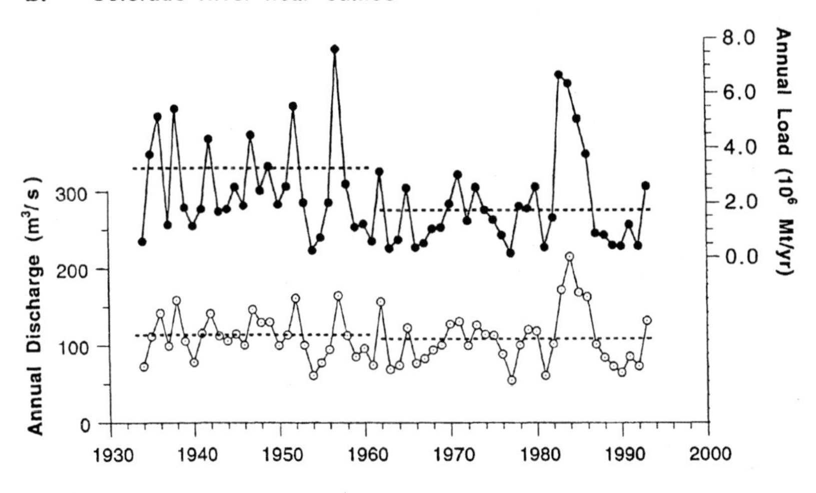

Fig. 4.4. Trends in annual suspended sediment load and discharge of (a) the Colorado River at Glenwood Springs and (b) the Colorado River near Cameo. Dashed lines indicate averages for individual periods. At both stations the differences in average annual suspended sediment load are statistically significant at the 0.025 level.

4.3. Changes in Channel Morphology

Rivers typically respond to reductions in discharge and sediment load by becoming smaller (Schumm, 1969), either as a result of deposition and narrowing in the main channel or filling in of subordinate channels (side channels), which then become part of the floodplain. Distinguishing between these processes is important in this case because the endangered fishes apparently show some preference for backwater and side-channel habitats (Osmundson et al., 1995). Thus, one of our main goals was to determine whether the river and associated side channels have just become smaller over time or whether some of these features have been lost and the river has become less complex. Our analysis of the 1937 and 1995 aerial photographs indicates a clear tendency towards the latter process. Since 1937, the total area of side channels and backwaters in the study reach has decreased by 31 percent, and the total area of islands has decreased by 20 percent (Table 4.1). The change in side-channel/backwater area is very systematic, with 25 out of 39 river mile (RM) reaches showing a decrease in area (Fig. 4.5). The change in island area is less systematic (Fig. 4.5) and due mostly to the loss of several large islands between RM 220 and RM 235. These islands appear to have been cut off from the river and used as gravel pits, probably during the construction of the interstate highway. The planform area of the main channel does not appear to have changed significantly since 1937 (Table 4.1), and individual reaches are equally divided between those that experienced an increase in area and those that experienced a decrease in area (Fig. 4.5).

Table 4.1. Summary of photogrammetric measurements of geomorphic features in the upper Colorado River.

	Total Planform Area		Change in Area (m^2)	Change in Area (%)
	1937	1995		
Main channel	4694000	4550000	-144000	3.1
Islands	3858000	3105000	-753000	19.5
Side channels	1054000	723000	-331000	31.4

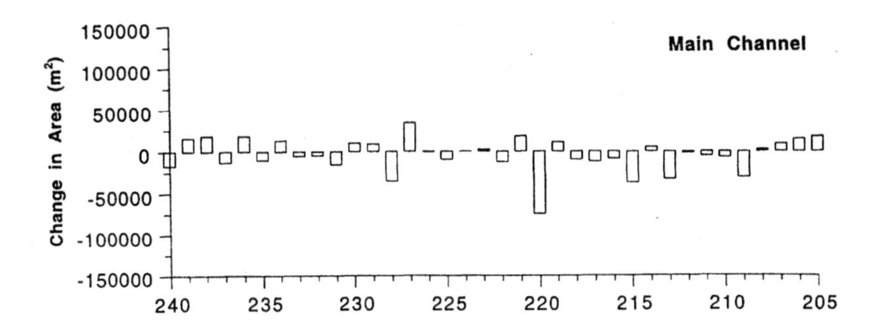

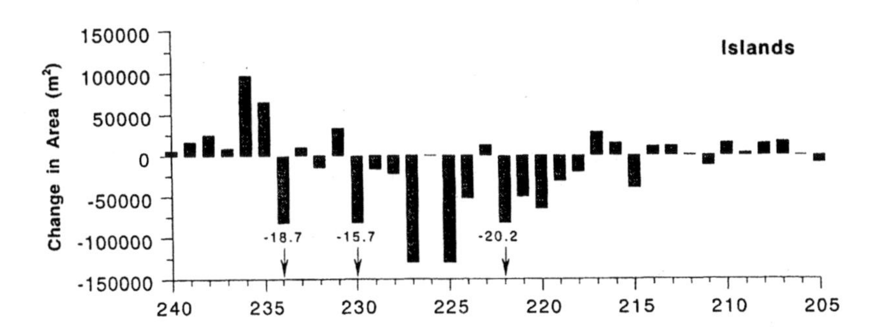

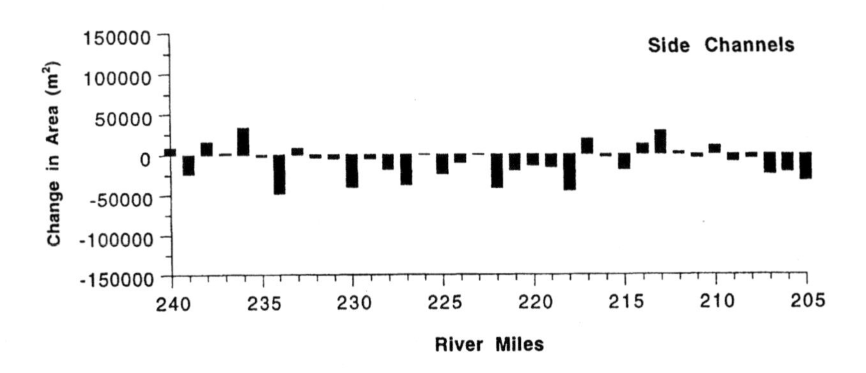

Fig. 4.5. **Changes in main channel area, island area, and side channel area for the period 1937 through 1995 for the reach between Rifle and DeBeque.**

Although there is certainly some error associated with these measurements, the error would have to be very large for the results to be insignificant. Essentially all of the side channels and backwaters would have to be overestimated in the early set of photographs and underestimated in the later set. The difference in discharge between the two sets of photographs is not much of a factor in this case, but even if it was, the higher flow occurs in the more recent set of photographs, so this would make the channel appear larger even if it had not changed. If we took this effect into account, the reduction in side channel area would be slightly greater than what we have indicated. These results suggest that the Colorado River has not only become smaller, but it has also become less complex as a result of side channels filling in. The process by which this occurs is easy to envision. Side channels are characterized by lower flow depths and lower flow velocities, thus even under natural conditions their sediment transport capacity is less than the main channel. If, as we have indicated above, the amount of sediment delivered to these reaches has not changed appreciably, but the river has lost some of its ability to carry this sediment, then whatever it cannot carry will be stored somewhere in the channel. Side channels are the likely sites of storage because they have a lower transport capacity. It is also true that flows through side channels are more ephemeral--side channels are topographically higher than the main channel, and they are not inundated as often. Some side channels may not flow for several years. This allows sediment to build up on the bed, and increases the likelihood that willow or tamarisk will colonize the deposits and permanently stabilize them. Vegetation establishment promotes further deposition until eventually the side channel has filled to the level of the floodplain. It is hard to say whether a 20 to 40 percent reduction in side-channel and backwater habitats is significant as far as the native fishes are concerned, or whether this would limit their re-introduction, but it seems clear that these features have been lost over time and the channel is indeed less complex now than it was in the past.

4.4. Existing Conditions

Our cross section surveys reveal subtle but recognizable trends in the bankfull width and depth of the channel (Fig. 4.6). The bankfull width averages 104 m for the reach as a whole, but appears to decrease from about 110 m near Parachute (RM 227-220, Fig. 4.6a) to about 75 m through DeBeque Canyon (RM 204-196, Fig. 4.6a). In contrast, the bankfull depth appears to increase slightly downstream, from about 2.5 m near Parachute to about 3.0 m through

DeBeque Canyon (Fig. 4.6b). The transitions in width and depth occur at about RM 208, which is near the town of DeBeque and the confluence with Roan Creek, the last major tributary to join the Colorado River before DeBeque

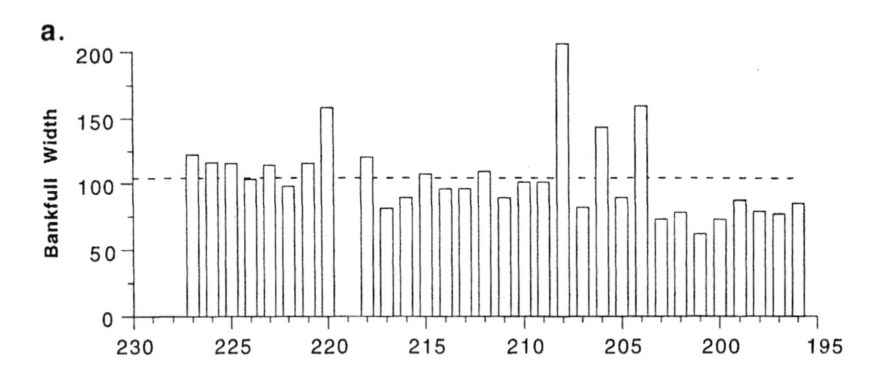

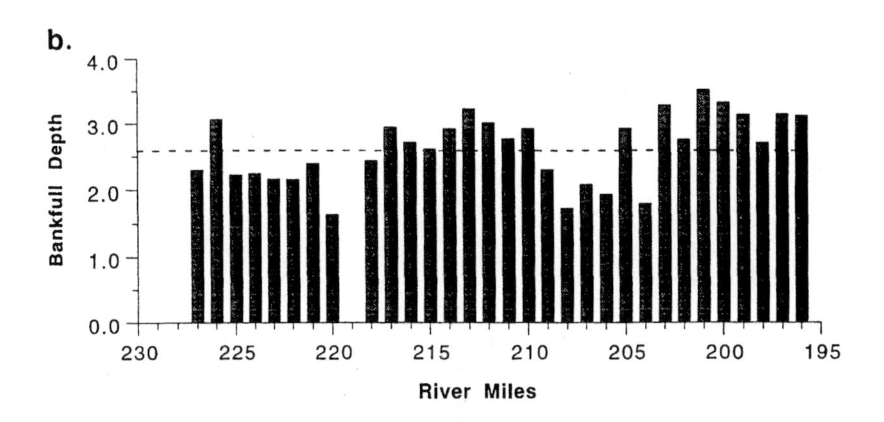

Fig. 4.6. Downstream variations in (a) bankfull width and (b) bankfull depth of the upper Colorado River. Data shown were obtained in the reach from approximately Parachute to the Cameo gaging station.

Canyon. We also note that about here the average gradient decreases from 0.0020 to 0.0016 (Fig. 4.7a). The reason for the change in gradient is not clear, but the decrease in slope more than offsets the increase in depth such that the bankfull dimensionless shear stress τ^*_b through DeBeque Canyon also decreases (Fig. 4.7b). Although this last result might imply that there is a change in sediment transport capacity, we would not attach too much significance to it because first, we see no evidence of aggradation, and second, the apparent difference in τ^*_b could easily be due to inaccuracies in our measurements of either the slope, or the median grain size, or both. In calculating τ^*_b, we used a reach-averaged value of 42 mm for D_{50}, which is about 20% less than what has been observed in reaches with similar gradients

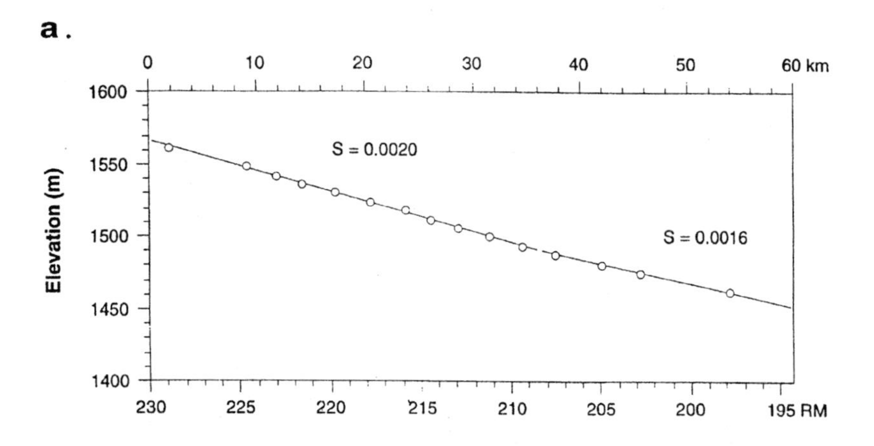

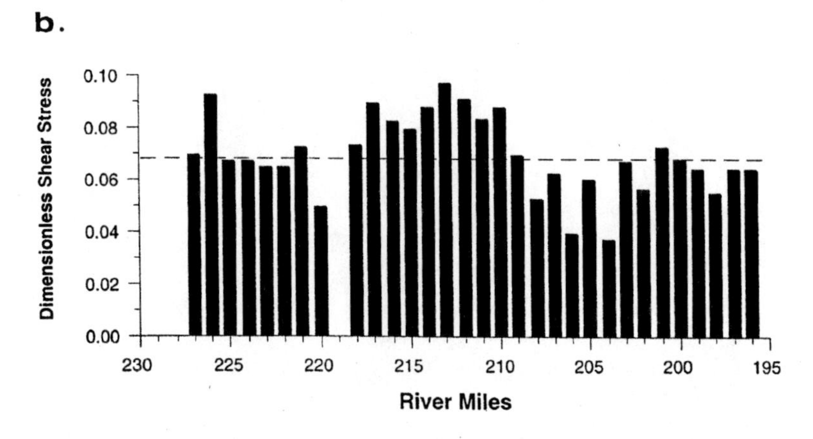

Fig. 4.7. **Plots of (a) the average channel slope and (b) the downstream variation in bankfull dimensionless shear stress τ^*_b of the upper Colorado River. Individual values of τ^*_b are based on field measurements between approximately Parachute to the Cameo gaging station.**

near Grand Junction (Van Steeter, 1996). If the bed material upstream from RM 208 was actually slightly coarser (say, 52 mm), then there would be no difference in τ^*_b between these reaches. As it stands, the reach upstream of RM 208 has an average τ^*_b of 0.073 and the reach below RM 208 has an average τ^*_b of 0.060, or twice the critical dimensionless shear stress.

To estimate the discharge required to initiate bed material transport (which we call the critical discharge, Q_c), we use Eqs. 3.1 and 3.2 to find the critical flow depth H_c, the Manning's equation to estimate the mean velocity U, and field measurements of the channel width W. Assuming uniform flow, we combine the continuity equation and the Manning's equation to get the following expression for the critical discharge:

$$Q_c = \frac{W\,H_c^{5/3}\,S^{1/2}}{n} \tag{4.3}$$

where, n is a roughness coefficient. To estimate the discharge that sets the width and depth of the channel, which we associate with the bankfull discharge Q_b, we followed the same steps, only we used the bankfull depth H_b and a slightly lower value of n . Based on results from previous flow measurements and modeling in reaches near Grand Junction (Van Steeter, 1996), we calculated Q_c using $n = 0.032$, and Q_b using $n = 0.030$.

A plot of individual cross section estimates of Q_c and Q_b is shown in Fig. 4.8. For the reach as a whole, Q_c averages 180 m^3/s, and Q_b averages 700 m^3/s. All but a handful of cross sections fall within 30% of these average

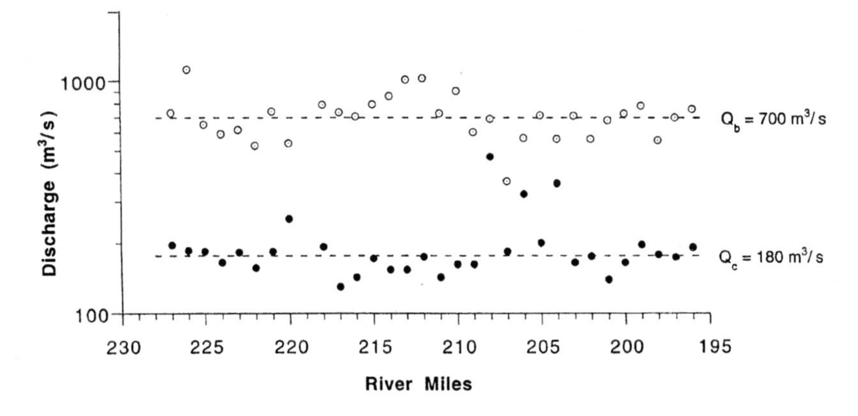

Fig. 4.8. Downstream variations in critical discharge Q_c and bankfull discharge Q_b.

values. The sections that deviate most from the average are anomalously wide or narrow (c.f. Fig. 4.6). Excluding these sites, our calculations indicate that Q_c lies between 120 m^3/s and 240 m^3/s, and Q_b lies between 525 m^3/s and 1025 m^3/s. Again, the variation in Q_c and Q_b might imply that the frequency of bedload transport or bankfull discharge is greater in some places than in others, but we do not think this is the case. If the discrepancies in Q_c (or Q_b) between individual cross sections were very large, then the channel would be aggrading or degrading. We see little evidence in the field that this is happening. Instead it is likely that most of the variation in Q_c and Q_b results from our assumptions about uniform flow, grain size, and slope. Whatever the case, our estimates of Q_c = 180 m^3/s and Q_b = 700 m^3/s give an indication of the flows that would cause widespread bedload transport and channel change. Using the Cameo gauge record for the period 1979-1993 as representative of present conditions, a discharge of 180 m^3/s would be exceeded, on average, about 56 days per year, and a discharge of 700 m^3/s would be exceeded about 3 days per year. The former number indicates that there might be a 2-month window of opportunity, centered around the peak of the hydrograph, when gravel substrates and spawning habitats are maintained in optimum conditions. This window is likely to also include the period of time when water temperatures conducive to spawning are in the range 18-22°C.

4.5. Comparison to Work in Other Reaches of the Colorado River

Conditions within reaches of the Colorado River near Grand Junction are very similar to those described above (Pitlick and Van Steeter, 1994; Van Steeter, 1996). Between 1937 and 1993, the total area of the main channel decreased by 15 percent, and the total area of side channels and backwaters decreased by 26 percent. When proportioned over the total length of the study reach (84 km), these changes equate to a 20 m decrease in the average width of the main-channel and a 7 m decrease in the average width of side-channels. We attribute most of the geomorphic changes in the Grand Junction area to the reduction in sediment loads described here, compounded by similar changes that occurred on the Gunnison River (the Gunnison River contributes about 40 percent of the flow and sediment load of the Colorado River below Grand Junction).

Field observations in reaches near Grand Junction, plus results from hydraulic modeling studies provide more direct information on discharges that initiate gravel transport, scour side channels, and thereby improve certain types of habitats. High snowmelt discharges in 1993 and 1995 produced widespread gravel transport and removed much silt from the channel bed. In both years flows remained near bankfull for several weeks, causing extensive re-working of gravel bars and scour of side channels. Below-average flows in 1994 caused few discernible changes, other than localized deposition of silt. Near-average flows in 1996 produced peak discharges about 20 percent less than the mean annual flood, but still high enough to entrain the bed material. Several times near the peak of the 1996 snowmelt we heard bedload moving across riffles as we floated the river. These observations confirm model-based estimates that discharges of at least 250 m^3/s are required to initiate gravel transport in the 15-mile reach and discharges of at least 400 m^3/s are required to initiate gravel transport in the 18-mile reach (the difference between the two reaches being the contribution from the Gunnison River). The former discharge is slightly higher than the estimated critical discharge for the Rifle-Palisade reach (180 m^3/s). The difference in discharges between reaches could be due to the way the estimates were obtained (using local vs. reach-averaged data), or the difference in channel capacity required because of the contribution from Plateau Creek which joins the Colorado River at the lower end of the Palisade-Rifle reach.

5. SUMMARY AND CONCLUSIONS

Native warm-water fishes of the upper Colorado River now live in an environment that is considerably different than it was a century or even 50 years ago. Irrigation diversions and dams limit their range of movement, non-native fishes compete for the same resources, and a variety of habitats used by the native fishes have been altered because of changes in the hydrology of the river. In this paper we have examined how changes in flow regime and sediment load have affected the geomorphology of a specific reach between Rifle and Palisade, Colorado, and have investigated the reach's suitability for providing additional habitat for the endangered Colorado squawfish and razorback sucker.

The natural flow regime of the upper Colorado River has been affected significantly by the construction and operation of reservoirs. Since the early 1960s, annual peak discharges of the Colorado River at Glenwood Springs have decreased by 28 percent, and annual peak discharges of the Colorado River near Cameo have decreased by 19 percent. The reduction in high flow frequency has in turn caused significant changes in the annual suspended sediment loads of the Colorado River. Since 1964, the average annual suspended sediment load at Glenwood Springs has decreased by 40 percent; at Cameo, the annual load has decreased by 26 percent. Our analysis of aerial photographs of a reach between these two gauges suggests that the principal geomorphic response to the change in sediment loads involved the filling in and abstraction of side channels; the area of the main channel has not changed appreciably since 1937.

Field studies of existing conditions within the Rifle-Palisade reach reveal consistent trends in the hydraulic geometry of the upper Colorado River. Reach-averaged values of the bankfull dimensionless shear stress vary between 0.060 and 0.073, which is about 2 to 2.5 times the critical value. We estimate that a discharge of about 180 m^3/s is required to reach the critical shear stress. Under the present hydrologic regime this discharge is exceeded 15 percent of the time, or about 56 days per year. The bankfull discharge is estimated to be about 700 m^3/s. This discharge is exceeded 3 days per year, or slightly less than 1 percent of the time. The former discharge would be important for flushing fine sediment from the bed and maintaining the quality of gravel substrates used for spawning, whereas the latter discharge would be important for re-working and forming new gravel bars, and expanding the amount of existing habitat.

Although we have indicated that about 30 percent of the side-channel and backwater habitat in this reach has been lost, a significant amount of potential habitat still remains. At present, the area of side channels and backwaters in the Rifle-Palisade reach is roughly equivalent to the amount in the reaches near Grand Junction. Potential spawning bars are likewise relatively abundant in this reach, and as we have shown, these habitats might be maintained in optimum conditions for periods of many weeks. Other native fishes are found in the reach (D. Osmundson, personal communication, 1996), so sources of food for the endangered fishes are available. These results, taken together, suggest that the range and availability of habitats used by adult Colorado squawfish could be expanded significantly by providing access to this reach, and that the idea of modifying the diversion structures to allow fish passage should be given further consideration.

6. ACKNOWLEDGMENTS

This work was supported by the U.S. Fish and Wildlife Service, Colorado River Fishery Project. We would especially like to thank Frank Pfiefer, Chuck McAda, and Doug Osmundson for their advice, and David Lewis for assistance in carrying out the field work. Paul Mosley and an anonymous reviewer provided helpful comments on an earlier version of the manuscript.

7. REFERENCES

Andrews, E.D., 1994. Marginal bedload transport in a gravel-bed stream, Sagehen Creek, California, Water Resources Research 30:2241-2250.

Andrews, E. D., 1984. Bed-material entrainment and hydraulic geometry of gravel-bed rivers in Colorado. Geological Society of America Bulletin 95: 371-378.

Butler, D. L., 1986. Sediment discharge in the Colorado River near DeBeque, Colorado. U.S. Geological Survey Water Resources Investigations Report 85-4266.

Graf, W. L., 1992. Science, public policy, and western American rivers. Transactions of the Institute of British Geographers 17: 5-19.

Harvey, M. D., R. A. Mussetter, and E. J. Wick, 1993. A physical process-biological response model for spawning habitat formation for the endangered Colorado Squawfish. Rivers 4:114-131.

Hibbert, A. R., 1979. Managing vegetation to increase flow in the Colorado River basin. U.S. Forest Service General Technical Report RM-66, Fort Collins.

Iorns, W. V., C. H. Hembree, D. A. Phoenix, and G. L. Oakland, 1964. Water resources of the upper Colorado River basin- Basic Data. U.S. Geological Survey Professional Paper 442.

Kondolf, G. M. and P. R. Wilcock, 1996. The flushing flow problem: Defining and evaluating objectives. Water Resources Research 32:2589-2599.

Laronne, J. B. and H. W. Shen, 1982. The effect of erosion on solute pickup from Mancos Shale hillslopes, Colorado, U.S.A. Journal of Hydrology 59:189-207.

Liebermann, T. D., D. K. Mueller, J. E. Kircher, and A. F. Choquette, 1989. Characteristics and trends of streamflow and dissolved solids in the upper Colorado River basin, Arizona, New Mexico, Utah, and Wyoming. U.S. Geological Survey Water Supply Paper 2358.

McAda, C.W. and L.R Kaeding, 1991. Movements of adult Colorado Squawfish during the spawning season in the upper Colorado River, Transactions American Fisheries Society 120:339-245.

Osmundson, D. B., P. Nelson, K. Fenton, and D. W. Ryden, 1995. Relationships between flow and rare fish habitat in the 15-mile reach of the upper Colorado River. Final Report U.S. Fish and Wildlife Service, Grand Junction.

Parker, G., 1979. Hydraulic geometry of active gravel rivers. Journal of the Hydraulics Division, American Society of Civil Engineers 105:1185-1201.

Pitlick, J., 1992. Flow resistance under conditions of intense gravel transport. Water Resources Research 28:891-903.

Pitlick, J. and M. M. Van Steeter, 1994. Changes in Morphology and Endangered Fish Habitat of the Colorado River. Colorado Water Resources Research Institute Completion Report No. 188, Fort Collins.

Quartarone, F., 1993. Historical accounts of upper basin endangered fish, Upper Colorado River Fishery Project. U.S. Fish and Wildlife Service, Denver.

Schumm, S. A., 1969. River metamorphosis. Journal of the Hydraulics Division, American Society of Civil Engineers 95:255-273.

Schumm, S. A. and G. C. Lusby, 1963. Seasonal variation of infiltration capacity and runoff on hillslopes in western Colorado. Journal of Geophysical Research 68:3655-3666.

Stanford, J. A., 1994. Instream flows to assist the recovery of endangered fishes in the upper Colorado River basin. U.S. Department of Interior National Biological Survey Biological Report 24, Washington, D. C.

Stockton, C. W. and G. C. Jacoby, 1976. Long term surface water supply and streamflow trends in the upper Colorado River basin based on tree ring analysis. Lake Powell Research Project Bulletin 18, University of Arizona, Tucson.

Tyus, H. M. and C. A. Karp, 1989. Habitat and streamflow needs of rare and endangered fishes, Yampa River, Colorado. U.S. Fish and Wildlife Service Biological Report, 89(14).

U.S. Fish and Wildlife Service, 1987. Recovery implementation program for endangered fish species in the upper Colorado River basin. U.S. Fish and Wildlife Service Final Report, Denver.

Van Steeter, M. M., 1996. Historic and current processes affecting channel change and endangered fish habitats of the Colorado River near Grand Junction, Colorado. Unpublished Ph.D. Thesis, University of Colorado, Boulder.

Wilcock, P. R. and J. B. Southard, 1988. Experimental study of incipient motion in mixed-size sediment. Water Resources Research 24:1137-1151.

Wilcock, P.R. and B.W. McArdell, 1993. Surface-based fractional transport rates: Mobilization thresholds and partial transport of a sand-gravel sediment. Water Resources Research 29:1297-1312.

Wilcock, P. R., A. F. Barta, C. C. Shea, G. M. Kondolf, W. V. G. Matthews, and J. Pitlick, 1996. Observations of flow and sediment entrainment on a large gravel-bed river. Water Resources Research 32:2897-2909.

A COMPARISON OF ONE- AND TWO-DIMENSIONAL HYDRODYNAMIC MODELS FOR EVALUATING COLORADO PIKEMINNOW SPAWNING HABITAT, YAMPA RIVER, COLORADO

R. A. Mussetter[1], M. D. Harvey[1], L. W. Zevenbergen[2], and R.D. Tenney[3]

ABSTRACT

Colorado pikeminnow spawning habitat formation and maintenance were investigated with one-dimensional (1-D) and two-dimensional (2-D) hydrodynamic models, HEC-2 and RMA-2V, respectively, at Mathers Hole in the lower Yampa River Canyon. The cross-section based, 1-D model adequately represented the overall reach hydraulics and the macro- and meso-scale requirements of a previously developed three-level physical process-biological response model (PRM) (Harvey et al., 1993). However, the micro-scale requirements were not as well represented. In contrast, the finite element-based, 2-D model represented the hydrodynamic requirements for all three levels of the PRM. The 2-D model predicted that optimum hydrodynamic conditions for spawning (i.e., bar cobbles near incipient motion) occurred between 14 and 128 cms, a range of discharges where spawning fish have been captured at the site.

KEYWORDS

Fish spawning habitat, hydrodynamic model, hydraulics, one- and two-dimensional modeling

1. INTRODUCTION

Historically, the cyprinids (Colorado pikeminnow, humpback chub, bonytail chub, roundtail chub) and catostomids (razorback sucker, flannel mouth

[1] Mussetter Engineering, Inc., P.O. Box 270785, Fort Collins, CO 80527.

[2] Ayres Associates, P.O. Box 270460, Fort Collins, CO 80527.

[3] Colorado River Water Conservation District, P.O. Box 1120, Glenwood Springs, CO 81602.

sucker) were the dominant native fishes in mainstream habitats of the Colorado River Basin, and were widely and abundantly distributed in all of the major rivers (Colorado, Gunnison, Green, White, and Yampa) in the basin (Tyus, 1986; Tyus and Karp, 1989). With the exception of the roundtail chub and flannelmouth sucker, all of the above-named species are now threatened with extinction due to the combined effects of habitat loss, regulation of flows, proliferation of non-native competitors, and other man-induced disturbances (Tyus, 1992).

Ongoing efforts are underway to better understand the habitat requirements for these species, with the intent to identify means of recovering the species. Although all of these species have been designated as federally-listed endangered species, more is known about the individual life-stage habitat requirements of the Colorado pikeminnow (*ptychocheilus lucius*) than about the other listed species (Tyus and Karp, 1989). Therefore, evaluations of the flow requirements for habitat formation and maintenance for individual life stages are more easily undertaken for the pikeminnow (Harvey et al., 1993). Establishment of physical process-biological response models should be carried out in the least hydrologically modified setting where the fish are most likely to be occupying the optimum habitat (Tyus, 1992). The Yampa River Basin, located in northwestern Colorado, is the least hydrologically modified in the upper Colorado River Basin and, therefore, is the best suited to evaluate flow-habitat relations (Tyus, 1992; Harvey et al., 1993).

Adult Colorado pikeminnow are distributed in the mainstem Yampa River from its mouth at the confluence with the Green River upstream for a distance of about 200 km. A major Colorado pikeminnow spawning migration to a limited number of sites within the Lower Yampa Canyon occurs in May and early June as adult fish migrate from the White, Green, and Upper Yampa Rivers (Tyus and McAda, 1984) (Fig. 1.1). Spawning occurs on the falling limb of the annual snowmelt hydrograph when the water temperature is between 17°C and 27°C. Depending on the magnitude of the runoff discharge, spawning can occur as early as June in low-water years and as late as August in high-water years (Tyus and Karp, 1989). For successful spawning, a clean, fines-free, cobble substrate is required (Hamman, 1981), and the substrate must remain stable and fines-free for a period of about 2 weeks between laying and fertilization of the eggs and larval emergence.

A three-level physical process-biological response model (PRM) for spawning habitat formation was developed from field measurements, 1-D hydraulic modeling (HEC-2), and analysis of incipient motion and sediment

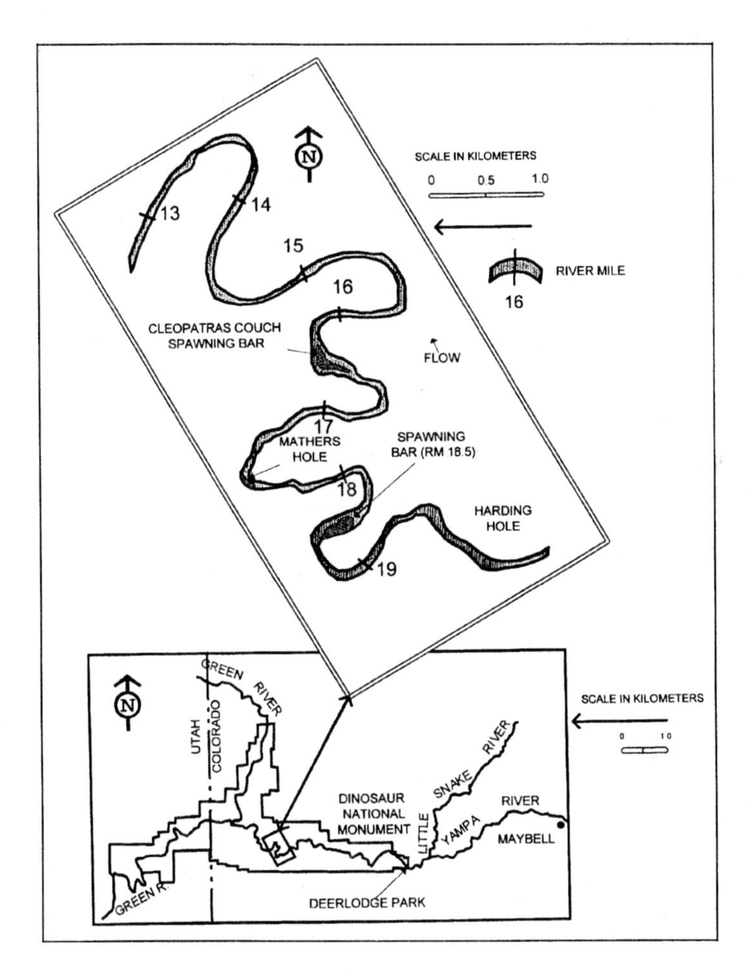

Fig. 1.1. General and detailed location maps of the Yampa River and the pikeminnow spawning bars in the lower Yampa Canyon, Dinosaur National Monument, Colorado.

transport at a known pikeminnow spawning bar (Cleopatras Couch), located at river mile (RM) 16.5 (RK 26.6) in the lower Yampa Canyon (Harvey et al., 1993). (The locations along the Yampa River that are discussed in this paper have historically been referred to by their river mile (RM). Due to their familiarity to those involved in the Upper Colorado River Recovery Program, reference to these sites will continue to be by RM through the remainder of the paper.) The model was subsequently verified at three other known spawning sites: (1) Mathers Hole at RM 17.5 (RK 28.2), (2) an unnamed site at RM 18.5 (RK 29.8) (both are in the lower Yampa Canyon) (Fig. 1.1) (Mussetter and Harvey, 1994), and (3) Three Fords West in Desolation Canyon on the Green River (Harvey and Mussetter, 1994). The model requires that, at a macro-scale level, the river must be confined so that high discharges do not disperse over a floodplain, and the geometry of the reach must result in backwater at higher flows. On a meso-scale, the backwatered and confined higher flows must cause a primary bar to be formed from a wide range of sediment sizes (silts to cobbles) that are in transport at those higher discharges. Increased local hydraulic gradient over the bar during the flow recession as a result of reduced downstream tailwater must result in dissection of the bar and creation of smaller-scale bars (tertiary bars) in the chute channels that are composed of fines-free gravels and cobbles. Finally, at a micro-scale level, the interactions between the hydraulic energy of the recessional flows and the tertiary bar sediments must produce a sand-free, clast-supported deposit in the secondary bars whose constituent cobbles are at a condition of incipient motion, defined by a dimensionless grain shear stress (τ'_*) between 1.0 and 1.5. The clast-supported nature of the spawning sediments is maintained by trapping of sand in the pool upstream of the primary bar, where the flow velocities are low at the range of lower discharges when spawning takes place.

The use of a 1-D, step-backwater, hydraulic model (HEC-2; USACOE, 1990) to describe the hydrodynamic conditions in a complex multidimensional situation at the spawning bars has been questioned (Stanford, 1993). Sufficient topographic data were available at the Mathers Hole site to permit development of a finite-element 2-D hydrodynamic model (RMA-2V; BYU, 1995) of the reach. The principal purpose of this paper is to present a comparison of the utility of the 1-D and 2-D models for evaluating the formation of Colorado pikeminnow spawning habitat.

2. DESCRIPTION OF MATHERS HOLE

Mathers Hole is located at RM 17.5 (RK 28.2) in a very sinuous (sinuosity = 2.6), bedrock- (Weber Sandstone) incised meandering reach of the Lower Yampa Canyon within Dinosaur National Monument in Colorado (Fig. 1.1). The left side of the canyon within the approximately 1,000-m-long project reach is composed of a 150-m-high Weber Sandstone cliff, and the right side of the canyon is composed of a strath terrace (Weber Sandstone overlain by coarse-grained fluvial sediments) that is about 12 m high. The strath terrace was not overtopped during the 1984 flood of record (915 cms). The coarse-grained (tertiary) spawning bar is located immediately upstream of a very sharp bend in the canyon (Lisle, 1986; O'Connor et al., 1986) and is attached to the left bank, which is the outside of the bend. The geomorphology, low-flow topography, and sedimentology of the project reach, as well as the locations of the 15 surveyed cross sections, are shown on Fig. 2.1. In addition to the cross sections, a detailed topographic survey of the entire reach was conducted. The median size (D_{50}) of the surface sediments ranges from about 100 mm at the head of the bar to about 60 mm at the downstream end. The pikeminnow spawning site is located just downstream of cross section 10, and the D_{50} of the sediments that compose the tertiary bar is 84 mm. Details of the hydrology of, and sediment supply to, the lower Yampa Canyon are provided in Harvey et al. (1993). Representative daily peak flows for dry, average, and wet years are 198, 312, and 283 cms, respectively, and the annual sediment transport through the canyon is estimated to be about 2 million tons (Elliott et al., 1984).

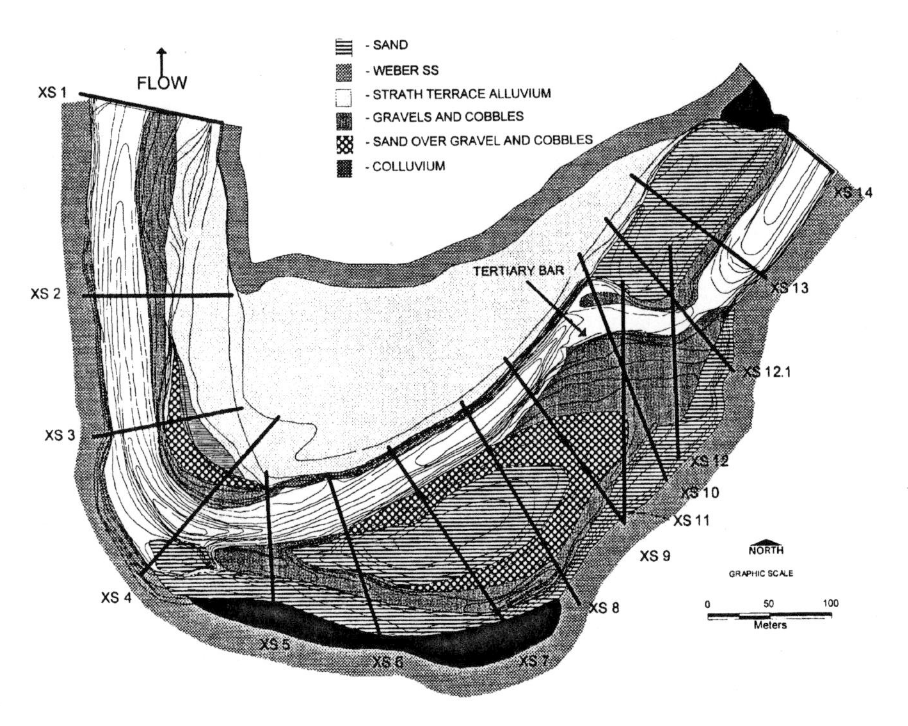

Fig. 2.1. Geomorphic map of the Mathers Hole study reach showing locations of surveyed cross sections and the tertiary spawning bar.

3. ONE-DIMENSIONAL HYDRAULIC ANALYSIS

The original hydraulic and incipient motion analyses for the Mathers Hole spawning bar reach were performed using the 1-D HEC-2 step backwater computer program (Mussetter and Harvey, 1994). The HEC-2 model was developed from 15 surveyed cross sections (Fig. 2.1), and was calibrated to the measured low-flow (~8.5 cms) water-surface profile, 1993 high-water marks (498 cms), and historic high-water marks that correspond to the 1984 flood of record (915 cms). The critical shear stress for a given discharge was estimated from the Shields (1936) relation:

$$\tau_c = \tau_{*_c} (\gamma_s - \gamma) D_{50} \qquad (3.1)$$

where τ_c is the critical shear stress, τ_{*c} is the dimensionless critical shear stress, γ_s is the unit weight of sediment (26,000 N/m^3), γ is the unit weight of water (9,810 N/m^3) and D_{50} is the median particle size. When the critical shear stress for the median size is exceeded, the bed is mobilized and all sizes up to about 5 times the median size are capable of being transported by the flow (Parker et al., 1982; Andrews, 1984). Reported values for τ_{*c} for the median size of the bed material range from 0.03 (Meyer-Peter and Muller, 1948; Neill, 1968) to 0.06 (Shields, 1936), but a value of 0.047 is generally used in engineering practice (Meyer-Peter and Muller, 1948). The lower value of 0.03 indicates true incipient conditions (Parker et al., 1982) while the higher value of 0.047 defines a low, but measurable, transport rate (Neill, 1968).

The bed shear stress due to grain resistance (τ') is a better descriptor of near-bed hydraulic energy and is, therefore, used in the incipient motion analysis (Einstein, 1950; Mussetter, 1989). The grain shear stress is estimated from the relation:

$$\tau' = \gamma Y' S \qquad (3.2)$$

where Y' is the portion of the total hydraulic depth associated with grain resistance (Einstein, 1950) and S is the energy slope. The value of Y' is computed by iteratively solving the semilogarithmic velocity profile equation:

$$\frac{V}{V'_*} = 5.75 + 6.25 \log \left(\frac{Y'}{k_s}\right) \qquad (3.3)$$

where V is the mean velocity, k_s is the characteristic roughness height of the bed (3.5 D_{84}; Hey, 1979), and V'_* is the shear velocity due to grain resistance given by:

$$V'_* = \sqrt{g\,Y'\,S} \tag{3.4}$$

For purposes of evaluating incipient motion, it is convenient to define the dimensionless grain shear (τ'_*), which is the ratio of the grain shear stress (τ') to the critical shear stress (τ_c):

$$\tau'_* = \frac{\tau'}{\tau_c} = \frac{\gamma\,Y'S}{\tau_{*c}\,(\gamma_s - \gamma)\,D_{50}} \tag{3.5}$$

If $\tau'_* < 1$, the shear stress is insufficient to mobilize the bed material, and if $\tau'_* > 1$, bed mobilization is indicated. Optimum conditions for spawning at the tertiary bar occurs when $1 < \tau'_* < 1.5$ using τ_{*c} of 0.03 (Harvey et al., 1993). Significant sediment transport occurs when $\tau'_* > 2$ using τ_{*c} of 0.03.

Four cross sections, (3, 6, 10, and 14) (Fig. 2.1) can be used to demonstrate the hydraulic characteristics of the Mathers Hole spawning bar reach. Conveyance weighting was used to estimate the variation in hydraulic conditions across each of the cross sections. Cross section 3 is located in a pool downstream of the sharp bend that causes upstream backwater at higher discharges. Incipient conditions ($1 < \tau'_* < 1.5$) occur at discharges between 153 and 599 cms, but the shear stress continues to increase as the discharge increases (Fig. 3.1). Cross section 6 is located in a pool immediately upstream of the sharp bend (Fig. 2.1). Incipient conditions occur at discharges between 156 and 291 cms, but shear stress does not increase significantly above 283 cms because of the backwater caused by the sharp bend (Fig. 3.2). Deposition occurs on the bar at discharges higher than about 283 cms.

Cross section 10 is located towards the downstream end of a riffle, and the hydraulics at this cross section were used to represent the tertiary bar which is located about 15 m downstream (Fig. 2.1). At this cross section, incipient conditions ($1 < \tau'_* < 1.5$) occur at all discharges above 18 cms (Fig. 3.3). The data in Fig. 3.3 demonstrate the backwater effect at the higher discharges, but they also indicate that significant sediment transport ($\tau'_* > 2$) does not occur on the riffle, which is a very unlikely condition. It is, therefore, apparent that the conveyance weighted, 1-D model results do not accurately represent the local hydraulics across the cross section at higher discharges at this location.

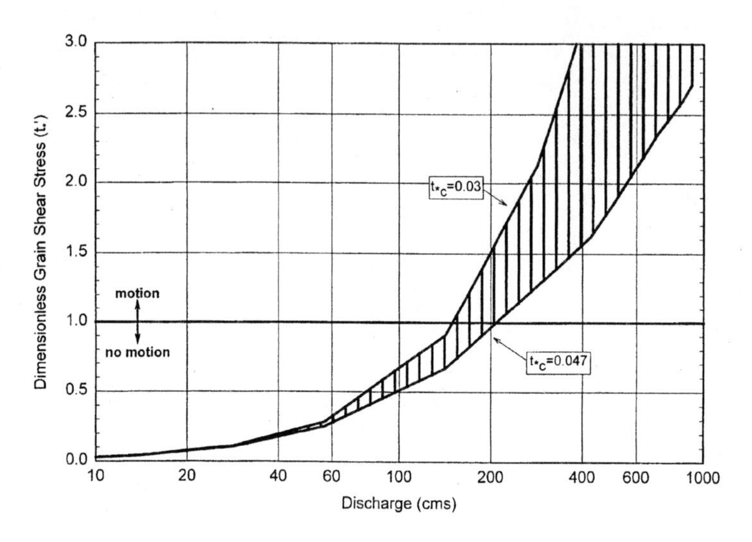

Fig. 3.1. **Variations in dimensionless grain shear stress (τ'_*) at Cross Section 3, for two values of dimensionless critical shear stress (τ_{*c}) at discharges ranging from 8.5 to 915 cms.**

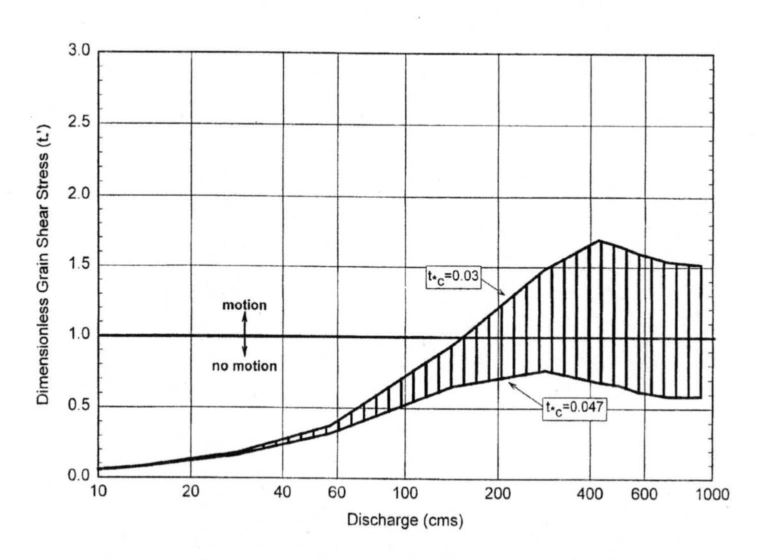

Fig. 3.2. **Variations in dimensionless grain shear stress (τ'_*) at Cross Section 6, for two values of dimensionless critical shear stress (τ_{*c}) at discharges ranging from 8.5 to 915 cms.**

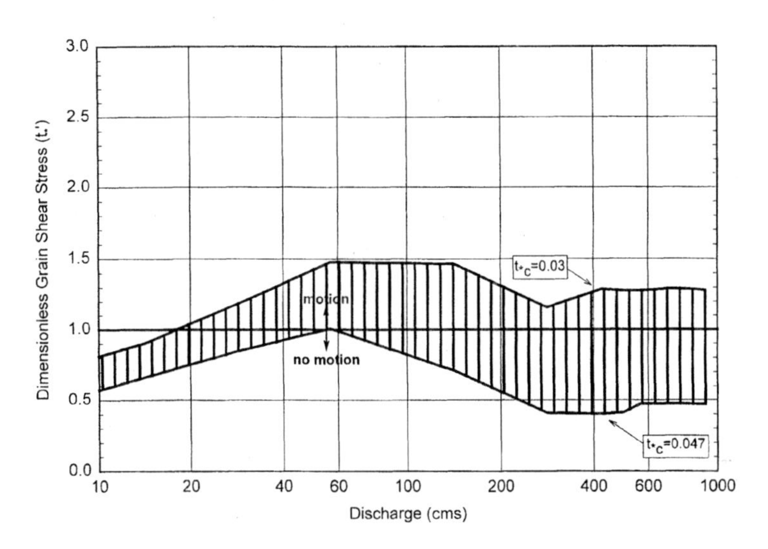

Fig. 3.3. Variations in dimensionless grain shear stress (τ'_*) at Cross Section 10, for two values of dimensionless critical shear stress (τ_{*c}) at discharges ranging from 8.5 to 915 cms.

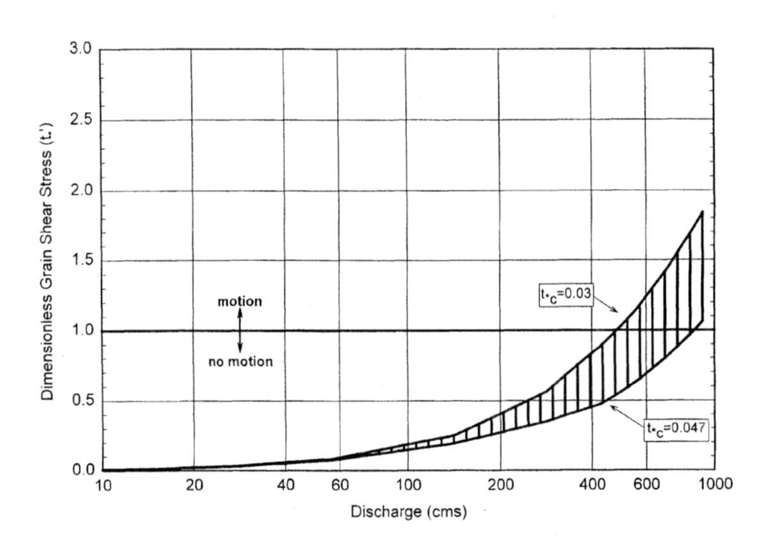

Fig. 3.4. Variations in dimensionless grain shear stress (τ'_*) at Cross Section 14, for two values of dimensionless critical shear stress (τ_{*c}) at discharges ranging from 8.5 to 915 cms.

Cross section 14 is located in a pool at the upstream end of the reach (Fig. 2.1). Incipient conditions occur at discharges between 482 and 859 cms, and the shear stress increases monotonically with increasing discharge (Fig. 3.4). Even though the analysis indicates that the bed material in the pool will be mobilized at a relatively infrequent discharge (approximately 20-year event), sediments already in transport that are delivered to the pool from upstream at a lower range of discharges can be transported through the pool and delivered to the spawning bar.

4. TWO-DIMENSIONAL HYDRAULIC ANALYSIS

The 2-D RMA-2V model uses elements and nodes to represent the geometry of the study reach in place of the cross sections that are used in the 1-D model (Fig. 4.1). Each element is assigned a Manning's n flow resistance value that

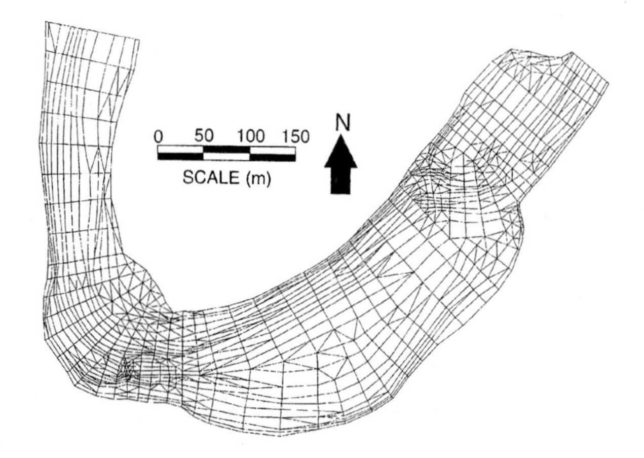

Fig. 4.1. Finite element network for 2-D model of Mathers Hole study reach.

varies with flow depth. The relationship between Manning's n and depth was estimated using Hey's (1979) gravel-bed river flow resistance equation that incorporates the Darcy-Weisbach friction factor and the D_{84} of the bed material (Equation 3.3, with $k_s=3.5D_{84}$). The representative D_{84} used in the analysis for this reach was 150 mm, based on the sediment sampling that was conducted during the field surveys. Output from the model includes the flow depth and depth-averaged vector velocity at each node in the finite element grid. The 2-D model estimates the magnitude and direction of the flow at each point in the grid allowing for the evaluation of flow separation and circulation, and estimation of the local bed shear stresses.

The same water-surface elevations and high-water marks were used to calibrate the 1-D and 2-D models. The predicted water-surface profiles from the two models are similar (Fig. 4.2), especially at discharges below 142 cms, where the flow is confined to the low-flow channel and the 1-D approximation is reasonable for the primary flow path.

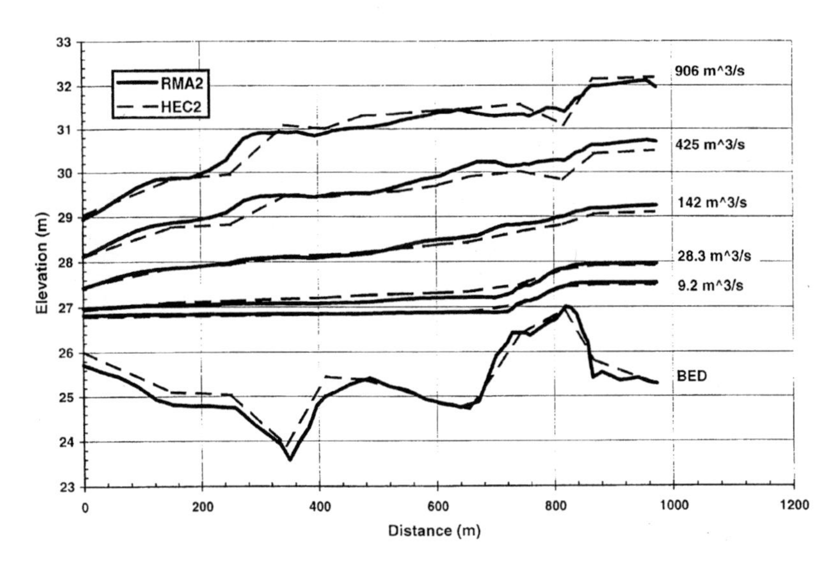

Fig. 4.2. Comparison of water-surface profiles along the Mathers Hole study reach for discharge between 9.2 and 915 cms, as predicted by the 2-D (RMA-2V) and 1-D (HEC-2) models.

The dimensionless grain shear stress was estimated for each node in the finite element model to evaluate the ability of the modeled flows (9.2, 14.2, 28.3, 142, 283, 425, and 915 cms) to mobilize the surface bed material. Dimensionless grain shear stress was computed using:

$$\tau'_* = \frac{\tau'}{\tau_c} = \frac{\gamma Y' S}{\tau_{*c}\ (\gamma_s - \gamma)\ D_{50}} = \frac{0.606\ V^2\ n^2}{\tau_{*c}\ D_{50}\ Y^{1/3}} \tag{4.1}$$

where Manning's n is a function of D_{84} and flow depth, computed using Hey's (1979) equation (Eq. 3.3). A dimensionless critical shear (τ_{*c}) of 0.03 and a representative D_{50} of 84 mm were used in the computations. In interpreting the results that are presented here, it should be recognized that local changes in the bed gradation are likely during the passage of the hydrograph as a result of winnowing of fines from the cobble/gravel matrix or by deposition of material brought into the area from upstream. The actual dimensionless shear stress in areas where the model results indicate that incipient conditions are exceeded by a large amount probably reduces back toward the 1.0 to 1.5 range due to a local increase in the D_{50} that results from the winnowing process.

Figure 4.3 shows the distribution of the dimensionless grain shear stress throughout the project reach, based on the 2-D model results for a

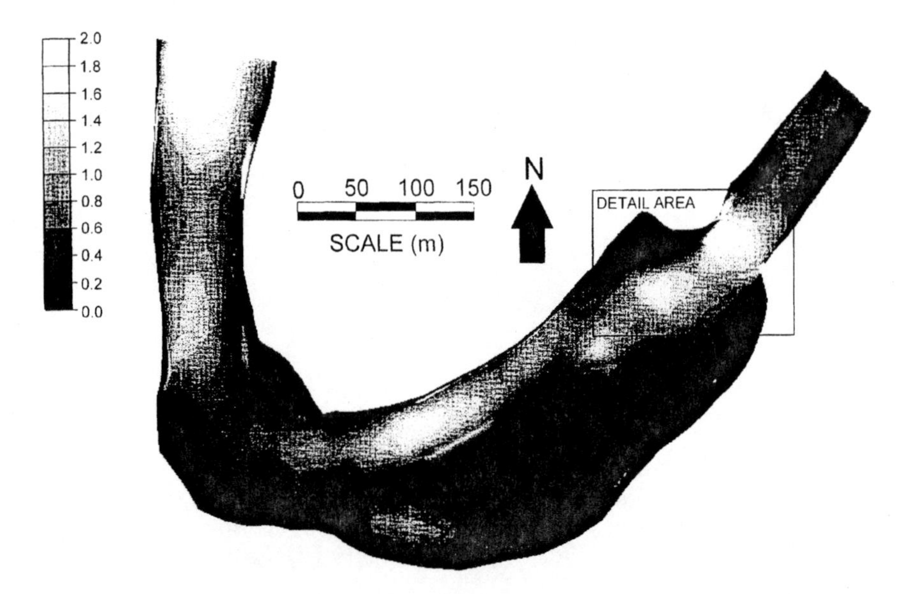

Fig. 4.3. Distribution of dimensionless grain shear stress (τ'_*) in the Mathers Hole reach at 283 cms.

discharge of 283 cms. At this discharge, nearly the entire bar, except for a small portion located between Cross sections 6 and 7, is inundated. Zones of highest shear stress are located downstream of Cross section 3, which is unaffected by downstream backwater, and between Cross sections 9 and 13, which are upstream of the backwater generated by the sharp bend at Cross section 4. The zones of highest shear stress at this discharge generally follow the location of the low-flow channel.

Dimensionless grain shear stress (τ'_*) versus discharge plots were constructed for the same cross sections (3, 6, 10, and 14) that were analyzed in the 1-D analysis so the model outputs could be directly compared. The comparative data for the range of incipient motion conditions ($1< \tau'_* <1.5$) for the four cross sections are presented in Table 4.1. At Cross section 3, downstream of the bend, the critical discharge range (283 to 453 cms) in the 2-D model is both higher and spans a wider range than in the 1-D model (153 to 219 cms). At cross section 6, upstream of the bend, the range of critical discharges and the spans are quite similar for the two models.

Table 4.1. Comparison of One- and Two-Dimensional Model Estimations of Dimensionless Grain Shear Stress (τ'_*) for Four Cross Sections at Mathers Hole.

Cross Section Number	D_{50} (mm)	Discharge (cms) ($\tau_{*c} = 0.03$)			
		($\tau'_* = 1.0$)		($\tau'_* = 1.5$)	
		1-D	2-D	1-D	2-D
3	69	153	283	219	453
6	84	156	119	283	255
10	84	18	20	>915	128
14	84	482	468	859	538

The most significant difference between the two models occurs in the vicinity of cross section 10. The 2-D model indicates that incipient conditions occur between about 20 to 128 cms (Table 4.1), which is also within the range of flows at which spawning pikeminnow have been captured at the tertiary bar located at the downstream end of the riffle represented by cross section 10 (Mussetter and Harvey, 1994). Further, the 2-D model also indicates that significant sediment transport ($\tau'_* > 2$) occurs at a discharge of about 482 cms. In contrast, the 1-D model indicated that significant transport did not take place over the range of modeled flows (Fig. 3.3). At Cross section 14 (upstream pool), the 2-D model indicates that incipient conditions occur over a narrower and lower range of flows (468 to 538 cms) than is indicated by the 1-D model (482 to 859 cms).

The 2-D model enabled a more detailed evaluation of the dynamics of the reach, especially in the location of the tertiary bar used by the pikeminnow for spawning. Contour maps of dimensionless grain shear stress at discharges of 915, 283, 142, 28.3, and 14.2 cms were constructed for the area outlined in Fig. 4.3 between cross sections 9 and 12. At 915 cms (Fig. 4.4), the dimensionless grain shear stress in the low-flow channel, and the area designated as the tertiary bar, are well below critical conditions due to downstream energy losses. Sediment in transport over the primary bar surface where the shear stresses are high ($\tau'_* > 2$) cascades over a slip face into the low-flow channel, thereby depositing on the riffle where the shear stresses are low ($\tau'_* < 1$). At 283 cms (Fig. 4.5), the shear stresses across the primary bar are only slightly above critical, and over the majority of the tertiary bar and the low-flow channel, shear stresses remain below critical because of the downstream backwater effect.

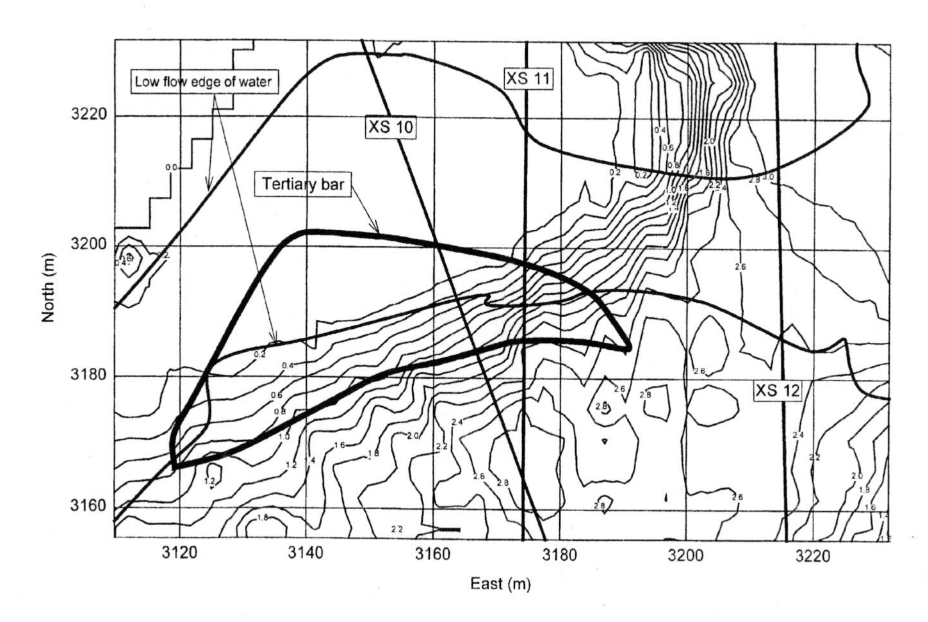

Fig. 4.4. Contour plot of dimensionless grain shear stress (τ'_*) in the vicinity of the tertiary bar at a discharge of 915 cms.

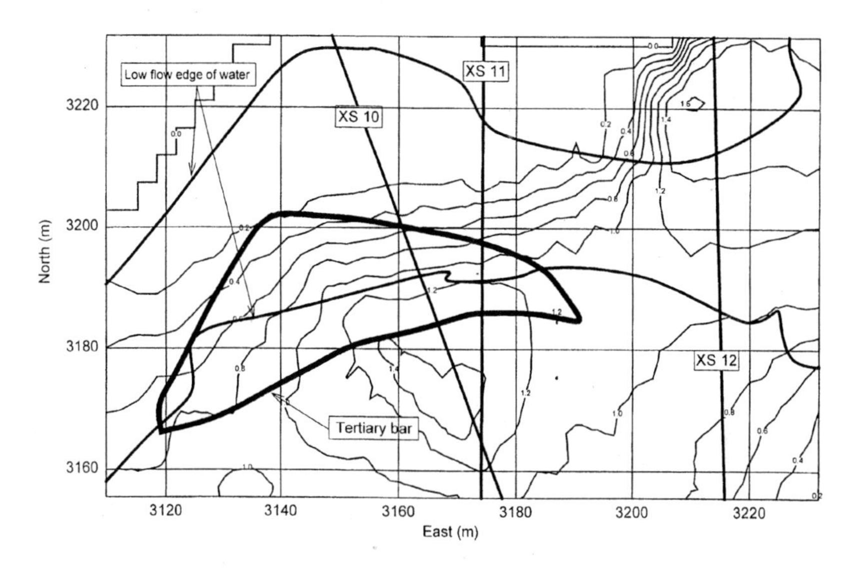

Fig. 4.5. **Contour plot of dimensionless grain shear stress (τ'_*) in the vicinity of the tertiary bar at a discharge of 283 cms.**

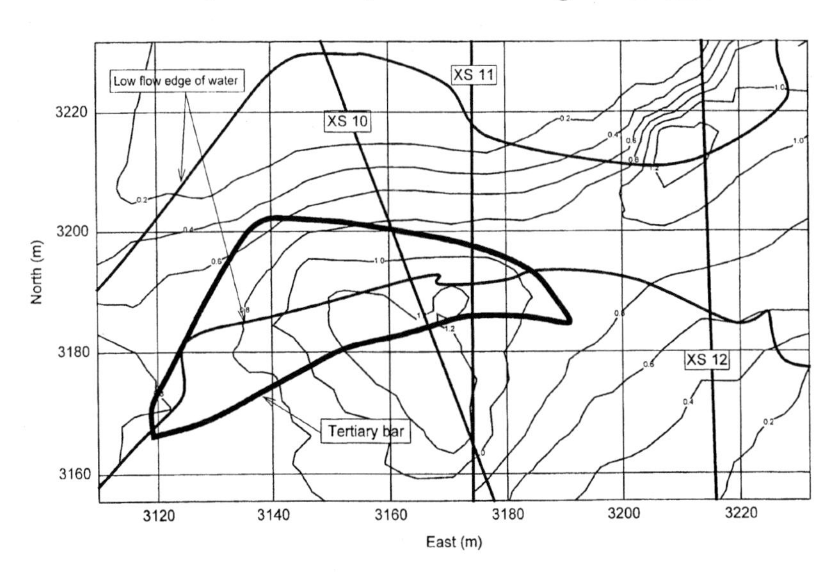

Fig. 4.6. **Contour plot of dimensionless grain shear stress (τ'_*) in the vicinity of the tertiary bar at a discharge of 142 cms.**

At a discharge of 142 cms, the shear stresses over the head of the riffle reduce to near the incipient condition (Fig. 4.6) and in the upstream pool shear stresses drop well below critical (Fig. 3.4), significantly reducing the supply of coarse-grained sediment to the primary bar. The portion of the tertiary bar that is at or above critical conditions, however, expands. The combination of a reduced upstream sediment supply and an expanded zone of mobile material on the tertiary bar provide the initial stage for creation of a fines-free, clast-supported deposit. At 28.3 cms (Fig. 4.7), the upstream supply of sediment is eliminated by the pool, the head of the primary bar is no longer inundated, and the flow path more closely follows the low-flow channel. The head of the riffle is mobile at this discharge, and the zone of above-critical conditions on the tertiary bar migrates towards the edge of the bar, which is believed to be the location at which spawning occurs. At 14.2 cms (Fig. 4.8), only a small portion of the tertiary bar meets the shear stress criterion for spawning ($1<\tau'_*<1.5$), and this correlates with the lower discharge limit for the presence of spawning pikeminnow.

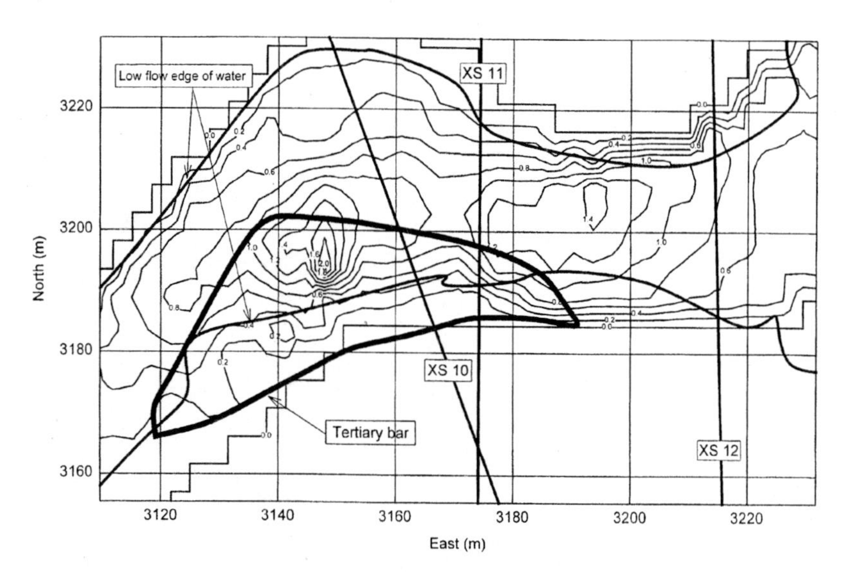

Fig. 4.7. Contour plot of dimensionless grain shear stress (τ'_*) in the vicinity of the tertiary bar at a discharge of 28.3 cms.

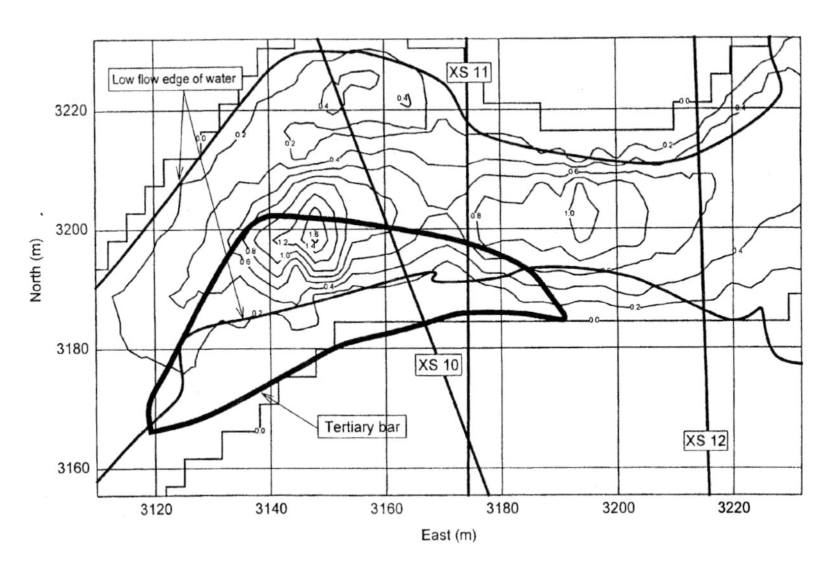

Fig. 4.8. Contour plot of dimensionless grain shear stress (τ'_*) in the vicinity of the tertiary bar at a discharge of 14.2 cms.

5. DISCUSSION

The physical process-biological response model (PRM) that was developed to describe the interactions among the physical setting, hydrodynamics, sediment transport, and spawning behavior of Colorado pikeminnow at known spawning locations in the lower Yampa Canyon was originally based on a 1-D hydraulic (HEC-2) analysis of the reach (Harvey et al., 1993; Mussetter and Harvey, 1994). However, the use of a 1-D model to describe the conditions in localized and complex multidimensional situations has been questioned (Stanford, 1993). Further, a 1-D analysis of spawning conditions at Mathers Hole indicated that the 1-D model may underestimate the shear stresses at key locations within the reach, especially at higher discharges (Fig. 3.3). The 2-D hydrodynamic model (RMA-2V) provides a detailed evaluation of the reach hydraulics, which, in turn, allows for a detailed evaluation of conditions that are responsible for formation of the spawning habitat.

At a reach scale, as exemplified by the comparison of the discharges required to generate critical shear stresses ($1<\tau'_*<1.5$) at four cross sections distributed through the reach (Table 4.1), it is apparent that the 1-D model adequately represents the overall reach hydraulics when compared to the 2-D model (Fig. 4.2). However, the 1-D model fails to adequately predict the shear stresses at the higher discharges at cross section 10, where the hydraulic conditions are very complex (Fig. 3.3). Critical discharges predicted by the 1-D model in the pool upstream of cross section 12 (Fig. 3.4) are higher than those predicted by the 2-D model; whereas, the 2-D model predicts higher critical discharges than the 1-D model in the pool in the downstream portion of the reach at cross section 3 (Fig. 3.1). The hydraulic conditions generated by the 1-D model at cross section 10, located towards the downstream end of the riffle, provide a reasonable representation of the hydraulics of the tertiary bar at the lower flows where the flows are more confined, but the 1-D model does not adequately define the upper range of the required conditions for spawning. The 1-D model does, however, generally support the macro-, meso- and micro-scale requirements of the PRM at the Mathers Hole spawning bar.

The major advantage of the 2-D model is the ability to more accurately represent the local hydrodynamic conditions at a wide range of discharges at specific locations within the overall reach, which, in turn, enables the genesis of the tertiary bar to be evaluated. At very high flows (Fig. 4.4), the entire riffle between the tertiary bar and cross section 12 is below critical conditions, and is likely depositional ($\tau'_* < 1$). At 283 cms, the shear stresses in this reach approach the critical condition (Fig. 4.5). At 142 cms (Fig. 4.6), the shear stresses at the head of the riffle are at, or below, critical indicating that the supply of coarse-grained sediment to the tertiary bar is very low, but the zone of above-critical conditions over the tertiary bar has expanded. At 28.3 cms, the required range of critical shear stresses ($1 < \tau'_* < 1.5$) for spawning are present over much of the tertiary bar (Fig. 4.7), and at 14.2 cms, they are restricted to a small portion of the tertiary bar. The lower limit for spawning at this location probably occurs at about 14.2 cms. The 2-D model indicates that the required range of shear stresses for spawning are present between about 14.2 and 128 cms, a range of discharges where fish in spawning condition have been captured at the site.

The ability to compute shear stresses at individual nodes within the finite-element network provides a means of more accurately evaluating shear stress conditions over the entire area of interest, rather than at the specific cross sections on which the 1-D model is based. Additionally, the 2-D model directly simulates the flow separation and circulation patterns along the reach. The separation and circulation zones are often incorrectly included as effective flow areas in a 1-D analysis. Because the 2-D model results apply at specific locations, the need to use cross-sectionally averaged values, or conveyance weighting techniques where the flow fields are strongly multidimensional, is eliminated. These advantages are offset to some extent by the increased topographic data requirements for generation of the finite-element network. For habitat investigations where knowledge of hydraulic conditions in localized areas is important, a 2-D model is probably the most appropriate analytical tool.

6. CONCLUSIONS

Overall reach hydraulics and the macro-, meso-, and micro-scale requirements of the physical process-biological response model for explaining the formation of spawning habitat for the Colorado pikeminnow can be adequately evaluated by a 1-D backwater model such as HEC-2. However, the 1-D models, which evaluate conditions at individual cross sections, do not describe the hydraulic conditions between cross sections, or the complex local hydraulic conditions at the cross sections, as well as the finite-element-based 2-D model. The genesis of the tertiary bar, and the specific range of discharges required to generate the optimum shear stresses for spawning on the tertiary bar, are much better defined by the RMA-2V model. Provided that adequate topographic data are available to develop the finite-element network, a 2-D hydrodynamic model is, therefore, better suited for habitat investigations.

7. ACKNOWLEDGMENTS

This study was conducted for the Colorado River Water Conservation District as part of their Endangered Species Habitat Investigations program. The authors thank Eric Kuhn for his support and encouragement to publish the results of the investigation.

8. REFERENCES

Andrews, E.D. 1984. Bed material entrainment and hydraulic geometry of gravel-bed rivers in Colorado. *Geol. Soc. America Bulletin 95*. 371-378.

Brigham Young University - Engineering Computer Graphics Laboratory. 1995. FastTABS Hydrodynamic Modeling Reference Manual.

Einstein, H.A. 1950. The bed load function for sediment transportation in open channel flows. U.S. Soil Conservation Service. *Tech. Bull. No. 1026*.

Elliott, J.G., J.E. Kircher, and P.V. Guerard. 1984. Sediment transport in the Lower Yampa River. U.S. Geological Survey. *Water Resources Investigation Report 84--4141*.

Hamman, R.L. 1981. Spawning and culture of Colorado squawfish in raceways, Progressive Fish Culturist. 43(2):173-177.

Harvey, M.D. and R.A. Mussetter. 1994. Green River endangered species habitat investigations. Report to the Colorado River Water Conservation District.

Harvey, M.D., R.A. Mussetter, and E. J. Wick. 1993. A physical process-biological response model for spawning habitat formation for the endangered Colorado squawfish. *Rivers*. 4(2):1-19.

Hey, R. 1979. Flow resistance in gravel-bed rivers. *Journal of the Hydraulics Division.* ASCE. 105 (HY4):14500.

Lisle, T.E. 1986. Stabilization of a gravel channel by large streamside obstructions and bedrock bends, Jacoby Creek, Northwestern California. *Geological Society of America Bulletin.* 97(8):999-1011.

Meyer-Peter, E., and R. Muller. 1948. Formulas for bed load transport. In Proceedings of the 2nd Congress of the International Association for Hydraulic Research. Stockholm. 2:39-64.

Mussetter, R.A. 1989. Dynamics of Mountain Streams. Doctoral Dissertation. Colorado State University, Dept. of Civil Engineering.

Mussetter, R.A. and M.D. Harvey. 1994. Yampa River Endangered Fish Species Habitat Investigations. For Colorado River Water Conservation District.

Neill, C.R. 1968. Note on initial movement of coarse uniform bed material. *Journal of Hydraulic Research.* 6:2:173-176.

O'Connor, J. E., R.H. Webb, and V.R. Baker. 1986. Paleohydrology of pool and riffle pattern development. Boulder Creek, Utah, *Geological Society of America Bulletin.* 97(4):410-420.

Parker, G., P.C. Klingeman, and D.G. McLean. 1982. Bedload and size distribution in paved gravel-bed streams. *Journal of the Hydraulic Division.* ASCE, 108(HY4): 17009.

Shields, A. 1936. Application of similarity principles and turbulence research to bed load movement. California Institute of Technology. Translation from German Original. 167.

Stanford, J.A. 1993. Instream flows to assist the recovery of endangered fishes of the Upper Colorado River Basin: Review and synthesis of ecological information, issues, methods and rationale. For U.S. Fish and Wildlife Service, *FLBS Open File Report 130-93.*

Tyus, H.M. 1992. An instream flow philosophy for recovering endangered Colorado River fishes. *Rivers.* 3(1):27-36.

Tyus, H. M. 1986. Strategies in the evolution of the Colorado squawfish (*Ptychocheilus lucius*). *Great Basin Naturalist.* 46:656-661.

Tyus, H.M. and C.A. Karp. 1989. Habitat use and streamflow needs of rare and endangered fishes, Yampa River, Colorado. U.S. Fish and Wildlife Service, *Biological Report 89.*

Tyus, H.M. and C.W. McAda. 1984. Migration, movements and habitat preferences of Colorado squawfish, *Ptychocheilus lucius*, in the Green, White, and Yampa Rivers, Colorado and Utah. *Southwestern Naturalist.* 29:289-299.

U.S. Army Corps of Engineers (USACOE). 1990. Water Surface Profiles, HEC-2 User's Manual. Hydrological Engineering Center, Davis, California.

Section 5

ENGINEERING GEOMORPHOLOGY

RIVERS BRING GEOMORPHOLOGISTS AND ENGINEERS TOGETHER

Pierre Y. Julien[1]

ABSTRACT

Equilibrium downstream hydraulic geometry and regime equations have fascinated generations of scientists around the world. Professor Stanley A. Schumm, who's retirement is honored with this symposium, combined empirical equations to determine qualitative effects of changes in water and sediment discharges on river width, depth, wavelength and slope. Over the years, the author has developed a method to provide quantitative estimates in hydraulic geometry. The comparison between qualitative and quantitative predictions is presented in this paper.

KEYWORDS

Fluvial geomorphology, river engineers, channel morphology

1. INTRODUCTION

The analysis of downstream hydraulic geometry is of foremost importance in river engineering and the contributions of hundreds of scientists could be referenced. Landmark contributions include those of Bose, Inglis, Lacey, Lane, Blench, Leopold, Schumm and Simons. The recent contributions of many geomorphologists and engineers should be cited along side of Shen, Harvey, Chang, Hey, Thorne, Griffiths, Yang, Richards, Baird, etc.

Dr. Stanley A. Schumm's most outstanding contribution to the understanding of river morphology was perhaps the definition of river adjustments to altered regime conditions. In 1969, he combined empirical equations and provided qualitative guidelines as to how the width, depth, meander wave length and slope of alluvial channels should change under increased/decreased water discharge and sediment input to a river. Schumm's

[1] Professor of Civil Engineering, Colorado State University, Engineering Research Center, Fort Collins, Colorado, 80523

contribution triggered research ideas among engineers seeking ways to derive regime equations from "physical" concepts in order to provide quantitative estimates of changes in hydraulic geometry. This paper promotes a free flow of information between engineers and geomorphologists. The collaboration between fluvial geomorphologists and river engineers is real and effective at Colorado State University. Drs. Stanley A. Schumm and Daryl B. Simons pursued experimental research at the Engineering Research Center leading to innovative ideas in the field of alluvial river mechanics.

Dr. Schumm certainly contributed to better understanding of the complex interaction of natural processes both at a specific river reach and at the entire watershed scale. His analysis of the effects of runoff and type of sediment load in terms of silt-clay content enabled the determination of changes in downstream hydraulic geometry defined as width, depth, slope and meander wavelength as a function of changes in water and sediment discharge.

The objective of this paper is first to review a landmark contribution of Dr. Schumm: how hydraulic geometry changes with water and sediment discharges. The author wishes to show how Schumm's concepts were developed into an engineering analysis of changes in downstream hydraulic geometry. The author suggests a method to determine quantitative changes in hydraulic geometry from changes in discharge, grain size and slope of noncohesive alluvial rivers. Similarities and differences between both governing equations and qualitative results are highlighted.

2. RIVER METAMORPHOSIS

Dr. Schumm (1969, 1971a) stated that during graded and steady time, channel morphology reflects a complex series of independent variables, but the discharges of water and sediment integrate most of the other independent variables. He added that it is the nature and quantity of sediment and water moving through a channel that largely determines the morphology of stable alluvial channels.

Although no good relationship could be determined between channel morphometry and sediment size, Dr. Schumm suggested that good correlations for width and depth were obtained when using an index for the type of sediment load. The index M representing the percentage of silt and clay in the channel was then combined with the mean annual water discharge Q in cubic feet per second to empirically define the channel width W, depth h in feet, the meander wavelength λ in feet and the channel slope S.

$$W = 2.3\frac{Q^{0.38}}{M^{0.39}} \tag{2.1}$$

$$h = 0.6M^{0.34}Q^{0.29} \tag{2.2}$$

$$\lambda = 1890\frac{Q^{0.34}}{M^{0.74}} \tag{2.3}$$

$$S = \frac{60}{M^{0.38}Q^{0.32}} \tag{2.4}$$

From these four empirical relationships, Dr. Schumm (1969, 1971b) investigated the possibility that channels could undergo a complete change in morphology (river metamorphosis) when changes in discharge and sediment load were of sufficient magnitude or exceeded threshold values. For instance, a suspended load meandering channel could be transformed into a braided channel. Arguing that the silt and clay in the channel M reflects the nature of sediment load moving through a channel, Schumm substituted M for $1/Q_s$ to determine the effects of sediment load on hydraulic geometry. The basic equations (2.1) through (2.4) provided a basis for the discussion of natural and man-induced changes in river morphology. The qualitative treatment indicated simply a direction of change, i.e. increase/decrease, rather than the quantitative magnitude of the change. The water and sediment discharges Q and Q_s were then related to the hydraulic geometry in width W, depth h, meander wavelength λ and slope S as

$$Q = \frac{Wh\lambda}{S} \tag{2.5}$$

and

$$Q_s = \frac{W\lambda S}{h} \tag{2.6}$$

To discuss the effects of changing discharge and sediment load on channel morphology, a plus/minus exponent was used to indicate how channel morphology would change as a result of a change in discharge and/or sediment load. The following relationships were then obtained:

$$Q^+ \sim W^+h^+\lambda^+S^- \tag{2.7}$$

$$Q_s^+ \sim W^+h^-\lambda^+S^+ \tag{2.8}$$

$$Q^- \sim W^-h^-\lambda^-S^+ \tag{2.9}$$

$$Q_s^- \sim W^-h^+\lambda^-S^- \tag{2.10}$$

An increase/decrease in discharge changes the dimensions of the channel and its slope, but an increase/decrease in bed material load at constant mean annual discharge changes not only channel dimensions but also the shape of the channel (width-depth ratio) and its sinuosity.

3. QUANTITATIVE ANALYSIS

There is an extensive literature on regime equations, and the reader may refer to Julien and Simons (1984) for detailed acknowledgments of numerous contributions. This presentation hereby focuses on the strength of the collaboration between engineers and geomorphologists at Colorado State University. Inspired by the collaborative work of Simons and Schumm at Colorado State University, Julien (1988, 1989) pursued research to quantitatively determine the changes in downstream hydraulic geometry of noncohesive alluvial channels from the stability of sediment particles under two-dimensional flow conditions. Under steady uniform bankfull flow conditions, the dominant discharge Q is

$$Q = WhU \tag{3.1}$$

where the mean velocity vector U is taken normal to the cross-sectional area.

The power form of the resistance equation enables the user to obtain closed-form relationships from

$$U = b\sqrt{8g}\left(\frac{h}{d_s}\right)^m h^{1/2}S^{1/2} \tag{3.2}$$

where the exponent m increases with decreasing relative submergence. It is clear that the Chezy equation corresponds to m=0 and Manning-Strickler's relationship corresponds to m=1/6.

The downstream bed shear stress τ_θ applied in straight open channels under steady uniform flow conditions is a function of the bed slope S, the mass density of water ρ, and the hydraulic radius R_h=Kh

$$\tau_\theta = K\rho g h S \tag{3.3}$$

For channels with large width-depth ratios, the parameter K approaches unity and the hydraulic radius R_h becomes equal to the flow depth h.

The stability of noncohesive particles in straight alluvial channels is described by the relative magnitude of the downstream shear force and the weight of the particle. The ratio of these two forces defines the longitudinal mobility factor, also called the Shields number τ_θ^*:

$$\tau_\theta^* = \frac{\tau_0}{(\rho_s - \rho) g d_s} \tag{3.4}$$

where ρ_s is the mass density of sediment particles. The critical value of the Shields number, $\tau_{\theta c}^* \approx 0.047$ for the median grain size d_{50}, identifies the beginning of motion of noncohesive particles in turbulent flows over rough boundaries. For values of the Shields number below the critical value ($\tau_\theta^* \leq \tau_{\theta c}^*$), the particles on the wetted perimeter of the alluvial channel are stable. Beyond this threshold ($\tau_\theta^* > \tau_{\theta c}^*$), the particles enter motion and the rate of sediment transport increases with the Shields number. Two significant concepts are associated with the Shields number: (1) the threshold concept described by $\tau_{\theta c}^*$ for the beginning of motion of noncohesive particles; and (2) the concept that beyond the threshold value, the sediment transport rate increases with the Shields number. Since the Shields number depends primarily on flow depth, it is thus associated with the vertical processes of aggradation and degradation in alluvial channels.

In straight alluvial channels without secondary circulation, a cross section is stable when threshold conditions exist simultaneously for all particles located on the wetted perimeter of the channel. A cosinusoidal cross-sectional shape is obtained with the maximum flow depth determined from the critical Shields number $\tau_{\theta c}^*$ and the bank slopes at the free surface inclined at the submerged angle of repose ϕ.

Secondary circulation in curved channels is generated through a change in downstream channel orientation. The streamlines near the surface are deflected toward the outer bank whereas those near the bed are deviated toward the inner bank. The near-bed velocity, the tangential bed shear stress and the drag on the bed particles are commonly directed toward the inner bank.

Flow in bends is analyzed in cylindrical coordinates. The relative magnitude of radial acceleration terms indicates that the centrifugal acceleration is counterbalanced by pressure gradient and radial shear stress, as suggested by Rozovskii (1961). The following formulation for the deviation angle λ is used:

$$\tan \lambda = b_r \left(\frac{h}{d_s} \right)^p \frac{h}{R} \tag{3.5}$$

where the values of p and b_r accommodate a wide spectrum of conditions pertaining to the secondary circulation in alluvial channel bends. For example, Rozovskii's approximation with D=11 corresponds to b_r=11 and p=0.

The downstream hydraulic geometry equations for noncohesive alluvial channels for hydraulically rough turbulent flows were derived by Julien and Wargadalam (1995) after combining (3.1), (3.2), (3.4), and (3.5). Several coefficients were found not to vary significantly and the system of equations reduced to the following with flow depth h in m, surface width W in m, average flow velocity U in m/s and friction slope S as:

$$h = 0.133 Q^{\frac{1}{3m+2}} d_s^{\frac{6m-1}{6m+4}} \tau_\theta^{*\frac{-1}{6m+4}} \tag{3.6}$$

$$W = 0.512 Q^{\frac{2m+1}{3m+2}} d_s^{\frac{-4m-1}{6m+4}} \tau_\theta^{*\frac{-2m-1}{6m+4}} \tag{3.7}$$

$$U = 14.7 Q^{\frac{m}{3m+2}} d_s^{\frac{2-2m}{6m+4}} \tau_\theta^{*\frac{2m+2}{6m+4}} \tag{3.8}$$

$$S = 12.4 Q^{\frac{-1}{3m+2}} d_s^{\frac{5}{6m+4}} \tau_\theta^{*\frac{6m+5}{6m+4}} \tag{3.9}$$

from the equilibrium or dominant flow discharge Q in m^3/s, the median grain size $d_s = d_{50}$ in meters, and the Shields parameter $\tau_\theta^* = K\gamma hS/(\gamma_s - \gamma)d_{50}$, given the resistance exponent m calculated from $m = 1/\ln(12.2\ h/d_{50})$. The details of the deviation can be found in Wargadalam (1993).

4. CALCULATION PROCEDURE

The recommend procedure for the calculation of the downstream hydraulic geometry starts with the user selection of three independent variables. To include the effects of sediment transport, the user may want to calculate four dependent variables: average flow depth h in m, surface width W in m, average flow velocity U in m/s and equilibrium slope S. These are a function of three known independent variables: discharge Q in m^3/s, median grain size ds in m, and dimensionless Shields number τ_θ^*. Equations (3.6) to (3.9) are solved with the five-step procedure outlined below, from Table 4.1.

Table 4.1. Downstream hydraulic geometry as a function of Q (m^3/s), d_s (m) and τ_θ^*.

	Coefficient a	Discharge Exponent b	Grain Size, Exponent c	Shields Number Exponent d
Flow depth h (m)	0.133	$\frac{1}{2+3m}$ $0.28 < b < 0.5$	$\frac{-1+6m}{4+6m}$ $-0.25 < c < 0.28$	$\frac{-1}{4+6m}$ $-0.25 < d < -0.14$
Top width W (m)	0.512	$\frac{1+2m}{2+3m}$ $0.5 < b < 0.57$	$\frac{-1-4m}{4+6m}$ $-0.42 < c < -0.25$	$\frac{-1-2m}{4+6m}$ $-0.28 < d < -0.25$
Mean flow velocity U (m/s)	14.7	$\frac{m}{2+3m}$ $0 < b < 0.14$	$\frac{2-2m}{4+6m}$ $0.14 < c < 0.5$	$\frac{2+2m}{4+6m}$ $0.43 < d < 0.5$
Slope S	12.4	$\frac{-1}{2+3m}$ $-0.5 < b < -0.28$	$\frac{5}{4+6m}$ $0.71 < c < 1.25$	$\frac{5+6m}{4+6m}$ $1.14 < d < 1.25$

Example: Calculate the downstream hydraulic geometry given

$Q = 104\ m^3/s,\ \ d_{50} = 0.056\ m,\ \ \tau_\theta^* = 0.047$
near the beginning of motion

1. Roughly estimate the flow depth, e.g. $h = 1$ m
2. From the flow depth and grain size calculate m from

$$m = \frac{1}{\ln\left(\frac{12.2h}{d_s}\right)} = 0.186$$

3. Calculate the exponents b, c, d for flow depth from Table 4.1, given m=0.186

$$h = aQ^b d_s^c \tau_\theta^* = 0.133\,(104)^{0.39}\,(0.056)^{0.023}\,(0.047)^{-0195} = 1.38\ m$$

4. Repeat steps 2 and 3 with calculated flow depth until convergence:

$m = 0.175$ gives $h = 1.49$m, and $m = 0.172$ gives $h = 1.51$m

5. Calculate the channel width W, flow velocity U, and slope S using the last value of m and the exponents of Q, d_s and τ_θ^* in Table 4.1, e.g., with $m = 0.172$,

$$W = 0.512\,(104)^{0.534}\,(0.056)^{-0.335}\,(0.047)^{-0.267} = 36.4\ m$$
$$U = 14.7\,(104)^{0.068}\,(0.056)^{0.329}\,(0.047)^{0.466} = 1.87\ m/s$$
$$S = 12.4\,(104)^{-0.397}\,(0.056)^{0.994}\,(0.047)^{1.199} = 2.86 \times 10^{-3}$$

The user may prefer to use a different set of known independent variables. Equations (3.6) to (3.9) must then be rearranged through algebraic transformation (Julien and Wargadalam, 1995) isolating each dependent variable on the left-hand side as a function of power functions of the three user-selected independent variables. For instance, geomorphologists may prefer to calculate flow depth h, width W, mean velocity U, and Shields number t^*_θ as explicit functions of discharge Q in m^3/s, median grain size d_s in m, and channel slope S. The recalibrated equations (5.1) through (5.3) can be solved with the procedure below corresponding to Table 4.2.

Table 4.2. Downstream hydraulic geometry as a function of Q (m^3/s), d_s (m) and slope S.

	Coefficient a	Discharge Exponent b	Grain Size Exponent c	Slope Exponent d
Flow depth h (m)	0.2	$\frac{2}{5+6m}$ $0.25<b<0.4$	$\frac{6m}{5+6m}$ $0<c<0.375$	$\frac{-1}{5+6m}$ $-0.2<d<-0.125$
Top width W (m)	1.33	$\frac{2+4m}{5+6m}$ $0.4<b<0.5$	$\frac{-4m}{5+6m}$ $-0.25<c<0$	$\frac{-1-2m}{5+6m}$ $-0.25<d<-0.2$
Mean flow velocity U (m/s)	3.76	$\frac{1+2m}{5+6m}$ $0.20<b<0.25$	$\frac{-2m}{5+6m}$ $-0.125<c<0$	$\frac{2+2m}{5+6m}$ $0.375<d<0.4$
Shields number τ_θ*	0.121	$\frac{2}{5+6m}$ $0.25<b<0.4$	$\frac{-5}{5+6m}$ $-1<c<-0.625$	$\frac{4+6m}{5+6m}$ $0.8<d<0.875$

5. VERIFICATION AND VALIDATION

The analytical formulations were tested with a comprehensive data set consisting of 835 field channels and 45 laboratory channels by Julien and Wargadalam (1995). The data set covers a wide range of flow conditions from meandering to braided, sand-bed and gravel-bed rivers with flow depths and channel widths varying by four orders of magnitude. The exponents of hydraulic geometry relationships change with the relative submergence. Four exponent diagrams illustrate the good agreement with several empirical regime equations found in the literature. They illustrate the results of the three-part analysis consisting of calibration, verification, and validation of the proposed hydraulic geometry equations. Field and laboratory observations are in very good agreement with the calculations of flow depth, channel width, mean flow velocity, and friction slope.

$$h = 0.2Q^{\frac{2}{5+6m}} d_s^{\frac{6m}{5+6m}} S^{\frac{-1}{5+6m}} \quad (5.1)$$

$$\tau_\theta^* = 0.121 Q^{\frac{2}{5+6m}} d_s^{\frac{-5}{5+6m}} S^{\frac{4+6m}{5+6m}} \tag{5.2}$$

$$U = 3.76 Q^{\frac{1+2m}{5+6m}} d_s^{\frac{-2m}{5+6m}} S^{\frac{2+2m}{5+6m}} \tag{5.3}$$

Example: Calculate the downstream hydraulic geometry given

$$Q = 104 m^3/s,\ d_{50} = 0.056 m,\ S = 2.87 \times 10^{-3}$$

1. Roughly estimate the flow depth, e.g. h = 1 m
2. From the flow depth and grain size, calculate m from

$$m = \frac{1}{\ln\left(\frac{12.2h}{d_s}\right)} = 0.186$$

meander wavelength

3. Calculate the exponents b, c, d for flow depth from Table 4.2, given m=0.186

$$h = 0.2\ (104)^{0.327}\ (0.056)^{0.182}\ (0.00287)^{-0.163} = 1.40\text{m}$$

4. Repeat steps 2 and 3 with calculated flow depth until convergence:

$m = 0.175$ gives $h = 1.48$m and $m = 0.173$ gives $h = 1.50$m

5. Calculate the channel width W, flow velocity U, and Shields number t_θ^* using the last value of m and the exponents of Q, d_{50} and S in Table 4.2, e.g., with m = 0.173,

$$W = 1.33\ (104)^{0.446} (0.056)^{-0.115} (0.00287)^{-0.223} = 54.2 \text{ m}$$
$$U = 3.76\ (104)^{0.223} (0.056)^{-0.057} (0.00287)^{0.388} = 1.29 \text{ m/s}$$
$$\tau_\theta^* = 0.121\ (104)^{0.331} (0.056)^{-0.828} (0.00287)^{0.834} = 0.047$$

6. COMPARISONS BETWEEN GEOMORPHIC AND ENGINEERING APPROACHES

From the second column in Table 4.1, the influence of an increase in discharge on hydraulic geometry is expected to be an increase in surface width, flow depth, flow velocity and a decreased slope, thus

$$Q \sim \frac{WhU}{S} \tag{6.1}$$

comparison with Dr. Schumm's relationship (2.5) is more than satisfactory.

Likewise, an increased sediment load is equivalent to an increased value of the Shields parameter, which through column 4 of Table 4.1 gives

$$Q_s \sim \tau_\theta^* \sim \frac{US}{Wh} \tag{6.2}$$

Comparisons with Dr. Schumm's relationship (2.6) show that the only difference is that the channel width appears on the denominator of (6.2) instead of the numerator of (2.6). The geomorphic approach would predict that an increased sediment load increases the width-depth ratio.

An increased width-depth ratio is probably the correct short-term response of a channel to an increased sediment load. Through aggradation within the reach, a river is forced out of banks and will braid until the channel slope is sufficiently steepened to reach equilibrium between incoming and outgoing sediment load. The long-term effect is likely to be that under a steeper slope, the flow velocity will increase and the cross-section areas Wh will decrease, thus yielding (6.2).

Finally, an estimated change in hydraulic geometry given a 100 percent increase in discharge in the channel described below Table 4.2 would result in the following changes

$$\frac{h_2}{h_1} \sim \left(\frac{Q_2}{Q_1}\right)^{0.33} = 2^{0.33} = 1.26$$

$$\frac{W_2}{W_1} \sim \left(\frac{Q_2}{Q_1}\right)^{0.45} = 2^{0.45} = 1.37$$

$$\frac{U_2}{U_1} \sim \left(\frac{Q_2}{Q_1}\right)^{0.22} = 2^{0.22} = 1.16$$

$$\frac{\tau^*_{\theta 2}}{\tau^*_{\theta 1}} \sim \left(\frac{Q_2}{Q_1}\right)^{0.33} = 2^{0.33} = 1.26$$

Under equilibrium conditions the depth would thus be expected to increase by 26 percent, the width by 37 percent, the velocity by 16 percent, and the Shields number should increase by 26 percent. Further verification of these quantitative assessments is deemed appropriate and future collaboration among engineers and geomorphologists will determine whether such equations are useful and/or whether they need to be modified.

7. SUMMARY AND CONCLUSIONS

The problem of defining the hydraulic geometry of alluvial channels has come a long way in recent decades. Dr. Schumm pioneered the understanding of how changes in discharge and sediment load in alluvial channels can change channel width, depth, velocity and slope. Through collaborative work among geomorphologists and engineers, quantitative assessments are now possible from (3.6) through (3.9). Future field applications will determine the reliability of the proposed downstream hydraulic geometry of noncohesive alluvial channels. This paper illustrates how the free flow of information and genuine collaboration between geomorphologists and engineers can contribute to advances in alluvial river mechanics. With his ability to test hypotheses through experimental research at the Engineering Research Center, Dr. Schumm definitely demonstrated leadership in his field and set the example for generations to follow.

8. REFERENCES

Julien, P.Y. 1988. Downstream hydraulic geometry of noncohesive alluvial channels. *Int. Conf. on River Regime*. John Wiley & Sons, Inc., New York, N.Y., 9-16.

Julien, P.Y. 1989. Geometrie hydraulique des cours d'eau a lit alluvial. Proc. IAHR Conf., Nat. Res. Council, Ottawa, Canada, B9-16.

Julien, P.Y. and D.B. Simons. 1984. Analysis of hydraulic geometry relationships in alluvial channels. Report CER83-89PYJ-DBS45, Colorado State University, April, 47p.

Julien, P.Y. and J. Wargadalam. 1995. Alluvial channel geometry: Theory and applications. J. Hydr. Engrg., Vol. 121, No. 4, April.

Rozovskii, I.L. 1961. Flow of water in bends of open channels. Translated by Y. Prushansky, Israel Program Sci. Translation, Jerusalem, Israel.

Schumm, S.A. 1969. River metamorphosis. J. Hydr. Engrg., ASCE, Vol. 95, No. HY1, pp. 255-273.

Schumm, S.A. 1971a. Fluvial geomorphology: Historical perspective. Chap. 4 in River Mechanics. Edited by H.W. Shen.

Schumm, S.A. 1971b. Fluvial geomorphology: Channel adjustment and river metamorphosis. Chap. 5 in River Mechanics. Edited by H.W. Shen.

Wargadalam, J. 1993. Hydraulic geometry equations of alluvial channels. Ph.D. dissertation, Colorado State University, Fort Collins, CO.

LIST OF SYMBOLS			
Symbol	Description	Symbol	Description
b	coefficient	Q_s	sediment discharge
b_r	streamline deviation coefficient	R	radius of curvature of river bends
d_s	sediment size	S	channel slope
g	gravitational acceleration	U	mean flow velocity
h	flow depth	W	channel width
K	hydraulic radius to flow depth ratio	λ	meander wavelength
m	exponent of the resistance equation	ρ	mass density of water
M	silt and clay index	ρ_s	mass density of sediment
p	streamline deviation exponent	τ_θ	Shields parameter
Q	discharge	τ_θ^*	shear stress

GEOMORPHIC DESIGN OF A SUBALPINE COLORADO DEBRIS FLOW CHANNEL

H.S. Pranger, II, A.B. Wilhelm, and J.O. Wilcox[1]

ABSTRACT

A 1,250-foot segment of perennial stream channel was designed at a coal mine located in a scenic, subalpine watershed in west-central Colorado. The morphology of the stream, Dutch Creek, has resulted primarily from recurrent debris flows. The channel substrate is composed principally of sandstone cobbles. A typical engineering approach to the design of Dutch Creek would result in a conspicuous, costly conveyance structure of limited durability. However, the unique site conditions afforded the opportunity to use a design approach that emphasizes geomorphic principles. Morphometric and sedimentologic properties of a reference reach of Dutch Creek provided criteria for the channel design. The channel sinuosity, channel gradient, channel width, floodplain width, channel depth, and channel materials of the reference reach were restored to the maximum constructable extent. Design detail beyond that presented here is unnecessary. The design only provides for an initial channel morphology. After construction, we expect rapid and continuous, although not catastrophic, channel adjustments to occur as step-pools develop. The channel will eventually blend inconspicuously into the surrounding landscape. Monitoring will be the key to verifying the success of this geomorphic channel design approach.

KEY WORDS

Cobbles, channel design, debris flow, stability, geomorphology, step-pools, monitoring

1. INTRODUCTION

In 1995, the State of Colorado Division of Minerals and Geology (DMG) requested technical assistance from the U.S. Office of Surface Mining Reclamation and Enforcement (OSM) regarding an unusual perennial stream at

---

[1] Mining Engineer and Hydrologist, U.S. Office of Surface Mining, Denver, Colorado.

the former Dutch Creek Mine. Dutch Creek, itself, is unusual because it is located in a scenic subalpine environment and debris flow is the primary channel-forming process. In a cooperative effort with the DMG, OSM agreed to provide a design for a 1250-foot reach of Dutch Creek. After evaluating the unusual characteristics of the site, OSM pursued a design approach that emphasizes geomorphic principles, rather than strict engineering principles.

2. SITE DESCRIPTION

2.1. Location and Project Area Description

The project area is located in scenic west-central Colorado, four miles west of Redstone, Colorado. Redstone is located approximately 21 miles west of Aspen, Colorado, and 26 miles south of Glenwood Springs, Colorado.

The Dutch Creek project area, where the designed channel would be located, lies between approximately 7,995 and 8,085 feet AMSL. Dutch Creek is a tributary of Coal Creek, which flows into the Crystal River at Redstone. The project area once contained refueling, washing, transportation, maintenance, and other mine facilities that supported five underground coal mines located at approximately 10,000 feet AMSL. The project area was constructed by leveling a debris fan near the mouth of Dutch Creek.

A concrete flume diverts Dutch Creek around the eastern edge of the project area. The flume is approximately 10 feet wide, four feet deep, and has a gradient of two percent. The flume's downstream end is perched approximately 60 feet above its confluence with Coal Creek, resulting in a waterfall. A 1,750-foot reference reach (see Harrelson et al., 1994 and Rosgen, 1994) of the native Dutch Creek channel, located immediately upstream of the project area, provided morphometric and sedimentologic characteristics for the channel design.

2.2. Geomorphic Setting

The Dutch Creek watershed lies between about 8,000 and 11,000 feet AMSL and has a total drainage area of approximately 4.1 square miles. The Mancos Shale that underlies this watershed has been thermally metamorphosed, giving it greater resistance to erosion than unmetamorphosed Mancos Shale. Accordingly, the Dutch Creek watershed is relatively steep. Cliff-forming sandstone units within the Mancos Shale are found near the watershed divides and provide material for recurrent debris flows.

Debris flow is the major channel-forming process for Dutch Creek and the entire Coal Creek watershed. Debris flows can occur during almost any rainfall event, provided sufficient rock is available for mobilization (Costa and Jarrett, 1981; Costa, 1984). Debris flows create channels that are much larger than could have been produced by water floods alone. Dutch Creek flows in a relatively small channel contained within the bottom of the much larger debris-flow channel. In effect, the bottom of the debris flow channel serves as a floodplain for Dutch Creek. Unless flooding, Dutch Creek flows within the inner channel.

In general, debris flows are relatively rare phenomena, occurring anywhere from about once every 10 to 10,000 years (Costa, 1984). However, debris flows occurred in Dutch Creek in 1977 and 1981. The 1977 debris flow caused $500,000 in damages to the project area.

Dutch Creek has a median particle diameter of 4.25 inches (Fig. 2.1), as determined by pebble counts (see Wolman, 1954). However, the substrate material grain size ranged from boulders greater than 8 feet in diameter to clay particles.

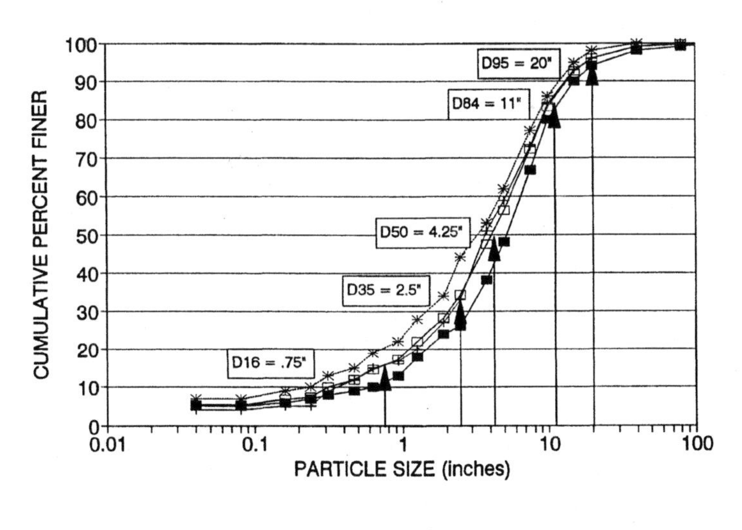

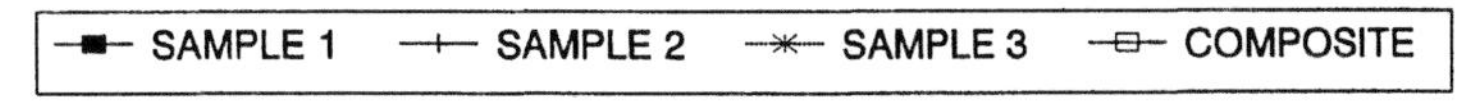

Fig. 2.1. Grain-size distribution for the reference reach of Dutch Creek.

2.3. Flow Characteristics

Dutch Creek is a perennial stream that flows primarily in response to snowmelt runoff and secondarily to rainfall events and groundwater inflow. Annual snowfall typically exceeds 200 inches at the project area. Rainfall in a high-elevation watershed such as this is typically far less intense than at lower elevations (Jarrett, 1990). Based on an extensive evaluation of USGS flow records, Jarrett (1987) found that Colorado streams above 2,300 meters AMSL (7546 feet) produce a maximum unit discharge of 1.1 cubic meters per second per square kilometer, therefore maximum expected flow should average 413 cfs at the mouth of Dutch Creek.

After heavy May and early June snowfall in 1995, the Dutch Creek watershed received moderate rains in mid-to-late June. Flow from one particular rain-on-snow event nearly topped the flume and was estimated to be 400 cfs, or near the maximum expected flow for Dutch Creek. Boulders up to five feet in diameter were carried through the flume during this event. By mid-July the flow had dropped to below 20 cfs.

3. DESIGN APPROACH

3.1. Design Concept

Ideally, stable channels are designed on reclaimed lands according to documented technical criteria that are widely applicable (Lidstone and Anderson, 1989). Engineering criteria, such as limiting permissible velocities, tractive stress and regime procedures, are most often used as stream channel design criteria at coal mines. Designing stream channels to engineering criteria alone may result in aggradation upstream and degradation downstream of the designed reach (Harvey, Watson and Schumm, 1985). An underlying assumption of these criteria is that channel stability is equivalent to complete channel rigidity during a large flood.

For example, a precisely engineered design of Dutch Creek would likely result in a costly and conspicuous riprapped diversion structure. The channel would probably be wider, straighter and steeper than the native channel. The design would likely include a synthetic filter fabric or geogrid beneath the riprap. An engineered channel would be composed of large, angular riprap containing rock materials foreign to the watershed. Also, the cost of importing rock riprap to the relatively remote project area would be prohibitive. Channel stability, and the aesthetic and dollar value of the reclaimed channel, would be reduced with a strictly engineered flood conveyance structure.

The approach used to design the Dutch Creek channel does not result in a standardized design based on standard criteria, but rather a site-specific design based on site-specific data. Engineering geomorphology has been identified as a solution to stream engineering problems (Schumm and Harvey, 1993). Integration of geomorphic and engineering studies on complex river problems increases the probability of project success (Schumm and Harvey, 1993; Jackson and Van Haveren, 1984; Toy et al., 1987). The geomorphic approach has been successfully applied to engineering, fish habitat restoration, and water resource projects (Rosgen, 1994), as well as coal mine stream reclamation projects (Lidstone, 1982; Bergstrom, 1985; Lidstone and Anderson, 1989).

Although the design reach can only include an approximation of the reference reach's highly variable geomorphic characteristics, the approximation should be adequate to ensure that only relatively small channel adjustments occur after construction (see Jackson and Van Haveren, 1984).

3.2. Data Acquisition

Field data were collected in July, 1995 from the reference reach and project area in order to refine an existing 1994 topographic map that was not geospatially referenced. Surveyed elevations were collected at 628 locations within and surrounding the reference reach and the project area. Survey benchmarks were fixed using Global Positioning System technology. Grain-size data were collected using Wolman counts from thirty sample lines covering 1,750 feet of the reference reach.

3.3. Design Procedure

The channel design was produced by manually constructing a contour map that incorporated the best combination of reference reach morphometric parameters. The reference reach's values of channel sinuosity, gradient, width, floodplain width and flow depth were assimilated into the designed reach to the maximum constructable extent. The overall alignment of the channel was somewhat compromised because of spatial constraints and to ensure a suitable channel gradient. However, the channel's aesthetic value was not compromised because of the alignment modification. A surface modeling software package was used to determine the balance of cut-and-fill material in the project area.

4. DESIGN EVALUATION

The designed channel, like the native channel, is composed of a relatively small stream channel located within the bottom of a much larger channel. The designed confluence of Dutch Creek and Coal Creek was located as far upstream on Coal Creek as possible to reduce the channel gradient to that of the reference reach.

Thalweg and gradient profiles of the reference and design reaches are presented in Figs. 4.1 and 4.2. As seen on Fig. 4.1, the reference reach appears to have a nearly uniform profile, with the gradient averaging 6.3 percent. Figure 4.2, on the other hand, indicates a highly variable channel gradient in the reference reach, due to at least 31 step-pools. The profile of the designed channel has been designed to a uniform gradient of 6.8 percent. The average gradient of the reference reach was slightly increased due to spatial

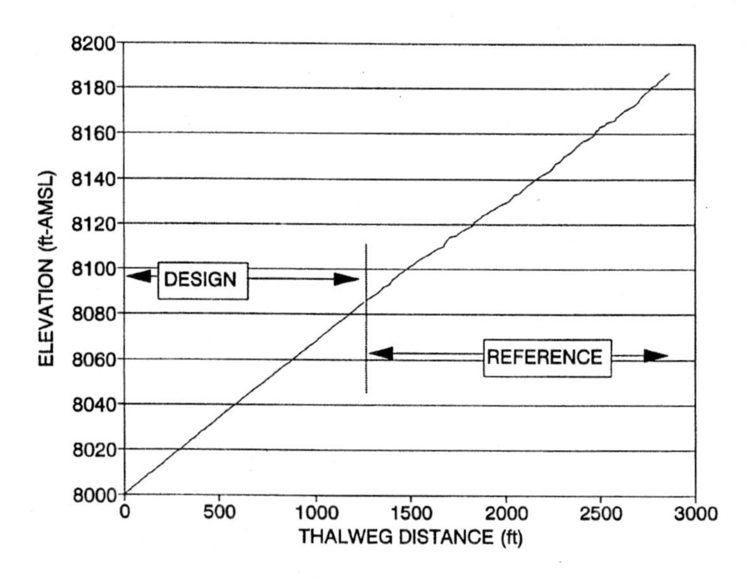

Fig. 4.1. Thalweg profile of Dutch Creek from the design reach through the reference reach.

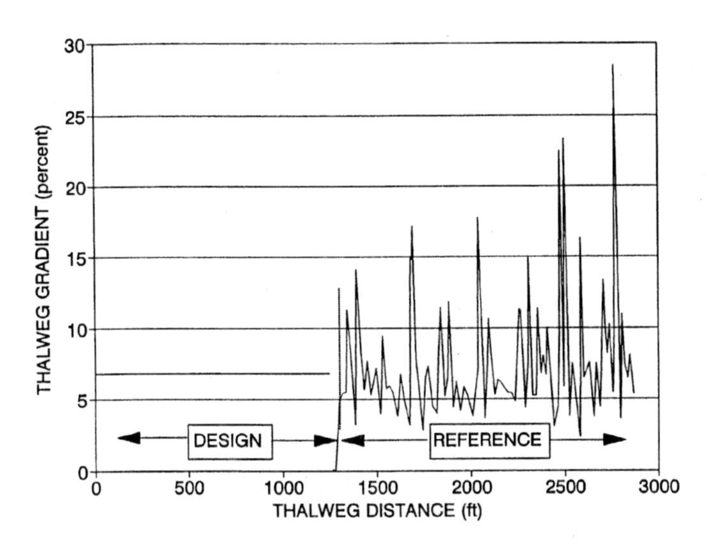

Fig. 4.2. Thalweg gradient in Dutch Creek from the design reach through the reference reach.

constraints in the project area. Step-pools could not be reasonably incorporated into the channel design because of equipment and cost limitations. A step-pool profile will develop in the designed reach as channel materials are mobilized under the perennial streamflow.

Channel and "floodplain" width profiles are presented in Fig. 4.3. The channel width in Fig. 4.3 is the top width of the inner channel. The "floodplain" width on Fig. 4.3 is the bottom width of the larger debris flow channel. The channel width profile of the design reach very closely approximates the width profile of the reference reach, both in mean value and in variability. The design reach "floodplain" is overall slightly wider and less variable than the reference reach. However, the "floodplain" profile was designed as shown on Fig. 4.3 for several reasons.

1. The flood capacity will be slightly greater in the design reach. The movement of floodplain substrate material during high flows will be reduced, which will mitigate the effect of the designed channel being slightly steeper than the reference reach.

2. The widest part of the floodplain in the design reach occurs where the channel makes its first major bend. Widening the channel there will reduce the forces impinging on the outer bank during high flow.

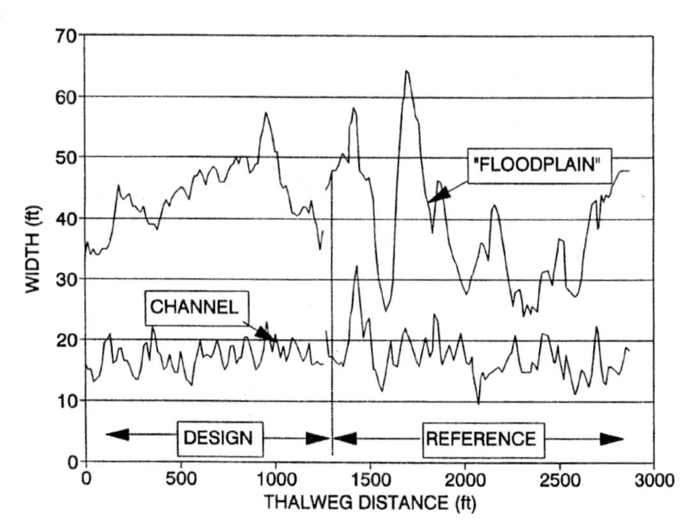

Fig. 4.3. Channel and floodplain width for Dutch Creek from the design reach through the reference reach.

3. By reducing the floodplain width continuously in the downstream direction, considerable earthmoving savings are realized because the cut depth increases in the downstream direction. At its narrowest point, the floodplain width is still well within design tolerance.

The channel initially will be constructed with a roughly parabolic cross-section with a maximum depth of two feet. The channel depth profile of the design reach will develop significant variability as step-pools form. The substrate material thickness will be two feet below the designed stream channel and 1 foot deep below the adjoining floodplain. Sufficient quantities and size of similar substrate material to line the design reach are found within the project area.

Figure 4.4 displays a perspective view of the proposed project area with no vertical exaggeration. The larger channel makes six distinct angular bends that are intended to resemble the two major bends of the reference reach. The sinuosity of the inner channel within the design reach is 1.13, as compared to the reference reach sinuosity of 1.05. The designed channel incorporates more angular bends in the larger channel than are found in the reference reach. The gradient reduction that was achieved through this relatively minor channel pattern manipulation was necessary to maintain the hydraulic characteristics of

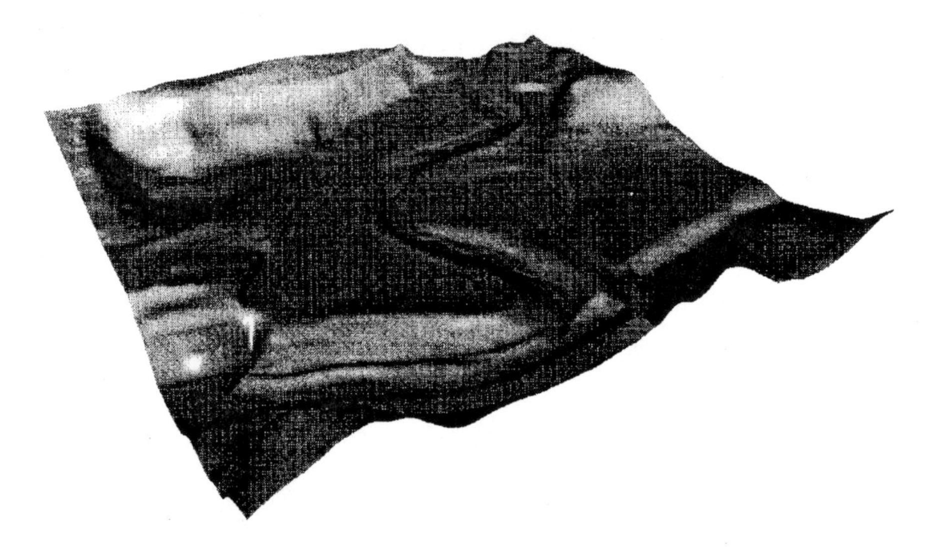

Fig. 4.4. Orthographic perspective view of the designed reach of Dutch Creek through the project area.

the native channel. Perspective views such as this provide a means to determine whether the proposed design is either suitable or too conspicuous for the site. A statistical summary of the morphometric characteristics of the reference and design reaches is presented in Table 4.1.

Table 4.1. Summary comparison of morphometric characteristics of the reference and design reaches of Dutch Creek.

Parameter	Reference Reach Mean	Std. Dev.	Design Reach Mean	Std. Dev.
Channel Slope (percent)	6.3	–	6.8	–
Channel Width (feet)	17.7	3.8	17.2	2.3
Floodplain Width (feet)	39.9	9.9	43.8	5.3
Channel Sinuosity	1.05	–	1.13	–

– indicates value not calculated or not applicable.

To evaluate the ultimate success or failure of the proposed design of Dutch Creek, a long-term monitoring program would be essential (Emmett and

Rosgen, 1995; Rosgen, 1994; Kondolf, 1995; Kondolf and Micheli, 1995). Stream channel monitoring can indicate the stability of a stream by evaluating whether the stream is: (1) aggrading; (2) degrading; (3) changing particle sizes of the bed material; (4) changing the rate of lateral migration through accelerated bank erosion; and/or (5) changing morphological type through evolutionary sequences (Rosgen, 1995). Permanent cross-sections with bench mark locations, bank and toe pins, and scour chains could be placed in the reference reach and the reconstructed channel to determine vertical and lateral stability. Pebble counts (Wolman, 1954) of channel deposits could be used to determine bed-load changes within the system. In addition, repeat annual photographs would enhance the understanding of the channel's development through time.

5. DISCUSSION AND CONCLUSIONS

The geomorphic design approach for stream channel reconstruction is straightforward. The approach is based on a basic premise: restoring the shape of a stream and the materials over which the stream is flowing essentially restores the stream to "native" geomorphic conditions. By restoring the shape of a stream, the hydraulic and gravitational forces causing erosion will be replaced. By restoring the materials of a stream, the forces resisting erosion will be reestablished. However, the question remains - to what extent must the shape and materials of a stream be approximated? The proposed stream design approximates to the maximum constructable extent the shape and materials of the native channel. The stream will inevitably change its shape and develop a step-pool morphology soon after construction. A key goal for the design is to shorten the time required for the channel to reach its "equilibrium morphology" (Jackson and Van Haveren, 1984).

Riprap-lined stream channels are often reconstructed assuming that the channel itself should neither contribute sediment to nor accumulate sediment from the larger stream system. However, the proposed channel design procedure assumes that erosion and deposition of sediment should be maintained and not eliminated in the larger native and reclaimed stream system. Dutch Creek is a high-energy stream transporting a substantial amount of bed-load, much of it in the typical riprap size range. Under these conditions, a design comprised of a simple approximation of the stream channel's morphometric and sedimentologic characteristics is sufficient to ensure restoration success without undue risk of catastrophic failure. Field studies of Dutch Creek's response after construction would be necessary to validate this geomorphic channel design approach in this debris flow dominated subalpine environment.

6. REFERENCES

Bergstrom, F.W. 1985. Determination of Longitudinal Profile for Stream Channel Reclamation. *Proceedings of the Second Hydrology Symposium on Surface Coal Mining in the Northern Great Plains*, Gillette, WY, February 26-27, 1985; 29-60.

Costa, J.E. and R.D. Jarrett. 1981. Debris Flows in Small Mountain Stream Channels of Colorado and Their Hydrologic Implications. *Bulletin of the Association of Engineering Geologists*, v. 18, No. 3, pp. 309-322.

Costa, J.E. 1984. Physical Geomorphology of Debris Flows. *Developments and Applications of Geomorphology*, edited by J.E. Costa and P.J. Fleisher, Springer-Verlag, Berlin, Heidelberg, Chapter 9; 269-317.

Emmett, W.W. and D.L. Rosgen. 1995. River Assessment and Monitoring, short course notes, *Wildland Hydrology Associates*. Pagosa Springs.

Harrelson, C.C., CL. Rawlins, and J.P. Potyondy. 1994. Stream Channel Reference Sites: *An Illustrated Guide to Field Technique*. USDA Forest Service, General Technical Report RM-245; 61pp.

Harvey, M.D., C.C Watson, and S.A. Schumm. 1985. Stream Channel Restoration Criteria. *Second Hydrology Symposium on Surface Coal Mining in the Northern Great Plains*, February 27-27, 1985, Gillette, WY;61-73.

Heede, B.H. 1986. Designing for Dynamic Equilibrium in Streams. *Water Resources Bulletin* 22;3;351-357.

Jackson, W.L. and B.P. Van Haveren. 1984. Design for a Stable Channel in Coarse Alluvium for Riparian Zone Restoration. *Water Resources Bulletin*. 20;5;695-703.

Jarrett, R.D. 1987. Flood Hydrology of Foothill and Mountain Streams in Colorado. Ph.D. Dissertation, Colorado State University, Fort Collins, 239 pp.

Jarrett, R.D. 1990. Hydrologic and Hydraulic Research in Mountain Rivers. *Water Resources Bulletin*. 26;3;419-429.

Kondolf, G.M. 1995. Learning From Stream Restoration Projects. Watersheds 94'. *Proceedings of the Fifth Biennial Watershed Management Conference*. University of California Water Resources Center Rep. 86;107-110.

Kondolf, G.M. and E.R. Micheli. 1995. Evaluating Stream Restoration Projects. *Environmental Management,* 19;1;1-15.

Lidstone, C.D. 1982. Stream Channel Reconstruction and Drainage Basin Stability.*Proceedings of the Hydrology Symposium on Surface Coal Mines in Powder River Basin*, November 3-5, 1982, Gillette, WY;43-57.

Lidstone, C.D. and B.A. Anderson. 1989. Considerations in the design of erosionally stable channels on reclaimed lands. *Proceedings of Symposium on Evolution of Abandoned Mine Land Technologies*, Riverton, WY, June 14-16, 1989;478-495.

Pranger, H.S. II, A.B. Wilhelm, J.O. Wilcox, and R.A. Welsh, Jr. 1996. Geomorphic Engineering Design of a Perennial-Stream Channel at a Subalpine Colorado Coal Mine. *Proceedings, 1996 Billings Reclamation Symposium - Planning, Rehabilitation and Treatment of Disturbed Lands*, March 17-23, 1996. Montana State University Reclamation Research Unit Publication No. 9603; 320-329.

Rosgen, D.L. 1994. A classification of natural rivers. *Catena,* 22;169-199.

Schumm, S.A. 1977. *The Fluvial System*. New York, John Wiley & Sons.

Schumm, S.A. and M.D. Harvey. 1993. Engineering Geomorphology. *Proceedings of the 1993 conference on Hydraulic Engineering*, sponsored by the Hydraulics Division, Am. Soc. Civil Engineers, 2;394-399.

Toy, T.J., C.C. Watson, D.I. Gregory, and S.C. Parsons. 1987. Geomorphic Principles, Stability and Reclamation Planning. *Fourth biennial Billings Symposium on Mining and Reclamation in the West and the National Meeting of the American Society for Surface Mining and Reclamation*, March 17-19, 1987, Billings, Montana.

Wolman, M.G. 1954. A method of sampling coarse riverbed material. *Trans. Am. Geophys. Union,* 35; 951-956.

EROSION AND SEDIMENT CONTROL PLAN FOR YOUNGS CREEK HYDROELECTRIC PROJECT: A GEOMORPHIC APPROACH

Jeffrey R. Laird[1]

ABSTRACT

An *Application for License for Major Unconstructed Project* for the Youngs Creek Hydroelectric Project northeast of Seattle, Washington, was submitted to the Federal Energy Regulatory Commission (FERC) in August 1990. In May 1992, the FERC granted approval of the license application and the project is presently under construction.

To facilitate acceptance, the project proponent entered into a voluntary agreement to comply with the Washington Department of Fisheries Hydroelectric Project Assessment Guidelines. Because of the potential impact on the local fisheries, a detailed Erosion and Sediment Control Plan (ESCP) was required. Specific considerations and methods for the ESCP were agreed upon and included assessment of sediment production and transport rates, sedimentation impacts on fish habitat, and evaluation of most-probable and worst-case sediment-producing events. Detailed, site-specific mitigation for construction activities were also to be outlined in the ESCP.

Geomorphic principles and techniques were used to identify and describe sources and rates of sedimentation in the drainage basin under existing conditions, as well as during and following project construction. Under existing conditions, 770 tons of sediment are estimated to be delivered to the stream system annually from the drainage basin. Under the most probable conditions during project construction, an additional 585 tons of sediment were estimated to be produced. Under a postulated worst-case condition involving a landslide during construction, an estimated 725 tons of sediment would be delivered to the stream. An additional 2.3 tons of sediment would be produced during operation of the project from road use. Most of the project-related (road erosion) sediment would be finer than 0.4 mm, which could affect fisheries since sediment with diameters of 0.85 mm or less are reported to be most damaging to fishery spawning gravel.

Mitigation and control methods based on accepted civil engineering practices were developed to reduce the generation and delivery of sediment during construction. The ESCP included construction scheduling, revegetation, drainage measures and site-specific sediment control structures. With a properly designed and implemented ESCP, the potential risks to fisheries from project construction and operation were determined to be low. This paper presents a summary of that ESCP.

---

[1] Shannon & Wilson, Inc., Seattle, Washington

KEYWORDS

Erosion, hydroelectric power, FERC, sediment transport, geomorphology, erosion control

1. INTRODUCTION

The Youngs Creek Hydroelectric Project is located on Youngs Creek, a tributary to Elwell Creek and the Skykomish River. The project site is located about 48 kilometers (km) northeast of Seattle, Washington and about 8 km south of Sultan, Washington. The proposed project is a 7.5 megawatt, run-of-the-river, small hydroelectric facility. The project includes a diversion weir, an intake structure, a 4.4-km-long steel penstock, a powerhouse and switchyard, and a 9.8-km-long single-circuit transmission line. The project will require the construction of about 2 km of permanent access road and 1 km of temporary haul road.

An Application for License for Major Unconstructed Project for the Youngs Creek Hydroelectric Project was submitted to the Federal Energy Regulatory Commission (FERC) in August 1990. To facilitate acceptance of the license application, the project proponent entered into a voluntary agreement with federal, state and local agencies, as well as the Tulalip Indian Tribes, to comply with the Washington Department of Fisheries Hydroelectric Project Assessment Guidelines. Because of the potential impact from project construction and operation on the local fisheries, a detailed Erosion and Sediment Control Plan (ESCP) was required under these guidelines. Specific considerations and methods for the ESCP were agreed upon which make this ESCP different from typical ESCPs. Some of these required considerations include (1) a description of the existing physical conditions of the drainage basin, (2) a detailed sedimentation study that addressed existing sediment yields and transport, and (3) the relative impacts of additional, construction-related sediment. The ESCP was to include a detailed, site-specific description of the construction methods and related erosion and sediment control mitigation and control measures. Finally, a risk assessment was to be conducted addressing the potential risk of project construction and operation to the fisheries.

2. EXISTING CONDITIONS

Existing conditions in the Youngs Creek basin were evaluated so that background sediment sources, characteristics, transport, and the relative impact of project construction, could be addressed.

2.1. Topography

Youngs Creek drains an area of approximately 37.8 km^2. The drainage basin upstream of the proposed intake site covers 21.5 km^2. Topography in the Youngs Creek drainage basin is varied, with moderate slopes interrupted locally by flat valley bottoms and relatively gentle ridge tops and plateaus. Peaks in the basin reach elevations up to 1,200 m.

Youngs Creek has three distinct channel reaches defined by channel gradient. The upper reach of Youngs Creek, from the basin divide to about 1,370 m upstream of the proposed intake, has an average gradient of 4.5 percent and is characterized by moderate valley slopes and a predominantly alluviated channel. The middle reach, which includes the project bypass reach, has an average gradient of 7.5 percent and extends down to the powerhouse site at elevation 180 m. The lower reach begins downstream of the proposed powerhouse and has an average gradient of 2 percent. This reach is characterized by low gradient side slopes and abundant alluvium in the channel and banks.

2.2. Geology

The project area is underlain by volcanic rocks and glacial deposits. The volcanic rocks originated from the Mt. Persis volcanic center 16 kilometers to the east of the project, approximately 42 to 47 million years ago (Tabor et al., 1982). Lava flows of andesite and dacite have formed plateaus in the upper basin (Tpp and Tpa, Fig. 2.1). Tuffs and tuff breccias are exposed in the creek bed and banks at the intake site, and exposures of these rocks continue in the creek bed downstream to the proposed powerhouse site.

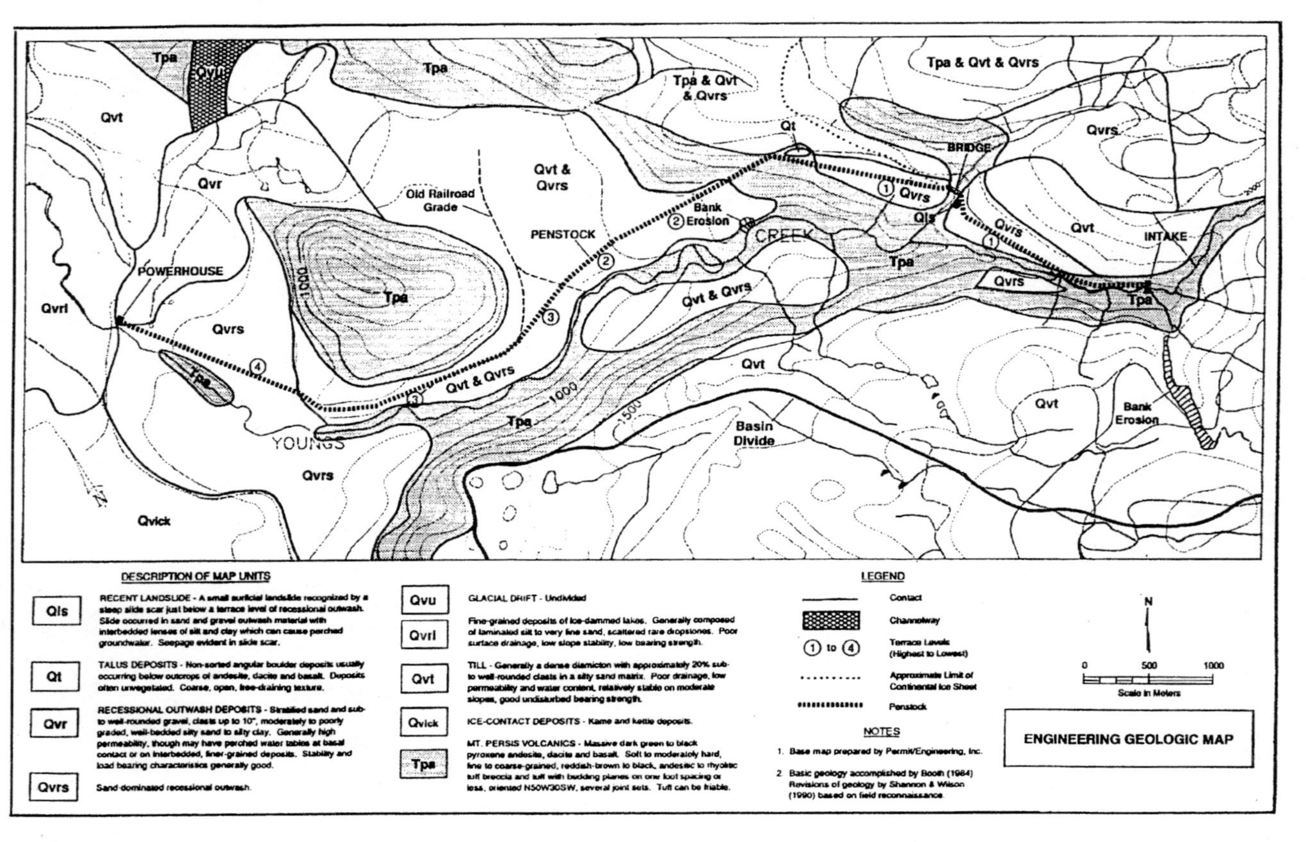

Fig. 2.1. Engineering Geologic Map of the Youngs Creek Site.

The Youngs Creek drainage basin was glaciated during the Vashon Stade of the Fraser continental glaciation in late Pleistocene, between 12,000 to 15,000 years ago. Glacial ice of the Puget Sound lobe covered all of the Youngs Creek drainage basin but the highest ridge, which forms the eastern divide (Booth, 1984). Glacial deposits in the basin include lodgement till (Qvt), sand- and gravel-dominated, fluvial recessional outwash (Qvr and Qvrs), lacustrine deposits (Qvrl), kame and kettle, and morainal ice-contact deposits (Qvick) (Fig. 2.1). More recent, Holocene soil deposits include stream alluvium, talus (Qt), landslide debris (Qls), bog deposits and colluvium (Fig. 2.1).

2.3. Hydrology

Discharges to Youngs Creek are supplied by surface and subsurface runoff derived from rainfall and snowmelt. Annual precipitation in the Youngs Creek basin averages about 1520 mm. Average monthly precipitation for the months of October through March in the project site vicinity is about 185 mm, while average monthly precipitation during the remaining, drier summer months is about 85 mm.

The average yearly flow at the diversion site is about 2.1 m^3/s. The highest average streamflows historically occur in December and May, at 3.1 and 3.0 m^3/s, respectively, while the lowest average streamflows historically occur in August and September, at 0.6 and 0.9 m^3/s, respectively. The 10-year flood has a probable one-day peak discharge of 25 m^3/s, while the 100-year flood has a probable one-day peak discharge of 34 m^3/s.

3. SEDIMENTATION STUDY

The major considerations relating to sedimentation for this project included (1) identifying existing sources and the rates of sediment delivery to the drainage system so that the potential, relative increases of construction-generated sediment over background rates could be evaluated, (2) evaluating the existing transportation and deposition of sediment below the project so that potential impacts and residence times of construction-generated sediment in the channel could be assessed, and (3) assessment of the effects of project construction and operation on basin sedimentation.

3.1. Existing Sediment Sources

3.1.1. Creep

Soil creep is an important contributor to the sediment budget of forested basins in humid mountains (Dietrich et al., 1982). Reported rates of creep for 30 to 65 percent gradient slopes in humid climates are 1 to 4 mm per year throughout the depth of colluvium (Selby, 1982). Creep rates 3 to 4 times higher than those reported for forested slopes have been measured on clear-cut slopes (Gray, 1973).

Sediment delivery by soil creep to streams in the upper and lower drainage basins was estimated using the following formula. Creep rates were adjusted from the reported creep rates based on the relative slope gradients and the area of clear-cuts in the basin. The product is multiplied by 2 to account for both banks of a channel.

$$W = (DD * h * vc * s * A) * 2 \tag{3.1}$$

where

W	=	weight of sediment supply (tons/yr)
DD	=	drainage channel density (m/km^2)
h	=	depth of soil (m)
vc	=	creep rate (m/yr)
s	=	unit weight of soil (tons/cubic m)
A	=	drainage basin area (km^2)

An estimated 103 tons of sediment per year is delivered by creep to the upper drainage channels, while sediment delivered by creep to the lower drainage channels is estimated at 93 tons per year (Table 3.1).

The grain-size distribution of sediment derived from creep which enters the drainage system should be similar to the parent-material, which consists of glacial deposits (till and outwash) and weathered volcanic bedrock. Averaged grain-size distributions of these soils indicate that 33 percent of the material is composed of sediment finer than 0.85 mm, while 54 percent of the material ranges from medium sand to fine gravel (0.85 to 19 mm), and the remaining 13 percent of the material is coarse gravel to boulders.

Table 3.1. Sediment budget for Youngs Creek Drainage Basin.

Sediment Source	Fine Sediment (<0.85 mm) (tons/year)	Medium Sand - Fine Gravel (0.85 to 19 mm) (tons/year)	Coarse Gravel - Boulders (>19 mm) (tons/year)	Total (tons/year)
Upper Basin (21.5 km^2)				
Soil Creep	34	56	13	103
Landslides	129	210	51	390
Road Erosion	46	0	0	46
Total	209	266	64	539
Lower Basin (16.3 km^2)				
Soil Creep	31	50	12	93
Landslides	40	65	15	120
Road Erosion	21	0	0	21
Total	92	115	27	234

3.1.2. Landslides

Field investigations and analysis of aerial photographs revealed no significant current or recent past landslide activity. However, landsliding is assumed to occur, though the landslides may be relatively small and infrequent. Reported annual rates of landslide sediment production on undisturbed and timber-managed slopes in study basins which had geology and climate similar to that of the Youngs Creek drainage basin range from 16 to 2,400 tons/km^2, respectively (Ice, 1985).

Average rates of landslide sediment production were determined by multiplying the reported rates referred to above by the slope area of each land-use type (undisturbed and timber-managed). Only slopes with gradients exceeding 30 percent were considered. Annual landslide sediment production in the upper and lower drainage basins was estimated at 390 and 120 tons, respectively (Table 3.1). The grain-size distribution of sediment derived from landslides is assumed to be the same as that of creep-derived sediment, as their parent sources are similar. This estimate gives an upper range for landslide sediment delivery because not all of the landslide sediment that is produced may be delivered to the stream channels.

3.1.3. Surface Erosion

Sediment production and delivery from roads is directly related to road use intensity and precipitation rates. Sediment rates are reported from studies conducted in the Clearwater Creek basin on the Olympic Peninsula in Washington (Reid and Dunne, 1984). These rates vary from 440 tons/km for heavily used roads to 0.5 tons/km for abandoned roads. The road erosion rates reported by Reid and Dunne were adjusted for the lower precipitation rate in the Youngs Creek basin by taking the ratio of the Youngs Creek basin annual precipitation (1,520 mm) to the Clearwater Creek basin annual precipitation (3,550 mm) to the 0.8 power. The various reported road erosion rates were reduced an additional 20 percent to account for losses to the forest floor (Irvin and Sullivan, unpublished report, cited in Duncan et al., 1987). These adjusted rates for various road use levels were then applied to road segments that deliver sediment to stream channels in the Youngs Creek basin.

Roads in the upper basin are estimated to deliver 46 tons of sediment to Youngs Creek annually, while lower basin roads are estimated to deliver 21 tons of sediment per year (Table 3.1). Sediment produced by road erosion is primarily clay, silt, and fine sand (Reid and Dunne, 1984).

3.2. Existing Sediment Transport

3.2.1. Channel Morphology and Grain-Size Distribution

The project bypass reach of Youngs Creek is a steep, bedrock and boulder channel in which gravel (sediment ranging from 6 to 76 mm in diameter) patches are limited to small, low-velocity areas. Downstream of the powerhouse, gravel beds are much more numerous, due both to lower gradients and to an abundant available supply of sand and gravel. Gravel and cobble bars are also much more numerous 1,370 m upstream of the diversion weir, where the average stream gradient is about 4.5 percent and gravel sources are more numerous.

Both Wolman point counts of the armor layer sediment and grab samples of sediment beneath the armor layer were utilized to characterize the sedimentation regime in Youngs Creek downstream of the diversion weir. The median grain size (D_{50}) of the bed armor layer increases downstream from 71 to 178 mm. The median grain size of the grab samples varied from 7 to 17 mm. Less than 2 percent of the sediment by weight was finer than 0.85 mm, the size below which sediment is generally considered to adversely affect salmonids.

3.2.2. Sediment Transport Rates

The low amounts of sand and gravel present in the bypass reach of Youngs Creek are a result of high sediment-transport capacity, limited sites of deposition, and low basin sediment yield. Youngs Creek is typical of high-gradient streams in that it is supply-limited; that is, its capacity to transport sediment exceeds the available supply.

To evaluate channel hydraulics, the section of Youngs Creek downstream of the diversion weir to Elwell Creek was divided into seven reaches, each with similar morphology. The flow depth corresponding to a given discharge was calculated with a form of the Manning's equation.

The ratio between particle settling velocity (V_s) and the shear velocity of water near the bed (u_*) determines whether a particle will be suspended. This balance is expressed as the Rouse number (Simons and Senturk, 1977). A Rouse number of 1 indicates a particle will travel as suspended load, particles with Rouse numbers between 1 and 2.5 will intermittently go into suspension and a Rouse number above 2.5 indicates that transport occurs entirely as bedload.

During the "bankfull" or 1.5-year recurrence-interval flood in Youngs Creek, the maximum grain size in suspension ranges from 0.25 to 0.81 mm through the different bypass reaches. This sediment would travel the 10 kilometers from the proposed intake to the mouth of the creek in about 1.7 hours. Particles with sizes between 0.56 to 2.02 mm would travel in intermittent suspension through the different bypass reaches and would move at a rate up to an order of magnitude more slowly.

Overall, transport capacity diminishes in the reaches below the proposed powerhouse due to flatter stream gradients. The maximum sizes of sediment in suspension, and in intermittent suspension, in these reaches during the 1.5-year flood are estimated to be 0.43 and 1.1 mm, respectively.

Along several sections of the bypass reach, the bed of Youngs Creek is armored with a layer of boulders and cobbles. Significant bedload transport occurs once flow exceeds a critical discharge needed to partially disrupt the armor layer. Critical discharges in the Youngs Creek bypass reach were determined with the method developed by Bathurst et al. (1987). These discharges range from 0.4 to 13.5 m^3/s and are exceeded an average of 329 days per year, and about once every 1.25 years, respectively. At least partial disruption of the armor layer and significant bed load transport probably occur in Youngs Creek within this range of flows during part of the year.

Bedload transport rates were calculated with the Meyer-Peter and Muller (1948) formula. The results suggest that, at most, 7.5 hours of annual flood flow would be sufficient to transport the estimated 290 tons per year of bedload sediment which enters the upper end of the project reach under natural conditions.

3.3. Effects on Sedimentation During Construction

The potential effects of construction-related erosion on Youngs Creek have been evaluated for two scenarios. The "most probable" case implies average climate and hydrology, along with successful design and maintenance of all erosion control measures. The "worst case" scenario would result from the most potentially damaging combination of climate and hydrology and a subsequent failure of erosion-control measures.

3.3.1. Most-Probable Case Scenario

In the most-probable case, the only significant, construction-related sediment delivery would be caused by surface erosion from the increased lengths and use-levels of roads.

Heavy-use annual road erosion rates, reported by Reid and Dunne (1984) from the Clearwater, Washington area, were corrected for rainfall differences between the project site and the study area and for forest floor losses, as discussed above. The adjusted heavy-use road erosion rate (47 tons/km) was applied to the approximately 12.2 km of new and existing construction access and haul roads in the drainage basin. The heavy road use would produce an additional 572 tons of fine-grained sediment over the six-month construction period; however, not all of this sediment would be delivered to the stream system. This represents a temporary 190 percent increase above the fine-grain sediment yield of 301 tons per year estimated for the entire Youngs Creek drainage basin. It is anticipated that the low drainage density of tributary channels in the vicinity of the roads, and the distance of the roads from Youngs Creek, will reduce the total amount of road-derived sediment, which actually reaches Youngs Creek. Most of this sediment would be transported downstream in suspension. It is unlikely that sediment derived from construction-related road erosion would lead to long-term deposition of fine sediment on the creek bed in the lower reaches of Youngs Creek.

3.3.2. Worst-Case Scenario

Runoff from a severe storm and slope disturbance during penstock construction could potentially initiate a landslide in an area of steep slopes that are mantled with loose to moderately dense sand and gravel outwash. A potential landslide site was assumed to be 12 m wide, 21 m long and 1.5 m deep, with a total mass of 730 tons. Sediment was routed down the creek assuming delivery of the entire landslide volume to Youngs Creek over a one-hour period during a 10-year peak flood discharge.

Following delivery into the creek, about 363 tons of the slide mass would go into suspension and intermittent suspension, rapidly moving downstream. Most of the suspended and intermittently suspended sediment derived from the slope failure would not drop out of suspension until sometime after reaching the lower gradient reaches downstream from the powerhouse. However, even through these reaches, sediment finer than 0.75 mm (approximately 70 percent of the total suspended and intermittently suspended landslide load) would be transported in intermittent suspension out of Youngs Creek during the 10-year flood stage. The remainder of this sediment would be transported as bedload.

The remaining 364 tons of sediment from the landslide would move downstream as bedload at minimum rates of 97 tons per hour. At these discharges, about 3 to 4 hours of flow would be required to remove the slide material from the point of slide entrance into the creek. While such sustained high flood flows probably do not occur in the drainage basin, smaller flows capable of bedload transport do occur more frequently on Youngs Creek. The volume of bed load in Youngs Creek would increase during the several years required for the bedload sediment to be transported out of the creek.

3.4. Effects on Sedimentation During Operation

3.4.1. Increased Sediment Yield

During operation, the 12.2 km of road, which will be heavily used during construction, will revert to light use, with a consequent reduction in erosion rates. The overall change from pre-project conditions will be the addition of 1.0 mile of lightly used road. The calculated increase in sediment yield from roads is 2.3 tons per year. This is a negligible increase over the 301 tons per year estimated as the pre-project basin fine-grained sediment yield (Table 3.1).

3.4.2. Effects of the Diversion Weir Impoundment

The impoundment can store about 377 tons of sediment. This is about 1.3 times greater than the average annual weight of bedload (290 tons), which is estimated to enter into the impoundment from upstream. Sluicing of the sediment trapped behind the diversion weir can occur during the short-duration, moderate-magnitude, rainfall-generated floods in the winter or during the spring months when longer-duration and higher-magnitude floods generated by melting of the snowpack occur. Temporary storage of bedload sediment behind the diversion weir is not anticipated to significantly impact the sedimentation regime downstream.

3.4.3. Effects of Reduced Flows in the Bypass Reach

A stream's sediment load over a period of years is transported predominantly during moderate flow events of intermediate frequency (Wolman and Miller, 1960). "Effective discharge" is defined as the flow which transports the most sediment over a period of years. Effective discharges reported in the literature for gravel-bedded streams with significant bed load transport have return periods ranging from 1.15 to 3.26 years (Pickup and Warner, 1976; Andrews, 1980).

The proposed maximum flow diversion is 3.5 m^3/s, so flow in the bypass reach during the 1.5-year "effective discharge" flood event would be 19.5 m^3/s, or 85 percent of the total unregulated peak flow. Because shear stress varies with depth and, assuming that sediment transport capacity varies to the 1.5 power of bed shear stress (Meyer-Peter and Muller, 1948), then using Leopold and Maddock's (1953) at-a-station relation of discharge (Q) to flow depth (d), where $Q^{0.4} \sim d$, a 15 percent reduction in discharge during the 1.5-year flood event would result in a 9 percent reduction in the sediment transport capacity.

As discussed previously, the existing sediment transport capacity of Youngs Creek is relatively high. The creek is easily capable of transporting most of its sediment load, as demonstrated by the general lack of gravel deposition in the bypass reach. It is, therefore, unlikely that a 15 percent reduction in flood flows, due to flow diversion, will produce noticeable changes in the volume or distribution of bed sediment in the bypass reach.

4. MITIGATION AND CONTROL MEASURES

Soil disturbance during construction will result in the potential for increased sediment delivery into Youngs Creek. Appropriate scheduling of construction and implementation of mitigation measures can minimize sediment influx to the creek. However, a detailed ESCP will only function if it is properly implemented, monitored and maintained. The Youngs Creek Hydroelectric Project ESCP detailed intended, station-to-station, project construction methods and practices and recommended specific concurrent erosion and sediment control measures.

4.1. Erosion and Sediment Control Plan Basic Principles

Basic principles of the Youngs Creek Hydroelectric Project ESCP included the following:

- Scheduling of major land disturbing activities during the dry season and performance of instream work during low-flow periods.
- Minimizing the area and duration of the construction disturbance.
- Using the proper type and size of equipment for the job.
- Protection of bare soil from rainfall and overland flow, and revegetation as soon after final grading as permitted by seasonal conditions.
- Reduction of the velocity of runoff from construction areas with proper control measures and minimizing the volume of construction runoff flowing over disturbed areas by planned diversions.
- Providing drainage facilities to control the runoff released from the construction areas.

- Trapping of sediment at the construction area with an emphasis on source isolation.
- Interception and diversion of water away from the construction areas whenever possible.
- Clearing only those areas which will be graded and stabilized in the current season.

4.2. Construction Scheduling

Prudent scheduling of clearing and construction activities (pre-construction planning) is considered to be the most effective method of mitigating erosion, sedimentation and slope instability (Controlling Sediment from Construction Sites, 1990). A separate document, entitled "Sediment-Related Construction Scheduling Guidelines" (Shannon & Wilson, 1991) was prepared for this and other small hydroelectric projects proposed in the Cascades. This study evaluated climate and streamflow characteristics in the west Cascades region to determine periods of high, moderate and low precipitation and streamflow. Construction activities related to small hydroelectric development and their potential to deliver sediment to a stream channel under various local site limitations (i.e., slope gradient, seepage) were assessed. From this, a matrix was developed for scheduling project construction activities (Table 4.1).

The construction activities with the most potential for introduction of sediment into Youngs Creek are the construction of the diversion weir, intake structure and intake laydown area, because they are in the existing floodplain. There is also potential for significant sedimentation during construction of the first 850 m of penstock and access road from the intake, because of steep slopes with seepage, the proximity to Youngs Creek, and the crossing of a tributary stream.

Table 4.1. Construction Activity Risk and Assigned Season.[1]

Risk of Sediment Reaching Stream [2]	Low	Medium	High
Local site limitations[3]	None	Moderate	Severe
Construction Activity			
Major Earthmoving	Extended Dry Season	Extended Dry Season	Dry Season
Minor Earthmoving	All Year	Extended Dry Season	Extended Dry Season
Bedrock Excavation	All Year	All Year	Extended Dry Season
Non-Earth Moving	All Year	All Year	All Year
In-Stream	----	----	Low Flow season

1 The assigned construction season should be based on the more restrictive of the two activity risks: sediment reaching stream or local site limitations.

2 Based on slope gradient and distance from channel.

3 Includes slope composition (rock or soil) and gradient, presence of seepage or streams; existing landslides; unstable geologic units.

4.3. Revegetation

The primary revegetation method recommended in the Youngs Creek Hydroelectric Project ESCP was reseeding with grasses and legumes using hydroseed equipment. Grasses with a fibrous root system provide quick stabilization of soils, while legumes can fix nitrogen. The recommended seed mix for the project followed an erosion control seed mix presently used by the U.S. Forest Service Mt. Baker Ranger District on forest lands to stabilize skid trails, fire lanes and roads (Table 4.2).

Table 4.2. Recommended seed mix for the Youngs Creek Hydroelectric Project.

Seed Variety	Percent (%) by Weight
Trifolium repens (White dutch clover) pre-inoculated	15
Lolium perenne (Perennial rye grass)	25
Phleum pratense (Timothy)	25
Lotus corniculatus (Birdsfoot trefoil)	15
Dactylis glomerata (Orchard grass)	20

The recommended application rate was 70 kg of seed mix per hectare, with a 20-10-10 fast release nitrogen fertilizer at a rate of 250 kg per hectare, wood fiber or straw mulch applied at rates of 2,200 kg per hectare, and tackifiers at a rate of 50 kg per hectare on slopes with gradients flatter than 2.5H:1V or 100 kg per hectare on slopes with gradients of 2.5H:1V and steeper.

4.4. Site-Specific Construction Activities Mitigation

The Youngs Creek Hydroelectric Project ESCP described, station-by-station, general construction methods and slope stabilization and erosion- and sediment-control measures. Plans and profiles of the project elements were provided showing topographic contours, stationing, proposed cuts and fills and erosion-control measures. In addition, typical and specific construction and erosion-control details were provided.

4.4.1. Access Roads

Approximately 135 m of new access road will be constructed to the powerhouse site. No road cuts, slope stability problems or significant erosion hazards are anticipated along this alignment. The road will be surfaced with a minimum 15 cm lift of crushed rock. An adjacent laydown/spoils area will be surrounded with drainage ditches that will discharge water into a straw-bale sediment barrier prior to release of the water onto low-sloped, vegetated ground.

Construction of 1,815 m of road, including a bridge across a tributary creek, will be required to gain access to the intake site. Construction will begin from an existing logging road and proceed upstream towards the intake site. The road will require extensive cuts and some fills. Construction will be limited to the dry season. Cuts across slope gradients steeper than 50 percent will be full-benched. Ditches and culverts with stabilized outfalls will be installed at slope depressions and approximately every 150 m. Gabion walls will be constructed to avoid excavation of long back slopes on steep soil slopes. Subdrains and subdrain dams will be installed in areas of seepage upslope of the road alignment to intercept subsurface water and divert it into roadside ditches. Blasting of rock slopes will be required and blast mats will be used to retain rock spoils. Road spoils will be end-hauled to spoils areas. Silt fences will be installed along the road cut where it is adjacent to the creek.

4.4.2. Diversion Weir and Intake Structure

A construction laydown area is proposed on the right bank just downstream of the intake site. The streamside of the fill placed for the laydown area will be riprapped. The laydown area will be sloped so that surface water will drain back toward the road and into straw-bale sediment barriers.

The intake structure will be founded in bedrock on the right bank. A temporary, sandbag cofferdam with an impervious liner on its streamside face will be constructed to divert creek flow around the intake structure site and toward the left bank. Sump pumps will be installed to pump water to a straw-bale sediment barrier. Concrete leachate from the construction site will be collected and pumped to a holding tank and properly disposed of off-site.

Following completion of the intake structure, the sandbag cofferdam will be re-positioned so that the creek flow will be diverted into the intake structure and discharged downstream. Removal of scattered patches of boulder and cobble alluvium and excavation of the upper layer of bedrock in the channel bed for the foundation will be necessary. Sump pumps will be installed to collect sediment-laden water and concrete leachate. Sediment will be filtered out from this water using straw-bale barriers and concrete leachate will be pumped to a holding tank for proper off-site disposal.

4.4.3. Penstock

The penstock and the access road to the intake site will follow the same alignment for the first 870 m downstream of the intake site. For 426 m downstream from the intake structure, the penstock will be anchored on a narrow bench excavated into bedrock downslope of the access road. The shallow soil overburden overlying the bedrock bank will be removed prior to blasting and excavation. Blast mats will be used to prevent flyrock from entering the creek. The penstock will be founded in a trench about 2 meters below the intake access road surface for the remainder of the distance to the tributary creek crossing.

Beyond the tributary creek crossing, the penstock will, in general, traverse gentle- to moderate-gradient, forested slopes to the powerhouse. A primitive access road will be constructed for penstock installation. The penstock will be placed in a 2-m-deep trench and surrounded with select, drainage backfill. Subdrain dams will be installed at appropriate intervals to divert subsurface water from the trench and out onto the surface. Trench spoils will be placed in the remainder of the trench and mounded and compacted over the trench. Original ground contours will be restored. Waterbars will be constructed to divert surface water off of the disturbed surface and the disturbed ground will be revegetated.

4.4.4. Powerhouse

A 6-m-deep excavation will be made in glacial recessional outwash for the powerhouse foundation. Following construction of the powerhouse, soil will be backfilled into the excavation. The tailrace invert will be about 20 feet beneath the ground surface and daylight on the right creek bank, where it will discharge into Youngs Creek.

Surface water will be diverted away from the powerhouse construction area with ditches. Sediment from surface water pumped out of the powerhouse excavation will be retained at straw bale barriers. Silt fences will be placed downslope of the construction area. Riprap will be placed around the tailrace outfall and along the right bank.

5. RISK ASSESSMENT

An assessment evaluating the risk of project-related sediment impacting anadromous fish populations was conducted as part of the Youngs Creek Hydroelectric Project ESCP. The assessment identified 1,585 m of habitat available for anadromous fish in Youngs Creek. Historical information and surveys concluded that only coho salmon and steelhead trout utilize the stream.

Sediment sampling transects were established in the bypass and anadromous fish reaches of Youngs Creek to evaluate existing potential spawning habitat sediment characteristics, in particular, the amount of fine sediment. The average median grain size was 17 and 31 mm and the average percent of sediment finer than 0.85 mm was 4.0 and 7.7 percent for the bypass and anadromous fish reaches, respectively.

High levels of fine sediment in gravel used for laying and raising anadromous fish eggs and embryos (redds) are clearly detrimental. However, a critical threshold for the amount of fines in a redd has not been determined. The Washington State Department of Fisheries recognizes this difficulty and suggests a qualitative threshold of "no resource loss due to introduction of additional sediment into the stream" (WDF,1987). This threshold is to be evaluated on three criteria: (1) proximity of the project to the resource; (2) existing and potential value of the resource; and (3) ability of the resource to recover.

Based on the existing sediment rates and transport characteristics of the Youngs Creek drainage basin, the relatively low amount and short duration of additional sediment from project construction, the distance of the project from the "at-risk" fish habitat, and the ability of the fisheries to recover from a sedimentation event, construction and operation of the project were determined to be a low risk.

In May 1992, the FERC granted approval of the license application and the project is presently under construction.

6. REFERENCES

Andrews, E. D. 1980. Effective and Bankfull Discharges of Streams in the Yampa River Basin, Colorado and Wyoming. Journal of Hydrology 46: 311-330.

Bathurst, J. C., W. H. Graf and H. H. Cao. 1987. Bed Load Discharge Equations for Steep Mountain Rivers. Sediment Transport in Gravel-Bed Rivers. John Wiley & Sons, New York. p. 453-492.

Booth, Derek B. 1984. Surficial Geology of the West Half of the Skykomish River Quadrangle, Snohomish and King counties, Washington. U. S. Geological Survey Open-File Report 84-213.

Controlling Sediment From Construction Sites. 1990. Seattle, Washington Short Course Proceedings. College of Engineering, University of Wisconsin.

Dietrich, W. E., T. Dunne, N. F. Humphrey and L. M. Reid. 1982. Construction of Sediment Budgets for Drainage Basins. U.S. Forest Service General Technical Report PNW-141, p. 5-23.

Duncan, S. H., R. E. Bilby, J. W. Ward and J. T. Heffner. 1987. Transport of Road-Surface Sediment Through Ephemeral Stream Channels. Water Resources Bulletin 23: 113-119.

Gray, D. H. 1973. Effects of Forest Clear-Cutting on the Stability of Natural Slopes: Results of Field Studies, Interim Report. University of Michigan, Department of Civil Engineering, Ann Arbor, Michigan. 002790-1-P. 119 p.

Ice, George C. 1985. Catalog of Landslide Inventories for the Northwest. National Council of the Paper Industry for Air and Stream Improvement, Inc., Technical Bulletin 456. 138 p.

Leopold, L.B. and T. Maddock, Jr., 1953, The Hydraulic Geometry of Stream Channels and Some Physiographic Implications: U.S. Geological Survey Professional Paper 252, p. 1-57.

Meyer-Peter, E., and R. Muller. 1948. Formulas for Bed-Load Transport. Proceedings, 2nd Meeting, International Association of Hydraulic Structures Research, Stockholm, Sweden. p. 39-64.

Pickup, G., and R. F. Warner. 1976. Effects of Hydrologic Regime on Magnitude and Frequency of Dominant Discharge. Journal of Hydrology 29: 51-75.

Reid, L. M., and T. Dunne. 1984. Sediment Production From Forest Road Surfaces. Water Resources Research 20: 1753-1761.

Selby, M. J. 1982. Hillslope Materials and Processes. Oxford University Press, Oxford, England.

Shannon & Wilson, Inc. 1991. Sediment-Related Construction Scheduling Guidelines. Report prepared for Hydro West Group, Bellevue, Washington. 18 p.

Simons, D. B., and F. Senturk. 1977. Sediment Transport Technology. Water Resources Publications, Fort Collins, Colorado. 807 p.

Tabor, R. W., and others. 1988. Preliminary Geologic Map of the Sauk River 30 by 60 minute Quadrangle, Washington. U.S. Geological Survey Open-File Map 88-692.

Washington Department of Fisheries. 1987. Hydroelectric Project Assessment Guidelines. draft report by the Washington State Department of Fisheries, Olympia, Washington. 99 p.

Wolman, M.G. and Miller, J.P. 1960. Magnitude and frequency of forces in geomorphic processes. Journal of Geology, v. 68, no. 1, p. 54-74.

RESPONSE AND RECOVERY OF THE MISSOURI RIVER DOWNSTREAM OF FORT PECK DAM MONTANA, WITH RESULTING BOUNDARY DISPUTE

W.R. Womack[1]

ABSTRACT

Following closure of Fort Peck dam in 1937, the Missouri River degraded and widened for about 47 miles downstream of the dam, triggering a complex response that affected the river for hundreds of miles downstream. Most of the degradation near the dam occurred between 1945 and 1955. Between 1949 and 1967, a reach about 180 river miles downstream widened and shallowed, apparently in response to degradation and bank erosion upstream. The episode of widening on the downstream reach was followed by entrenchment after 1967, and channel widths decreased to near pre-dam values. As the channel began to narrow, a series of mid-channel bars formed, many of which eventually attached to the bank on both sides, forming a terrace below the floodplain. When oil was discovered in the 1980s underlying the floodplain just upstream of the North Dakota border, the State of Montana initiated legal proceedings to claim portions of the young terrace and the underlying oil. River law in Montana and other states awards the state all islands and "accumulations of land" formed within the beds of streams. Although the district court decided in favor of the landowners, the Montana Supreme Court overturned the verdict, accepting the state's argument that the terrace consists of land accreted to islands rather than land accreted to the banks. This case clearly illustrates the problems created by river laws that do not account for natural process or temporary changes. Following the reasoning accepted by the Montana Supreme Court, the state would be awarded the terrace along both banks. As the river continues to migrate, large areas of accreted land would then become state property, and eventually the state would own a major portion of the floodplain. A similar battle over oil occurred along the Yellowstone River near Sidney, where the state claimed a large point-bar, again following the reasoning that the point-bar originated as an "island". Since point-bars do form initially in the channel opposite the accreting bank, the state was able to produce air photos that illustrated the point-bar's origin as an "island or accumulation of land" within the channel.

[1] Principal, Womack and Associates, 5825 Lazy Lane, Billings, Montana 59106.

KEYWORDS

River metamorphosis, complex response, dam effects, river islands, river law

1. INTRODUCTION

River changes and boundary disputes go hand-in-hand. When a river migrates, property boundaries shift with the river, through a process known as accretion. When a river abruptly changes course (as a result of a meander cutoff, for example), the boundary remains fixed, and the process is known as avulsion. Although accretion and avulsion have specific technical meanings that refer to river processes, in the legal realm these terms are also defined by rates of change, i.e., if land forms along a river "slowly and imperceptibly," by definition the process is accretion, and vice versa for avulsion. The laws of Montana and other states further confuse the issue by awarding to the state all "islands and accumulations of land" formed in the beds of streams. Since lateral accretion does not occur on the banks of streams, but rather in the beds of streams, problems of interpretation are inevitable. This paper addresses changes to the Missouri River downstream of Fort Peck dam, and the legal battle that followed when classical river law attempted to cope with a river system undergoing fundamental and human-induced change.

Following closure of Fort Peck dam (Fig. 1.1) in 1937, the Missouri River degraded and widened for about 47 miles downstream of the dam, as is common downstream of major dams (Williams and Wolman, 1984). These changes triggered a complex response that affected the river for more than 180 river miles downstream over a period of more than 50 years. Near the downstream end of the altered reach, the river widened and shallowed considerably, forming many bars and eroding large areas of both banks. Thanks to a fortuitously located oil well, the response of the river led to a boundary dispute that eventually reached the Montana Supreme Court.

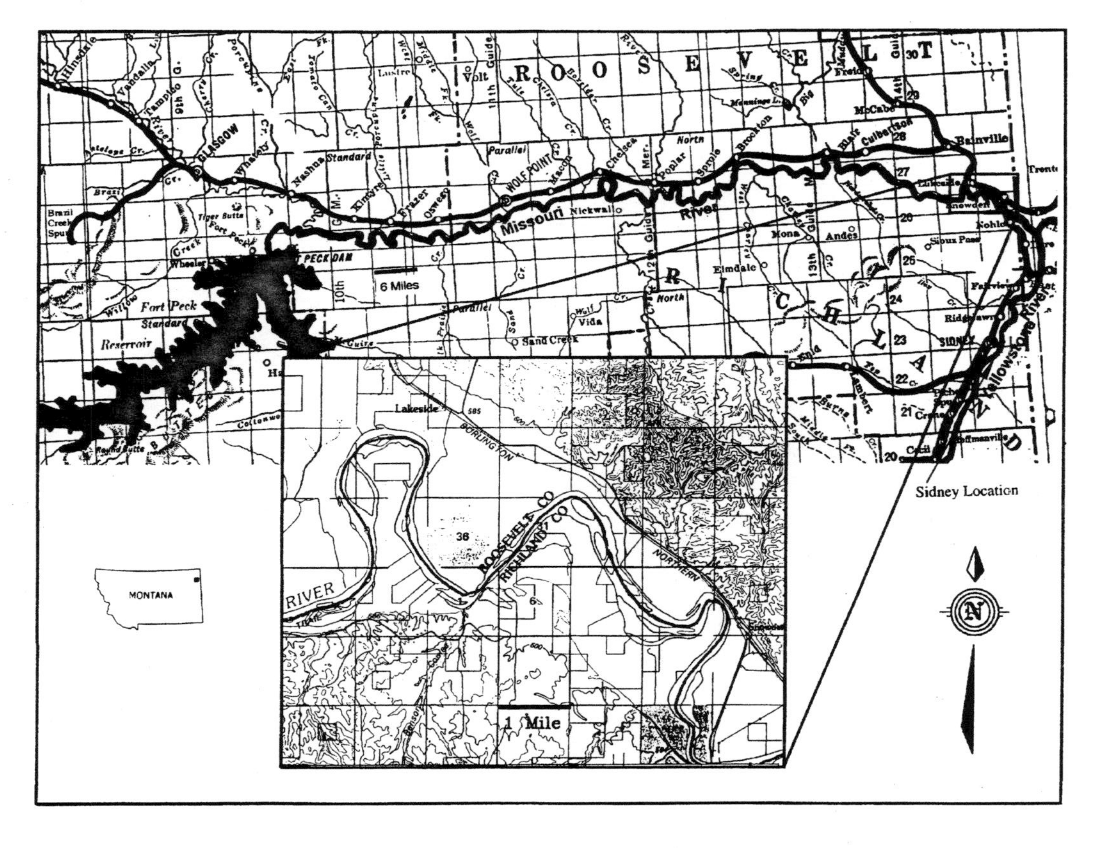

Fig. 1.1. Location map.

2. EFFECTS OF FORT PECK DAM

Fort Peck Dam caused significant changes to the flow regimen of the Missouri, including rapid variations in stream flow, much higher than normal winter flows, and reduced peak flows. Obviously the water emerging from the dam carried little sediment load, so its erosive power was enhanced.

2.1. Alterations to Flow

Williams and Wolman (1984) documented the following changes: the average daily discharge increased from 7,063 to 9,888 cfs. The average annual peak discharge was slightly smaller (24,000 cfs as compared to 27,000 cfs pre-dam). In general, the high flows are higher and the low flows are lower after closure because the flow is curtailed for parts of the day and relatively large flows are released at other times. For example, on December 18, 1988, the low flow was 2,880 cfs and the high flow was 15,960 cfs.

2.2. Changes to Reach Immediately Downstream of Dam

The streambed degraded by 4.9 feet at a gaging station 8 miles downstream of the dam (Williams and Wolman, 1984). Maximum degradation of 5.7 feet was recorded in 1973, 14 miles downstream. Degradation 47 miles downstream was 0.8 foot in 1973. However, cross-section data compiled by the Corps of Engineers (Corps) during the period 1936 to 1994 indicate that degradation was irregular except in the first few miles downstream of the dam. Channel width in the degraded reach increased by as much as 55 feet between 1936 and 1973, although some reaches experienced little change in width. The Corps data were recorded in English units, and that format will be used hereafter.

2.3. Changes Downstream of Culbertson, Montana

The Missouri enters a narrow gorge at Culbertson, about 140 river miles downstream of Fort Peck dam. It emerges from the gorge about 160 river miles downstream of Fort Peck, just upstream of the North Dakota border, and forms three prominent meander loops (Fig. 1.1). This reach, which is about 125 river miles downstream of the degraded reach described by Williams and Wolman (1984), underwent profound changes during and immediately following the period of degradation. The changes were investigated using historical surveys, aerial photographs, and the database compiled by the Corps of Engineers for study of the river's response to Fort Peck Dam.

The 1894 survey (Fig. 2.1) shows the three meanders to be in roughly the same positions as today, and the river appeared to be rather narrow compared to its width in later years, with a single channel. Aerial photographs taken in 1938 (Fig. 2.2) exhibit the same pattern.

Beginning between 1938 and 1949 and continuing into the late 1960s, the river widened and shallowed. The width increased by as much as three times and numerous mid-channel bars formed. Survey data collected by the Corps clearly illustrate this trend. However, the aerial photographs provide a particularly graphic illustration of the river changes. The river is visibly wider and there are obviously many more bars in the photos taken in 1949 (Fig. 2.3), 1956 (Fig. 2.4), and 1967 (Fig. 2.5). These changes correspond in time to the period of degradation below Fort Peck, where most of the changes occurred between 1945 and 1955 (Williams and Wolman, 1984).

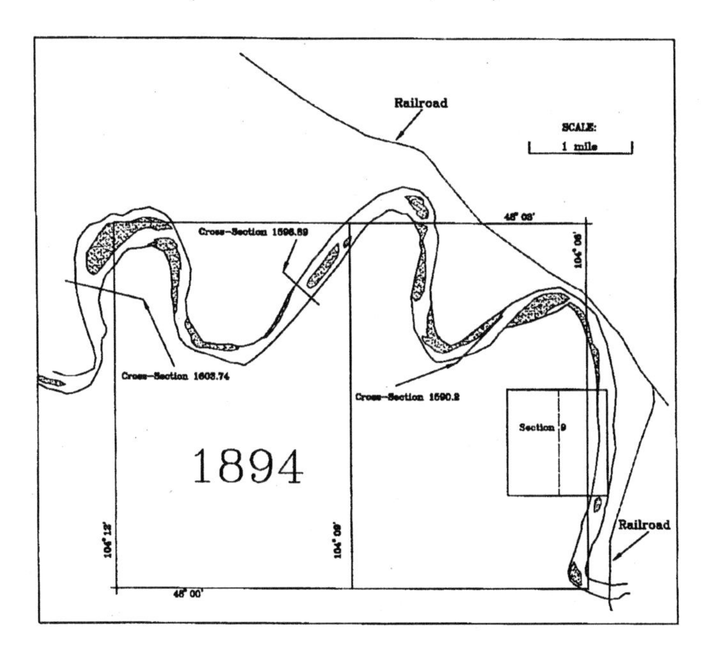

Fig. 2.1. **Missouri River Channel in 1894, based on map published by the Missouri River Commission, 1894.**

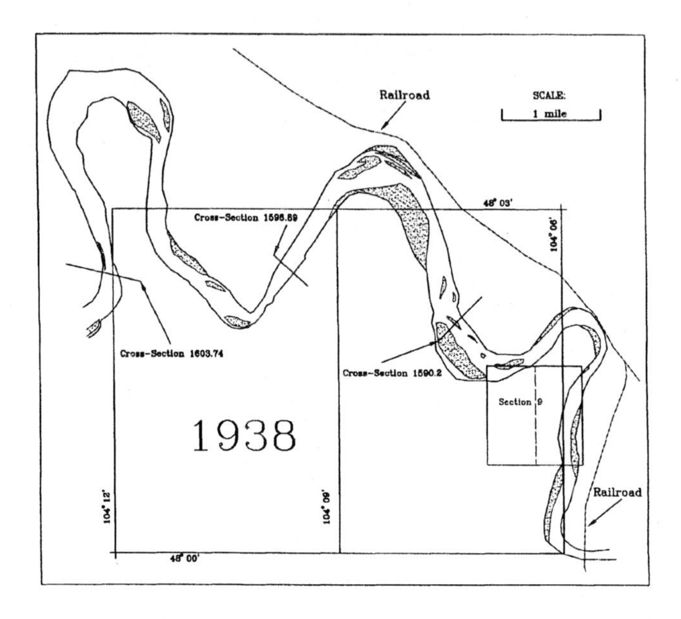

Fig. 2.2. Missouri River Channel in 1938, based on aerial photographs.

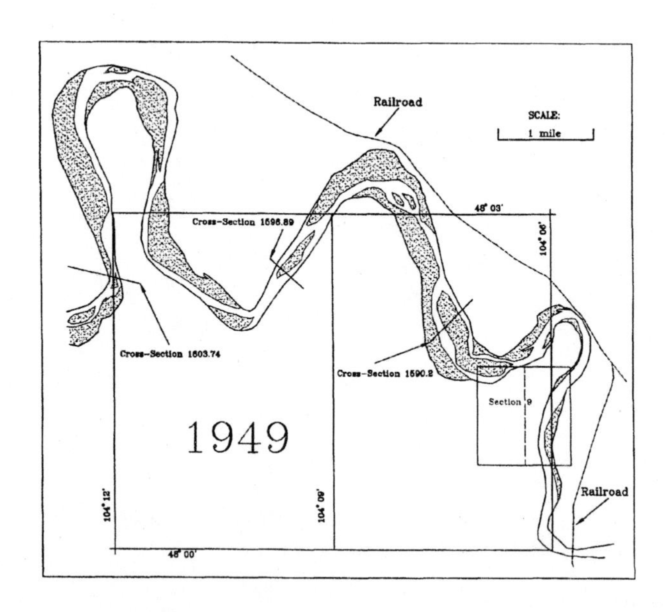

Fig. 2.3. Missouri River Channel in 1949, based on aerial photographs.

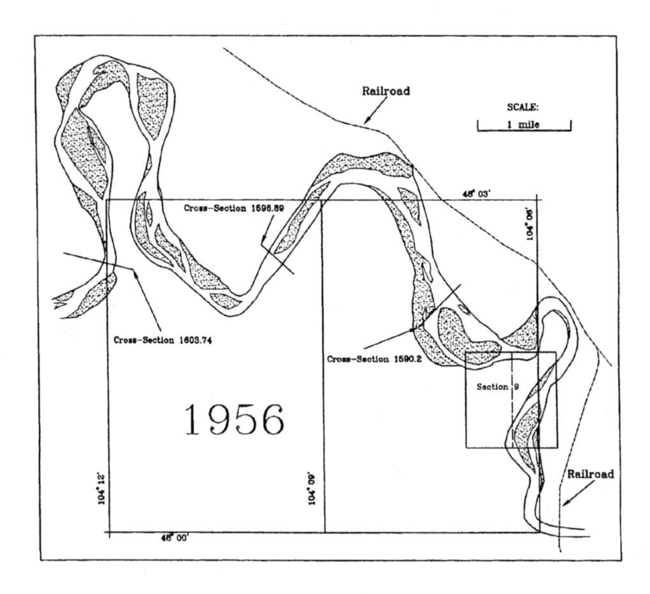

Fig. 2.4. Missouri River Channel in 1956, based on aerial photographs.

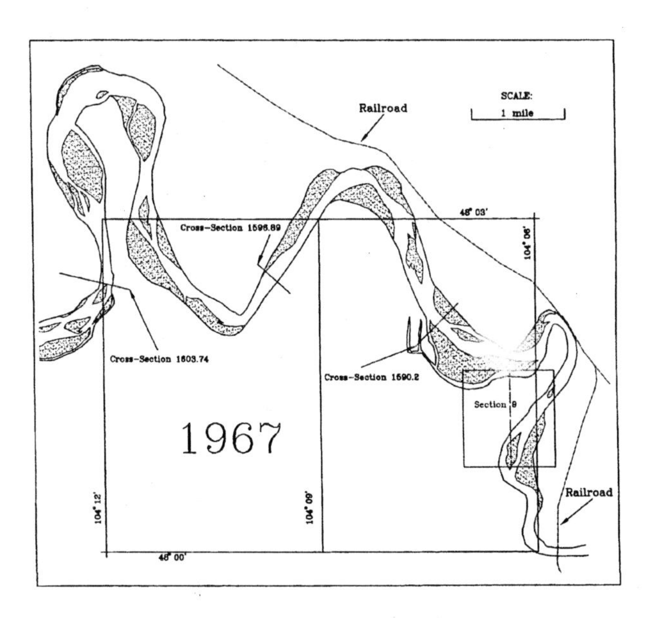

Fig. 2.5. Missouri River Channel in 1967, based on aerial photographs.

Following the period of widening, the river apparently adjusted to the changes in regime, and began to narrow and deepen, taking on a channel pattern similar to the predam morphology (Figs. 2.6 and 2.7). During this period many of the mid-channel bars attached to the banks, forming low terraces along both banks (Figs. 2.8 and 2.9).

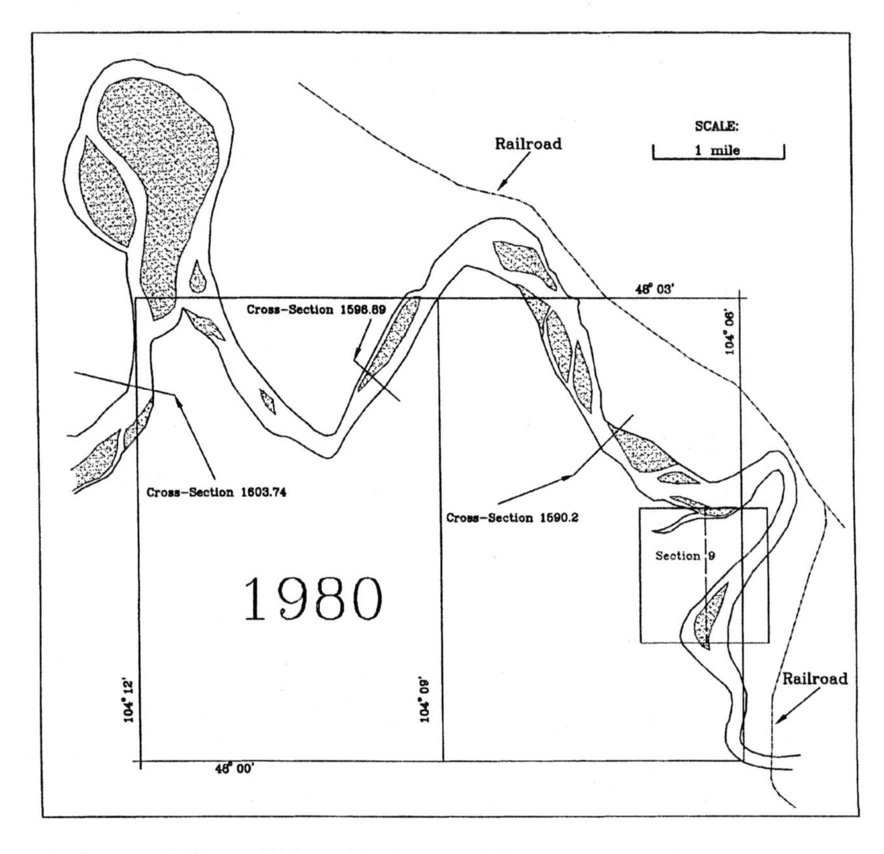

Fig. 2.6. Missouri River Channel in 1980, based on aerial photographs.

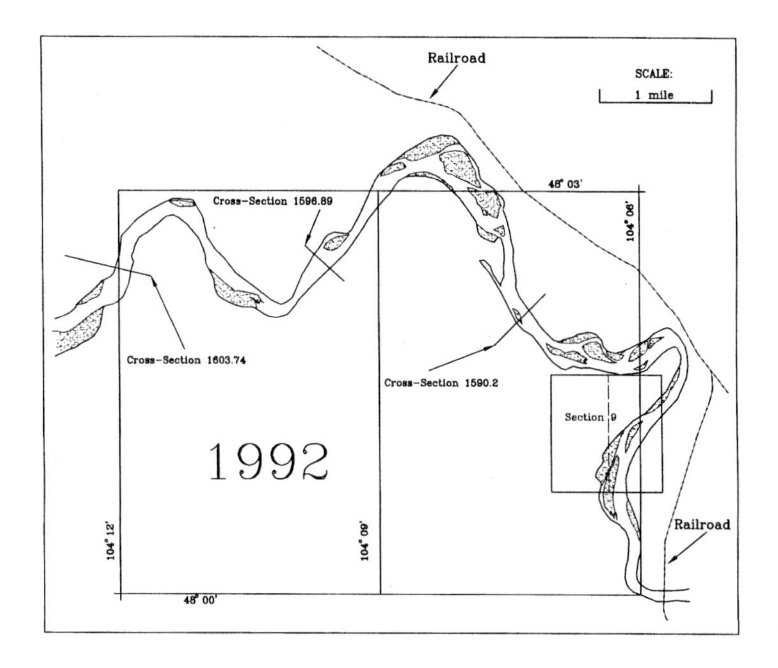

Fig. 2.7. Missouri River Channel in 1992, based on aerial photographs.

Fig. 2.8. Oblique aerial photograph of the accreted mid-channel bars near the South Boundary of Section 9; view northeast.

Fig. 2.9 Ground photograph of mid-channel bars accreted to the riverbank near the Boundary of North and South Half of Section 9; view south.

2.4. Analysis

The response and recovery in the study reach downstream of Culbertson seems to be very similar to the "river metamorphosis" process documented by Nadler and Schumm (1981) and Bradley (1983). The river apparently responded to an increase in sediment load triggered by bank erosion and degradation in the reach immediately downstream of Fort Peck Dam. Changes this far downstream from a major dam are unusual, but the pattern of widening and aggradation appears to be similar to changes recorded on the Colorado River near Needles, California, after the closure of Hoover Dam (Williams and Wolman, 1984). Zhou and Pan (1994) documented major changes to the Yellow River in China hundreds of kilometers downstream of Sanmenxia Reservoir.

Figure 2.10 illustrates changes in width measured from aerial photographs of the study reach, showing dramatic widening following 1938 and narrowing following 1967. The cross sections follow section line boundaries in the area of the boundary dispute discussed in Section 4. Near the west boundary of Section 9 (and extended northward, Section 4), Township 26N, Range 39E, Richland County, the channel widened from about 1,400 feet to about 3,600 feet between 1938 and 1956, and narrowed following 1967 to about 900 feet by 1992. Along the south boundary of Section 9, the river widened from about 600 feet in 1938 to about 2,000 feet in 1956, and narrowed to about 1,100 feet by 1992. Along the east-west half-section line of Section 9, the change was 600 feet to about 1,300 feet between 1938 and 1956, narrowing to about 1,000 feet by 1992.

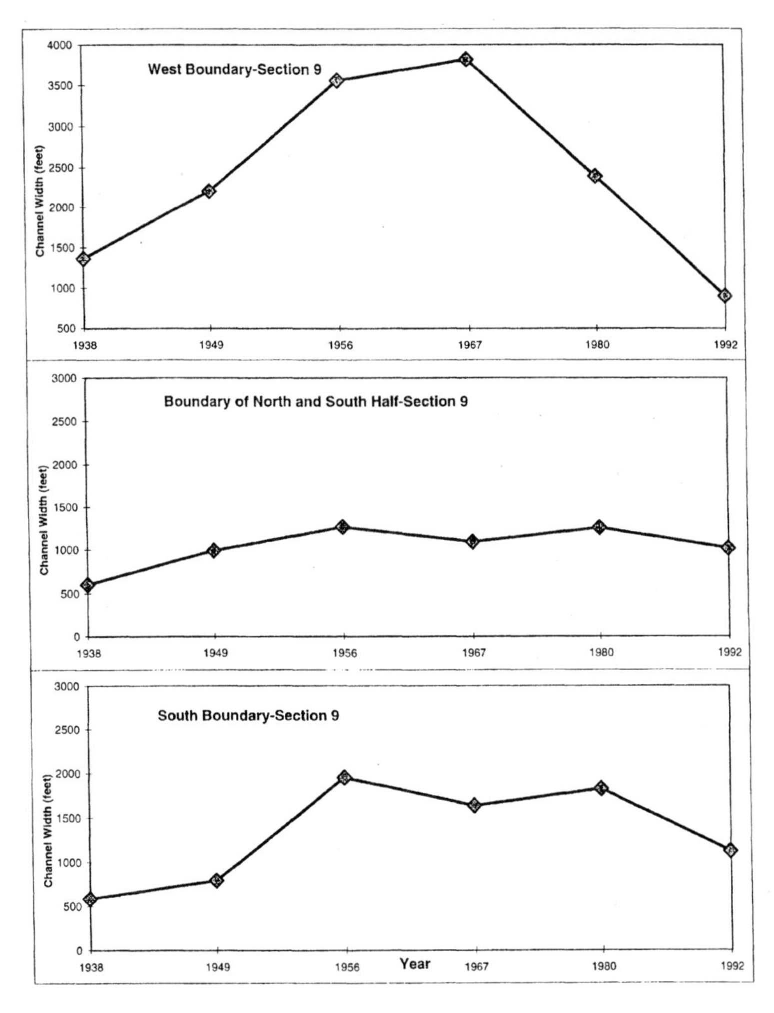

Fig. 2.10. Changes in channel width in the study reach; note the decreasing trend back to pre-dam conditions.

Figures 2.11 through 2.13 present thalweg elevation and average bed elevation recorded by the Corps downstream of Fort Peck during the period

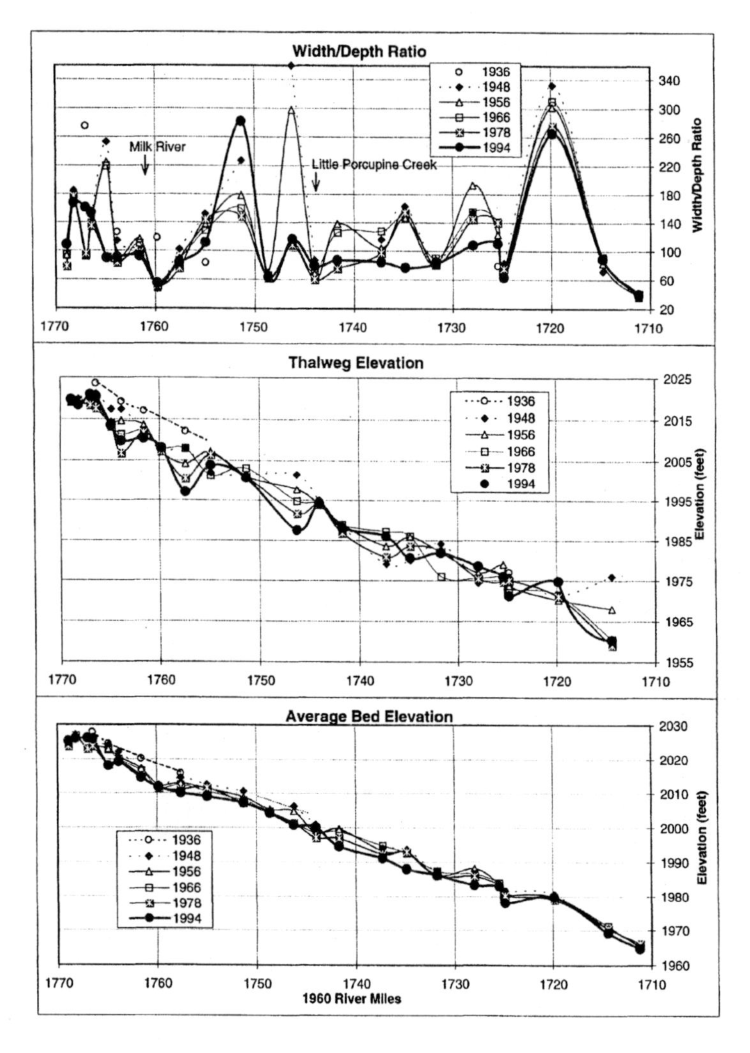

Fig. 2.11. **Elevation and width/depth ratios for River Miles 1770-1710, 1936-1994.**

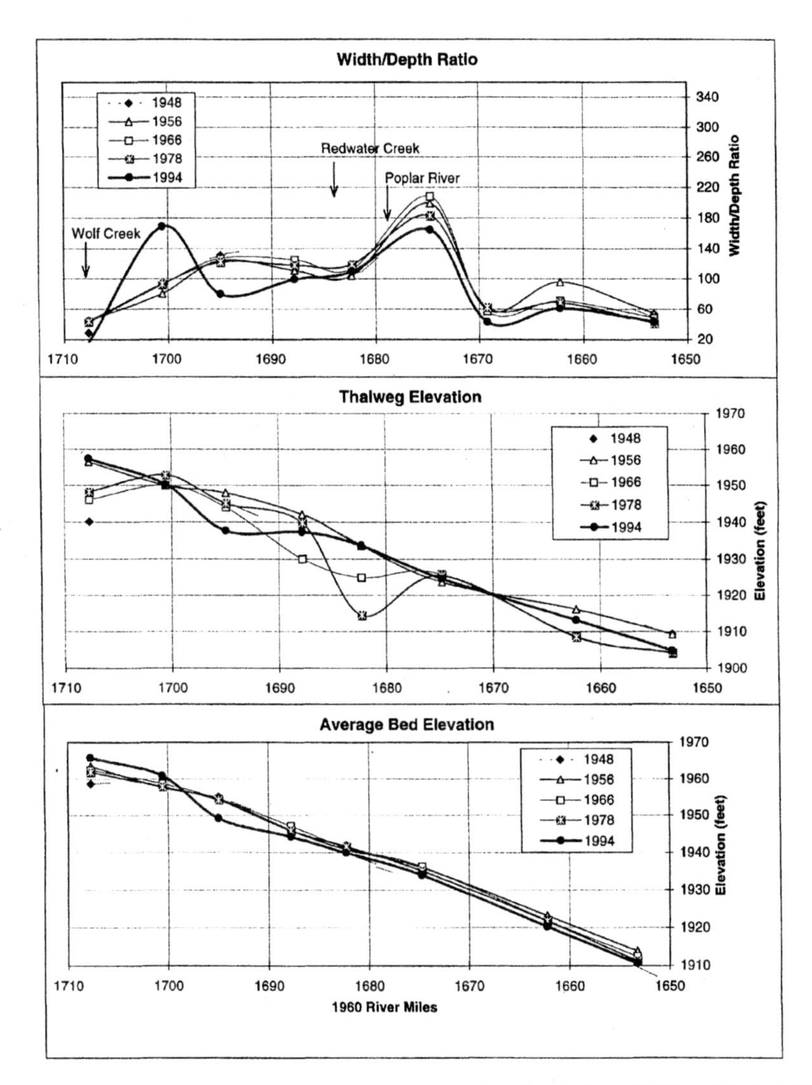

Fig. 2.12. Elevation and width/depth ratios for River Miles 1710-1650, 1948-1994.

1936 to 1994. Width/depth ratios calculated from the Corps data, using top width (normalized to a discharge of 26,000 cfs) divided by average depth, have also been included on Figs. 2.11 through 2.13. The river has been divided into

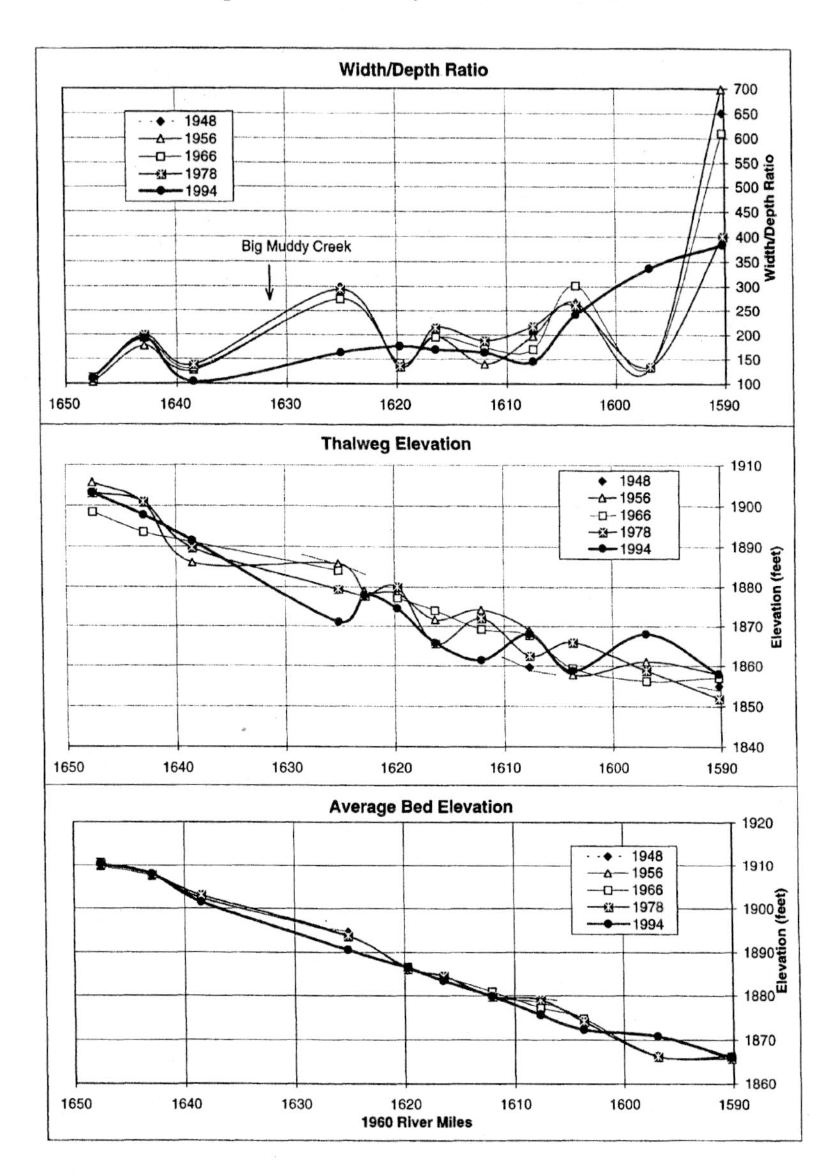

Fig. 2.13. Elevation and width/depth ratios for River Miles 1650-1590, 1948-1994.

three reaches of approximately 60 river miles each for convenience of presentation. The study reach is at the downstream end of the Corps study area (Fig. 2.13), beginning at about 1960-river-mile 1605 and extending downstream to about mile 1585, beyond the cross section at mile 1590.2, which is the farthest downstream measured by the Corps.

The cross-section data are rather sparse through the study reach, and the positioning of the sections may not adequately represent river behavior. Also, very few data were recorded in the years immediately preceding and following dam closure. In 1936, for example, cross sections were measured regularly for only about 10 river miles below the dam (Fig. 2.11). In 1948, regular cross-section measurements were extended downstream to about river mile 1682, about 87 river miles below the dam. Only after 1956 were cross sections measured at regular intervals downstream into the study reach.

Following closure of the dam, both the thalweg elevation and average bed elevation decreased considerably downstream to about river mile 1755 (Fig. 2.11). The average bed elevation continued to degrade until at least 1978 downstream to about river mile 1725 (about 40 river miles below Fort Peck). Williams and Wolman (1984) reported degradation for about 47 miles downstream of the dam. However, the degradation was irregular, and thalweg elevations exhibited periodic increases and decreases at many cross sections downstream of river mile 1755 after 1948 (Fig. 2.11). The river initially widened through some portions of the degraded reach, but width/depth ratios in the years following 1956 appear to be consistently lower than in 1948.

Downstream of river mile 1725, aggradation and degradation occurred with no clear pattern, apparently expressing complex response of the river to the changes upstream (Petts, 1979). The width/depth ratios and thalweg elevation showed considerable local variation, although average bed elevations appear to have undergone little change.

The term "complex response" (Schumm, 1973) refers to the complex reaction of the river system to an initial triggering event. In the case of the Missouri downstream of Fort Peck dam, the reach immediately downstream of the dam responded to changes in flow regimen by degrading and widening. However, farther downstream the change in sediment load triggered a much more complicated series of events, so that some reaches were undergoing aggradation and others degradation at the same time.

The only Corps cross sections within the study reach are at 1960 river miles 1603.74; 1596.89; and 1590.2. Their locations are shown on the figures drawn from air photos, and the three points at the right end of Fig. 2.13 represent elevations and width/depth ratios. The sections at 1603.74 and 1596.89 exhibited relatively little change between 1956 and 1978. Between

1978 and 1994, a cutoff just downstream of river mile 1603.74 may have caused the channel to degrade at 1603.74, and aggrade at 1596.89. The width/depth ratio at mile 1590.2 decreased abruptly after 1966, reflecting abandonment of the south channel as the river narrowed (compare Figs. 2.5 and 2.7).

Figure 2.14 illustrates relative changes in thalweg elevation occurring in the intervals between measurements (which vary from 8 to 12 years) taken from the Corps data presented on Figs. 2.11 through 2.13. There were consistently negative elevation changes in the degrading reach immediately downstream of the dam up until about 1978. However, downstream of about river mile 1755, positive elevation changes begin to appear as early as the interval 1948-1956, and gains and losses in thalweg elevation with time become erratic. Large fluctuations in thalweg elevation, approaching 20 feet at some cross sections, occurred farther downstream. Downstream of about river mile 1650, the elevation gains and losses between measurements exhibit an interesting regularity after 1956, and suggest that pulses of aggradation and degradation may move progressively downstream, leading to a complex response of the river system.

After 1978 most of the elevation changes in the reach downstream of river mile 1650 were negative, indicating a general entrenchment of the lower reach. It seems possible that the river may have widened as the normal pattern of flow and sediment load was altered, and then narrowed and degraded as the complex response of the system attenuated with time.

In the upper reach below Fort Peck dam, the average width/depth ratio was about 135 in 1936, increasing to 151 in 1948 and decreasing gradually to about 110 after 1978 (Fig. 2.15). In the middle reach the average width/depth ratio in 1948 was about 80, increasing to about 98 in 1966 and decreasing gradually to 87 in 1994. On the lower reach, including the study area, the maximum average width/depth ratio was about 384 in 1948, decreasing gradually to 243 in 1994. Clearly the river has narrowed with time throughout its length between Fort Peck and the North Dakota border. The width/depth ratios are very high, particularly on the lower reach, similar to those documented by Schumm et al. (1994) on the Mississippi River below Cairo, Illinois.

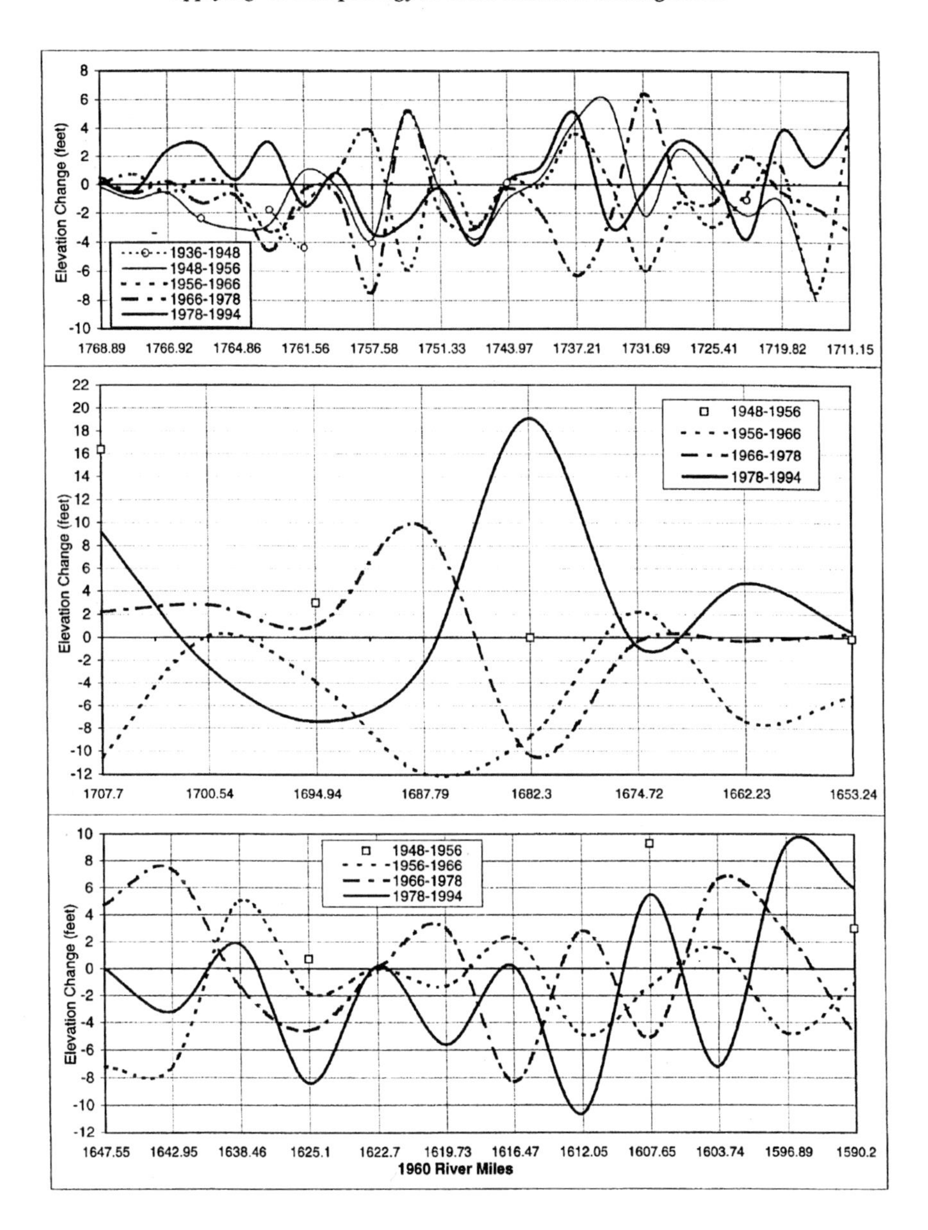

Fig. 2.14. Relative changes in thalweg elevation, 1936-1994.

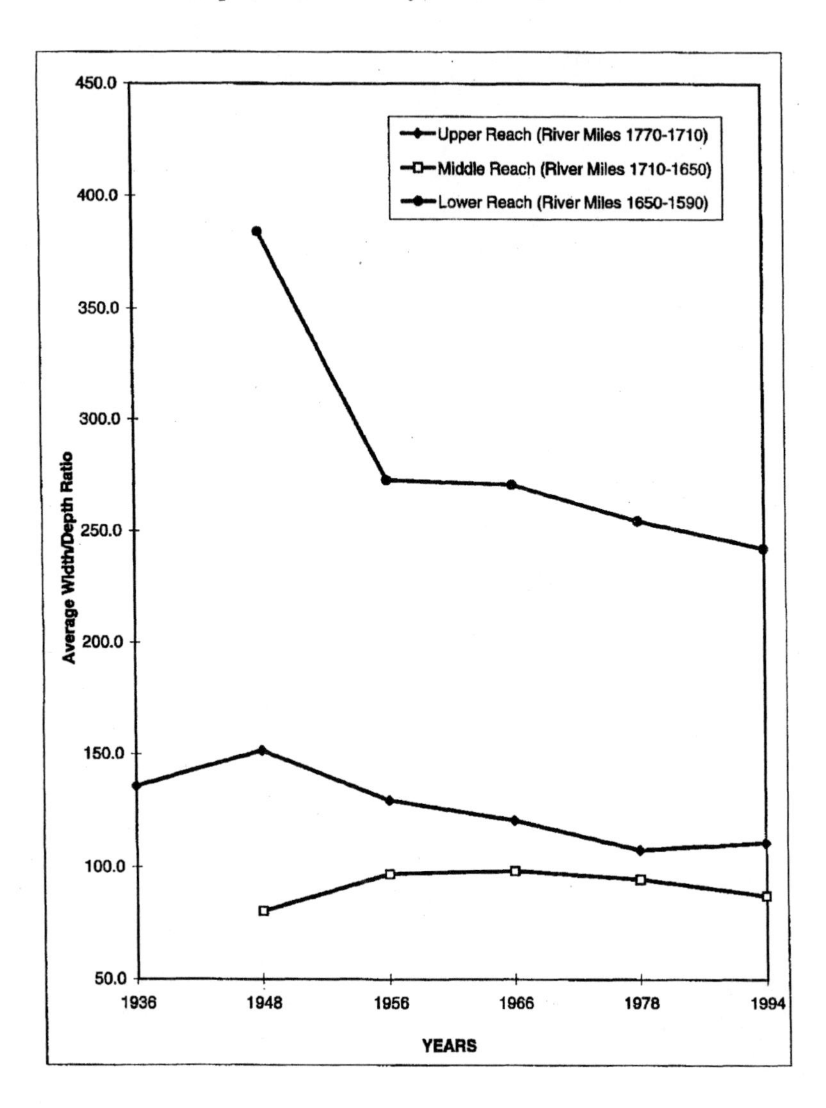

Fig. 2.15. Average width/depth ratios for the upper, middle, and lower river sections.

Tributaries seem to have little influence on river behavior. With the exception of the Milk River, which flows into the Missouri just below Fort Peck dam, tributaries downstream of the dam are prairie streams with very small discharges compared to the main stem Missouri.

Discharge data from the gaging stations at Wolf Point and Culbertson (Fig. 2.16), for the period 1929 through 1995, reflect the changes discussed by Wolman and Miller (1984). The total annual discharge shows a slight upward trend until 1983, then a decline. The annual peak discharge shows a strong downward trend over the whole period, although the annual variations are rather large. There appears to be no obvious pattern in these data to explain the response and recovery of the river in the study reach.

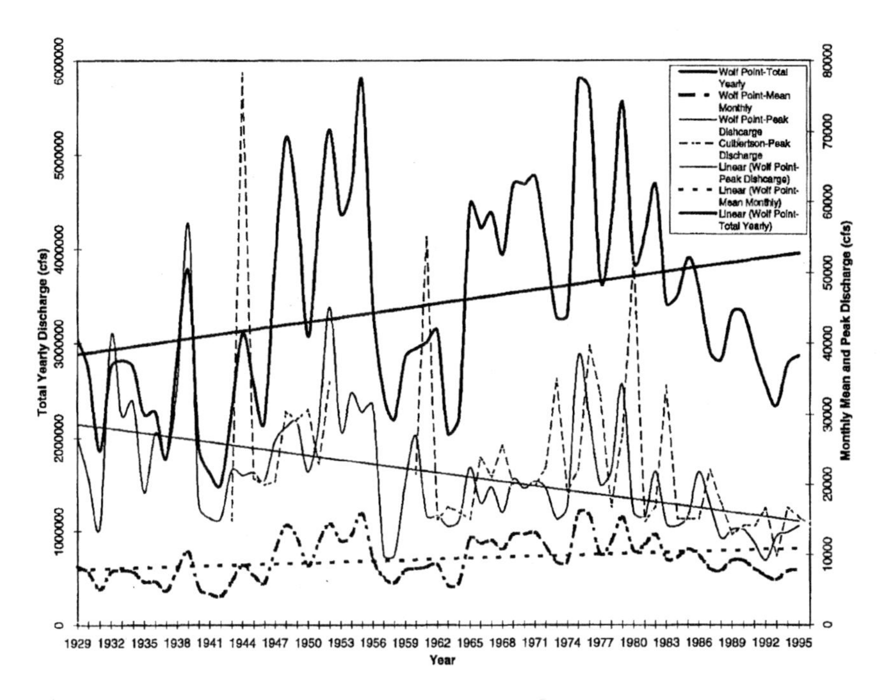

Fig. 2.16. Discharge data from Wolf Point and Culbertson Gaging Stations on the Missouri River.

In summary, there does not appear to be an explanation for the geomorphic changes that occurred downstream of Fort Peck dam, other than the closure of the dam itself. The changes in the reach immediately downstream of the dam have been well documented (Corps of Engineers, 1952; Williams and Wolman, 1984) and are clearly attributable to the changes in flow regime and increase in erosive power that the dam imposed on the river. This study has shown that during the same period the shape of the river was altered hundreds of miles downstream of the dam.

These changes do not conform to the normal technical concepts of accretion and avulsion. Accretion and avulsion are normally defined in the context of a stable meandering river system, where in the course of migration the river erodes the outer banks of meander bends and accretes point-bar deposits along the inner banks, undergoing periodic avulsive changes when meanders are cut off. Although a stable river changes its position, it maintains its basic cross section and planform within a relatively narrow defining range of channel parameters such as gradient, sinuosity, and width/depth ratio. During the period of change in the study reach the geometry of the channel, as expressed by the width/depth ratio, underwent drastic change.

Inherent to the process of erosion and accretion is that the river migrates through its floodplain, reworking the floodplain deposits and replacing them with point-bar deposits (Chorley et al., 1984). On the other hand, when avulsion occurs, the river "jumps" to a new channel, bypassing a portion of the floodplain and leaving it intact. In this case, the river underwent a fundamental change in form, attacking both banks and creating numerous mid-channel bars, which eventually joined the banks on both sides to form a low terrace. The terrace-forming process does not meet the technical definition of accretion, because it did not involve meander migration and point-bar formation. Nor is it avulsion, because an abrupt channel change did not leave behind identifiable land. If anything, it is probably closest to accretion, because the lands in question were initially destroyed and then replaced by deposition against the riverbanks.

3. RIVER BOUNDARY LAW

The most important defining terms in river boundary law are "accretion" and "avulsion". Chorley et al. (1984) state: "*Of interest is the fact that if there is a political or administrative boundary line associated with the stream, the former is often held to move with the channel if the channel shifts slowly due to a process of bank erosion and lateral accretion. However, the boundary will often be held to remained fixed if the change is abrupt and avulsive in nature, for example, when a neck or chute cutoff occurs.*"

Beck (1967), in a study of accretion law on the Missouri River, provided a brief description of accretion, avulsion, and other legal terms involved in river migration:

- Accretion is generally defined as the process where the action of water causes the gradual and imperceptible deposit of soil in a certain place so that it becomes dry, "fast" land.
- When water gradually recedes baring land in the process, the proper descriptive term is dereliction.
- If soil is lost by the gradual encroachment of water, the process is generally referred to as erosion.
- When the water location changes suddenly, as for example, when a river leaves its old bed forming an entirely new one, the inundating process is referred to as avulsion and the baring process as reliction, as contrasted with dereliction above. Generally today, however, the entire swift change process is referred to as avulsion, whereas reliction is used in the situation described above where dereliction should be used.

This discussion illustrates the problems that can arise when legal definitions are applied to river processes (including natural or man-caused changes). For example, in the case of McCafferty vs. Young in Lewis and Clark County, Montana, it was concluded that when the river channel moved one quarter-mile in less than 100 years, the migration was "perceptible" and, therefore, avulsive, and the change in the course of the river did not affect location of the county line which was formerly located in the center of the river's old channel (Beck, 1967). This reasoning bears no relationship whatever to the processes of river migration and concepts of accretion and avulsion used by the technical community. The following table summarizes the technical and legal definitions of these terms, in the context of river migration.

The main difference between the technical and legal concepts is that the technical definitions are process-oriented and the legal definitions are time-oriented. The geomorphologist views migration of a river through its floodplain as erosion and accretion, no matter how rapidly the river moves.

Term	Technical Definition	Legal Definition
Erosion	Reworking of floodplain deposits by removal on the outside of meanders	Removal of soil by gradual encroachment of water
Accretion	Deposition of point-bars on the inside of meanders	Gradual and imperceptible deposition of soil to become dry land
Avulsion	Abandonment of river channel in favor of a shorter path, say across a meander neck or to the sea	Swift change of river location.

Land ownership was the responsibility of the Federal government up until statehood. According to the "equal footing" doctrine of U.S. law promulgated in 1844, after statehood the beds of rivers became state property. Under the laws of Montana, North Dakota, and other states, all islands and "accumulations of land" that rise from the beds of rivers are state land. If channels between islands and riverbanks are abandoned by the river, the boundary between state property and the adjacent landowner's property becomes the centerline of the abandoned channel.

The legal definition of islands is significant because point-bars, which are classic lateral accretion deposits, arise within the beds of streams rather than on the banks (Morisawa, 1985). The point-bar sometimes originates as a mid-channel bar, separated from the accreting bank by a back channel. The upstream end of the back channel becomes plugged, and the bar joins the bank. Even though the land formed by this process is clearly accreted, if viewed early in the process (say on an aerial photograph), it will appear as a midchannel bar, subject to legal interpretation as an "island."

4. SECTION NINE BOUNDARY DISPUTE

The Missouri River passes Section 9, Township 26N, Range 39E, in Richland County, Montana, about 2.5 miles upstream of the North Dakota boundary (Fig. 1.1). An oil well was successfully completed in Section 9 in 1984. The State of Montana Oil and Gas Conservation Commission assigned the west half of Section 9 as the "spacing and pooling unit" for the well. The Montana Department of State Lands subsequently filed for quiet title of portions of the unit that had been eroded by the river over the years.

In the years following 1967, lands rose from the riverbed in the areas claimed by the state, as the river narrowed and recovered from the impact of Fort Peck Dam (compare Figs. 2.5 and 2.7). By 1987, low terraces had formed in the areas of Section 9 eroded earlier, particularly at the north end and at the southeast corner of the west half section. The state contended that the terraces and bars were islands or lands accreted to islands.

The landowners' attorneys had originally intended to contend that the rate of change was too rapid to be "imperceptible", and the boundaries should remain at their original locations because the movement was avulsive. However, true avulsion was obviously not involved, and they eventually argued that the changes in the river had occurred due to the temporary changes caused by Fort Peck Dam, and that the terraces formed against the banks following recovery should be viewed as accreted lands. The landowners' attorneys also argued that since point-bars initially form within river channels and are obviously accreted, the fact that the bars originated within the channel was irrelevant.

The landowner prevailed in the Richland County District Court in Sidney, but the verdict was subsequently overturned by the Montana Supreme Court in 1992, which agreed with the state's contention that the bars which became the terrace were "islands." The Montana Supreme Court correctly understood that the changes were not avulsive. The bars were determined to be islands because they had become vegetated before the back channels were abandoned. The terrace at the north end of Section 9 had small trees, and the terrace along the east side had grass and willows.

Although the Montana Department of State Lands has not as yet chosen to do so, the Supreme Court decision opens the door to sweeping claims along the Missouri River. Figure 4.1 is a map based on the 1991-1992

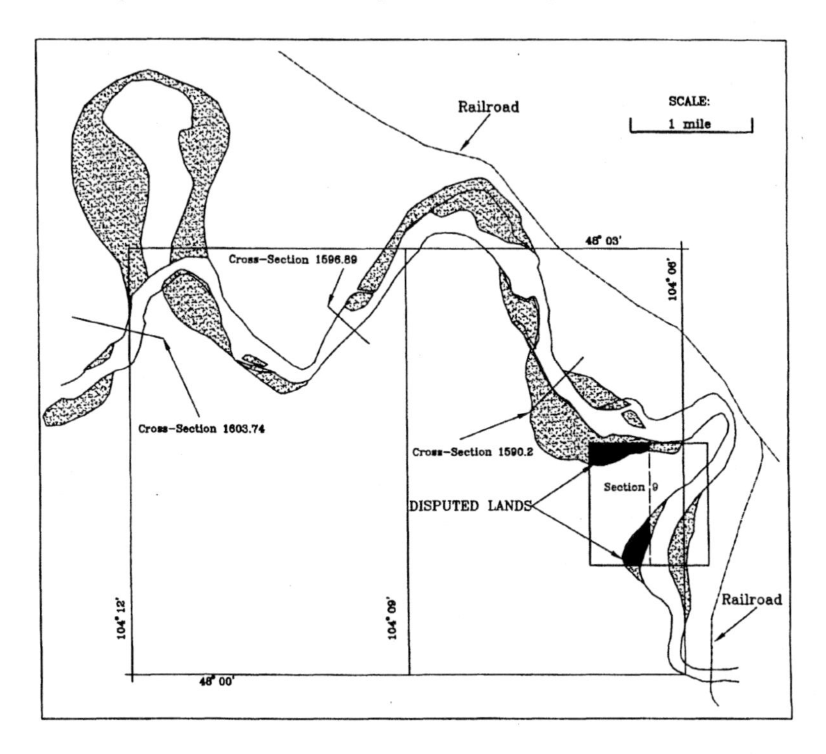

Fig. 4.1. Stippled areas are lands that rose from the bed of the river and which belong to or could be claimed by the State. Disputed lands in Section 9 were claimed by the State.

aerial photographs, with similar terrace areas outlined upstream and downstream of Section 9. As the figure shows, large areas of both banks could be claimed by the state following the reasoning that prevailed at Section 9. An interesting potential consequence is that the state would own the accreted lands regardless of the direction of river migration. This appears to violate the principle of "compensation" normally implicit in the process of river migration and accretion. As noted by Williams and Wolman (1984), channel widening downstream from dams may result in erosion of both banks without compensatory deposition. The extensive river widening that occurred in this reach allows the state to gain land at the expense of all riparian landowners because the normal process of erosion and accretion was altered.

5. DISCUSSION

At least three fundamental problems appear in Montana river boundary law, some of which extend to river law in general. The first is the emphasis on rates of change; i.e., the use of words such as "perceptible" and "imperceptible". Accretion and avulsion have specific technical definitions, none of which have anything to do with speed. The second problem is the notion that accreted lands do not rise in the river channel. Lateral accretion occurs in the channel, typically as point-bars develop in the channel and accrete to the bank (Morisawa, 1984). In Montana, if it can be demonstrated that a body of land originated in the stream (i.e., with a channel separating it from the bank), it is classified as an "island" and granted to the state. An example case is discussed in the following paragraph. The third obvious problem is that river law in general has difficulty in coping with unusual processes like river metamorphosis or temporary alterations to river systems caused by dams, land use changes, etc.

Up to now, the state of Montana has only claimed oil lands along the rivers. In a case on the Yellowstone River near Sidney (Fig. 1.1), the state successfully claimed a large area of oil-bearing land that originated as a mid-channel bar and eventually accreted to the bank. As shown by aerial photographs, the disputed land originated as a mid-channel bar sometime between 1938 and 1949. As river migration continued, the bar joined the left bank of the river between 1967 (aerial photo Fig. 5.1a) and 1980 (aerial photo Fig. 5.1b), and is by all appearances a classic point-bar. Further study of the aerial photograph on Fig. 5.1b shows that this process has been repeated several times on this particular bend. The outlines of old point-bars and back

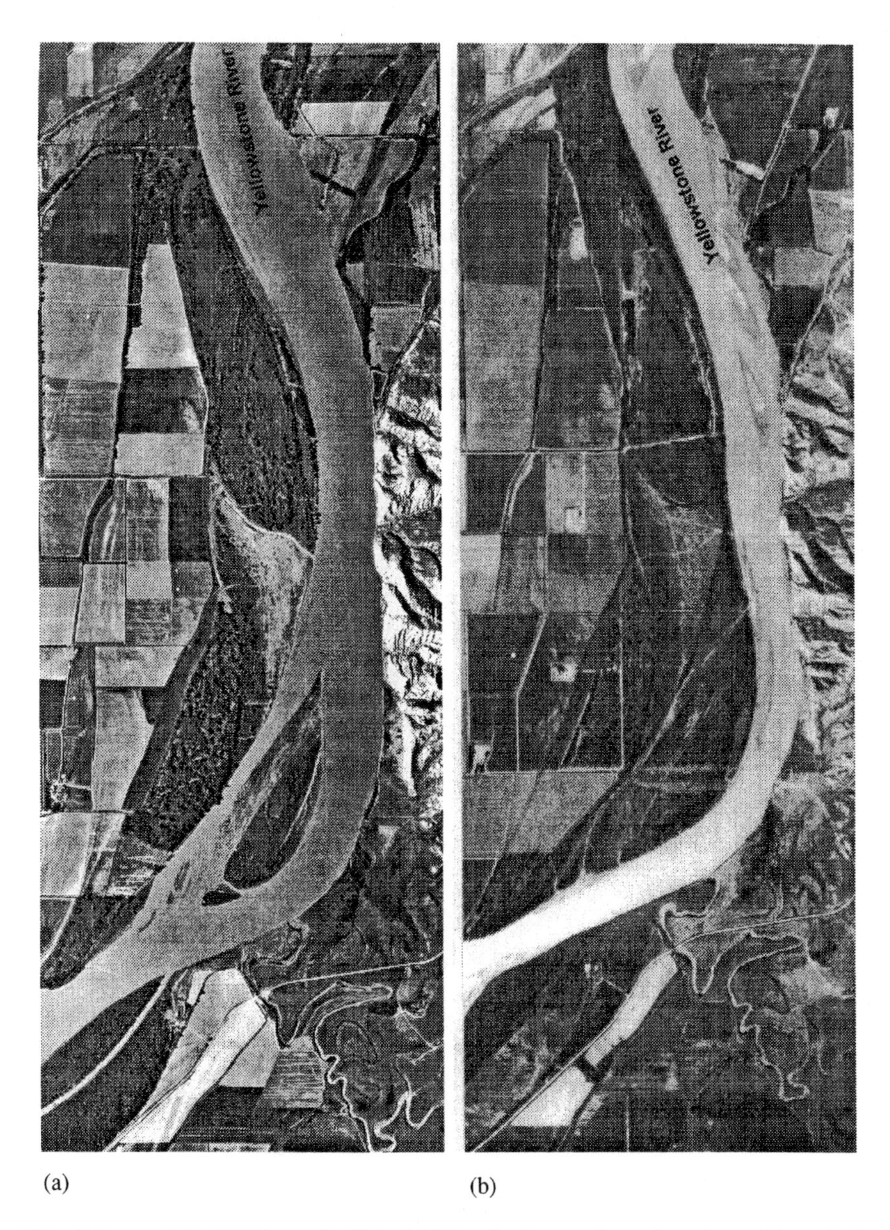

Fig. 5.1. (a) 1967 and (b) 1980 photographs showing disputed bar/island in the Yellowstone river near Sidney, Montana. Note by 1980, the disputed area has accreted to the left bank.

channels are clearly delineated by vegetation. Some lands currently under cultivation at this location apparently originated as mid-channel bars. Similar lands downstream along the Yellowstone River have been claimed by the adjacent landowners by quiet title action, and the state has shown no interest. Those sites apparently are not underlain by oil.

Clearly, Montana River law is inconsistent with natural process and is inconsistently applied. Technical professionals obviously need to be involved in educating attorneys and judges on river process, but cases like these typically draw expert witnesses who will argue both sides. As long as dams are built and people alter rivers, there is likely to be a steady flow of work for technical experts. And as long as there is oil, the land disputes are likely to be particularly heated.

6. ACKNOWLEDGMENTS

The Corps of Engineers provided data on discharge, thalweg and average bed elevations, and channel width. Carla Cary van Siclen worked many hours analyzing the Corps database.

7. REFERENCES

Allen, P.M., R. Hobbs, and N.D. Maier. 1989. Downstream Impacts of a Dam on a Bedrock Fluvial System: Bull. AEG, v. 26, no. 2, p. 165-190.

American Society of Civil Engineers. 1978. Environmental Effects of Large Dams, 225p.

Arizona Dept. Transportation. 1979. Salt River Bed Study Report: unpublished.

Beck, R.E. 1967. The Wandering Missouri River: a Study in Accretion Law: 43 North Dakota Law Review 429.

Beck, S., D.A. Melfi, and K. Yalamanchili. 1983. Lateral Migration of the Genesee River, New York in *River Meandering*: ASCE, New York, p. 510.

Bradley, J.B. 1983. Transition of a Meandering River to a Braided System Due to High Sediment Concentration Flows in *River Meandering*: ASCE, New York, 89p.

Brice, J.C. 1975. Airphoto Interpretation of the Form and Behavior of Alluvial Rivers: Final Report for the U.S. Army Research Office.

Brice, J.C. 1977. Lateral Migration of the Middle Sacramento River, California: NTIS Report PB-271662/AS, 56 p.

Burke, T.M. 1983. Channel Migration on the Kansas River in *River Meandering*: ASCE, New York, p. 250.

Camp, L.S. 1977. Land Accretion and Avulsion: the Battle of Blackbird Bend: 56 Nebraska Law Review.

Chorley, R.J., S.A. Schumm, and D.E. Sugden. 1984. *Geomorphology*: Metheun, New York, 574 p.

Emmett, W.W. 1974. Channel Aggradation in Western United States as Indicated by Observations at Vigil Network Sites: Zeit. Geomorph. Supplementary v. 21, p. 52-62.

Fraser, J.C. 1972. Regulated Discharge and the Stream Environment: River Ecology and Man, p. 263-285.

Gottschalk, L.C. 1964. Reservoir Sedimentation in Chow, V.T., *Handbook of Applied Hydrology*: McGraw-Hill.

Lagasse, P.F. 1994. Variable Response of the Rio Grande to Dam Construction in *The Variability of Large Alluvial Rivers* (S.A. Schumm and B.R. Winkley, eds.), ASCE, p. 395-422.

Mahkaveev, N.T. 1970. Effect of Major Dam and Reservoir Construction on Geomorphological Processes in River Valleys: Geomorphology no. 2, p. 106-110.

Morisawa, M. 1985. *Rivers-Form and Process*: Longman, New York.

Nadler, C.T. and S.A. Schumm. 1981. Metamorphosis of the South Platte and Arkansas Rivers, Eastern Colorado: Physical Geography, v. 1, p. 95.

Petts, G.E. 1979. Complex Response of River Channel Morphology Subsequent to Reservoir Construction: Progress in Physical Geography, v. 3, no. 3, p. 329-362.

Rahn, P.H. 1977. Erosion Below Main Stem Dams on the Missouri River: Bull. AEG, v. 14, no. 3, p. 157-181.

Schumm, S.A. 1973. Geomorphic Thresholds and Complex Response of Drainage Systems: Proceedings of the 4th Annual Geomorphology Symposium Binghamton, p. 299-310.

Schumm, S.A., I.D. Rutherfurd, and J. Brooks. 1994. Pre-Cutoff Morphology of the Lower Mississippi River in *The Variability of Large Alluvial Rivers* (S.A. Schumm and B.R. Winkley, eds.): ASCE, p. 13-44.

Serrad, J.E. and J.K. Howard. 1983. Sacramento River Bank Erosion (problem statement) in *River Meandering*: ASCE, New York, p. 271.

Strand, R.I. 1977. Sedimentation in *Design of Small Dams*: Bureau of Reclamation.

U.S. Army Corps of Engineers. 1952. Report on Degradation Observations, Missouri River Downstream from Fort Peck Dam: Fort Peck, Montana, 71 p.

USDA Agricultural Research Service. 1969. Summary of Reservoir Sediment Deposition Surveys made in the United States through 1965: Misc. Pub. No. 1143.

Williams, G.P. and M.G. Wolman. 1984. Downstream Effects of Dams on Alluvial Rivers: USGS PP 1286, 83 p.

Wolman, M.G. and L.B. Leopold. 1957. Floodplains: USGS PP 282-C.

Zhou, Z. and X. Pan. 1994. Lower Yellow River in *The Variability of Large Alluvial Rivers* (S.A. Schumm and B.R. Winkley, eds.), ASCE, p. 363-394.

SIZE OF PAVEMENT IN GRAVEL-BED RIVERS

Michael A. Stevens[1]

ABSTRACT

Since the advent of Wolman's method of sizing riverbed gravels, there is a lack of rigor and some confusion in the reporting and use of bed-material descriptors of gravel and cobble size. The Wolman size is not the same as that used earlier in research and in fieldwork; for example Lane and Carlson, 1953. If we are diligent in reporting exactly how the gravels were sampled and sized, then the Kellerhals-Bray equivalence criteria can be employed to convert most reported sizes to whatever the need may be. To judge how to use any information on gravel size, one must know how the samples were taken. Then, it is imperative that all data be reported as "percent finer by weight" or "percent finer by number." The early work in the transport of gravels used "percent finer by weight." The particle diameter to be used in equations was to be determined from size by weight, often the procedure being dissimilar for different authors. Some of the problems arise because people are using the Wolman Count (percent finer by number) in expressions such as Shield's initiation of motion relation and the Meyer-Peter and Muller bed-load transport equation. Both require "percent finer by weight" which is not given directly by the Wolman Count unless the surface material of the bulk sample is the same as the material in the entire sample. The Kellerhals-Bray conversion is needed to convert the Wolman Count to the suitable "percent finer by weight" if that is not the case.

KEYWORDS

Gravel-bed rivers, sampling, gravel transport, degradation, pebble count, size equivalence

1. INTRODUCTION

The bed of a gravel-bed river is commonly covered with a layer of somewhat uniform and coarse particles underlain by a subpavement, a mixture of

[1] Consultant, P.O. Box 3263, Boulder, Colorado 80307.

particles that includes the coarse sizes but also fines which are not found in abundance, if at all, in the pavement. To sample the pavement, it is common to use the Wolman (1954) method whereby one picks up a stone along a transect at equal intervals until approximately 100 stones are collected (see Church et al., 1987). The intermediate axis of each stone in the sample is measured and the frequency of stone size by number is obtained. The Wolman method is equivalent to the grid-by-number method that has been employed by many. Because of the work of Kellerhals and Bray (1971) and Church et al. (1987), some may be using the Wolman count of pavement stones as equivalent to the sieve-by-weight of a bulk sample, but this is not the size needed for many analyses.

By the Wolman count method, the pavement of the Uncompahgre River between Montrose and Delta, Colorado, is approximately 2 inches (50 mm). In this paper, it is shown that it is more correct to consider a size that is approximately 4 inches (100 mm), twice the Wolman count size, for the Uncompahgre pavement. The critical question is whether this river could degrade under flow conditions resulting from a proposed hydroelectric plant. The pavement size is a very important issue for this project.

2. CONCEPTS

The method chosen for sampling in gravel-bed rivers depends on the purpose of the study. Principal concerns are: (1) *gravel transport*, (2) *degradation*, (3) *flood levels*, (4) *substrate suitability*, and (5) *mining*. Each concern could warrant a different method of collecting and processing samples of the gravel bed and materials beneath it. In the Uncompahgre study, the concern is degradation.

For the pavement, the water exerts a fluid shear stress on the bed that is resisted by the submerged weight of the surface particles. The fluid shear stress is the force promoting movement. Therefore, it is necessary to consider the submerged weight of particles per unit area of bed surface to determine if movement occurs. When all particles have the same density, volume is a surrogate for submerged weight in particle frequency analysis.

For gravel-bed river pavements, one could do no better than to collect all surface stones over the entire area of the bed comprising one population (that is, over one facies of the bed surface), sieve and weigh them all, and report the results as size versus percent finer by weight. This is the area-by-weight method.

Only two field methods of sampling seem practical for pavements. One is the grid or Wolman count method and the other is collecting all surface stones within a measured area. The Wolman method is the only one practical for sampling under water and one can sample a much larger region than by picking up all the stones in one area. The area-by-weight method is better for handling such issues as imbrication and texture.

Kellerhals and Bray (1971) first put forth the concepts whereby size equivalence can be had for samples collected under a variety of methods. The same model (Fig. 2.1), a volume of densely packed cubes of three sizes, is employed here. The numbers of cubes in the volume are 4,608, 576, and 72 for the small, intermediate, and large cubes, respectively. The surface layer is comprised of the same three sizes of cubes in numbers as indicated in Table 2.1.

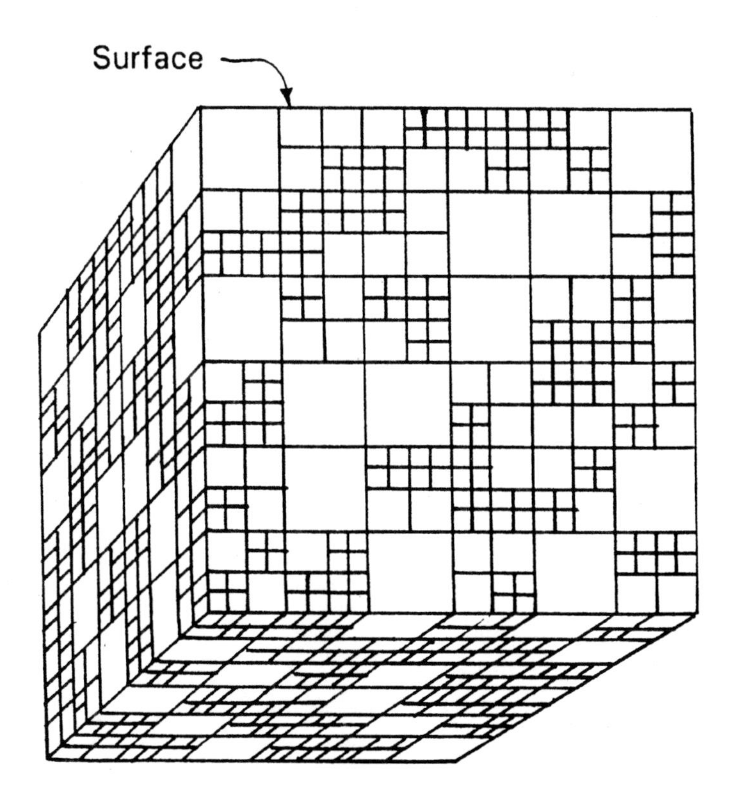

Fig. 2.1. Sample of Densely Packed Cubes of Three Sizes.

Table 2.1. Mean size of the cubes on the surface (Fig. 2.1), area-by-weight method.

Size Di	Number n_i	Area of Each Stone	Volume of Each Stone	Area $n_iD_i^2$	Volume, $n_iD_i^3$	Percent in Class
4	12	16	64	192	768	57.1
2	48	4	8	192	384	28.6
1	192	1	1	192	192	14.3
Total	252			576	1344	100.0

According to weight (volume times submerged unit weight), the surface is comprised of 57.1 percent size-4 cubes, 28.6 percent size-2 cubes and 14.3 percent of size-1 cubes. This gradation is the best estimate of the capability of the surface particles to resist motion.

The mean size D_m that gives the same weight per unit area as the mix of cubes is

$$n_m D_m^3 = 1344$$

with n_m being the number of mean-size stones. These n_m cubes cover the same surface area so

$$n_m D_m^2 = 576$$

Solving these two expressions yields

$$D_m = 1344/576 = 2.33$$

slightly larger than the intermediate size, and

$$n_m = 576/2.33^2 = 106 \quad \text{or} \quad n_m = 1344/2.33^3 = 106.$$

Suppose the flow removes all the smallest cubes from the surface. This leaves 192 holes in the surface. How does the representative mean size change? In Table 2.2, there are only 60 cubes left to resist the same shear stress as before. Actually, the holes could affect the magnitude of the fluid shear stress, increasing it slightly but that is ignored here.

Table 2.2. Mean size of the cubes on the surface with the smallest removed (Fig. 2.1), area-by-weight method.

Size D_i	Number n_i	Area of Each Stone	Volume of Each Stone	Area $n_i D_i^2$	Volume, $n_i D_i^3$	Percent in Class
4	12	16	64	192	768	66.7
2	48	4	8	192	384	33.3
1	192	1	Holes	192	0	0
Total	252			576	1152	100.0

In the model, the holes count for area but not for weight or volume (if volume is used instead of weight). The assumption is that there is a homogeneous distribution of all sizes of particles and holes. Now, the volume of the cubes on the surface is 786 + 384. As the force per unit area is assumed the same, the area is still 576. Thus, the removal of the fines reduces the representative mean size to

$$D_m = (768+384)/576 = 1152/576 = 2.00$$

This is a modest reduction of 14 percent considering the number of cubes forming the surface changed from 252 to 60. The new gradation is two-thirds by weight large cubes and one-third intermediate cubes. It is the larger particles (represented by 768/576) that are most effective in resisting motion and not the smallest ones (384/576).

The removal of cubes from the surface affects the stability of the bed as defined by Shields' parameter

$$\tau_* = \tau_o / (\gamma_s - \gamma) D_m$$

Here

τ_o = force of water on the bed divided by the bed area = fluid stress on the bed

γ_s = unit weight of the cubes, and

γ = unit weight of water

A low value of Shield's parameter (low stress, large D_m, or both) means that the bed does not move. A high value indicates that there is considerable bedload movement.

The surface is less stable with holes in it. Few cubes must carry the same load. The Shield's parameter must be greater if there are holes. As the stress is assumed unchanged, D_m must be smaller with holes than without holes. In the extreme, with only one large cube on the surface it is likely to move as its D_m (= 64/576) = 0.11; with only one small cube D_m = 1/576 = 0.00017 and will certainly move.

In this analysis of stability, there are two issues; the weight of the particle, represented here by its volume, and the surface area over which the water shear is distributed. Then, the desired gradation is that given by the Kellerhals-Bray "area-by-weight".

The Kellerhals-Bray formal conversion factor for different methods of sampling gravel is defined by the expression

$$f_{ci} = f_{oi}D_i^x / \Sigma f_{oi}D_i^x$$

where the summation is over all class sizes and

f_{ci} = converted fraction of the observed sample to the desired distribution;

f_{oi} = observed proportion of the sample in the *i-th* size class with the mean size D_i; and

x = integer depending on the method of collecting and analyzing the sample.

The Kellerhals-Bray values for x for all types of sampling are listed in Table 2.3. A value of x = 0 means that no conversion is required, and the size distribution obtained by the one method gives the same distribution as by the other. It is important to notice that the grid-by-number method (or Wolman method) is equivalent to sieve-by-weight , but only if the surface material of the bulk sample is the same as all the rest of the material in the bulk sample volume.

The issue of holes has been addressed by Diplas and Sutherland (1988) and Diplas and Fripp (1992) in context with using wax or clay for sampling areas. Such void fillers collect particles differently from grid and transect methods, the likely choice for fieldwork. Some different equivalence criteria are needed for void-fillers samples. Use of wax is more common in the laboratory where particles are smaller and the bed can be dried for sampling.

Table 2.3. Kellerhals and Bray x-values for finding the equivalency of samples or converting from one to the other.

Conversion from:	Conversion to:				
	Sieve-by-Weight	Grid-by-Number	Grid-by-Weight	Area-By-Number	Area-by-Weight
Sieve-by-weight	0	0	3	-2	1
Grid-by-number	0	0	3	-2	1
Grid-by-weight	-3	-3	0	-5	-2
Area-by-number	2	2	5	0	3
Area-by-weight	-1	-1	2	-3	0

3. EXPERIMENTS

In order to test the validity of the Kellerhals-Bray concepts of equivalence, Church et al. (1987) conducted an experiment whereby "... a bulk mixture of known size distribution between 5.66 and 45.3 mm and total weight of 20.2 kg was prepared. This was placed in a square box and mixed by inverting the box and shoveling the material. The mixing process is supposed to have been random and the surface of the gravel, when the box was opened, not to have any preferred texture or fabric. A grid-by-number sample was collected as follows. A 55-mm wire grid was placed over the mixture and 25 stones were collected from the surface under the grid points (a glass rod was used to make the selection procedure objective). The stones were size-graded by 1/2 ϕ intervals and replaced and the gravel remixed. The procedure was repeated until a total of 400 stones had been drawn.... The gravel was mixed once more and the surface spray-painted. The painted stones were removed, graded, then counted and weighted to yield area-by-weight samples. Typically, 800-1000 clasts were exposed. The entire experiment was repeated three times, using the same mixture."

The pertinent experimental values for the entire mixture in the box (sieve-by-weight) and the surface combined sample are given in Table 3.1. The combined surface sample was large, weighing 15,923 grams. The number of particles sampled by grid was 1200.

Table 3.1. Church et al. (1987) tests of the Kellerhals-Bray formal conversion factors for gravel samples.

Class Size (mm)	Mean[1] Size (mm)	Mixture in Box	Combined Surface Sample Fractions		
			Grid-by-Number	Area-by Weight	
		Sieve by Weight	Observed	Observed	Converted[2]
32.0-45.3	38.07	0.1304	0.0975	0.2407	0.1954
22.5-32.0	29.81	0.2212	0.1983	0.2856	0.3112
16.0-22.5	19.03	0.1971	0.2175	0.1875	0.2179
11.3-16.0	13.39	0.1869	0.1975	0.1555	0.1392
8.00-11.3	9.47	0.1859	0.2342	0.1096	0.1168
5.66-8.00	6.73	0.0785	0.0550	0.0211	0.0195

[1]The mean size is the geometric mean of the upper and lower class sizes.
[2]The conversion is from grid-by-number to area-by-weight.

In accordance with the cube model and hydraulics theory, the observed area-by-weight fractions (Table 3.1) are taken as the best estimate of the surface gradation. The findings are as follows:

1. Approximately 53 percent (using area-by-weight) of the surface particles are larger than 22.5 mm. The grid-by-number sample has only 30 percent by number larger than 22.5 mm.

2. As Kellerhals and Bray had shown, the grid-by-number fractions of the surface sample are approximately equivalent to the sieve-by-weight fractions of the entire mixture in the box. There was a slight bias with the grid-by-number sample giving a finer distribution than that of the mixture population.

3. The grid-by-number distribution of the surface sample is significantly different from the area-by-weight distribution, especially at the

extremes. Any derived D_i value is smaller in grid-by-number sampling than in the area-by-weight sampling. Therefore, the grid-by-number size distribution, by the theoretical construction developed here, is not a good measure of the surface pavement gradation when considering the resistance of the pavement to movement.

4. The grid-by-number distribution, when converted by the Kellerhals-Bray criteria, is a close estimate of the observed area-by-weight fractions for the surface.

5. The conversion methods for surface samples may yield very little information about the material below when that is different from the surface. A stable armor may vary even more so from what is underneath. Livsey's photograph of pavement resulting from degradation (see Vanoni, 1975) illustrates what the difference between armor and the parent material underneath can be. That means the grid-by-number sample of the pavement gives only a limited indication of what is in the subpavement, certainly not the sieve-by-weight information as in the box experiments. In the field what's in the box is not what's on the surface.

4. AREA SAMPLES

In theory, the surface samples should be taken from a measured area. The reason is that the total weight of the surface stones should be divided by the measured area. This is in keeping with the concept that it is the force of fluid per unit area of bed that is the important factor. It is possible that the sum of the areas of the sampled particles could be greater than the measured area.

In accordance with the hydraulics of the problem, the surface sampling for degradation studies should be conducted with the following in mind (Fig. 4.1). A stone sitting on top of others as shown in Fig. 4.1a should be discarded as it is likely to be move under almost any stress. The marked stone in Fig.4.1b should be kept. The stones below the holes in Figs. 4.1c, 4.1d, and 4.1e should not be taken for the pavement as they do not resist the fluid shear on the bed. The imbricated stones in Fig. 4.1f should all be taken. They have a high weigh-per-exposed area, indicative of their higher-than-average stability. For transport studies, the small stones in Fig. 4.1a and 4.1e would be included in the sample because at times they could be the only stones moving.

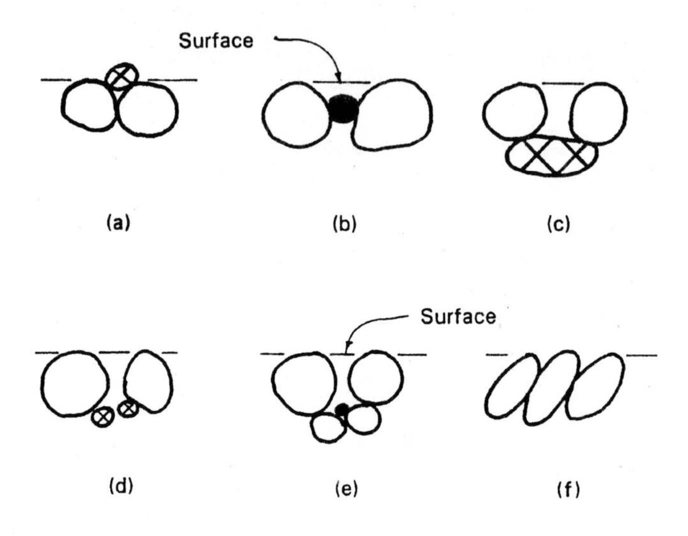

Fig. 4.1. Stones accepted or rejected in sampling the pavement.

The recommendations just cited should give basically the same pavement sample as collected by Church et al. (1987) with their sprayed paint and labeled "exclusive." When all painted stones beneath the surface layer are taken as part of the sample, they called the sample "inclusive." If the underlying painted stones are not collected, the sample is called "exclusive." The data in Table 2.3 are "exclusive."

5. VOLUME SAMPLING

On the basis of the conceptual model used here (Fig. 2.1), there is little likelihood of getting a meaningful volumetric sample of the surface in the field. The surface is the fluid mechanics equivalent of the substrate in biology. The area sample cannot be considered a volume sample as the conversions of the box experiment samples signify.

For the conceptual model, one can remove a volume equal to the surface area times the depth of the largest cube. Compared to the area-by-weight sample, this "surface volume" is finer than the sieve-by-weight. Obtaining a volumetric sample of the surface under water in the field is neither practical, desirable, or needed.

6. TRANSECT-BY-NUMBER

The Wolman count is a transect-by-number method of sampling the bed surface, equivalent to the grid-by-number. Suppose the grid is formed with a continuous wire, which is then marked at each intersection. When extended into a straight line, the wire becomes a transect. Stones collected from a single population (facies) at the grid intersections would be the same as those collected under the markers on the transect wire. Thus, the Wolman count is equivalent to transect-by-number, which in turn is equivalent to grid-by-number. No conversions among these three are necessary. In keeping with the terminology of Kellerhals and Bray, the Wolman count is hereafter referred to as grid-by-number. For fieldwork, the Wolman sampling method and the conversion to area-by-weight is the choice for pavement studies.

7. EXAMPLES

In addition to the experiments of Church et al. (1987), there are other direct comparisons of grid-by-number, transect-by-number and area-by-weight of gravel-bed and canal pavements. Some are presented here (Table 7.1).

Table 7.1. Relation among different surface samples.

Reference	Sample		D_{16}	D_{50}	D_{84}
Kellerhals and Bray (1971) after Ritter and Helley (1969)	Grid-by-number	observed	12	21	32
	Area-by-weight	observed	15	28	48
		converted[1]	16	26	38
Kellerhals (1967)	Transect-by-number	observed	100	200	280
	Grid-by-number	observed	80	170	260
	Area-by-weight	observed	110	210	240
		converted[2]	140	230	310
		converted[1]	120	210	290
Uncompahgre Study (1994)	Transect-by-number	observed	22	56	125
	Area-by-weight	converted[2]	49	114	218

[1] Converted from grid-by-number to area-by-weight.

[2] Converted from transect-by-number to area-by-weight.

Kellerhals and Bray (1971) had used the information of Ritter and Helley (1969) to show that the latter's grid-by-number and area-by-weight samples could be made equivalent by using the Kellerhals-Bray conversion.

Previously Kellerhals (1967) had employed different methods to sample a gravel bar with material from 0.5 to 18 in. (10 to 460 mm). Here (Table 7.1), for the grid-by-number, the 10 samples collected by two people are combined to yield a sample of 500 stones. The area-by-weight was a collection of all stones over two square yards (1.7 m^2). The transect sample of 100 stones was collected by pacing. The transect yielded larger stones than the grid samples. The grid-by-number and transect-by-number samples convert well to area-by-weight except for the largest fractions.

8. SIGNIFICANCE

The grid-by-number frequency distribution for the pavement in Reach 10 of the Uncompahgre River is given in Table 7.1 along with the Kellerhals-Bray conversion to area-by-weight. The median size D_{50} for the grid-by-number count is 56 mm whereas for area-by-weight it is 114 mm, twice as large. Based on D_{50}'s, one sample is twice as stable as the other.

Gessler's method (1971) of determining pavement stability is one that employs the complete distribution and not some single representative size. Also, it includes the concepts of fluid shear in determining stability. It is suitable for testing the significance of the difference between the grid-by-number and area-by-weight gradations. Using the gradations of Table 7.1, a discharge of 2,000 ft^3/s, depth of 5.2 ft, and slope of 0.0058, the Gessler bed stability factor is 0.55 for the grid-by-number distribution and 0.70 for the other. The Gessler interpretation is that the first would degrade slightly without slope reduction and the other is stable.

9. WHERE TO SAMPLE

Often the bed of a gravel-bed river is covered in different areas by bars of finer material on top of the pavement. This is illustrated in the channel bed topography of Lisle and Madej (1992). The most likely place for pavement to be continuously exposed is on the riffles. It is best to sample here for degradation studies.

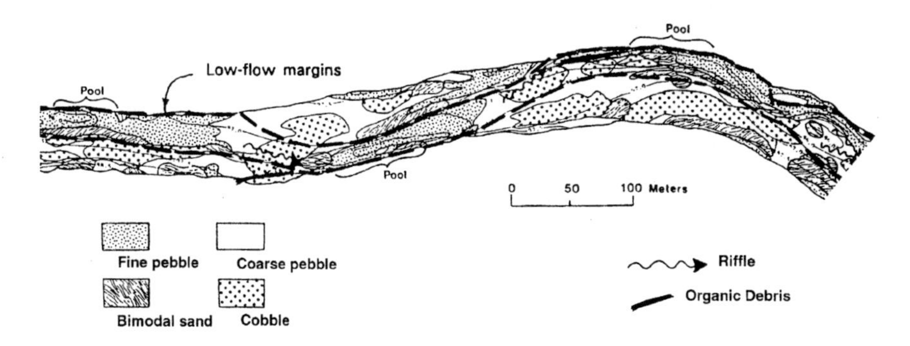

Fig. 9.1 Map of a streambed with bed materials, riffles, and pools identified (from Lisle and Madej, 1992).

10. USAGE

It is necessary to use the representative grain size, which is appropriate for the equation or graph for the particular purpose being pursued. For example, Meyer-Peter and Muller (1948) want us to compute the representative grain size in a very specific manner for their bed-load equation.

In his experiments on the initiation of motion, Shields used uniform sand of one size. We have no instructions for mixtures of sizes. Since the Wolman Count is such a practical way of sampling gravel-bed rivers, what is needed is an initiation-of-motion graph with the grid-by-number D_m as the representative size, taking into account the density of the stones. Such a graph could be derived from the published information and the Kellerhals-Bray equivalence criteria.

One can convert the area-by-weight representative grain size for pavement recommended here to any other of the Kellerhals-Bray sizes for use in various initiation-of-motion equations.

11. CONCLUSIONS

The conceptual model of Kellerhals and Bray and hydraulics principles are suitable for calculating the representative gradation of the pavement in gravel-bed rivers for the purpose of determining the pavement stability to resist the force of the fluid shear. The Kellhals-Bray method equivalence criterial specifies both how to sample (surface or bulk volume, grid or area) and how to determine frequency (by number or by weight). Hydraulic principles dictate that the area-by-weight gradation is the best estimate of the size distribution of the pavement when the interest is in degradation.

The grid-by-number method does not give the correct estimate of the particle sizes of the pavement unless the Kellerhals-Bray conversion is made to area-by-weight.

An attempt to sample the surface volume (area times depth equal to the largest particle) will most likely result in a sample, which is finer than the surface pavement.

The Wolman count method is a practical method for sampling gravel-bed pavement under water and over a large area. Named here as grid-by-number in accordance with the Kellerhals-Bray terminology, it is argued that the transect-by-number method is directly equivalent to sampling by grid so that no conversion is required to achieve a grid-by-number sample. Hence the transect-by-number sample can be converted to area-by-weight with the Kellerhals-Bray conversion factor.

At best, the grid-by-number sampling of pavement gives only an indication of the coarsest sizes in the subpavement. The grid-by-number conversion of surface samples to sieve-by-weight is not applicable for either the pavement (equivalent to a substrate) or subpavement unless they have the same size distribution.

Due to the potentially high variability of the surface material with respect to space and time in gravel-bed rivers, it is best to sample riffles when degradation is the prime concern. Unless riffle pavement moves, the reach cannot degrade. Also, the pavement on riffles is more likely to be the same for a range of flows, so it can be sampled at low flow.

It is necessary to report where (riffle, pool, bar, other) and how the samples were collected, the number of stones in the sample, percent finer by number or percent finer by weight, and other particulars of the sampling program, including its purpose.

12. REFERENCES

Church, M.A., McLean, D.G., and Wolcott, J.F. 1987. Riverbed gravels: sampling and analysis. In *Sediment transport in Gravel-Bed Rivers,* C.R. Thorne, J.C. Bathurst, and R.D. Hey, eds., John Wiley & Sons, New York, 43-79.

Diplas, P. and Sutherland, A.J. 1988. Sampling techniques for gravel sized sediments. *J. of Hydraulic Engineering,* ASCE, 114:5, 484-501.

Diplas, P. and Fripp, J.B. 1992. Properties of various sediment sampling procedures. *J. of Hydraulic Engineering,* ASCE, 118:7, 955-970.

Gessler, J. 1971. Beginning and ceasing of sediment motion. *River Mechanics,* H.W. Shen ed., Ft. Collins, Colorado, C 7, 1.

Kellerhals, R. 1967. Stable channels with gravel-paved beds. *J. Waterways and Harbors Div.,* ASCE, 93:1, 63-84.

Kellerhals, R. 1987. Discussion of Church, M.A., McLean, D.G., and Wolcott, J.F., 1987. Riverbed gravels: sampling and analysis. In *Sediment transport in Gravel-Bed Rivers,* C.R. Thorne, J.C. Bathurst, and R.D. Hey, eds., John Wiley & Sons, New York, 84-85.

Lane, E.W., and Carlson, E.J. 1953. Some factors affecting the stability of canals constructed in coarse granular material. IAHR Proceedings, Minnesota International Hydraulics Convention.

Lisle, T.E., and Madej, M.A. 1992. Spatial variation in armoring in a channel with high sediment supply." C 13, *Dynamics of gravel-bed rivers,* P. Billi, R.D. Hey, C.R. Thorne, and P. Tacconi, eds., John Wiley & Sons Ltd., New York, 277-293.

Meyer-Peter, E. and Muller, R. 1948. Formulas for bed-load transport. Report on Second Meeting, IAHR, Stockholm, Sweden, 39-64.

Ritter, J.R. and Helley, E.J. 1969. "Optical method for determining particles sizes of coarse sediment." *Techniques of Water Resources Investigations*, US Geological Survey, Book 5, C 3.

Vanoni, V.A. 1975. Fundamentals of sediment transportation. in C II, *Sedimentation Engineering,* V.A. Vanoni, ed., Manual No. 54, ASCE, New York, p.182.

Wolman, M.G. 1954. A method of sampling coarse riverbed material. *Trans.,* Am. Geophys. Union, 35, 951-956.

APPENDIX

Publications and Books by Stanley A. Schumm

<u>PUBLICATIONS</u> - (Abstracts are not included)

1955 - Schumm, S.A., The relation of drainage basin relief to sediment loss: Pub. Internat. Assoc. Hydrology, I.U.G.G., v. 1, p. 216-219.

1956 - Schumm, S.A., Evolution of drainage systems and slopes in badlands at Perth Amboy, New Jersey: Geol. Soc. America Bull., v. 67, p. 597-646.

1956 - Schumm, S.A., The movement of rocks by wind: Jour. Sed. Petrology, v. 26, p. 284-286.

1956 - Schumm, S.A., The role of creep and rainwash on the retreat of badland slopes: Am. Jour. Sci., v. 254, p. 693-706.

1957 - Schumm, S.A., and Hadley R.F., Arroyos and the semiarid cycle of erosion: Am. Jour. Sci., v. 256, p. 161-174.

1959 - Langbein, W.B., and Schumm, S.A., Yield of sediment in relation to mean annual precipitation: Am. Geophys, Union Trans., v. 39, p. 1076-1084

1960 - Schumm, S.A., The effect of sediment type on the shape and stratification of some modern fluvial deposits: Am. Jour. Sci., v. 258, p. 177-184

1960 - Schumm, S.A., The shape of alluvial channels in relation to sediment type: U.S. Geol. Survey Prof. Paper 352-B, p. 17-30

1961 - Schumm, S.A., Effect of sediment characteristics on erosion and deposition in ephemeral-stream channels: U.S. Geol. Survey Prof. Paper 352-C, p. 31-70.

1961 - Schumm, S.A., and Hadley, R.F., Progress in the application of landform analysis in studies of semiarid erosion: U.S. Geol. Survey Circ. 437, 14 p.

1961 - Hadley, R.F., and Schumm, S.A., Sediment sources and drainage-basin characteristics in upper Cheyenne River basin: U.S. Geol. Survey Water-Supply Paper 1531-B, p. 137-196.

1961 - Schumm, S.A., The dimensions of some stable alluvial channels: U.S. Geol. Survey Prof. Paper 424-B, p.26-27.

1961 - Schumm, S.A., and Lichty, R.W., Recent flood-plain formation along Cimarron River in Kansas: U.S. Geol. Survey Prof. Paper 424-C, p. 112-114.

1962 - Schumm, S.A., Erosion on miniature pediments in Badlands National Monument, South Dakota: Geol. Soc. America Bull., v. 73, p. 719-724.

APPENDIX – Publications and Books by Stanley A. Schumm

1963 - Schumm, S.A., and Lusby, G.C., Seasonal variation of infiltration capacity and runoff on hillslopes in western Colorado: Jour. Geophys. Research v. 68, p. 3655-3666.

1963 - Schumm, S.A., The disparity between present rates of denudation and orogeny: U.S. Geol. Survey Prof. Paper 454-H, 13p.

1963 - Schumm, S.A., A tentative classification of alluvial river channels: U.S. Geol. Survey Circ. 477, 10 p.

1963 - Schumm, S.A., Sinuosity of alluvial rivers on the Great Plains: Geol. Soc. America Bull., v. 74, p. 1089-1100.

1963 - Schumm, S.A., and Lichty, R.W., Channel widening and flood-plain construction along Cimarron River in southwestern Kansas: U.S. Geol. Survey Prof. Paper 352-D, p. 71-88.

1964 - Schumm, S.A., Seasonal variations of erosion rates and processes on hillslopes in western Colorado: Zeitschr. Geomorph., Supplementband, v. 5, p. 215-218.

1964 - Schumm, S.A., and Chorley, R.J., The fall of threatening rock: Am. Jour. Sci., v. 262, p. 1041-1054.

1964 - Schumm, S.A., A method of estimating hydrologic response to climate change: Oondoona, Sydney Univ. Geog. Soc., v. 6., p. 5-12.

1965 - Schumm, S.A., and Lichty, R.W., Time, space and causality in geomorphology: Am. Jour. Sci., v. 263, p. 110-119.

1965 - Schumm, S.A., Geomorphic research: Applications to erosion control in New Zealand: Soil and Water (Soil Conserv. and Rivers Control Council), v. 1, p. 21-24.

1965 - Schumm, S.A., Quaternary paleohydrology, in The Quaternary of the United States: Princeton, Princeton Univ. Press, p. 783-794.

1966 - Schumm, S.A. and Chorley, R.J., Talus weathering and scarp recession in the Colorado plateaus: Zeitschr. Geomorph., v. 10, p. 11-36.

1966 - Schumm, S.A., The development and evolution of hillslopes: Jour. Geol. Education, v. 14, p. 98-104.

1967 - Schumm, S.A., Rates of rock creep on hillslopes in western Colorado: Science v. 155 p. 560-561.

1967 - Schumm, S.A., Paleohydrology: Application of modern hydrologic data to problems of the ancient past, International Hydrology Symposium (Fort Collins, U.S.A.) Proc. v. 1, p. 185-193.

1967 - Schumm, S.A., Meander wavelength of alluvial rivers: Science, v. 157, p. 1549-1550.

1967 - Schumm, S.A., On the movement of surface markers, p. 160; Erosion measured by stakes, p. 161-162; On vegetation as a marker , p. 168; in Field

Methods for the Study of Slope and Fluvial Processes, Revue de Geomorphologie Dynamique, v. 17, p. 152-188.

1968 - Schumm, S.A., Aerial photographs and water resources: in Aerial Surveys and Integrated Studies: UNESCO, Natural Resources Research Series 6, p. 70-79.

1968 - Schumm, S.A., Weather modification and the landscape: Yale Scientific Magazine v. 43, p. 10, 11, 14, 16, 18, 23.

1968 - Schumm, S.A., Speculations concerning paleohydrologic controls of terrestrial sedimentation: Geol. Soc. America Bull, v. 79, p. 1572-1588.

1968 - Scheidegger, A.E., Fairbridge R.W. and Schumm, S.A., Badlands: in Encyclopedia of Geomorphology, p. 43-48: Rheinhold Book Corp., New York. 1295 p.

1968 - Schumm, S.A., River adjustment to altered hydrologic regimen, Murrumbidgee River and paleochannels, Australia: U.S. Geol. Survey Prof. Paper 598, 65 p.

1969 - Schumm, S.A., River metamorphosis: Jour. Hyd. Div., Proc. American Soc. Civil Engineers, HY 1, p. 255-273.

1969 - Schumm, S.A., Geomorphic Implications of climatic changes: in Water, Earth and Man (edited by R.J. Chorley) p. 525-534: London, Methuen and Co., 588 p.

1969 - Schumm, S.A., A geomorphic approach to erosion control in semiarid regions: Am. Soc. Agricultural Eng. Trans., v. 12, no. 1, p. 60-68.

1970 - Schumm, S.A., Experimental studies on the formation of lunar surface features by fluidization: Geol. Soc. America Bull., v. 81, p. 2539-2552.

1970 - Schumm, S.A., Bird, J.B., and Starkel, L., Report of work group on classification of hillslopes: Zeit. Geomorph. Supplementband 9, p. 85-87.

1971 - Schumm, S.A. and Kahn, H.R., Experimental study of channel patterns: Nature, v. 233, p. 407-409.

1971 - Schumm, S.A., Fluvial Geomorphology: The Historical Perspective, in River Mechanics, v. 1 (edited by H.W. Shen) Chap. 4, 30 p.

1971 - Schumm, S.A., Fluvial Geomorphology: Channel Adjustment and River Metamorphism, in River Mechanics, v.1 (edited by H.W. Shen) Chap. 5, 22p.

1972 - Schumm, S.A. and Khan, H.R., Experimental Study of channel patterns: Geol. Soc. America Bull., v. 83, p. 1755-1770.

1972 - Schumm, S.A., Fluvial Paleochannels: Soc. Econ. Paleontologist and Mineralogists, Spec. Pub. 16, p. 98-107.

1972 - Schumm, S.A., Khan, H.R., Winkley, B.R. and Robbins, L.G., Variability of river patterns: Nature (Phy. Sci.), v. 237, p. 75-76.

APPENDIX – Publications and Books by Stanley A. Schumm

1972 - Schumm, S.A., Paleohydrology, in Encyclopedia of Geochemistry and Environmental Sciences, Van Nostrand Reinhold Co., p. 876-880.

1972 - Schumm, S.A., Parker, R.S., Shepherd, R.G., Edgar, D.E. and Mosley, M.P., Anomalous behavior of model rivers: in International Geography 1972, Univ. of Toronto Press, p. 60-61.

1973 - Schumm, S.A. and Shepherd, R.G., Valley floor morphology: evidence of subglacial erosion? Area: Inst. of British Geographers, v. 5, p. 5-9.

1973 - Schumm, S.A. and Parker, R.S., Implications of complex response of drainage systems for Quaternary alluvial stratigraphy: Nature, Physical Sci., v. 243, p. 99-100.

1973 - Schumm, S.A. and Stevens, M.A., Abrasion in place: A mechanism for rounding and size reduction of coarse sediments in rivers: Geology, v. 1, p. 37-40.

1974 - Shepherd, R.G. and Schumm, S.A., Experimental study of river incision: Geol. Soc. America, Bull., v. 85, p. 257-268.

1974 - Schumm, S.A., Geomorphic thresholds and complex response of drainage systems, in Fluvial Geomorphology (edited by M. Morisawa) Publications in Geomorphology, SUNY Binghamton, New York, p. 299-310. Reprinted 1981, Allen and Unwin, London.

1974 - Schumm, S.A., Sediment yields of drainage basins: Encyclopedia Britannica, 15th Edition, v. 16, p. 474-480 (reprinted 1984).

1974 - Schumm, S.A., Structural origin of large Martian channels: Icarus, v. 22, p. 371-389.

1974 - Schumm, S.A., Geomorphology: Yearbook of Science and Technology, McGraw-Hill, N.Y., p. 13-23.

1975 - Patton, P.C. and Schumm, S.A., Gully erosion, northwestern Colorado: A threshold phenomenon: Geology, v. 3, p. 88-90.

1975 - Stevens, M.A., Simons, D.B. and Schumm, S.A., Man-induced changes of middle Mississippi River: Jour. Waterways, Harbors and Coastal Eng. Div., ASCE Proc., v. 101, p. 119-133.

1975 - Woolsey, T.S., McCallum, M.E. and Schumm, S.A., Modeling of diatreme emplacement by fluidization: Physics and Chemistry of the Earth, v. 9, p. 24-42, Pergamon Press, Oxford.

1976 - Schumm, S.A. and Beathard, R.M., Geomorphic thresholds: An approach to river management: in Rivers 76. American Soc. Civil Engineers, New York, v. 1, p. 707-724.

1976 - Stevens, M.A., Simons, D.B. and Schumm, S.A., Geomorphic study of Pool 25: in Rivers 76, American Soc. Civil Engineers, N.Y., v. 2, p. 1655-1679.

1976 - Schumm, S.A., Episodic erosion: A modification of the geomorphic cycle: in Theories of Landform Development (editors W.N. Melhorn and R.C.

Flemal). Pub. in Geomorphology, SUNY, Binghamton, N.Y., p. 69-85. Reprinted 1982 by Allen and Unwin, London.

1977 - Loronne, J.B. and Schumm, S.A., Evaluation of the Storage of Diffuse Sources of Salinity in the Upper Colorado River Basin: Colorado State University Environmental Research Center, Completion Report 79, 111 p.

1977 - Womack, W.R. and Schumm, S.A., Terraces of Douglas Creek, northwestern Colorado. An example of episodic erosion: Geology, v. 5, p. 72-76.

1977 - McCallum, M.D., Woolsey, T.S. and Schumm, S.A., A fluidization mechanism for subsidence of bedded tuffs in diatremes and related volcanic vents: Bull. Volcanologique, v. 34-4, p. 1-16.

1977 - Mosley, M.P. and Schumm, S.A., Stream junctions - A probable location for bedrock placers: Economic Geology, v. 72, p. 691-694.

1977 - Schumm, S.A., Applied fluvial geomorphology: in Applied Geomorphology (J.R. Hails, editor). Elsevier, Amsterdam, p. 119-156.

1978 - Ethridge, F.G. and Schumm, S.A., Reconstructing paleochannel morphology and flow characteristics: methodology, limitations, and assessment: in A.D. Miall, ed., Fluvial Sedimentology: Canada Soc. Petrol. Geol., Memoir No. 5, p. 703-722.

1979 - Begin, Z.B. and Schumm, S.A., Instability of alluvial valley floors: A method for its assessment: Am. Soc. Agr. Eng. Trans., v. 22, p. 347-350.

1979 - Schumm, S.A. and Meyers, D.F., Morphology of alluvial rivers of the Great Plains: Great Plains Agr. Council Pub. 91, p. 9-14.

1979 - Schumm, S.A., Geomorphic thresholds: the concept and its applications: Inst. British Geogr. Trans., v. 4, p. 485-515.

1980 - Schumm, S.A., Plan form of alluvial rivers: International Workshop on Alluvial Rivers, University of Roorkee (India), Proc., p. 4-21-4-33.

1980 - Schumm, S.A., Kirk Bryan Award - Response: Geol. Soc. America Bull. (Part 2), p. 1101-1102.

1980 - Shen, H.W., Schumm, S.A. and Doehring, D.O., Stability of stream channel patterns: National Acad. Sci., Transportation Research Record 736, p. 22-28.

1980 - Schumm, S.A., Some applications of the concept of geomorphic thresholds in Coates, D.R. and Vitek, J.D. (editors), Thresholds in Geomorphology: George Allen and Unwin, London, p. 473-485.

1980 - Schumm, S.A. Bradley, M.T., and Begin, Z.B., Application of Geomorphic Principles to Environmental Management in Semiarid Regions: Colorado State University Colorado Water Resources Institute, Completion Report 93, 43 p.

APPENDIX – Publications and Books by Stanley A. Schumm

1980 - Begin, Z.B., Meyer, D.F. and Schumm, S.A., Knickpoint migration due to base-level lowering: Jour. Waterways, Coastal, and Ocean Div., Am. Soc. Civil Eng. Proc., v. 106, p. 369-387.

1980 - Begin Z.B., D.F. Meyer and Schumm, S.A., Sediment production of alluvial channels in response to base-level lowering: American Soc. Agr. Eng. Trans., v. 23, p. 1183-1187.

1980 - Bergstrom, F. and Schumm, S.A., Episodic behavior in badlands: Internat. Assoc. Hydrologic Sciences Pub. 132 (Christchurch Symposium), p. 478-592.

1981 - Begin, Z.B., Meyer, D.F. and Schumm, S.A., Development of longitudinal profiles of alluvial channels in response to baselevel lowering: Earth Surface Processes and Landforms, v. 6, p. 49-68.

1981 - Patton, P.C. and Schumm, S.A., Ephemeral-stream processes: Implications for studies of Quaternary valley fills, Quaternary Res., v. 15, p. 24-43.

1981 - Schumm, S.A., Evolution and response of the fluvial system: sedimentologic implications, Soc. Economic Paleontologists and Mineralogists Spec. Pub. 31, p. 19-29.

1981 - Nadler, C.T. and Schumm, S.A., Metamorphosis of South Platte and Arkansas Rivers, eastern Colorado: Physical Geogr., v. 2, p. 95-115.

1982 - Schumm, S.A., Bean, D.W. and Harvey, M.D., Bed-form dependent pulsating flow in Medano Creek, southern Colorado: Earth Surface Processes and Landforms, v. 7, p. 17-18.

1982 - Schumm, S.A. and Harvey, M.D., Natural erosion in the USA: American Soc. Agronomy, Spec. Pub. 45, p. 15-22.

1982 - Laronne, J.B. and Schumm, S.A., Soluble-mineral content in surficial alluvium and associated Mancos Shale: Water Res. Bull., v. 18, p. 27-35.

1982 - Schumm, S.A, Costa, J.E., Toy, T., Knox, J., Warner, R., and Scott, J., Geomorphic assessment of uranium mill tailings: in Uranium Mill Tailings Management, Nuclear Energy Agency, Paris, p. 69-87.

1982 - Schumm, S.A., Costa, J.E., Toy, T., Knox, J., and Warner, R., Geomorphic hazards and uranium tailings disposal: Proc. Management of water from uranium mining and milling: Internat. Atomic Energy Agency (Vienna), p. 111-124.

1982 - Parker, R.S. and Schumm, S.A., Experimental study of drainage networks in Badland geomorphology and piping (Eds. R. Bryan and A. Yair) Geobooks Norwich U.K., p. 153-168.

1983 - Harvey, M.D., Watson, C.C. and Schumm, S.A., Channelized streams: An analog for the effects of urbanization: Internat. Symp. Urban Hydrology Hydraulics and Sediment Control, Proc. (Lexington, Ky) p. 401-409.

1983 - Schumm, S.A. and Chorley, R.J., Geomorphic controls on the management of nuclear waste: U.SD. Nuclear Regulatory Comm. NUREG/CR-3276, 137 p.

1983 - Nelson, J.D., Volpe, R.L., Wardwell, R.E., and Schumm, S.A., Design considerations for long-term stabilization of uranium mill tailings impoundments: U.S. Nuclear Regulatory Comm. NUREG/CR-3397, 163 p.

1983 - Burnett, A.W. and Schumm, S.A., Alluvial river response to neotectonic deformation in Louisiana and Mississippi, Science, v. 222, p. 49-50.

1984 - Begin, Z.B. and Schumm, S.A., Gradational thresholds and landform singularity. Significance for Quaternary studies. Quat. Res., v. 21, p. 267-274.

1984 - Watson, C.C., Harvey, M.D. and Schumm, S.A., Neotectonic effects on river pattern: in River meandering (ed. C.M. Elliott) Amer. Soc. Civil Engs. N.Y., p. 55-66.

1984 - Schumm, S.A., River morphology and behavior: Problems of extrapolation: in River Meandering (ed. C.M. Elliott) Amer. Soc. Civil Engs., N.Y., p. 16-29.

1984 - Winkley, B.R., Schumm, S.A., Mahmood, K., Lamb., M.S. and Linden, W.M., New developments in the protection of irrigation, drainage and flood control structures on rivers. Internat. Comm. Irrigation and Drainage 12th Congress (Fort Collins) Trans Vol. C, p. 69-111.

1985 - Schumm, S.A., Explanation and extrapolation in geomorphology; Seven reasons for geologic uncertainty: Japanese Geomorph. Union, Trans., V. 6, p. 1-18.

1985 - Schumm, S.A., Patterns of alluvial rivers: Annual Rev. Earth Planetary Sci., v. 13, p. 5-27.

1986 - Schumm, S.A., and Phillips, Loren, Composite channels of the Canterbury Plain, New Zealand: A martian analog?, Geology, v. 14, p. 326-329.

1986 - Schumm, S.A., Alluvial river response to active tectonics: in Active Tectonics, National Academy Press, Washington, D.C., p. 80-94.

1987-Schumm, S.A. and Ash, Donald, Paleohydrology: in Oliver, J.E., and Fairbridge, R.W. (eds.) Encyclopedia of Climatology, Van Nostrand-Reinhold, New York, p. 675-682.

1987 - Paine, A.D.M., Meyer, D.F. and Schumm, S.A., Incised channel and terrace formation near Mount St. Helens, Washington, Internat. Assoc. Hydrololgic Sci. Publ. 165, p. 389-390.

1987 - Harvey, M.D. and Schumm, S.A., Response of Dry Creek in California to land use change, gravel mining and dam closure. Internat. Assoc. Hydrologic Sci. Pub. 165, p. 451-460.

1987 - Jin, Desheng and Schumm, S.A., A new technique for modeling river morphology: in Gardner, V. (ed.). International Geomorphology, Wiley, Chichester, p. 681-690.

1987 - Gregory, D.I. and Schumm, S.A., The effect of active tectonics on alluvial river morphology: in Richards, K.S. (ed.) River channels: Environment and process: Blackwells, London, p. 41-68.

1987 - Phillips, Loren and Schumm S.A., Effect of regional slope on drainage networks: Geology, v. 15, p. 813-816.

1987 - Schumm, S.A. and Brackenridge, G.R., River response in Ruddiman, W.F. and Wright, H.E., Jr., North America and adjacent oceans during the last deglaciation: Geological Soc. America, v. K-3, p. 221-240.

1988 - Schumm, S.A., Variability of the fluvial system in space and time: in Rosswall, Thomas, Woodmansee, R.G. and Risser, P.G. (eds). Scales and global change: Wiley, New York, p. 225-249.

1988 - Schumm, S.A., Geomorphic hazards: Problems of prediction: Zeitschrft Geomorph. Supplementband, 67, p. 17-24.

1989 - Schumm, S.A. and Thorne, C.R., Geologic and geomorphic controls on bank erosion: in Ports, M.A. (ed.), Hydraulic Engineering: Proc. 1989 National Conference on Hydraulic American Soc. Civil Engs., p. 106-111.

1989 - Schumm, S.A. and Gellis, A.C., Sediment yield as a function of incised channel evolution: in Brush L.M. (ed.), Taming the Yellow River Silt and Floods, Kluwer Academic Publishers, N.Y., p. 99-109.

1990 - Buchanan, J.P. and Schumm, S.A., Niobrara River: in Wolman, M.G. and Riggs, H.C. (eds) Surface Water Hydrology, Geol. Soc. America, Boulder, CO, p. 314-321.

1991 - Rogers, R.D. and Schumm, S.A., The effect of sparse vegetative cover on erosion and sediment yield: Jour. Hydrology, v. 123, p.

1991 - Gellis, Allen, Hereford, Richard, Schumm, S.A., and Hayes, B.R., Channel evolution and hydrologic variations in the Colorado River basin: Factors influencing sediment and salt loads, Jour. Hydrology, v. 124, p. 317-344.

1992 - Schumm, S.A., The variability of large alluvial rivers: Significance for river engineering: Nile 2000, Conf. on protection and development of the Nile and other major rivers; Proc., v. 1, p. 3-3-1 to 3-3-15.

1993 - Schumm, S.A., River response to baselevel change: Implications for sequence stratigraphy: Jour. Geology, v. 101, p. 279-294.

1993 - Schumm, S.A. and Harvey, M.D., Engineering geomorphology: Hydraulic Engineering 1993, Hydraulics Div. American Soc. Civil Engs., p. 394-399.

1993 - Germanoski, Dru and Schumm, S.A., Change in braided river morphology resulting from aggradation and degradation: Jour. Geology, v. 101, p. 451-466.

1993 - Orlowski, L.A., Grundy, W.D., Mielke, P.W., Jr., and Schumm, S.A., Geological applications of multi-response permutation procedures: Mathematical Geol., v. 25, p. 483-500.

1993 - Wood, L.J., Ethridge, F.G., and Schumm, S.A., The effects of rate of base-level fluctuation on coastal-plain, shelf and slope depositional systems: an experimental approach: in Posamentier, H.W., Summerhayes, C.P., Haq,

B.V., and Allen, G.P., Sequence stratigraphy and facies associations: Internat. Assoc. Sedimentologists, Spec. Pub. 18, p. 43-53.

1993 - Jorgensen, D.W., Harvey, M.D., Schumm, S.A., and Flam, Louis, Morphology and dynamics of the Indus River: Implications for the Mohen jo Daro site in Shroder, J.F., Jr. (ed.) Himalaya to the Sea: Routledge, London, p. 288-326.

1994 - Wood, L.J., Ethridge, F.G., and Schumm, S.A., An experimental study of the influence of subaqueous shelf angles on coastal plain and shelf deposits: American Assoc. Petroleum Geologists, Memoir 58, p. 381-391.

1994 - Koss, J.E., Ethridge, F.G., and Schumm, S.A., An experimental style of the effects of base-level change on fluvial, coastal plain, and shelf systems: Jour. Sedimentary Research, v. B64, p. 90-98.

1994 - Schumm, S.A., Erroneous perceptions of fluvial hazards: Geomorphology, v. 10, p. 129-138.

1994 - Schumm, S.A., Ethridge, F.G., Origin, evolution, and morphology of fluvial valleys: Soc. Sedimentary Geology (SEPM) Special Paper 51, p. 11-27.

1994 - Schumm, S.A., Rutherfurd, I.D., and Brooks, John, Pre-cutoff morphology of the lower Mississippi River: in Schumm, S.A. and Winkley, B.R. (eds.) The Variability of Large Alluvial Rivers: American Soc. Civil Engineers Press, N.Y., p. 13-44.

1994 - Schumm, S.A. and Galay, V.J., The River Nile in Egypt: in Schumm, S.A. and Winkley, B.R. (eds.) The Variability of Large Alluvial Rivers: American Soc. Civil Engineers Press, N.Y., p. 75-100.

1994 - Harbor, D.J., Schumm, S.A., and Harvey, M.D., Tectonic control of the Indus River in Sindh, Pakistan: in Schumm, S.A. and Winkley, B.R. (eds.) The Variability of Large Alluvial Rivers: American Soc. Civil Engineers Press, N.Y., p. 161-175.

1994 - Harvey, M.D. and Schumm, S.A., Alabama River: Variability of overbank flooding and deposition: in Schumm, S.A. and Winkley, B.R. (eds.) The Variability of Large Alluvial Rivers: American Soc. Civil Engineers Press, N.Y., p. 313-337.

1994 - Schumm, S.A. and Ethridge, F.G., Origin, evolution and morphology of fluvial valleys: in Incised-valley systems SEPM Special Pub. 51, p. 13-27.

1995 - Schumm, S.A. and Rea, D.K., Sediment yield from disturbed earth systems: Geology v. 23, p. 391-394.

1995 - Schumm, S.A., Boyd, K.F., Wolff, C.G., and Spitz, W.J., A groundwater sapping landscape in the Florida Panhandle, Geomorphology, v. 12, p. 281-297.

1995 - Boyd, K.F. and Schumm, S.A., Geomophic evidence of deformation in the northern part of the New Madrid seismic zone, U.S. Geological Survey Professional Paper 1538-R, 35 p.

APPENDIX – Publications and Books by Stanley A. Schumm

1995 - Ouchi, S., Ethridge, F.G., James, E.W., and Schumm, S.A., Experimental study of subaqueous fan development, in Hartley, A.J. and Prosser, D.G., (eds.) Characterization of deep marine classis systems, Geological Survey, London Spec. Pub. No. 94, p. 13-28.

1995 - Orlowski, L.A., Schumm, S.A., and Meilke, P.E., Jr., Reach classification of the lower Mississippi River, Geomorphology, v. 14, p. 221-234.

1996 - Osterkamp, W.R. and Schumm, S.A., Geoindicators for river and river-valley monitoring, in Berger, A.R. and Iams, W.J. (eds.) Geoindicators, Balkema, Rotterdam, p. 97-114.

1996 - Schumm, S.A., The variability of large alluvial rivers: Significance for river engineering, in Nakato, T. and Ettema, R (eds.) Issues and Directions in Hydraulics, Balkema, Rotterdam, p. 135-144.

1996 - Schumm, S.A., Erskine, W.D., and Titleard, J.W., Morphology, hydrology, and evolution of the anastomosing Ovens and King Rivers, Victoria, Australia, Geological Soc. America Bulletin, v. 108, p. 1212-1224.

1997 - Schumm, S.A., Drainage density, Problems of prediction and application, in Staddart, D.R. (ed.) Process and form in geomorphology, Routledge, London, p. 15-45.

1997 - Schumm, S.A. and Spitz, W.J., Geological influences on the lower Mississippi River and its alluvial valley, Engineering Geology, v. 45, p. 245-261.

1997 - Spitz, W.J. and Schumm, S.A., Tectonic geomorphology of the Mississippi valley between Osceola, Arkansas, and Friars Point, Mississippi, Engineering Geology, v. 46, p. 259-280.

1997 - Lagasse, P.F., Schumm, S.A., and Zevenbergen, L.W., 1997. Quantitative techniques for stream stability analysis. In Holly, F.M., Jr. and Alsaffan, Adnan (eds.), Managing water: Coping with scarcity and abundance, Amer. Soc. Civil Engs., New York, p. 147-153.

1998 - Ethridge, F.G., Wood, Leslie, and Schumm, S.A., Cyclic variables controlling sequence development: Problems and perspectives in relative role of eustasy, climate, and tectonism in continental rocks: SEPM, Spec. Pub. 59, p. 17-29.

1998 - Schumm, S.A. and Lagasse, P.F., Alluvial fan dynamics - Hazards to highways: in Abt, S.R., Young-Pezestk, and Watson, c.C. (Eds.) Water Resources Engineering >98, v. 1, Am. Soc. Civil Engs., New York, p. 298-303.

1999 - Schumm, S.A., Causes and controls of channel incision: in Darby, S.E. and Simon, Andrew (eds.) Incised River Channels: Wiley Chichestor, p. 20-33.

1999 - Holbrook, J. and Schumm, S.A., Geomorphic and sedimentary response of rivers to tectonic deformation: a brief review and critique of a tool for recognizing subtile epeirogenic deformation in moder ane ancient steetings: Tectonicophysics, v. 305, p. 287-306.

1999 - Harvey, M.D. and Schumm, S.A., Indus River dynamics and the abandonment of Mohenjo Daro: In Meadows, A and Meadows, P.S. (eds.), The Indus River: Biodiversity, Resources, Human Kind, Oxford Univ. Press, p. 333-348.

1999 - Jones, L.S. and Schumm, S.A., Causes of avulsion: An overview. Internat. Assoc. Sedimentologists, Special Pub. 28, p. 171-178.

BOOKS:

1969 - Schumm, S.A. and Bradley, W.C. (editors), United States Contributions to Quaternary Research: Geol. Soc. America Special Paper 123, 305 p.

1972 - Schumm, S.A., River Morphology: Dowden, Hutchinson and Ross, Inc., 429 p.

1973 - Schumm, S.A. and Mosley, M.P., Slope Morphology: Dowden, Hutchinson and Ross, Inc., 468 p.

1977 - Schumm, S.A., Drainage basin morphology: Dowden, Hutchinson and Ross, Stroudsburg, PA, 352 p.

1977 - Schumm, S.A., The Fluvial System: John Wiley, N.Y., 338 p.

1980 - King, P.B. and Schumm, S.A., The physical geography (geomorphology) of W.M. Davis: Geobooks, Norwich, U.K., 174 p.

1984 - Schumm, S.A., Harvey, M.D. and Watson, C.C., Incised Channels: Morphology dynamics and control: Water Resources Pub., Littleton, CO. 200 p.

1985 - Chorley, R.J., Schumm, S.A. and Sugden, D.E., Geomorphology: Methuen, New York, 605 p.

1987 - Schumm, S.A., Mosley, M.P., and Weaver, W.E., Experimental fluvial geomorphology: John Wiley, N.Y., 413 p.

1991 - Schumm, S.A., To interpret the Earth, Ten ways to be wrong: Cambridge Univ. Press., Cambridge, 133 p.

1994 - Schumm, S.A. and Winkley, B.R. (editors), The Variability of Large Alluvial Rivers: American Society of Civil Engineers Press, N.Y., 467 p.

1999 - Schumm, S.A., Dumont, Jean, and Holbrook, John. The Mobile Earth: Alluvial Rivers and Active Tectonics. Cambridge University Press.

2000 - Schumm, S.A., Dumont, J., and Holbrook, J.. *Active Tectonics and Alluvial Rivers.* Cambridge University Press.

Pike's Peak, Colorado - September 1946